International Conference on Sustainable Materials (ICoSM 2020)

IOP Conference Series: Materials Science and Engineering
Volume 957

Genting Highlands, Malaysia
30 March 2020

Editors:

Mohamed Syahmie Mohamad Rasidi
Sharifah Shahnaz Syed Bakar

ISBN: 978-1-7138-4784-7
ISSN: 1757-8981

Printed from e-media with permission by:

Curran Associates, Inc.
57 Morehouse Lane
Red Hook, NY 12571

Some format issues inherent in the e-media version may also appear in this print version.

Institute of Physics
Dirac House, Temple Back
Bristol BS1 6BE UK

Phone: 44 1 17 929 7481
Fax: 44 1 17 920 0979

techtracking@iop.org

Additional copies of this publication are available from:

Curran Associates, Inc.
57 Morehouse Lane
Red Hook, NY 12571 USA
Phone: 845-758-0400
Fax: 845-758-2633
Email: curran@proceedings.com
Web: www.proceedings.com

TABLE OF CONTENTS

Extended Preface

International Conference on Sustainable Materials (ICoSM 2020)

International Conference on Sustainable Materials (ICoSM 2020) aims to provide a platform for scientists, researchers and students in the related field to discuss current research progress, challenges and practical solutions that could be adopted to keep abreast in this challenging world. This conference also is targeted to build the substantial cooperation, networking and promotes intercultural dialogue and cooperation amongst the academicians, scientist, researchers and engineers in relation to the Sustainable Materials. Sustainable materials is one of the most dynamic research area that promotes in finding new materials by applying new knowledge in sciences to enhance current technology and enrich quality of life. Thus, sustainable materials and related field are among the areas of research that have been given priority for strengthening the research activities in order to achieve sustainable development in this advancing world.

Topics of interest include, but are not limited to, ceramic materials and geopolymer; electronic materials; bio-materials; magnetic materials; polymers and rubber; composite metallurgy; modelling and simulation; lithium battery, fuel cells materials and energy storage; ionic, mixed and electronic transport phenomena; mineral processing; extractive metallurgy; and recycling technology.

The ICoSM2020 was scheduled to be held on 30th March 2020 at Awana Resort World Genting, Pahang, Malaysia. However on 18th March 2020, Malaysian government has issued a Movement Control Order (MCO) and banned all meetings/ mass gathering throughout the MCO period due to the novel coronavirus (COVID-19) outbreak. This virus outbreak has affected globally until now. The committees have waited for reduced spread out of the virus and condition improvement to resume the conference as planned but the movement controlled order is still on going where we need to adapt to physical distancing and hygiene practice. As a result, the conference was postponed to 3rd September 2020 with the new e-conference format. The virtual e-conference meeting was agreed to be commenced as we must be dynamic and soaring the materials spirit to ensure the knowledge and finding can be shared among the community. The participants also has showed full commitment by submitting their manuscript accordingly. Therefore, the committees have agreed to carry out the conference virtually.

The organizing committee is located at Universiti Malaysia Perlis, Malaysia. The total of 68 presenters from all over the country were divided among 3 parallel session where presentation period for each participants is 10 minutes whereas the keynote presentation were given 30 minutes time. Presentation were conducted online via google meet. Question and answer session was carry out immediately after each talk. Further discussion were continued personally by contacting the speaker directly afterwards. The session was accomplished smoothly with few hiccups that have been handled well by the committees and helpful and understanding participants. Technically, the e-conference ICoSM 2020 was a success and fruitful event of the year 2020.

IOP Conf. Series: Materials Science and Engineering **957** (2020) 011001 doi:10.1088/1757-899X/957/1/011001

The e-conference ICoSM 2020 was held virtually on 3rd September 2020 from 8.30 am and finished on the same day. The conference was officiate by Vice Chancellor of Universiti Malaysia Perlis Professor Ir. Ts. Dr. R Badlishah Ahmad and followed by thirty minutes keynote speech by 4 honorable speakers from established universities in Malaysia, Assoc. Prof. Dr. Noraiham Mohamad from Universiti Teknikal Malaysia; Associate Professor Dr. Khairel Rafezi Ahmad from Universiti Malaysia Perlis; Assoc. Prof. Dr. Noor Hana Hanif Abu Bakar from Universiti Sains Malaysia; and Professor Dr. Mariyam Jameelah Ghazali from Universiti Kebangsaan Malaysia. All participants joined the session the opening session via link http://meet.google.com/wwp-ibyv-oqb .

The parallel sessions were conducted as soon as the completion of keynote session. Three parallel session consisting of invited speakers and normal presenter where each presentation was given 10 minutes time followed by question and answer session. All presenters were given link for each session via programme book that were emailed to participants and uploaded into the ICoSM 2020 official website at https://icosm.unimap.edu.my/ . The link for parallel session 1 http://meet.google.com/msj-tcck-fsd ; parallel session 2 http://meet.google.com/xti-wriu-bjy ; and parallel session 3 http://meet.google.com/uxf-uppe-yto . Each session has some delays due to unstable connection and technical faulty such as participants unable to share their presentation slides but the organizing committees and participants were very helpful and considerate towards the situations while handling these difficulties. Technically, the overall presentation was a success.

IOP Conf. Series: Materials Science and Engineering **957** (2020) 011001 doi:10.1088/1757-899X/957/1/011001

Pictures taken during ICoSM 2020

IOP Conf. Series: Materials Science and Engineering **957** (2020) 011001 doi:10.1088/1757-899X/957/1/011001

IOP Conf. Series: Materials Science and Engineering **957** (2020) 011001 doi:10.1088/1757-899X/957/1/011001

REC
ICOsM PPKB is presenting
08:55
UNIVERSITI
MALAYSIA
PERLIS
UNIVERSITI MALAYSIA PERLIS
UNIVERSITI MALAYSIA PERLIS

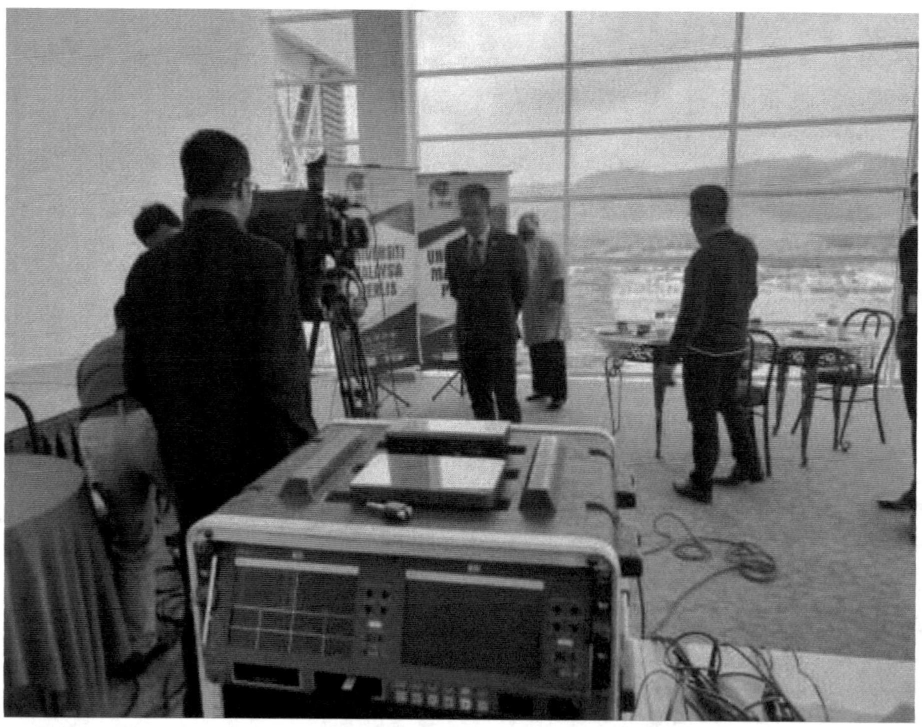

IOP Conf. Series: Materials Science and Engineering **957** (2020) 011001 doi:10.1088/1757-899X/957/1/011001

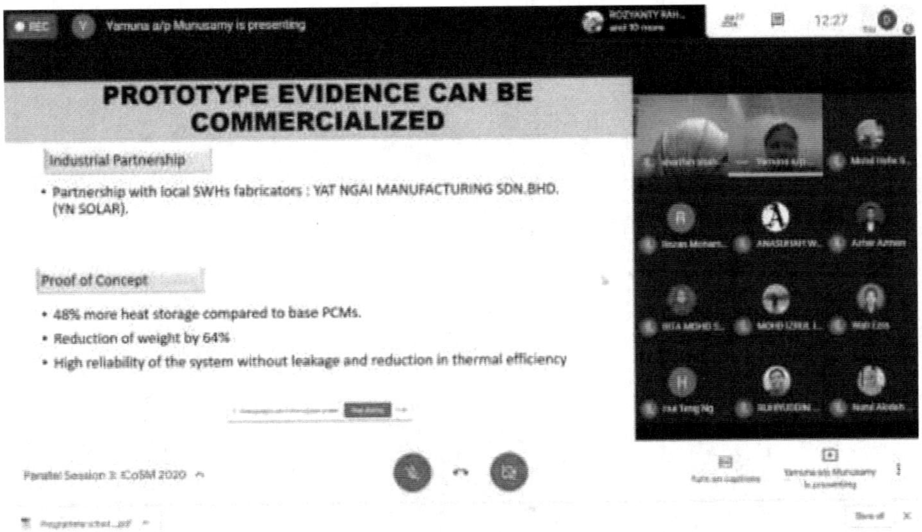

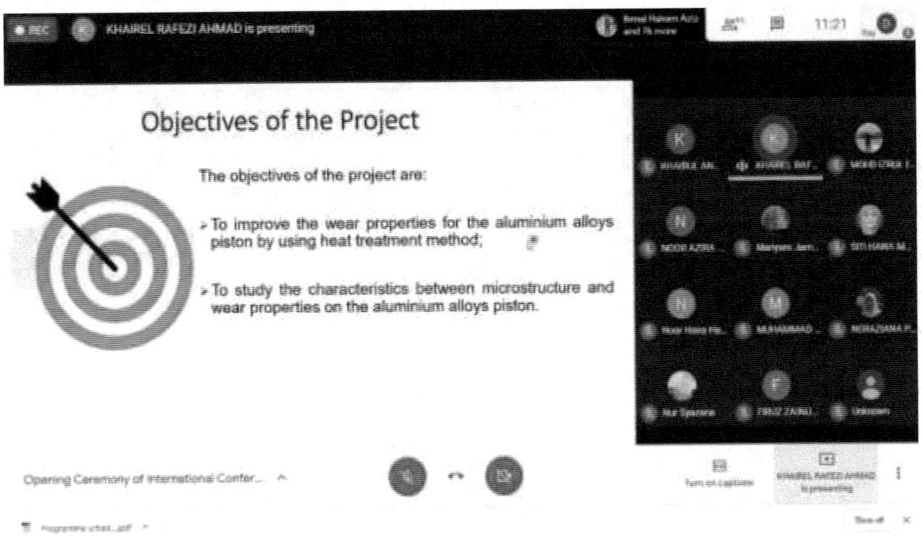

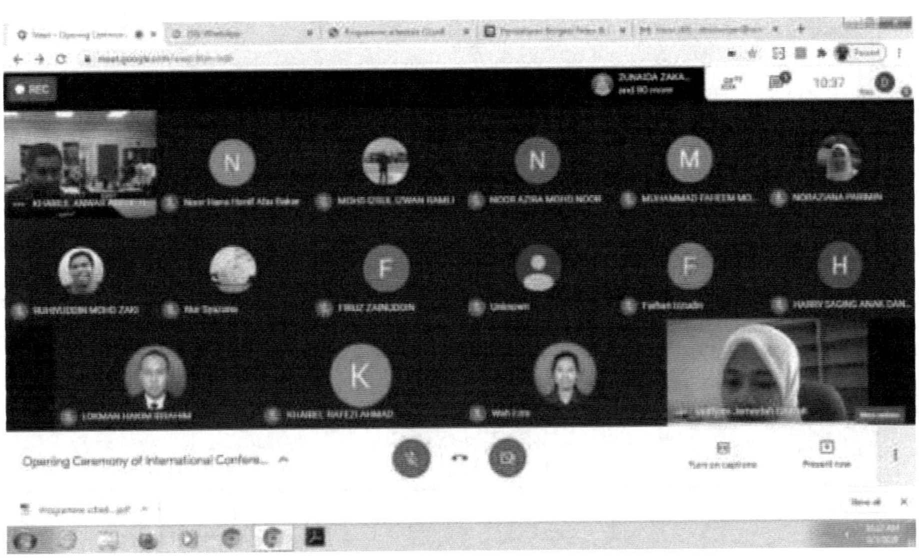

ICoSM 2020
ICoSM 2020

IOP Conf. Series: Materials Science and Engineering **957** (2020) 011002 doi:10.1088/1757-899X/957/1/011002

Peer review declaration

All papers published in this volume of IOP Conference Series: Materials Science and Engineering have been peer reviewed through processes administered by the Editors. Reviews were conducted by expert referees to the professional and scientific standards expected of a proceedings journal published by IOP Publishing.

- **Type of peer review: Single-blind**

The manuscripts were distributed among the reviewer according to their expertise via email. TWO reviewers were assigned for each manuscript. Reviewers were given MAX 14 days to review the manuscript before submitting the revised manuscript together with the reviewer form.

- **Conference submission management system:** Manually – participants need to register and upload their manuscript into Google form (Google drive)

- **Number of submissions received:** 87 papers

- **Number of submissions sent for review:** 87 papers

- **Number of submissions accepted:** 68 papers

- **Acceptance Rate (Number of Submissions Accepted / Number of Submissions Received X 100):** (68/87)*100 = 78%

- **Average number of reviews per paper:** 2 reviews

- **Total number of reviewers involved:** 70 reviewers

- **Any additional info on review process:** None

- **Contact person for queries:**

  - ✓ Mohamad Syahmie Mohamad Rasidi
    Email: syahmie@unimap.edu.my Mobile: +60199889846

  - ✓ Sharifah Shahnaz Syed Bakar
    Email: shahnaz@unimap.edu.my Mobile: +60143214884

IOP Conf. Series: Materials Science and Engineering **957** (2020) 012001 doi:10.1088/1757-899X/957/1/012001

Oxide Growth Behaviour of Fe-Ni-Cr Alloy at High Temperature Oxidation

Noraziana Parimin[1,*], and *Esah* Hamzah[2]

[1]School of Materials Engineering, Universiti Malaysia Perlis, 02600 Arau, Perlis Malaysia
[2]Faculty of Mechanical Engineering, Universiti Teknologi Malaysia, 81310 Skudai, Johor, Malaysia

Abstract. The isothermal oxidation test has been investigated on two types of Fe-Ni-Cr alloy, namely Fe-33Ni-19Cr and Fe-40Ni-24Cr alloys. Both alloys were undergoing an isothermal oxidation test at temperature of 500°C for 500 hours exposure time. The weight change per surface area of the oxidized samples has been recorded to calculate the oxidation kinetics of both alloys. The oxide growth behaviour of oxidized samples has been examined using scanning electron microscope (SEM) equipped with energy dispersive x-ray (EDX) spectrometer. The oxide phase formed on the sample surface has been analyzed using x-ray diffraction (XRD) technique. The results show that both alloys were followed a parabolic rate law, indicating a diffusion-controlled oxide growth mechanism. In addition, the oxidation kinetics indicating an increasing weight gain trend as the exposure time increase. Several oxide phases had formed on the oxidized surface of both alloys, consists of Cr-rich, Ti-rich, Fe-rich and spinel oxide structure. The surface morphology of both alloys demonstrated a continuous oxide scale formed on the alloy surface. Additionally, Fe-33Ni-19Cr alloy recorded a formation of Ti-rich oxide, whereas, Fe-40Ni-24Cr alloy displayed a formation of overgrown Nb-rich oxide particle which. Roles of the precipitates in oxidation mechanism give new insights into the alloy optimization.

1 Introduction

Ni-based alloy is characterized by the high phase stability of the face centre cubic austenite matrix, and are used in nuclear reactors, gas turbines, petrochemical, aerospace and heat-treating industries due to their favorable strength and excellent oxidation resistance at elevated temperature [1]. Ni-based alloy has the ability to form protective surface oxide scales at high temperatures that provides them with resistance to further high temperature oxidation [1-2]. For that reason, the oxidation behaviour of alloy has been extensively studied. Because of nickel oxide grows relatively fast, alloying element are normally added to improved its high temperature oxidation resistance, as well as mechanical properties. The element contains on the alloy is purposely added to give the topmost benefit to the alloy system in high temperature condition. Sufficient amount of each element will provide

* Corresponding author: noraziana@unimap.edu.my

protection against high temperature oxidation by forming compact, continuous slow growing oxides and thermodinamically stables compounds [3-4].

Fe-Ni-Cr alloys are a family of Ni-based alloy, which was developed for high temperature oxidizing condition. Fe-Ni-Cr alloys forms a protective surface films of Cr_2O_3, which have excellent oxidation resistance in many severe environment [5-10]. Research related to chromia forming alloy reported that Cr will enhance the formation of two oxide scale composed of Cr_2O_3 and spinel type oxide $MnCr_2O_4$. Chromia is a barrier against further oxidation due to its very low diffusion coefficients for oxygen and metals, hence offer a high oxidation resistance [8]. The study of oxidation related research on these alloys has been carried out by many researcher [1-21].

Fe-33Ni-19Cr and Fe-40Ni-24Cr alloys are usually used in power generation and thermal processing application. In these applications, the alloys are often subjected to the repeated thermal cyclic, which results a temperature gradient on heating and cooling, that affect the material properties. Therefore, microstructural stability and oxidation resistance behavior for related applications become a high priority [2]. Both alloys have higher Cr content, which are 19 and 24 % of Cr, which act as the oxide forming element. These alloys were chosen in this study due to the circumstances that typically 15-25% Cr has been identify to give best oxidation protection to the component by forming a slowly growing surface oxide scale of Cr_2O_3 during high temperature service [3]. Fe content in the alloy would favour a formation of hematite phase, which have the beneficial impact on the oxide scale spallation resistance [11-13].

In addition, alloying element has been added to both alloys to enhanced the properties during service. Minor alloying element, which are Mn, Ti, Si, Al and Nb, may significantly affect the oxidation behavior of the alloy [3-4, 14]. Si and Al will precipitates as internal oxides, while Mn and Ti may incorporate into the scale, forming a stable oxide layer. Additionally, the precipitates formation were found to have obvious influences on oxidation rate, which depends on the complex function of the alloy structure. The formation of precipitates at the scale-metal interface depends on the alloying element and working conditions of the oxidized samples. The present of the precipitates may decrease the adherence of the oxide scale to the metallic structure and contributed to the preferred site for pit formation [15-18]. Therefore, in this study, the behavior and growth of oxide scale on two types of Ni-based alloys has been investigated in terms of oxidation kinetics and oxide morphology of the oxidized samples.

2 Methodology

The materials used in this study were two types of Ni-based alloy, which are Fe-33Ni-19Cr (ASTM-B-409: ASME-SB-409, UNS N08810) and Fe-40Ni-24Cr (ASTM-B-409: ASME-SB-409, UNS N08120) alloys. The chemical composition of Fe-33Ni-19Cr and Fe-40Ni-24Cr alloys has been analyzed using optical emission spectrometer (OES). The composition of both alloys has been presented in Table 1. Samples for the high temperature oxidation tests were cut using wire electrical discharge machine (EDM) to dimensions approximately 10 mm x 10 mm x 3 mm. The samples for oxidation tests were ground using silicon carbide grinding paper to P600 grit finished. The surface area of each samples has been recorded before the oxidation test. All samples were weighted on METLER AT400 analytical balance with accuracy of ± 0.1 mg, before and after the oxidation test to measure the weight change. Both alloys were isothermally oxidized at 500°C for 500 hours. The oxidized samples were characterized in terms of oxidation kinetics, oxide phase analysis and oxide surface morphology. The oxidation kinetics was calculated form the weight change of the oxidized samples over a surface area. The oxide phase was analyzed using X-ray diffraction (XRD) model SIEMENS coupled with BRUKER D5000 XRD instrument software. The oxide scale morphology was examined using scanning electron microscope (SEM) coupled with energy dispersive x-ray (EDX) analyzer model JOEL JSM-6460 LA and PHILIPS XL40.

Table 1. Chemical composition of Fe-33Ni-19Cr and Fe-40Ni-24Cr alloys in weight percent (wt. %) with balance Fe.

Alloy / Element	Ni	Cr	C	Al	Ti	Si	Mn	P	Cu	N	Nb
Fe-33Ni-19Cr	33	19	0.078	0.534	0.489	0.315	0.556	0.007	0.082	0.02	-
Fe-40Ni-24Cr	40	24	0.048	0.080	0.029	0.441	0.702	0.014	0.114	0.20	0.44

3 Results and Discussion

3.1 Oxidation Kinetics

The weight change per surface area as a function of time of Fe-33Ni-19Cr and Fe-40Ni-24Cr alloys isothermally oxidized at 500°C for 500 hours is shown in Fig. 1. The oxidation kinetics of both alloys follows similar trend with increasing weight gain, as the exposure time increase. At the beginning of the oxidation test, both alloys recorded a rapid weight gain until 100 hours exposure time. The weight gains continuously increase as the oxidation exposure increase until 500 hours. Overall, Fe-33Ni-19Cr alloy exhibited slightly lower weight gain compared to Fe-40Ni-24Cr alloy. Similar weight change trend has been recorded by other researchers on oxidation study of Fe-Ni-Cr alloy [5, 10].

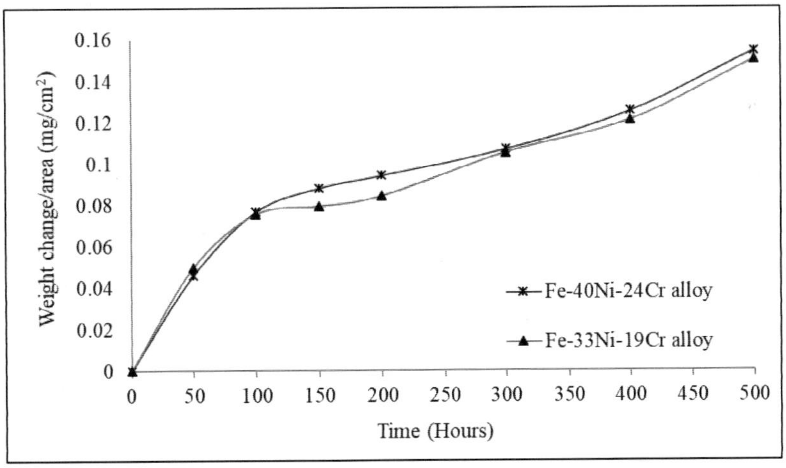

Fig. 1. Oxidation kinetics of Fe-33Ni-19Cr and Fe-40Ni-24Cr alloys at 500°C for 500 hours.

The rate law of the oxidized alloy was identified by a double log plot as shown in Fig. 2 (a). Equation (1) was used to fit the data where x is a weight change per surface area, t is a time and C is a constant. The m value indicating the rate data of the respecting curve, which value of 1, 2 and 3 for linear, parabolic and cubic oxidation, respectively. The double log plots indicate that both alloys produced good fitting results. The fitting parameter R^2 with a value approaching one denotes the increasing consistence of the data with the fitting. From the graphs, the m value for both alloys was recorded at 2.32 and 2.14 for Fe-33Ni-19Cr and

IOP Conf. Series: Materials Science and Engineering **957** (2020) 012001 doi:10.1088/1757-899X/957/1/012001

Fe-40Ni-24Cr alloys, respectively, indicating the oxidation kinetics approximately follows a parabolic rate law, m = 2. The analysis form the double log plots was confirmed that the oxidation rate law for both alloys were followed a parabolic rate law, thus lead to further identify the parabolic rate constant for both alloys. Our previous studies has found that the exposure of both alloy at others oxidizing temperatures recorded a similar parabolic rate law [15-16, 19-21]. Fig. 2 (b) shows the square of weight change as a function of time, with equation (2) was used to fit the data, where x is a weight change per surface area, t is a time, C is a constant and Kp is a parabolic rate constant. Both alloys show good fitting R^2 results. The parabolic rate constant recorded a value of 11.17 x 10^{-9} and 11.80 x 10^{-9} for Fe-33Ni-19Cr and Fe-40Ni-24Cr alloys, respectively. The parabolic rate constant for Fe-33Ni-19Cr alloy recorded a lower value, indicating a good oxidation resistance.

$$\log x = 1/m \log t + C \tag{1}$$

$$x^2 = Kp\, t + C \tag{2}$$

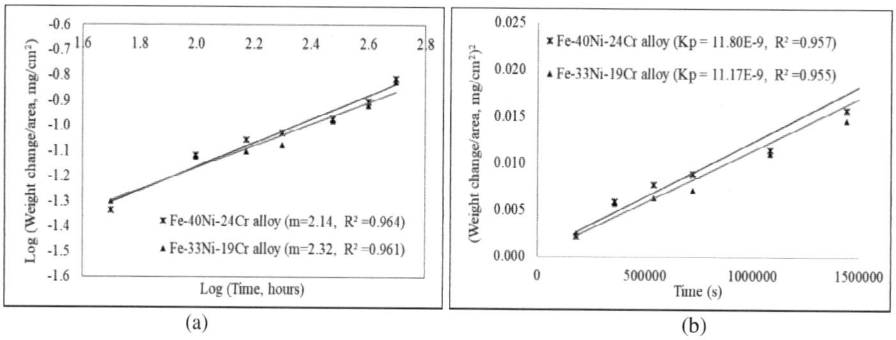

Fig. 2. Weight change as a function of exposure time of Fe-33Ni-19Cr and Fe-40Ni-24Cr alloys at 500°C: (a) double log plots indicating parabolic rate law (m=2) and (b) square of weight change indicating parabolic rate constant, Kp value.

3.2 Oxide Phase Analysis

The oxide phase analysis has been done by XRD technique for both alloys. The analysis recorded that the main oxide phases formed on the alloy surface of Fe-33Ni-19Cr and Fe-40Ni-24Cr alloys composed of several oxide phases consists of Cr-rich, Fe-rich, Ti-rich and spinel oxide structures. The oxides phases are Cr-rich oxides (Cr_2O_3, $Cr_{1.3}Fe_{0.7}O_3$, $(Cr_{0.88}Ti_{0.12})_2O_3$), Fe-rich oxides ($Fe_2O_3$ and Fe_3O_4) Ti-rich oxides (TiO_2 and $(Ti_{0.97}Cr_{0.03})O_2$) and spinel oxides structure ($MnCr_2O_4$, $MnFe_2O_4$, $FeCr_2O_4$, $NiCr_2O_4$ and $NiFe_2O_4$). The presence of Cr-Ti-rich oxides, such as $(Cr_{0.88}Ti_{0.12})_2O_3$ and $(Ti_{0.97}Cr_{0.03})O_2$ oxides are believe to reduce the effect of Cr evaporation, whereby the formation of Cr evaporation lead to the development of oxide exfoliation. According to [14], the $(Cr,Ti)_2O_3$ has a lower Cr vapor pressure comparable with $LaCrO_3$ due to the same hexagonal structure in between Ti and La. Oxidized Fe-33Ni-19Cr alloy were formed several oxide phase which are Cr-rich, Fe-rich, Ti-rich and spinel oxides structure that contributes the most to the oxide formation due to rapid diffusion of these metal ion during high temperature exposure. In addition, the formation of hematite (Fe_2O_3) phase have significant impact on oxide exfoliation resistance. According to other researchers [11-13], the formation of hematite would reduce the coefficient thermal expansion (CTE) mismatch in between hematite and spinel oxide due to the similar CTE volume. On the other hand, the formation of Mn-Cr spinel oxides phase may also enhance the oxide spallation resistance due to the incorporated of Mn and Cr oxide which can lower the Cr_2O_3 activity by the formation in $MnCr_2O_4$ oxide [3-4]. While, oxidized Fe-40Ni-24Cr alloy formed similar oxide scale with Fe-33Ni-19Cr alloy, with addition of Nb-rich oxide composes of NbO_2.

IOP Conf. Series: Materials Science and Engineering **957** (2020) 012001 doi:10.1088/1757-899X/957/1/012001

3.3 Oxide Surface Morphology

Fig. 3 shows SEM image of oxide surface morphology of Fe-33Ni-19Cr alloy after exposure for 300 hours. Fig. 3 (a) shows low magnification (1000x) of SEM image, demonstrated the uniform and continuous oxide scale formed on the alloy surface with disclose of undulation surface texture of oxide layer. The undulation surface texture was originating from the grinding effect to provides a groove marks on the sample surface, which act as oxide nucleating foundation to enhance the nucleation process. The undulation surface texture indicated that thin oxide scale was formed on the alloy surface. Whereas, high magnification (10,000x) SEM image in Fig. 3 (b) displays the uniform scale-shape oxide formed on the alloy surface. The elemental EDX analysis at the matrix area (A) recorded the presence of Fe, Ni, Cr and Mn which are identified as the main elements and alloying element of the alloy. The presence of oxygen indicates the thin oxide layer developed on the alloy surface. The elemental EDX analysis suggested that area A composes of mixed oxide phases. While, EDX analysis at spot B indicates the enrichment of element O and Ti, suggested that this oxide particle composed of Ti-rich oxide. According to other researcher [14], the formation of Ti-rich oxide was said to give significant effect on oxidation protection and mitigated oxide exfoliation.

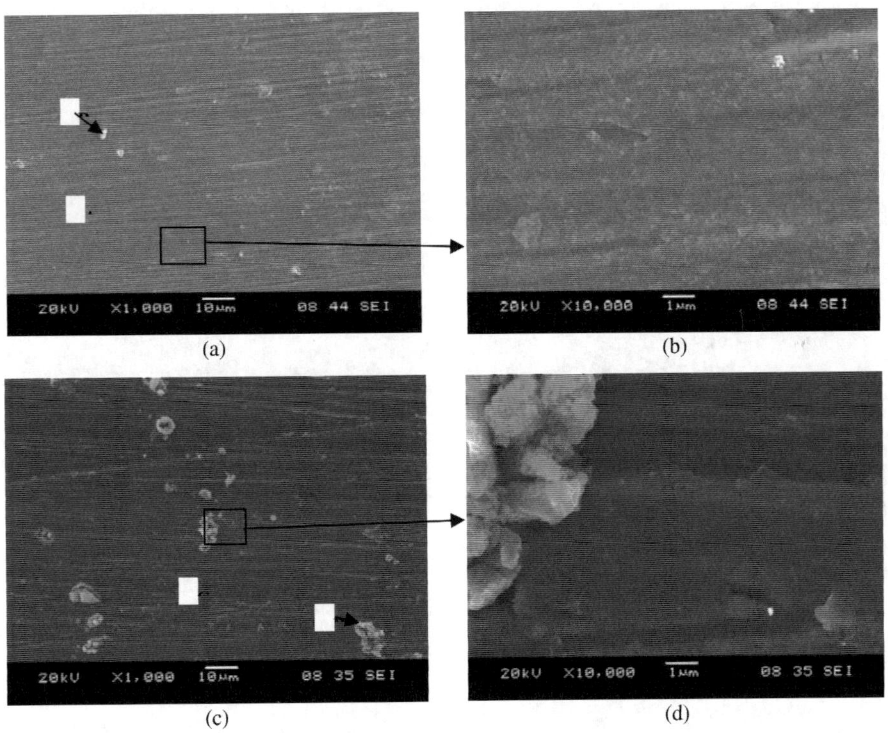

Fig. 3. SEM images of oxide morphology after exposure for 300 hours: (a-b) Fe-33Ni-19Cr alloy, (c-d) Fe-40Ni-24Cr alloy.

Fig. 3 (c) shows low magnification (1000x) SEM image of oxide surface morphology of Fe-40Ni-24Cr alloy after exposure for 300 hours, indicated the formation of continuous oxide layer with overgrown oxide particles distributed along the oxide surface. The Elemental EDX analysis on the area C recorded the enrichment of element Fe, Ni, Cr, Mn and O, indicating the formation of mixed oxide phases on the alloy surface. The elemental EDX analysis on the overgrown oxide particle at D recorded the enrichment of element Nb and O, suggesting the formation of Nb-rich oxide as analyzed by XRD technique. Whereas, the close-up SEM image (10,000x) in Fig. 3 (d) shows the protruding structure of overgrown Nb-rich oxide was well attached to the oxide structure. Other researcher [17-18] has recorded similar Nb-rich precipitates formed on oxidized of Ni-based alloy.

Fig. 4 (a) shows low magnification (1000x) SEM image of oxide surface morphology of Fe-33Ni-19Cr alloy after exposure for 500 hours showing a uniform and continuous oxide structure was formed on the alloy surface. The close-up SEM image (10,000x) on Fig. 4 (b) shows the similar oxide structure with indication of Ti-rich oxide particles formed at the isolated area on the alloy surface. Fig. 4 (c) shows the formation of uniform oxide structure with overgrown Nb-rich oxide particles on Fe-40Ni-24Cr alloy after exposure for 500 hours. The Nb-rich oxide particles was overgrown on the oxidized Fe-40Ni-24Cr alloy as isolated oxide particle. It is found that a crack was developed inside the overgrown oxide particles due to the effect of extensive growth of protruding oxide structure. The Nb element was added to alloy system as an alloying element to enhance the mechanical properties of the alloy. However, it is caused in the formation of Nb-rich oxide precipitates which encourages the occurrence of pitting. The presence of pitting on the oxide surface contains of Nb-rich oxide has been reported by other researcher [12], due to the massive difference in the composition between alloy matrix and precipitates. Whereas, Fig. 4 (d) shows a large area of the oxide spallation on the oxidized Fe-40Ni-24Cr alloy surface, mark as area E.

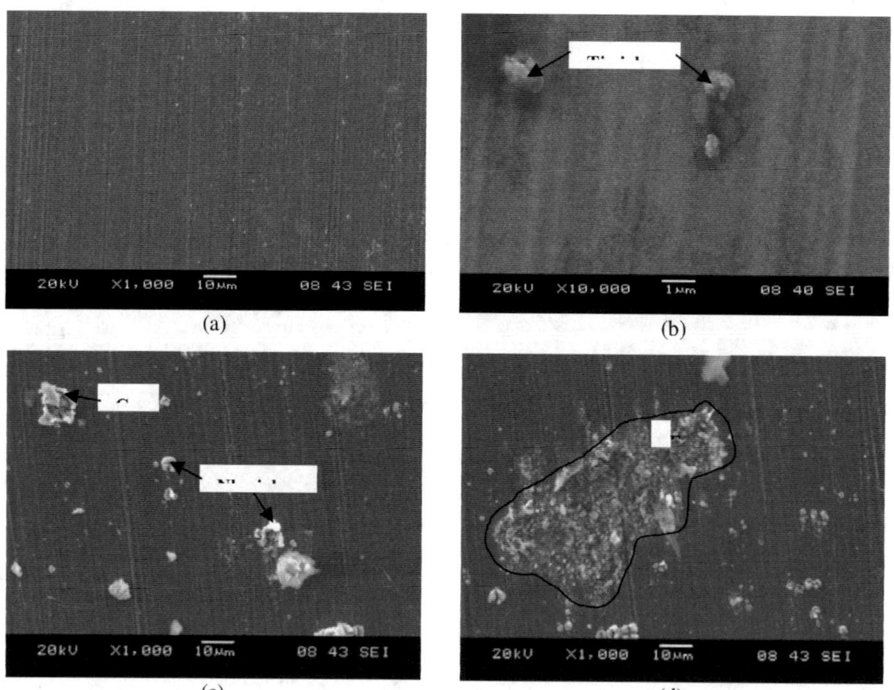

Fig. 4. SEM images of oxide morphology after exposure for 500 hours: (a-b) Fe-33Ni-19Cr alloy, (c-d) Fe-40Ni-24Cr alloy.

4 Summary

High temperature oxidation of Fe-33Ni-19Cr and Fe-40Ni-24Cr alloys had been study in order to investigate the oxidation kinetics, oxide phases and oxide scale behavior. The kinetics of isothermal oxidation of both alloys follow a parabolic rate law, indicating the diffusion-controlled oxide growth rate. Fe-33Ni-19Cr alloy recorded lower parabolic rate constant, therefore has good oxidation resistance. Several oxide phases were formed on the alloy surface. The oxide morphology formed on oxidized samples of both alloys are mostly uniform with mixed oxides phases detected. On the other hand, oxidized Fe-40Ni-24Cr alloy displayed a formation of overgrown Nb-rich oxide precipitates. The formation of Nb-rich oxide precipitates is significantly detrimental due to the induces of crack formation around the precipitates.

The authors would like to thank the Ministry of Education Malaysia for the research fund under the Fundamental Research Grant Scheme (FRGS) (Project Code. FRGS/1/2016/TK05/UNIMAP/02/4).

References

1. B.R. Barnard, P.K. Liaw, R.A. Buchanan, D.L. Klarstrom, Mater. Sci. & Eng.: A **527** (16-17) (2010)
2. M. Fulger, D. Ohai, M. Mihalache, M. Pantiru, V. Malinovschi, J. Nucl. Mater. **385** (2) (2009)
3. J. Zurek, D.J. Young, E. Essuman, M. Hänsel, H.J. Penkalla, L. Niewolak, W.J. Quadakkers, Mater. Sci. & Eng.: A **477** (1-2) (2008)
4. G.R. Holcomb, D.E. Alman, Scrita Mater. **54** (10) (2006)
5. S. Liu, J. Shen, X.H. Guo, L.L. Liu, Y. Niu, Corros. Sci. **135** (2018)
6. S. Swaminathan, S.M. Hong, M. Kumar, W.S. Jung, D.I. Kim, H. Singh, I.S. Choi, Surf. & Coat. Tech. **362** (2019)
7. M.S. Elbakhshwan, S.K. Gill, A.K. Rumaiz, J. Bai, S. Ghose, R.B. Rebak, L.E. Ecker, Appl. Surf. Sci. **426** (2017)
8. A. Col., V. Parry, C. Pascal, Corros. Sci. **114** (2017)
9. S. Bsat, B. Xiao, X. Huang, S. Penttila, Oxid. Met. **89** (2018)
10. B. Adam, L. Teeter, J. Mahaffey, M. Anderson, L. Arnadottir, J.D. Tucker, Oxid. Met. **90** (2018)
11. L. Tan, T.R. Allen, J,T. Busby, J. Nucl. Mater. **441** (1-3) (2013)
12. L. Tan, K. Sridharan, T.R. Allen, R.K. Nanstad, D.A. McClintock, J. Nucl. Mater. **374** (1-2) (2008)
13. L. Tan, K. Sridharan, T.R. Allen, J. Nucl. Mater. **348** (3) (2006)
14. L. Tan, X. Ren, K. Sridharan, T.R. Allen, Corros. Sci. **50** (7) (2008)
15. N. Parimin, E. Hamzah, IOP Conf. Ser.: Mater. Sci. Eng. **701** (2019) 012022
16. N. Parimin, Singh, IOP Conf. Ser.: Mater. Sci. Eng. **701** (2019) 012063
17. L. Tan, Corros. Sci. **50** (11) (2008)
18. G.S. Was, P. Ampornrat, G. Gupta, S. Teysseyre, E.A. West, T.R. Allen, K. Sridharan, J. Nucl. Mater. **371** (1-3) (2007)
19. N. Parimin, Jemahri, IOP Conf. Ser.: Mater. Sci. Eng. **701** (2019) 012062
20. N. Parimin, Z. Zulnuraini, S.A.C. Sakdun, Solid State Phenomena **280** (2018)

21. N. Parimin, E. Hamzah, A. Amrin, Solid State Phenomena **280** (2018)

Influence of Solution Treatment Temperature on the Microstructure of Ni-based HR-120 Superalloy

Noraziana Parimin[1,*], and *Esah* Hamzah[2]

[1]School of Materials Engineering, Universiti Malaysia Perlis, 02600 Arau, Perlis Malaysia
[2]Faculty of Mechanical Engineering, Universiti Teknologi Malaysia, 81310 Skudai, Johor, Malaysia

Abstract. The solution treatment process has been carried out on Ni-based HR-120 superalloy (Fe-40Ni-24Cr) at six different temperatures, namely 950°C, 1000°C, 1050°C, 1100°C, 1150°C and 1200°C. All samples were soaking for 3 hours at desired temperature, followed by water quench. The main objective of the solution treatment process is to vary the grain size of the alloy. The effect of solution treatment process on the alloy has been investigated in term of microstructure, grain size, phase present and hardness test, by using optical microscope, scanning electron microscope (SEM) equipped with energy dispersive x-ray (EDX) spectrometer, x-ray diffraction (XRD) technique and Vickers hardness test. As a results, the grain size of the solution-treated alloy were increase as the solution treatment temperature increase from 27.27 μm to 40.86 μm for solution-treated alloy at 950°C to 1200°C, respectively. Three precipitate phases were detected during phase analysis which are NbC, TiC and (Nb,Ti)C on the solution-treated alloy. In addition, as the grain size increase when the solution treatment temperature increase, the hardness value was decrease.

1 Introduction

Ni-based superalloys are often the choice in high temperature applications in aerospace, nuclear, chemical and power industries applications because of their excellent high temperature mechanical properties [1-4], where substantial cooling is needed. This extracts forfeit in terms of thermal efficiency. Materials that allowed higher operating temperatures, high temperature strength and excellent oxidation resistance, results in significant growths in efficiency. Heat resistant Fe-Ni-Cr superalloys can potentially meet this need. Our previous studies has found that the exposure of Fe-Ni-Cr alloy at several oxidizing temperatures recorded excellent oxidation resistance [5-7].

The excellent properties of heat resistant Fe-Ni-Cr superalloys at high temperature conditions is achieved by their good resistance to industrial environments and high strength economical alloy, providing cost-effective solutions to many maintenance problems in the heat-treating and industrial heating industries. Along with potential application and comparable materials composition, HR-120 superalloy were selected as candidate materials

* Corresponding author: noraziana@unimap.edu.my

for solution treatment study especially for thermal related applications. HR-120 superalloy, which is Fe-40Ni-24Cr alloy is a heat-resistant alloy with good oxidation resistance [8-10], and excellent strength at elevated temperature. It is used for heat treating equipment, waste incinerators, radiant tubes, muffles, retorts and furnace components [11].

HR-120 superalloy is a solid-solution strengthened, austenitic and heat resistant alloy [12-13]. The alloy is solution-annealed at a temperature ranging from 1175 to 1230°C and rapidly cooled [11]. The material is mainly composed of face center cubic austenitic grains with annealing twins [14]. Annealing twins occurs during cooling due to the change in crystal system because of re-arrangement of unstable crystal structure and transform into another more stable form. Annealing twins are fairly common in FCC metals [15]. The HR-120 superalloy is composed of austenitic phase with annealing twins which usually found in alloy with high nickel content [16-17]. HR-120 superalloy is a heat resistant alloys with excellent metallurgical stability at high temperature exposure. The comparable equiaxed austenitic structure is stable at prolonged high temperature with excellent oxidation resistant properties.

Extensive investigations have been conducted on heat treatment of Fe-Ni-Cr superalloy to optimize the influence of test parameters on physical and mechanical properties. Heat treatment process has great effect on the mechanical properties and the microstructure of superalloy. The effect of microstructure on superalloy has been widely studied including heat treatment, aging treatment, solution treatment, thermomechanical treatment and surface treatment [18-24]. In order to develop an understanding on the grain alteration through solution treatment, the solution treatment process on HR-120 superalloy, which is Fe-40Ni-24Cr alloy has been carried out. The finding from this research will give better understanding of the microstructure and hardness of the alloy.

2 Methodology

The materials used in this study is a Ni-based HR-120 superalloy, which is Fe-40Ni-24Cr alloy (ASTM-B-409: ASME-SB-409, UNS N08120). The chemical composition of this alloy has been analyzed using optical emission spectrometer (OES), presented in Table 1.

Table 1. Chemical composition of Fe-40Ni-24Cr alloy in weight percent (wt. %) with balance Fe.

Alloy / Element	Ni	Cr	C	Al	Ti	Si	Mn	P	Cu	N	Nb
Fe-40Ni-24Cr	40	24	0.048	0.080	0.029	0.441	0.702	0.014	0.114	0.20	0.44

The alloy were undergoing a solution treatment process at six different temperature, namely 950°C, 1000°C, 1050°C, 1100°C, 1150°C and 1200°C, soaking for 3 hours followed by water quench. The solution-treated samples was hot mounted in acrylic transparent resin and was ground using silicon carbide grinding paper P320, P500, P600, P800 and P1000 grit followed by polishing using diamond paste of 9 μm and 3 μm, and then final polishing with colloidal silica 0.04 μm to obtain a smooth and mirror like finish. The polished sample was etched in solution of 2 parts glycerol ($C_3H_8O_3$), 3 parts hydrochloric acid (HCl) and 1 part nitric acid (HNO_3) for 5-120 seconds to reveal the microstructure. The average grain diameter was measured based on at least 5 optical micrographs using Heyn linear intercept procedure ASTM E112. The solution-treated samples were characterized in terms of microstructure, grain size, phase analysis and hardness test. The microstructure of solution-treated samples were examined using optical microscope model NIKON equipped with i-Solution image analysis software and scanning electron microscope (SEM) coupled with energy dispersive x-ray (EDX) analyzer model JOEL JSM-6460 LA and PHILIPS XL40. The phase analysis was examined using x-ray diffraction (XRD) technique model SIEMENS coupled with BRUKER D5000 XRD instrument software. The hardness test was examined using Vickers hardness tester model Matsuzawa-DVK II series.

3 Results and Discussion

3.1 Microstructure of Solution-Treated Alloy

Fig. 1 shows the optical micrographs of the Fe-40Ni-24Cr alloy for all solution-treated samples. The optical image shows an austenitic matrix and primary precipitates of carbides present at the grain. In the Fe-40Ni-24Cr alloy, the addition of titanium, niobium and carbon have a tendency to form simple MC-type carbides, which are TiC, NbC or (Nb,Ti)C during high temperature exposure [24-26]. Fig. 2 shows SEM image of as-received Fe-40Ni-24Cr alloy with two area of EDX analysis. The large precipitates detected at spectrum 1 indicates the enrichment of element Nb and C as presented in chemical composition of EDX analysis as shown in Table 2. This nearly spherical morphology suggested the formation of NbC precipitate. Whereas, EDX analysis at spectrum 2 indicates the presence of major element Nb, Ti and C. The results suggested that at this area composes of precipitates of TiC and/or (Nb,Ti)C.

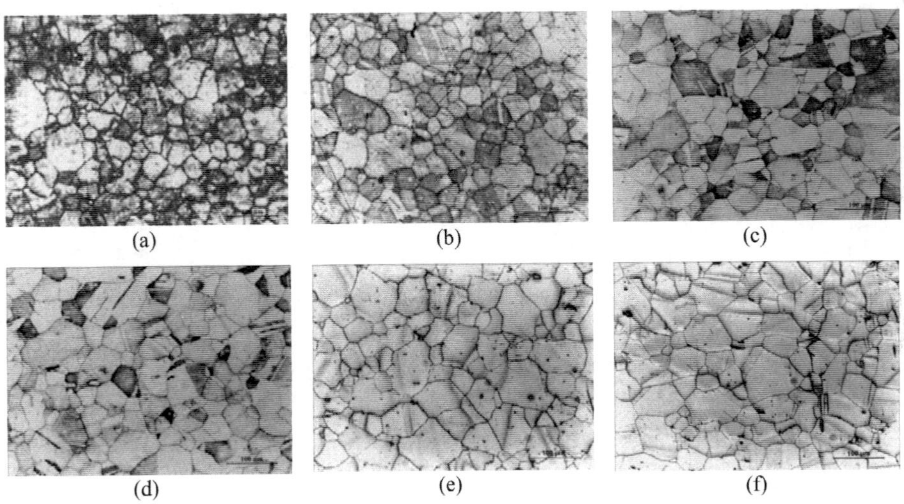

Fig. 1. Optical micrographs of solution-treated Fe-40Ni-24Cr alloy at different temperature: (a) 950°C, (b) 1000°C, (c) 1050°C, (d) 1100°C, (e) 1150°C and (f) 1200°C.

Fig. 3 shows the average grain sizes of solution-treated Fe-40Ni-24Cr alloy at different solution treatment temperatures. Results indicate the grain size increased with increase in solution treatment temperature. Solution-treated sample at 950°C for Fe-40Ni-24Cr alloy resulted finest grain size which is 27.27 μm, followed with linear increase in size up to 40.86 μm for solution-treated sample at 1200°. The grain size of as-received Fe-40Ni-24Cr alloy was 38.13 μm. The as-received sample had been annealed by the material supplier at the temperature of 1177° during manufacturing process. Effect of grain size coarsening is primarily caused by the carbon content in the grain boundary carbides dissolve into the alloy matrix, which leads to atypical grain growths occurred, thus reducing the pinning effect and the obstacle of grain growth. Similar observation was recorded by other researcher [23, 27], which discovered the irregular grains growth at solution-treated temperature above 1150°C for Fe-23Ni-12Cr and Fe-38Ni alloy.

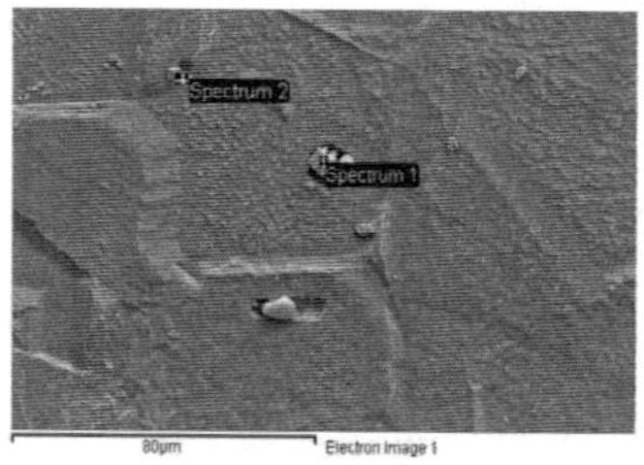

Fig. 2. SEM image of the as-received Fe-40Ni-24Cr alloy.

Table 2: EDX analysis of the as-received Fe-40Ni-24Cr alloy (atomic %).

Spectrum	Nb	Ti	C
1	53.78	-	46.22
2	16.16	67.54	16.30

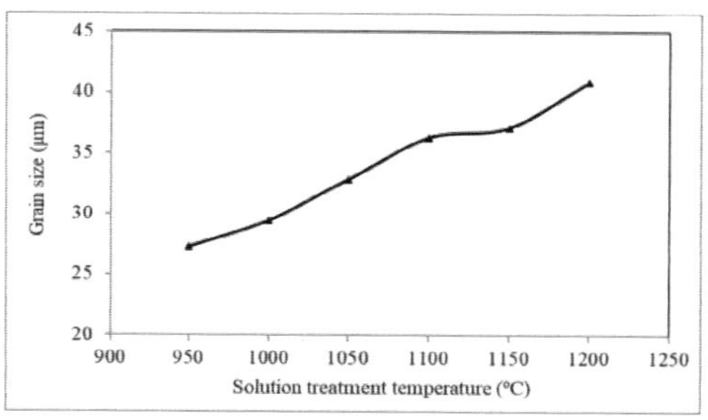

Fig. 3. Effect of solution treatment on the grain size of Fe-40Ni-24Cr alloys.

3.2 Phase Analysis of Solution-Treated Alloy

Fig. 4 shows the x-ray diffraction patterns of solution-treated samples of Fe-40Ni-24Cr alloy at 1100°C. This sample was chosen to represent other samples due to the similar pattern of all solution-treatment samples. The most intense peaks belong to the base alloy austenite phases. In addition, less intense peaks were also identified which belong to the NbC, TiC and (Nb,Ti)C phases. The NbC and (Nb,Ti)C peak were overlap on each other. The mixed presence of TiC and (Nb,Ti)C precipitates is supported by EDX analysis (Fig. 2, spectrum 2) which recorded the enrichment of Ti, Nb and C elements, indicating mixed existence of both carbides. According to other researchers [23-26], addition of Ti, Nb and C in the alloy system would favour a formation of MC carbides during high temperature exposure which are, TiC, NbC and/or (Nb,Ti)C.

3.3 Hardness of Solution-Treated Alloy

The hardness test results of solution-treated Fe-40Ni-24Cr alloy is shown in Fig. 5. The results indicate that increasing in the solution treatment temperatures was decrease the hardness of the alloy. The hardness value of Fe-40Ni-24Cr alloy was highest at solution-treated 950°C which is 207.66 HV followed by linear decrease of the hardness until solution-treated at 1200°C. However, the decrease in hardness value at higher solution treated temperature at 1100°C to 1200°C is small compared to lower solution treatment temperatures from 950°C to 1050°C showing large value range. The decrease in hardness value as temperature increases is due to the grain become larger. Similar observation has been recorded by other researcher [27]. The hardness value for as-received Fe-40Ni-24Cr alloy is 193.67 HV. Hardness is a resistance to plastic deformation, usually by indentation. Solution treatment process at high temperature will recovered great amount of imperfection and dislocation. When sufficient thermal energy is supplied during recovery, its allow the arrangement of dislocation into lower energy configuration, hence lowered the internal energy of the recovered metal. Prolong the exposure alloy at a temperature at which recrystallization take place will cause a new grain structure transforms completely. At this stage, the ductility of alloy usually increase significantly. Then, the grain growth would take place by the grain boundary migration through atomic diffusion across boundary resulting coarsening of grains. Consequently, as the grain enlarge, the capability of metal to resist plastic deformation will decrease due to reducing of grain boundary area which act as an obstacle for permanent plastic deformation, hence resulting lower hardness value.

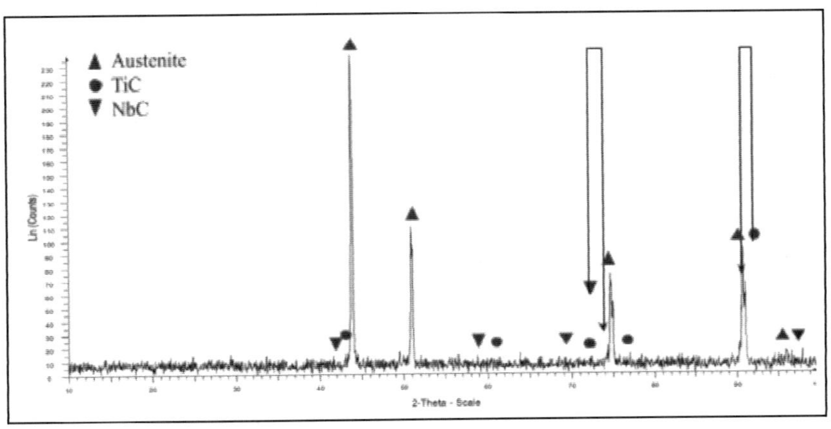

Fig. 4. XRD pattern of solution treatment of Fe-40Ni-24Cr alloy at 1100°C.

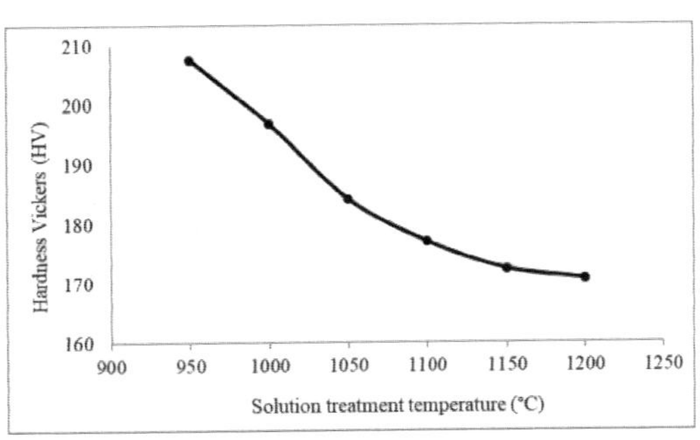

Fig. 5. Effect of solution treatment on the hardness of Fe-40Ni-24Cr alloys.

4 Summary

The solution treatment process on Fe-40Ni-24Cr alloy produced varies grain size. Fine grain size recorded at solution-treated 950°C which is 27.27 μm, whereas, coarse grain size recorded at solution-treated 1200°C, which is 40.86 μm. Three precipitate phases were detected consists of NbC, TiC and (Nb,Ti)C. The hardness of the solution-treated samples was decreasing, as the solution treatment temperature increase.

The authors would like to thank the Ministry of Education Malaysia for the research fund under the Fundamental Research Grant Scheme (FRGS) (Project Code. FRGS/1/2016/TK05/UNIMAP/02/4).

References

1. C.N. Athreya, K. Deepak, D.I. Kim, B. Boer, S. Mandal, V.S. Sarma, J. Alloys & Comp. **778** (2019)
2. D. Saber, I.S. Emam, R. Abdel-Karim, J. Alloys & Comp. **719** (2017)
3. J. Jiang, G. Xiao, Y. Wang, Y. Liu, J. Alloys & Comp. **784** (2019)
4. Z. Hao, R. Cui. Y. Fan, J. Lin, Int. J. of Mech. Sci. **150** (2019)
5. N. Parimin, Jemahri, IOP Conf. Ser.: Mater. Sci. Eng. **701** (2019) 012062
6. N. Parimin, E. Hamzah, A. Amrin, Solid State Phenomena **280** (2018)
7. N. Parimin, Z. Zulnuraini, S.A.C. Sakdun, Solid State Phenomena **280** (2018)
8. N. Parimin, E. Hamzah, IOP Conf. Ser.: Mater. Sci. Eng. **701** (2019) 012022
9. N. Parimin, Singh, IOP Conf. Ser.: Mater. Sci. Eng. **701** (2019) 012063
10. N. Parimin, E. Hamzah, A. Amrin, J. of Adv. Reser. In Fluid Mech. & Ther. Sci. **49** (2018)
11. HAYNES International, *HAYNES® HR-120® alloy, (H-3125l)* (U.S.A: Haynes Internatinal, 2019)
12. X. Ledoux, S. Mathieu, M. Vilasi, Y. Wouters, P. Del-Gallo, M. Wagner, Oxid. Met. **80** (2013)
13. L. Couture, F. Ropital, F. Grosjean, J. Kittel, V. Parry, Y. Wouters, Oxid. Met. **80** (2013)
14. Y. He, L. Chen, P. Liaw, R. Mcdaniels, C. Brooks, R. Seeley, D. Klarstrom, Int. J. of Fatigue **24** (2002)
15. F.C. Campbell, *Elements of Metallurgy and Engineering Alloys* (Ohio: ASM International, 2008)
16. K.E. Tillous, J. Dulcy, T. Belmonte, Oxid. Met. **82** (2014)
17. L.J. Chen, P.K. Liaw, H. Wang, Y.H. He, R.L. McDaniels, L. Jiang, B. Yang, D.L. Klarstrom, Mech. of Mater. **36** (2004)
18. A. Munitz, S. Salhov, G. Guttmann, N. Deromiw, M. Nahmany, Mater. Sci. & Eng. A **742** (2019)
19. C.Li, M. Seyring, X.Li, Y. Zhong, Z. Ren, M. Rettenmayr, Met. & Mater. Trans. A **50** (2019)
20. A. Munitz, L. Meshi, M.J. Kaufman, Mater. Sci. & Eng. A **689** (2017)
21. L. Tan, J.T. Busby, H.J.M. Chichester, K. Sridharan, T.R. Allen, J. of Nuc. Mater. **437** (2013)
22. L. Tan, X. Ren, K. Sridharan, T.R. Allen, Cor. Sci. **50** (2008)

23. D. Cai, Y. Mei, N. Pulin, L. Wenchang, Mater. Lett. **57** (2003)
24. I.A. Sustaita-Torres, S.H. Rodríguez, M.P. Guerrero-Mata, M. Garza, E. Valdés, F. Deschaux-Beaume, R. Colás, Mater. Chem. & Phy. 1**33** (2012)
25. B. Piekarski, Mater. Charac. **61** (2010)
26. R. Dehmolaei, M. Shamanian, A. Kermanpur, Mater. Charac. **59** (2008)
27. J.F. Zhang, Y.F. Tu, J. Xu, J.S. Zhang, J.L. Zhang, J. Iron & Steel Res. Int. **15** (2008)

Effect of Chemical Treatment on Tensile Properties of Oil Palm Empty Fruit Bunch (OPEFB)-based All Cellulose Composite (ACC) Films

Nur Liyana Izyan Zailuddin[1], *Azlin Fazlina* Osman[1,*], *Rozyanty* Rahman[1].

[1] Center of Excellence Geopolymer and Green Technology (CEGeoGTech), School of Materials Engineering, Universiti Malaysia Perlis, Arau, Perlis 02600, Malaysia.

Abstract. This study explored the potential of using oil palm empty fruit bunch (OPEFB) in the production all-cellulose composite (ACC) films. The isolation process of the raw OPEFB fiber was carried out using chemical process to extract the OPEFB nanocellulose. The ACC films from the OPEFB and microcrystalline cellulose (MCC) were prepared using dimethylacetamide (DMAC) and lithium chloride solvent system whereby the partially dissolved cellulose was transformed into the matrix phase surrounding the remaining non-dissolved fiber. The ACC films containing 1, 2, 3 and 4 % (wt/vol) OPEFB cellulose and 3 % (wt/vol) microcrystalline cellulose (MCC) were prepared and the effects of 2-ethylhexyl acrylate chemical treatment of the OPEFB cellulose on tensile properties of the ACC film were investigated. Results indicate that the chemical treatment using 2-ethylhexyl acrylate has reduced the hydroxyl group composition in the cellulose and allowing greater dissolution of the cellulose during the formation of the ACC film. As a result, the tensile strength and modulus of elasticity of the ACC film were significantly enhanced. However, both untreated and treated ACC films experienced the reduction in both properties when the cellulose concentration was increased from 1 % to 4% (wt/vol), due to the saturation of the cellulose particles and non-homogeneity of the ACC system.

1 Introduction

Synthetic plastics are daily used materials by billions number of human being from all over the world. However, these petroleum-based plastics can cause several negative impacts to the environment and ecosystem [1]. Due to these ecological concerns, there is an increasing effort to replace the conventional synthetic plastics with the natural based plastics (biodegradable and biorenewable) by undertaking significant number of research and development and translating them into commercial products [1- 4]. Some natural polymers that have been used for packaging application include starch, cellulose and chitin [1, 2, 4]. The addition of natural elements from lignocellulosic fibers (flax, ramie, sisal) as an

* Corresponding author: azlin@email.unimap.edu.my

alternative to the traditional glass fibers can have a much gentler effect to the surrounding [4, 5]. These natural fibers are also economical and have good mechanical properties [5].

In Malaysia, the plantation of palm oil produces large amount of residues like oil palm empty fruit bunch (OPEFB), oil palm shell and other biomass wastes [2, 6]. These lignocelluloses wastes can be generated into something useful for example to produce biomass feedstock, bioconversion and also utilize in composite fields [2, 6, 7]. Lignocelluloses like OPEFB consists a vast amount of cellulose which can be utilized in the production of biocomposite films. Our research group has investigated the utilization of OPEFB derived cellulose in the production of all-cellulose composite (ACC) [2, 6]. ACC is a composite whereby both the matrix and fiber are from the same component; in this case they are from cellulose sources. ACC can be produced by "surface selective dissolution" whereby part of the cellulose fiber is partially dissolved using cellulose solvents which is regenerated to form the matrix surrounding the non-dissolved fiber (which act as the reinforcement for the composite). Part of the cellulose fiber that are dissolved, becomes the matrix of the composite [8, 9]. The main advantage of ACC process is that it simplify the composite's method and can produce composite with good mechanical properties [8, 9].

In the production of polymer composite, processing method and route may affect the morphology and properties of the resultant composite material [10]. This fact also applies to the ACC material. Suitable cellulose solvents are required in order to properly dissolved the cellulose since the dissolution of cellulose can be quite difficult due to the structure of the cellulose itself which consist of their inter and intra hydrogen bond [2, 6, 11, 12]. Some solvents that are known to participate in the dissolution of cellulose include N-methylmorpholine-N-oxide (NMMO) [11], NaOH/urea solution [12] and N-dimethylacetamide/lithium chloride (DMAC/LiCl) [2, 6]. Furthermore, pre-treatment in the form of chemical, need to be introduced to optimize the cellulose dissolution process. In addition, ideal nanocellulose content should be determined to optimize the performance of the resultant regenerated cellulose composite film.

In this communication, we report the production of the ACC films from the OPEFB-derived cellulose and microcrystalline cellulose (MCC). DMAC/LiCl solvent system was used, whereby the partially dissolved cellulose were transformed into the matrix phase surrounding the remaining non-dissolved fiber. The effect of the cellulose chemical treatment by 2-ethylhexyl acrylate on tensile properties of the ACC films were studied and reported herein.

2 Materials and Method

The raw materials used in this study were oil palm empty fruit bunch (OPEFB) which was used as received from Malaysian Palm Oil Board (MPOB) Bangi, Selangor and microcrystalline cellulose (MCC) from cotton linters (Aldrich Chemistry) with the average size of 51 μm. The chemicals used for the production of ACC films were; sodium hydroxide (NaOH), ethanol (C_3H_3OH) and acetone (C_3H_6O), which were supplied by HmbG® Chemicals. Sodium Chlorite ($NaClO_2$) was supplied by Sigma-Aldrich (Germany) and was used in the bleaching treatment of the OPEFB fiber. Sulfuric acid was used during the hydrolysis step of OPEFB and has a molar mass of 98.10 g/mol. 2-ethylhexyl acrylate was used to treat the OPEFB nanocellulose and this chemical was supplied by Merck. N,N-Dimethylacetamide (DMAC) was also manufactured by Merck, Germany and was used to dissolve the cellulose. Lithium chloride (LiCl) was supplied by Across, Belgium and was used together with DMAC to dissolve the cellulose.

2.1 Chemical Isolation Processes of OPEFB Fiber to Obtain the Nanocellulose

Firstly, the OPEFB fiber was treated in 4 % NaOH solution at the temperature of 80 °C for 1 hour under stirring conditions. Then, the filler was subjected to distilled water and filtered in order to remove the alkaline. This step was repeated in a count of four times to assure substantial removal of the hemicellulose. The bleaching treatment was employed on the OPEFB fiber after it was dispersed in distilled water. The resultant OPEFB suspension was then bleached using the diluted $NaClO_2$ and mechanically stirred for 1 hour at 80 °C. The step was repeated three times to assure the removal of the lignin and hemicellulose. The treated fiber was subsequently washed with distilled water and filtered. The acid hydrolysis using 65 % H_2SO_4 was conducted under stirring condition at a temperature of 45 °C. The suspension of OPEFB was washed with cold distilled water in order to stop the reaction. The OPEFB suspension was then centrifuged at the speed of 7500 rpm for about 15 minutes and was repeated about 10 times. The produced nanocellulose was homogenized for 30 seconds at 5000 rpm before being dried in an oven at the temperature of 50 °C for one day.

2.2 Chemical Treatment of OPEFB Nanocellulose using 2-ethylhexyl acrylate

The chemical treatment of the nanocellulose was conducted by employing 3 % (v/v) of 2-ethylhexyl acrylate which was dissolved in ethanol and stirred. Then, OPEFB nanocellulose was added into the solution and was stirred for a while. After stirring, the solution was left overnight. The next day, the treated OPEFB nanocellulose was filtered and dried in an oven at a temperature of 70 °C.

2.3 Preparation of All Cellulose Composite (ACC) Films

Prior to the dissolution of cellulose, the OPEFB nanocellulose and MCC were activated in the following solvents firstly; in acetone (for 1 hour and dried in oven) followed by DMAC (for 1 hour and dried in oven). This activation step was conducted to ensure the cellulose swell before the dissolution process. For the dissolution step, the activated OPEFB nanocellulose and MCC were dissolved in DMAC for 10 minutes. Next, LiCl at 8 % (wt/vol) was added to the mixture. The mixture was constantly stirred for 30 minutes until the LiCl completely dissolved in the solution. The cellulose solution was then poured onto a glass plate and left to develop overnight. The next day, the resultant composite films were washed with distilled water to remove the residual of DMAC/LiCl before letting it dried at room temperature for a day.

2.4 Fourier Transform Infrared (FTIR) Analysis

Perkin Elmer spectrum FTIR spectrometer was used to analysis the functional groups of raw OPEFB fiber, OPEFB nanocellulose and acetic acid-treated OPEFB nanocellulose. The wavenumber range was between 500 - 4500 cm^{-1}. The spectra were recorded with 16 scans and resolution of 4 cm^{-1}.

2.5 Tensile Test

Instron Universal Testing Machine, model 5590, was used to conduct the tensile test which was based on ASTM D882. The cross-head speed of the testing was at 10 mm/min. Five replicates were tested for each samples. The mean values of the tensile strength, elongation at break and modulus of elasticity were recorded.

3 Results and Discussion

Fig. 1 exemplifies the FTIR spectra of the untreated and 2-ethylhexyl acrylate-treated OPEFB nanocellulose. The peaks at 3340 cm^{-1} (untreated nanocellulose) and 3338 cm^{-1} (treated nanocellulose) are assigned to the stretching vibrations of –OH group. The bands at 2888 cm^{-1} for both nanocelluloses correspond to the stretching vibration of C-H group [13-15]. The peaks between 1035 – 1036 cm^{-1} indicate the C-O-C pyranose ring stretching vibration in the nanocellulose [15].

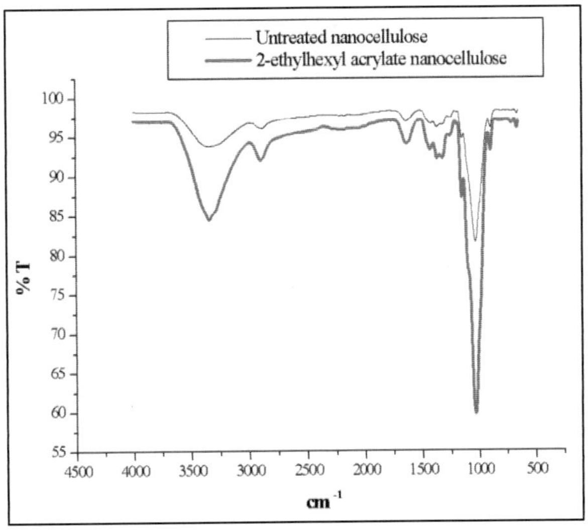

Fig.1. FTIR spectra of the untreated and 2-ethylhexyl acrylate-treated nanocellulose.

The band that is observed on 2-ethylhexyl acrylate-treated OPEFB nanocellulose at 1630 cm^{-1} can be described as the characteristic of carbonyl group (-C=O) [13]. Furthermore, the intensity and strength of the treated nanocellulose peaks as opposed to the untreated nanocellulose suggest that the chemical treatment have led to the substitution of hydroxyl group (OH) from nanocellulose with the carbonyl group (-C=O) originated from acrylate compound [14]. Fig. 2 illustrates the proposed molecular interaction between the nanocellulose and the 2-ethylhexyl acrylate treatment.

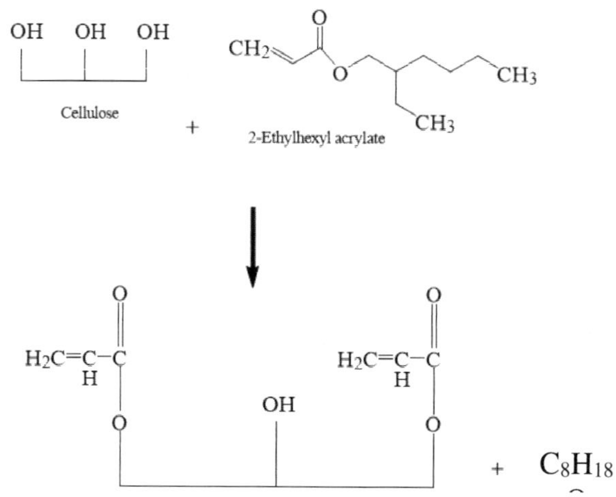

Fig. 2. Proposed molecular interaction between the nanocellulose and 2-ethylhexyl acrylate.

Fig. 3 demonstrates the tensile strength of both untreated and 2-ethylhexyl acrylate-treated ACC films incorporating fix amount of MCC (3 % (wt/vol)) plus OPEFB in the various amounts of 1, 2, 3, 4 % (wt/vol). Both ACC films exhibit similar results whereby the tensile strength decreased as the OPEFB contents increased from 1 % (wt/vol) to 4 % (wt/vol). Although both types of ACC films tend to show similar trend, 2-ethylhexyl acrylate-treated ACC films exhibit greater tensile strength than the untreated counterparts. Apparently, the 1 % (wt/vol) of 2-ethylhexyl acrylate treated OPEFB ACC film displays the highest tensile strength with the value of 14.8 MPa. As compared to the untreated film at 1 % (wt/vol), the increment of 85 % was obtained. In general, the treated ACC films showed better tensile strength than the untreated ACC films. The reason for this could be due to the treatment of the OPEFB nanocellulose with 2-ethylhexyl acrylate that reduced the intra and inter molecular bonding of the nanocellulose by replacing its hydroxyl group with the carbonyl group of the 2-ethylhexyl acrylate. This induced greater dissolution of the nanocellulose in the solvent, thereby more homogeneous composite structure was formed. This can be correlated with the findings of Zailuddin et al. [6] by which the presence of butylmethacrylate (BMA) treatment has led to the increment of the interfacial interactions between the dissolved cellulose matrix and the non-dissolved cellulose fiber in the biocomposite films.

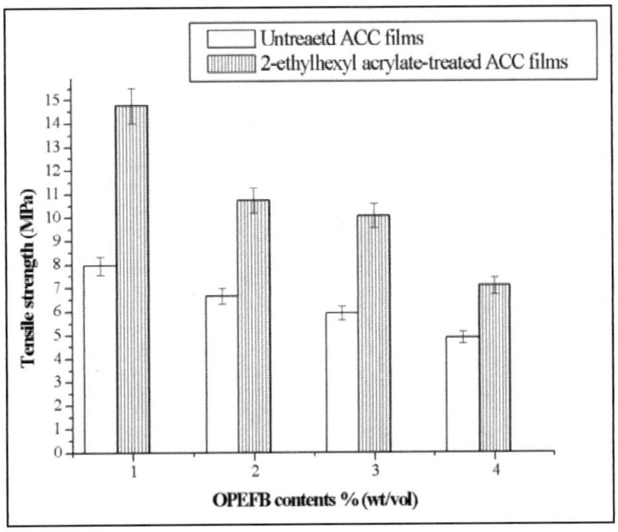

Fig. 3. The effect of OPEFB nanocellulose contents on the tensile strength of the untreated and 2-ethylhexyl acrylate-treated ACC films.

The effect of modulus of elasticity of the untreated and treated ACC films consisting consistent amount of MCC at 3 % (wt/vol) and different amounts of OPEFB nanocellulose (1, 2, 3, 4 % (wt/vol)) is illustrated in Fig. 4. The modulus of elasticity for 2-ethylhexyl acrylate-treated films is higher than the untreated composite films. However, it is observed that both ACC films exhibit similar trend of results in which the modulus of elasticity decreased as the OPEFB contents increased from 1 % (wt/vol) to 4 % (wt/vol). The modulus of elasticity for OPEFB content at 1 % (wt/vol) of 2-ethylhexyl acrylate-treated ACC film is 534.9 MPa while for the 1 % (wt/vol) untreated OPEFB ACC film is 179.8 MPa. The chemical treatment using 2-ethylhexyl acrylate has led to the improvement of the interfacial interactions between both filler and matrix phases of the composite film, hence the stiffness of the material increased. Apparently, both types of ACC films exhibit reduction in the modulus of elasticity when the OPEFB content increased from 1 % to 4% (wt/vol) due to cellulose particle aggregations or saturation that induced the formation of stress concentrated area and non-homogenized composite structure.

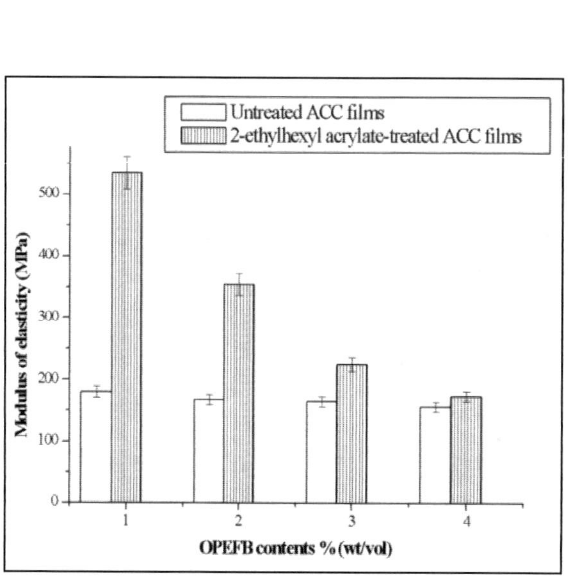

Fig. 4. The effect of OPEFB nanocellulose contents on the modulus of elasticity of the untreated and 2-ethylhexyl acrylate-treated ACC films.

4 Conclusion

In this study, we have investigated the use of 2-ethylhexyl acrylate for chemical treatment of the OPEFB cellulose and analyzed its efficiency to improve the tensile properties of the OPEFB cellulose-derived ACC film. Based on the FTIR spectra, it can be proved that the chemical treatment resulted in the substitution of hydroxyl group (OH) from nanocellulose with the carbonyl group (-C=O) from the acrylate compound. This reduction in hydroxyl group of the cellulose assists in better dissolution of cellulose during the formation of the ACC film. Both treated and untreated ACC films show the highest value of tensile strength and modulus of elasticity when 1 % (wt/vol) of OPEFB was used to form the composite. However, both types of ACC films exhibit reduction in tensile strength and modulus of elasticity when the OPEFB content increased from 1 % to 4% (wt/vol). This could be due to cellulose particle aggregations and saturation that induced the formation of stress concentrated area along the composite structure. Nevertheless, the treated ACC films showed higher tensile strength and modulus of elasticity as compared to the untreated ACC films due to improved interfacial interactions between the dissolved cellulose matrix and the non-dissolved cellulose fiber in the biocomposite films.

References

1. S.Y. Rong, N.M. Mubarak, F. A. Tanjung, J. Environ. Chem. Eng., **5**, 6132–6136 (2017).
2. N. L. I. Zailuddin, A.F. Osman, S. Husseinsyah, Z. Ariffin, F. H. Badrun, *Mechanical Properties and X-Ray Diffraction of Oil Palm Empty Fruit Bunch All-Nanocellulose Composite Films,* Proceedings of the Second International Conference on the Future of Asean (ICOFA) 2017 – Volume 2 – Science and Technology, Springer (2017).
3. S. N. F. S. Adam, A.F. Osman, R. Shamsudin, Tensile properties, biodegradability and bioactivity of thermoplastic starch (TPS)/ bioglass composites for bone tissue engineering Sains Malaysiana, **47**, 1303-1310 (2018).

4. I. Ibrahim, A.F. Osman, T.L. Ping, Effects of ultrasonication process on crystallinity and tear strength of thermoplastic starch/cellulose biocomposites, IOP Conference Series: Materials Science and Engineering, 701 (1), 012045 (2019).
5. M. R. Sanjay, G. R. Arpitha, L. Laxmana Naik, K. Gopalakrishna, B. Yogesha, Natural Resources, **7**, 108-114 (2016).
6. N. L. I. Zailuddin, S. Husseinsyah, F. N. Hahary, H. Ismail, Polymer-Plastics Technology and Engineering, **56**, 109-116 (2017).
7. A. Shariff, N.S.M. Aziz, N. Abdullah, J. Phys. Sci. **25**, 97–112 (2014).
8. N. Soykeabkaew, T. Nishino, T. Peijs, Composites: Part A, **40**, 321–328 (2009).
9. T. Huber, J. Mussig, O. Curnow, S. Pang, S. Bickerton, M.P. Staiger, Journal of Material Science **47**, 1171–1186 (2012).
10. A.F. Osman, K. Jack, G. Edwards, D. Martin, Effect of Processing Route on the Morphology of Thermoplastic Polyurethane (TPU) Nanocomposites Incorporating Organofluoromica, Advanced Materials Research, **832**, 27-32 (2014).
11. H. Zhao, J. H. Kwak, Y. Wang, J. A. Franz, J. M. White, J. E. Holladay, Carbohydr Polym, **67**, 97-103 (2007).
12. C-H Kuo, C-K Lee, Carbohydr Polym, **77**, 41–46 (2009).
13. N.S. Lani, N. Ngadi, A. Johari, M. Jusoh, J. Nanomater. 1-9 (2014).
14. S. Ramli, M. S. Ja'afar, M. A. Abd Sisak, N. Zainuddin, I.A. Rahman, Malaysian Journal of Analytical Sciences, **19**, 275-283 (2015).
15. B. Soni, E.B. Hassan, B. Mahmoud, Carbohydr. Polym. **134**, 581-589 (2015).

IOP Conf. Series: Materials Science and Engineering **957** (2020) 012005 doi:10.1088/1757-899X/957/1/012005

Isolation and characterization of microcrystalline cellulose extracted from banana fiber in poly(lactic acid) biocomposite produced from solvent casting technique

Luqman. M[1,*], *Tamil Arasi* Vesuanathan[1], and *Mohd Nazry* Salleh[1]

[1]Center of Excellence Geopolymer & Green Technology, School of Material Engineering, Universiti Malaysia Perlis,02600 Jejawi, Perlis, Malaysia

Abstract. The purpose of this project is to extract Microcrystalline Cellulose (MCC) from Banana fiber to produce Poly(Lactic Acid)/Microcrystalline Cellulose, (PLA/MCC) biocomposite film using a solvent casting technique. The initial phase is the production of MCC from Banana fiber by involving three steps; alkaline treatment, bleaching and acid hydrolysis. MCC produced from the different concentration of Nitric acid in the acid hydrolysis process, were used to analyze the morphological and crystalline properties using SEM and XRD respectively. MCC produced from the higher concentration of Nitric acid shows good morphological properties and higher % crystallinity. Then, MCC compounded into PLA with different filler loading to produce PLA/MCC biocomposite film using a different solvent which is Chloroform and Dichloromethane. Results show that lower MCC loading, induces good filler matrix interaction and this evidences by the improvement of the Tensile strength and Young's Modulus, as well as shows improvement in loss factor and storage modulus which studied from DMA. Moreover, the addition of MCC slightly improved the thermal stability of PLA. From using a different solvent to produce PLA/MCC biocomposite film, Dichloromethane solvent improves the tensile strength and Young's modulus of biocomposite film, while Chloroform reduces the tensile properties of biocomposite film.

1 Introduction

Poly (lactic acid), (PLA) thermoplastic is a biocompatible polymer derived from renewable resources such as corn, sugar beets, potato, cheese and whey by fermentation processes [2]. PLA can be processed similarly like polyolefins such as Polyethylene and Polypropylene, moreover, PLA have good stiffness and strength [1,2]. PLA is mainly used in commodity areas, for instance, in packaging, textile and composite with the aid of new technologies and for large scale of production [3]. PLA have better properties, but its low deformation at break, high modulus, hydrophilic properties and low heat resistance have limited its application

[*] Corresponding author: luqman@unimap.edu.my

primarily to rigid thermoformed packaging [4]. In this study, PLA based biodegradable composites are prepared with microcrystalline cellulose as the reinforcing phase.

Banana plantation in Malaysia covers 26,000, hectare and it is the second most widely cultivated fruit with the total production of 530,000 metric, ton yearly. For every cycle of banana production, four times of waste are generated. Somehow, the leaves can use for wrapping, but the waste stems getting accumulated in banana growing areas. Thus, recycling consideration of banana stem will solve the country's agriculture disposal problem in eco-friendly way. These agriculture wastes are least cost and also containing cellulose richly, which about 60-65% [5]. In this study, banana fibers chosen to be the raw material to obtain microcrystalline cellulose.

Microcrystalline cellulose, (MCC) is purified, partially depolymerized, non-fibrous form of cellulose that occurs as white. Generally, MCC is odourless, tasteless crystalline powder composed of porous particles and consist of linear homopolysaccharide of β-(1,4) linked anhydro D-Glucose units [6]. MCC which is extracted from natural fiber widely being used in various field such as pharmaceutical, cosmetics, food, plastic composites because of its, compatibility, hydrophilicity, acid-insolubility and biodegradability. There are few methods to extract MCC from lignocellulosic material. In this study, MCC from banana fiber extracted using two steps including cellulose extraction using alkaline pretreatment and MCC preparation through acid hydrolysis method.

MCC in PLA as reinforcing filler has a better dispersion and have ability to improve its properties when using suitable amount of MCC to compound. Moreover, the techniques used to produce PLA/MCC composite also affect the properties of composite produced. Technique used in this study is solvent casting technique; which is a simple technique that used to produce polymer composite with uniform thickness distribution, high optical purity with low haze [7]. Solvents used in this study are Dichloromethane and Chloroform; both solvents are good solvents in order to dissolve PLA [8].

2 Experimental

2.1 Materials

The Poly (lactic acid), PLA is supplied by TT Biotechnology Sdn. Bhd., Penang. Hydrogen Peroxide is supplied by Bendosen Company, while Nitric acid, and Sodium Hydroxide supplied by HmbG Chemicals. Banana fiber obtained from locally in Perlis. The fibers were cut into small pieces and dried under sunlight for 1 week, then crushed to have fine powders.

2.2 Microcrystalline preparation

40g of Banana fibers were soaked in 1000 ml of 1 M Sodium Hydroxide solution overnight. Then the mixture was stirred and heated at 70-80ºC for 4 hours. The treated banana stems were bleached using 10% hydrogen Peroxide in 15% Sodium Hydroxide solution at 90ºC for 45 min. Bleached banana fibers were washed using distilled water until becomes neutral. Acid hydrolysis process was conducted on bleached fibers using 0.5M, 1.0M and 1.5M of Nitric acid in the reflux system, respectively. The hydrolysis process was conducted for 2.5 hours at 70-80ºC. Then, the suspension obtained was filtered and washed to obtained microcrystalline cellulose (MCC). The obtained MCC was dried in an oven at 70ºC for 1 day and ground to obtain the fine powder.

2.3 Preparation PLA/MCC biocomposite

Poly (Lactic Acid), PLA pellets were dissolved completely in 45ml of Dichloromethane solvent and stirred at room temperature. Subsequent PLA pellets dissolve completely, 5 php of Microcrystalline cellulose powder which hydrolyzed using 0.5M Nitric acid was slowly added and stirred for 40min. Then, 45 ml of PLA/MCC mixture was cast on the glass mold of 15cm X 15cm and make sure no air bubble present. Following that, the PLA/MCC composite was let to be dried in room temperature for 24 hours. The steps were repeated to form a composite film of PLA/MCC by using different MCC loading which hydrolyzed using different concentration of Nitric acid followed by another solvent, which is Chloroform. The formulation of PLA/MCC biocomposite film with different concentration of Nitric acid hydrolyzed MCC, different loading of Microcrystalline cellulose powder and the different solvent are shown in Table 1.

Table 1. Formulation of PLA/MCC biocomposite produced from Dichloromethane and Chloroform solvents.

Material	PLA/MCC - Dichloromethane solvent	PLA/MCC - Chloroform solvent
PLA, php	100	100
Concentration of Nitric acid, M	0.5, 1.0, 1.5	0.5, 1.0, 1.5
MCC loading, php	5, 10	5, 10

*php= part per hundred polymer

2.4 Fourier Transform Infrared (FTIR) Spectroscopy

Fourier Transform Infrared (FTIR) spectroscopy was used to identify the changes in the functional group of raw fiber and MCC powder prepared using different acid concentration. It can prove that obtained powders from the isolation process were MCC. Raw fiber and MCC powder tested from Perkin Elmer Spectrometer 2000 using Attenuated total reflection (ATR) method. The scan range was set from 600-4000cm^{-1} at room temperature.

2.5 Scanning Electron Microscope (SEM)

Scanning electron microscope (SEM) machine model JEOL JSM-6460 LA was used to study morphology of MCC powder and tensile fracture surface of biocomposite film. The sample was mounted onto aluminium stubs and coated with a thin layer of palladium to provide a conductive coating. The system was running at 15kV and studied the morphology of samples at a magnification of 500x, 1000x and 1500x.

2.6 X-ray Diffractometer (XRD)

X-ray diffraction (XRD) analysis is used to study the crystallinity behavior of raw banana fiber and MCC powder. An X-ray diffractometer model of Bruker D2 Phaser was used and the test carry out using Ni-filtered Cu Kα radiation with the scan of 10° to 50° (2θ angle range) and scan rate is 2°/min. The crystallinity of Raw Banana fiber powder and MCC powder were calculated using Crystallinity Index (CRI) shows Equation 1.

$$\text{CRI}\ (\%) = \frac{I_{002} - I_{am}}{I_{am}} \times 100\% \qquad (1)$$

2.7 Tensile Properties

The tensile properties of PLA/MCC biocomposite were identified using Universal Tensile Machine (UTM) Instron 5590. Tensile properties such as Tensile strength, Young's Modulus and Break elongation for each sample were obtained by following ASTM D 882 method. The test was conducted at crosshead speed of 10 mm/min and the sample dimension was 80mm×15mm.

2.8 Dynamic Mechanical Analysis (DMA)

Dynamic Mechanical Analysis (DMA) is used to identify the thermomechanical properties of PLA/MCC biocomposite film formed using two different MCC loading. Perkin Elmer DMA 8000 model was used to test the sample with 50mm×5mm dimension in tension mode. The heating rate used is 2°C/mm, the frequency used is 1 Hz and the temperature range set from 30°C-120°C.

3 RESULT AND DISCUSSION

3.1 Properties of MCC powder produced from different concentration of Nitric acid in acid hydrolysis process

3.1.1 Fourier Transform Infrared (FTIR) Spectroscopy

Fig. 1(a) shows the FTIR spectra of Raw fiber and peak at 3289cm^{-1} is assign for O-H stretching vibration from cellulose. The peak at wavelength of 2917 cm^{-1} indicates the C-H stretching while peak at 1011cm^{-1} indicates C-O stretching of primary alcohol of cellulose. Peak at wavelength of 1604 cm^{-1}corresponded to C=O vibration of aromatic skeletal of lignin. In Fig. 1(b-d), same peaks can be observed as in Raw fiber which are at 3317-3319 cm^{-1}, 2890-2893cm^{-1}, and 1014-1016 cm^{-1}.expect peaks at 1325-1326 cm^{-1} and 1644 cm^{-1}. Thus, peak at 1644 cm^{-1}in MCC powder spectra belongs to the strong absorption of water by cellulose which proves that low amount of hemicellulose presence in MCC powder. While peak between 1325-1326 cm^{-1} indicates the presence of crystalline cellulose. Moreover, in the MCC spectra, elimination of 1604cm^{-1}peak evidences the removal of aromatic skeleton of lignin.

3.1.2 X-Ray Diffraction

Fig. 2 shows the X-ray diffraction patterns and the relative degrees of cellulose crystallinity. From Fig. 2, it can be observed, all samples the peaks were at 2θ = 16°, 22.5°, and 36°, where 2θ = 22.5° is belongs to (002) crystallographic peak. From Fig. itself, it is very obvious that, the crystallinity have increased when compared the peaks intensity of MCC powders with its Raw fiber. The crystallinity index,% for all samples were calculated and summarized in the Table 2. The Crystallinity index of 1.5M Nitric acid hydrolyzed MCC is higher compared to other samples which is 51.3% and the Raw fiber has the lower crystallinity % which is 37.5%. The higher the concentration of acid, the higher crytsallinity of MCC powder produced.

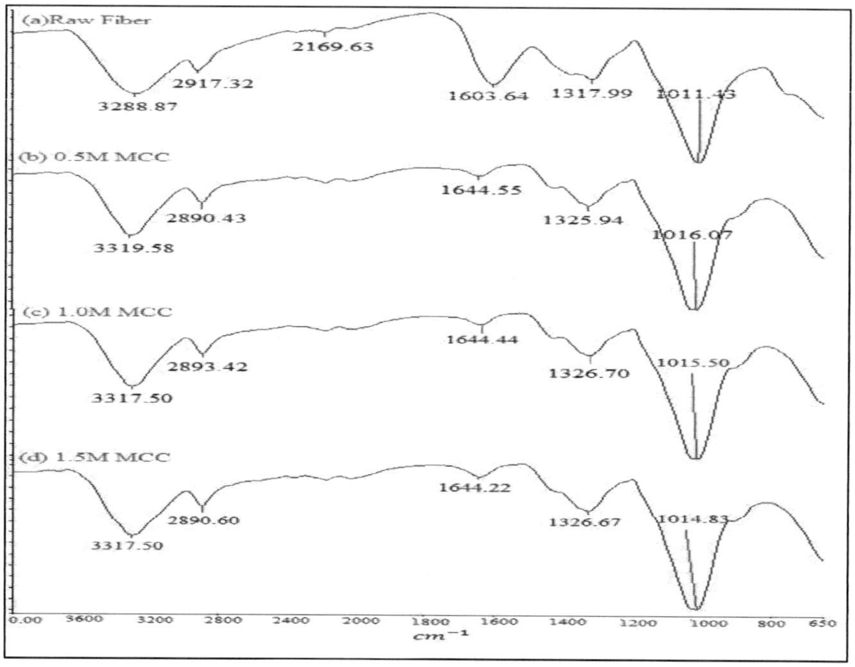

Fig. 1. FTIR analysis of Raw fiber and MCC produced from different concentration of acid.

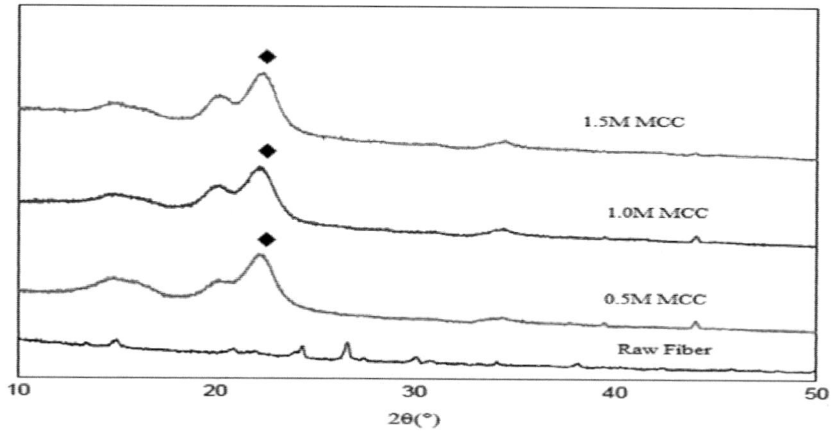

Fig. 2. X-Ray Diffraction results of Raw fiber and 0.5 M, 1.0M, 1.5M Nitric acid hydrolyzed MCC powders.

Table 2. Crystallinity Index of MCC powder.

Sample	Crystallinity (%)
Raw fiber	37.5
0.5M MCC	47.0
1.0M MCC	48.6
1.5M MCC	51.3

Fig. 3(a) and Fig. 3(b) shows SEM micrograph of 0.5M and 1.5M Nitric acid hydrolyzed MCC at ×500 magnification respectively. 0.5M MCC have very rough surface with irregular shape and the particle seems long in length. 1.5M MCC have rough surface with almost regular shapes and the MCC particles seems short in length.

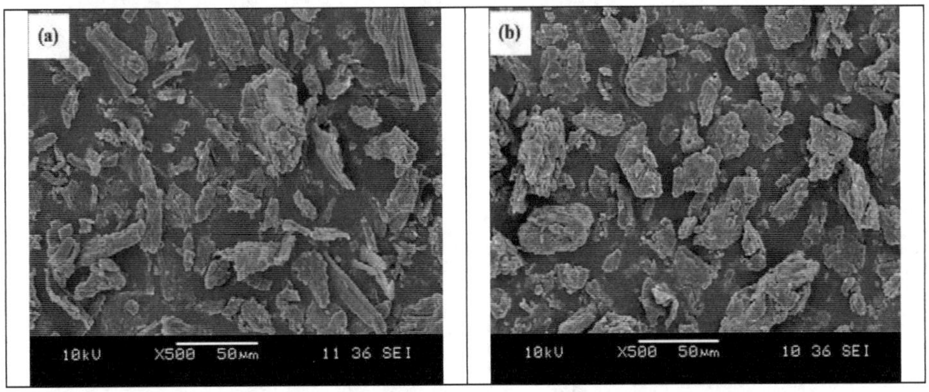

Fig. 3. SEM Micrograph of (a)0.5M and (b)1.5M MCC at magnification of x500.

3.2 Effect of different loading of MCC on the properties of PLA/MCC biocomposite film

3.2.1 Tensile Strength

Fig. 4. depicts the effect of different MCC loading on the tensile strength of PLA/MCC biocomposite film. 5 php MCC filled films have the highest tensile strength compared to 10 php MCC filled films. For instance, the maximum tensile strength is holds by 1.5M/5php MCC/DC film. Lower MCC content in PLA can have better dispersion and better interfacial bonding with matrix. This will eventually increase the tensile strength of biocomposite film. Higher MCC content in PLA tend to flocculates and deteriorates the properties of biocomposite film, which proven by the lower tensile strength of 10php MCC filled PLA/MCC film.

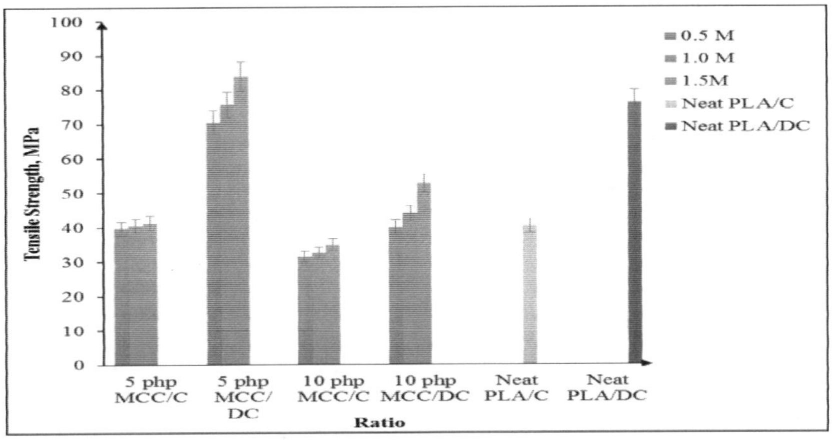

Fig. 4. Effect of different MCC loading in the tensile strength of PLA/MCC biocomposite.

Fig. 5 shows the Young's modulus of PLA/MCC biocomposite films with different MCC loading. The highest value for Young's modulus is shown by 1.5M/5php MCC/ DC. As explained earlier, lower MCC loading will have good dispersion and interfacial bonding with PLA matrices. Thus, higher force is needed to break the samples. The lower Young's modulus is shown by 1.5M/10php MCC/C film. 10 php of MCC loading in PLA matrix tend agglomerate and unable to distribute the stress well under tension. Thus, higher MCC loaded film possess lower Young's modulus.

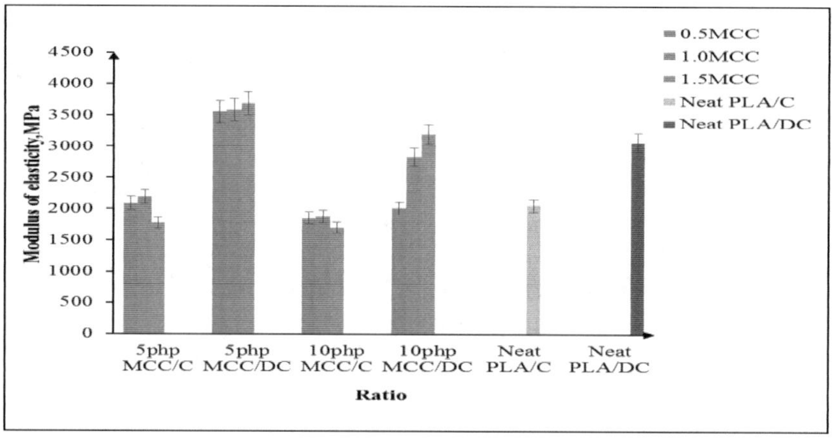

Fig. 5. Effect of different MCC loading inYoung's Modulus of PLA/MCC biocomposite.

3.2.2 Thermomechanical Properties

Fig. 6 shows the loss factors of PLA/MCC biocomposite films with different MCC loading. Tan delta can be defined as dimensionless parameter which measures the ratio of loss modulus to storage modulus [9]. When compares the MCC filled PLA films with Neat PLA/DC, the tan delta peak slightly shifted to higher temperature. Thus, the addition of MCC restricts the movement of molecules and there are some interactions presence between PLA and MCC in filled PLA [1]. Tan delta peaks of Neat PLA/DC which accurately measured at 70.3°C is lower than 1.5M/5MCC/DC (72.0°C) and 1.5M/10MCC/DC (73.3°C) composites. From Fig. 6.0, the tan delta of three composites exhibit in following order, Neat PLA>1.5M/10MCC/DC> 1.5M/5MCC/DC. Thus film containing 5 php MCC possess low tan delta which proves low loss of energy and good interfacial interaction between PLA matrix and MCC filler.

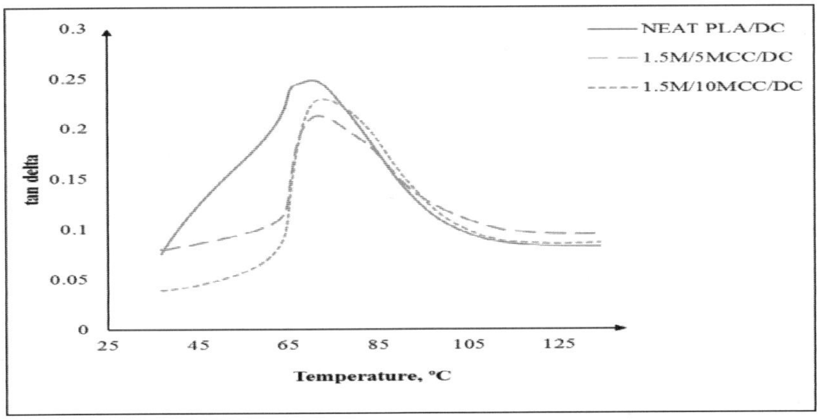

Fig. 6. Temperature dependence on loss factor of PLA/MCC biocomposite film.

3.3 Effect of different solvent on the properties of PLA/MCC biocomposite film

Fig. 7 shows the tensile strength of 1.5M/5MCC/C, 1.5M/10MCC/C, 1.5M/5MCC/DC, and 1.5M/10MCC/DC. From the Fig. it is known that, the tensile strength of 1.5M/5MCC/DC is higher compared to other ratios. Dichloromethane solvent have lower boiling point, thus have the shorter evaporation time. So, when the PLA pellets are dissolved by the Dichloromethane solvent, it will evaporate in short period of time and will not provide the chances for the MCC to flocculate in the PLA matrix. This will result in PLA/MCC film with good dispersion of filler among the matrix which leads to higher tensile strength. PLA/MCC biocomposite film produced from Chloroform solvent have lower tensile strength compared to film produced from Dichloromethane solvent. Chloroform with longer evaporation time results in the flocculation of MCC in the PLA matrix, especially in the film of using 10php MCC. The flocculation deteriorates the dispersion of MCC in the PLA matrix and reduce the properties of PLA/MCC biocomposite film. This can be proven from the lower tensile strength of 1.5M/5MCC/C and 1.5M/10MCC/C compared other two films

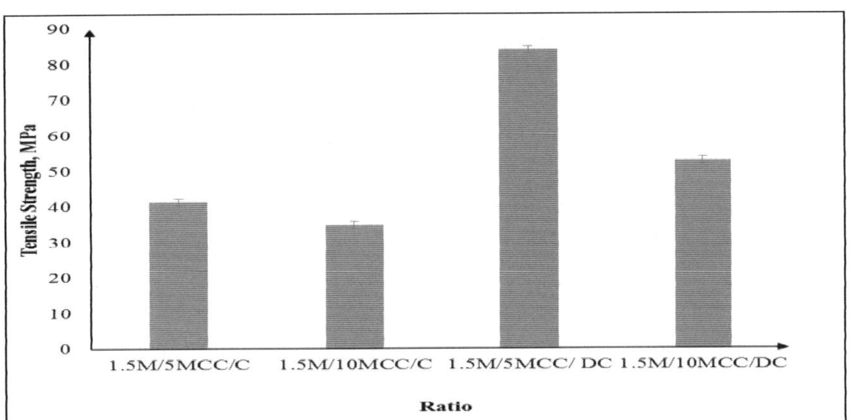

Fig. 7. Effect of different solvent on the tensile strength of PLA/MCC biocomposite film.

Fig. 8 shows the effect of using different solvent on the break elongation of PLA/MCC biocomposite. Fig. 8.0 reveals that the break elongation of Chloroform film is higher than

Dichloromethane films. This can be relate with the crystallinity of sample, where the break elongation will be reduced with increase in crystallinity. Good dispersion of MCC in PLA matrix increases the crystallinity of PLA biocomposite films as well as decreases the break elongation.

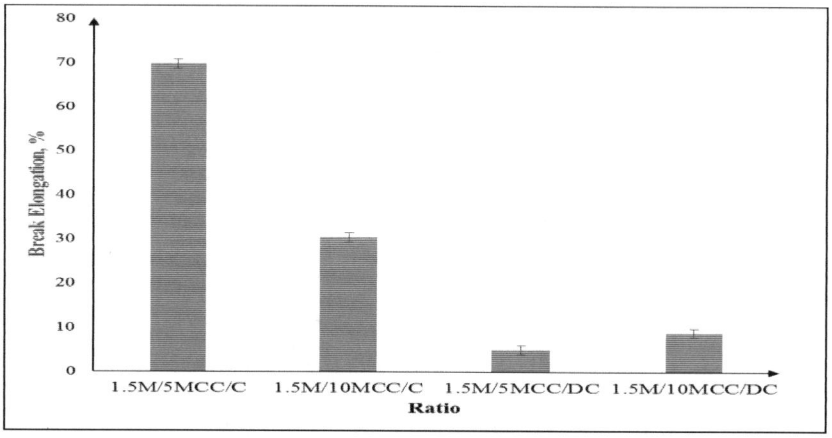

Fig. 8. Effect of different solvent on the Break Elongation of PLA/MCC biocomposite film.

3.3.1 Morphology of PLA/MCC biocomposite film

Fig. 9(a) and Fig. 9(b) shows the SEM micrograph of tensile fractured morphology of PLA/MCC/C and PLA/MCC/DC PLA biocomposite film at magnification of ×500 respectively. In Fig. 9(a), fractured surface of 1.5M/5MCC/C seen to be very rough. The rough surface of the fractured sample is the effect of using Chloroform, which induced a greater chain mobility of the polymer causes a rough surface of the film due to its slow evaporation rate [8]. Moreover, from Fig. 9.0(a) it can be clearly observed the MCC pullout and holes. This indicates that, there are least interfacial bonding between the PLA matrix and MCC filler.

Fig. 9(b), shows the morphology of 1.5M/5MCC/DC film's tensile fractured surface at ×500 magnification. The fractured surface of 1.5M/5MCC/DC is smoother than 1.5M/5MCC/C in Fig. 9(a). Fig. 9(b) reveals that, there are interfacial interactions between the PLA matrix and MCC surfaces. It also can be concluded that, there are least level of pulled out of MCC from PLA matrix. In Fig. 9(b) also have the evidence for the adherence of PLA matrix on the MCC surface. Thus, it can be concluded that, Dichloromethane providing good interfacial bonding between PLA matrix and MCC filler.

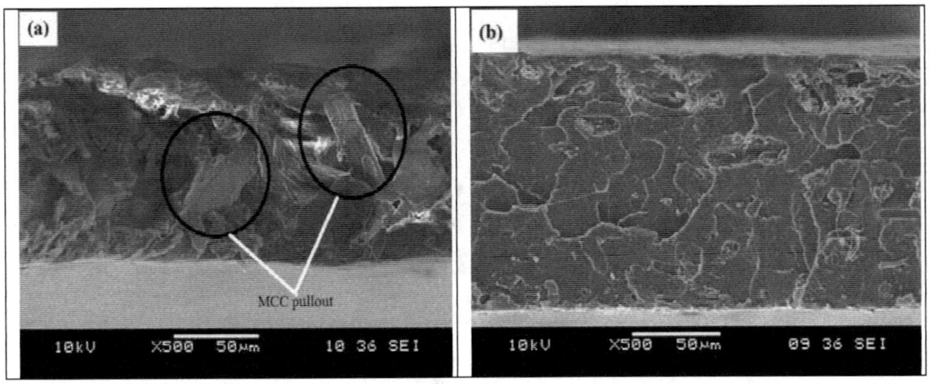

Fig. 9. Morphology of fractured surface of (a) PLA/MCC/C and (b) PLA/MCC/DC biocomposite film.

4 Conclusions

MCC produced from higher concentration of Nitric acids shows higher Crystallinity Index and have smaller size as well as particles with almost regular shape. Incorporation of lower loading of MCC in PLA matrix, improves the tensile strength and Young's modulus in biocomposite. Moreover, MCC in PLA slightly improves the thermal stability and reduces the loss factor. From using different solvents, Dichloromethane have improved the tensile strength and Young's modulus but reduced the break elongation of biocomposite film. PLA/MCC biocomposite film formed by Chloroform solvent shows lower tensile properties. Dichloromethane solvent induces good interaction between PLA and MCC, which can be observe by SEM micrograph, while Chloroform solvent induces poor interaction between PLA and MCC.

This work was financially supported by RACER/1/2019/TK05/UNIMAP/2.

References

1. Mathew, A. P., Oksman, K., & Sain, M., Mechanical Properties Of Biodegradable Composites From Poly Lactic Acid (PLA) And Microcrystalline Cellulose (MCC), *Journal Of Applied Polymer Science,* **97(5)**, 2014-2025 (2005)
2. Byun, Y., Whiteside, S., Thomas, R., Dharman, M., Hughes, J., & Kim, Y. T., The Effect Of Solvent Mixture On The Properties Of Solvent Cast Polylactic Acid (PLA) Film, *Journal Of Applied Polymer Science*, **124(5)**, 3577-3582 (2012)
3. Drumright, R. E., Gruber, P. R., & Henton, D. E., Polylactic Acid Technology, *Advanced Materials*, **12(23)**, 1841-1846 (2000)
4. Pillin, I., Montrelay, N., & Grohens, Y., Thermo-Mechanical Characterization Of Plasticized PLA: Is The Miscibility The Only Significant Factor, *Polymer*, **47(13)**, 4676-4682 (2006)
5. S. Mekhilef, R. Saidur, A. Safari, And W. E. S. B. Mustaffa, Biomass Energy In Malaysia: Current State And Prospects," *Renewable And Sustainable Energy Reviews*, **15(7)**, 3360-3370 (2011)

6. Deepa, B., E. Abraham, B.M. Cherian, A. Bismarck And J.J. Blaker Et Al., Structure, Morphology And Thermal Characteristics Of Banana Nano Fibers Obtained By Steam Explosion, *Bioresour. Technol.*, **102**, 1988-1997 (2011)

7. Siemann, U., Solvent Cast Technology: A Versatile Tool For Thin Film Production. *Scattering Methods And The Properties Of Polymer Materials*, Page 307 (2005)

8. Hughes, J., The Effects Of Solvent Mixture On The Thermal And Mechanical Properties Of Solvent Cast Poly-Lactic Acid (PLA) Film. *Master Thesis, Clemson University*, USA (2011)

9. Zugenmaier, P., Materials Of Cellulose Derivatives And Fiber-Reinforce Cellulose Polypropylene Composites: Characterization And Application. *Pure And Applied Chemistry*, **78(10)**, 1843-1855 (2006)

Photocatalytic Behaviour of TiO_2-geopolymer Paste under Sunlight

Siti Norsaffirah Zailan[1,2], *Norsuria* Mahmed[1,2,*], and *Mohd Mustafa* Al Bakri Abdullah[1,2]

[1]School of Materials Engineering, Universiti Malaysia Perlis, Pusat Pengajian Jejawi 2, 02600 Jejawi, Perlis, Malaysia
[2]Center of Excellence Geopolymer & Green Technology (CeGeoGTech), School of Materials Engineering, Universiti Malaysia Perlis (UniMAP), Taman Muhibbah, 02600 Jejawi, Arau, Perlis, Malaysia

Abstract. The photocatalytic degradation of methylene blue (MB) of the geopolymer paste with addition nanoparticles titanium dioxide (TiO_2) was investigated under illumination of sunlight radiation. In this work, TiO_2-based geopolymer paste was prepare by dry mixing of Class F fly ash (FA) and different amount of nano-TiO_2 (0, 2.5, 5.0, 10.0 & 15.0 wt%). Then, the alkali activator, which is combination of sodium hydroxide solution with sodium silicate was added to the dry mix FA-TiO_2 with ratio of 2.5. All the sample was cured at room temperature and aged for 28 days. The photocatalytic behaviour of nano-TiO_2 was evaluated and compared through photocatalytic testing and surface degradation testing by measuring the photodegradation of MB under sunlight rays. The experimental results revealed that the addition of nano-TiO_2 to the geopolymer paste shows excellent result on photocatalytic degradation of MB after exposure to sunlight up to 150 minutes.

1 Introduction

Recently, in highly polluted area, surface of the building faces the serious problem of tending to become dusty and dull in appearance. Thus, in order to maintain the surface cleaning of building, paint or high consumption of chemical detergents is used, consequently, a high cost was being issued [1]. The release of high amounts carbon dioxide (CO_2) and concrete dust during manufacturing process of Ordinary Portland cement (OPC) is a major contribution to the air pollution. Hence, air pollutions have becoming a hot topic has been attracted attention researcher and industries, which in turn geopolymer become an alternative material to OPC by replacing cements-based binder with industrial by-product such as fly ash (FA) [2-3].

In order to pursue the concept of green environment blending materials, the properties of geopolymer materials was expanded by adding a photocatalyst for self-cleaning purpose. The photocatalyst that most commonly used in cementitious materials is nanoparticles of titanium dioxide (TiO_2) due to its good photocatalytic behaviours. Properties of self-cleaning is developed by addition nanoparticles of photocatalytic material in the geopolymer structure

* Corresponding author: norsuria@unimap.edu.my

[3]. There is a researcher reported that nano-TiO_2 by playing a role of photocatalyst on the surface of cementitious materials, was activated due to the presence of ultraviolet (UV) radiation from the sunlight. In addition, there are two significant advantages by adding photocatalytic materials onto cementitious materials which is, reduce the air pollution and keep the surface of the concrete buildings free from dirt and preserves its colour, even in polluted conditions and industrial areas [4]. Thus, the building aesthetic appearance is maintained and also effectively reduces costs of maintenance in cleaning process.

Based on the literature, research work on the application of self-cleaning onto the geopolymer field are still lacking. This works aims to investigate the potential effect of nanoparticles titanium dioxide as a photocatalyst materials on the photocatalytic behaviour of self-cleaning based-geopolymer paste when directly subjected to sunlight rays.

2 Experimental Method

2.1 Materials and Chemicals

Fly ash as raw materials used in this study is supplied by Cement Industries of Malaysian Berhad (CIMA) in Perlis, Malaysia. The high purity (99.9%) of titanium dioxide (TiO_2) nanoparticles which has a particle size ranging from 30nm to 50nm were obtained from Hongwu Nanometer was used as photocatalyst. In this work, sodium hydroxide (NaOH) solution and sodium silicate (Na_2SiO_3) were purchased from Formosa Plastic Corp., and Taiwan South Pacific Chemical Industries Sdn. Bhd., Malaysia, respectively, and deionized water was used as alkali activator. Methylene blue (MB) used as dye was supplied from Sigma-Aldrich and ethanol (EtOH) solution with a 96% purity was obtained from Altia.

2.2 Mix Design and Sample Preparation

In the preparation of TiO_2-geopolymer paste, the class F fly ash was sieved at 150 µm. The 12M solution of sodium hydroxide (NaOH) was prepared by diluted the 480g of NaOH pellet in 1 litre deionized water, then was allowed to cool at room temperature for 24 hours. The alkali activator solution is a combination of NaOH and sodium silicate (Na_2SiO_3) at ratio 2.5 by mixing 71.43g of 12 M NaOH solution with 178.57g of Na_2SiO_3 solution. Fly ash was dry-mixed first with 2.5 wt% of nano-TiO_2 powder. Then, dry mixture was mixed with alkali activator at ratio of solid-to-liquid ratio of 2.0 that was conducted by mechanical mixer for 15 minutes until a homogeneous paste was obtained. Then, the slurry was casted into disc specimen with a diameter 90 mm and 10 mm thickness (European Standard ISO 679) under curing conditions of room temperature for 28 days. The sample was labelled as Sample $T_{2.5}$. The preparation of the TiO_2-geopolymer paste were repeated by varying the percentage by mass of nano-TiO_2, which are 5.0, 10.0 and 15.0 wt%. These samples were labelled as Sample T_5, T_{10} and T_{15}, respectively. A geopolymer paste without the addition of nano-TiO_2 also was prepared and labelled as Sample T_0. The details about mix proportion of TiO_2-geopolymer paste is illustrated in Table 1.

Table 1. Mix design of TiO_2-geopolymer paste.

Name of Sample	Mix Design					
			Alkali Activator		Mass Ratio	
	TiO_2-nanoparticles (wt%)	Fly Ash (g)	NaOH 12M (g)	Na_2SiO_3 (g)	S/L	Na_2SiO_3/NaOH
T_0	0	500	71.43	178.57	2.0	2.5
$T_{2.5}$	2.5	487.5				
T_5	5.0	475				
T_{10}	10.0	450				
T_{15}	15.0	425				

2.3 Photocatalytic Testing

The ability of TiO_2-geopolymer paste to be functioned as self-cleaning was determined through photocatalytic test. A photocatalytic testing was carried to study the performance of the geopolymer paste with different wt% of nano-TiO_2. The test was conducted by immersed disc specimen with a diameter 90 mm and thickness of 10 mm in 150 ml of MB solution with concentration 10 mg/L. Then, the specimen is directly exposed under sunlight for 150 minutes as shown in Fig. 1.

In photocatalytic testing, an ultraviolet-visible spectrometer (Lambda 25, Perkin Elmer, USA) was used to measure the degree of degradation methylene blue (MB) solution. The degradation degree of the TiO_2-geopolymer paste was obtained using the Eq. 1:

$$A_t = \frac{I_a}{I_o} = \frac{I_o - I_t}{I_o} \quad (1)$$

where,
A_t is degradation ratio
I_o is base absorbency
I_t is absorbency after time t

Fig. 1. Disc of TiO_2-geopolymer paste immersed in methylene blue under sunlight.

2.4 Surface Degradation Testing

The decomposition of organic compound was evaluated through surface degradation testing. Experiment was carried out by applied directly dye the MB solution with the concentration of 30 mg/l on the surface of TiO_2-geopolymer paste disc specimens. Then, the specimens were exposed directly to the sunlight as a natural source for 150 minutes and visually observed in every 30 minutes.

3 RESULT AND DISCUSSIONS

3.1 Photocatalytic behaviour of TiO2-geopolymer paste under sunlight

The ability of TiO_2-geopolymer paste as a photocatalyst material was evaluated through photocatalytic test. Fig. 2 shows the degradation of MB solution under natural source radiation as a function time of TiO_2-geopolymer paste.

The photocatalytic behavior of specimens was evaluated in terms of the efficiency of MB solution degradation. Fig. 2 shows the MB solution was degraded proportional with time. As can see, the colour gradually changed from blue to colourless from 30 min to 150 min. After 90 min exposed to sunlight, it was clearly seen that the MB solutions start to degrade. The role of nano-TiO_2 as a photocatalyst materials in the geopolymer paste was proved by level of MB degradation was increased as a function of time.

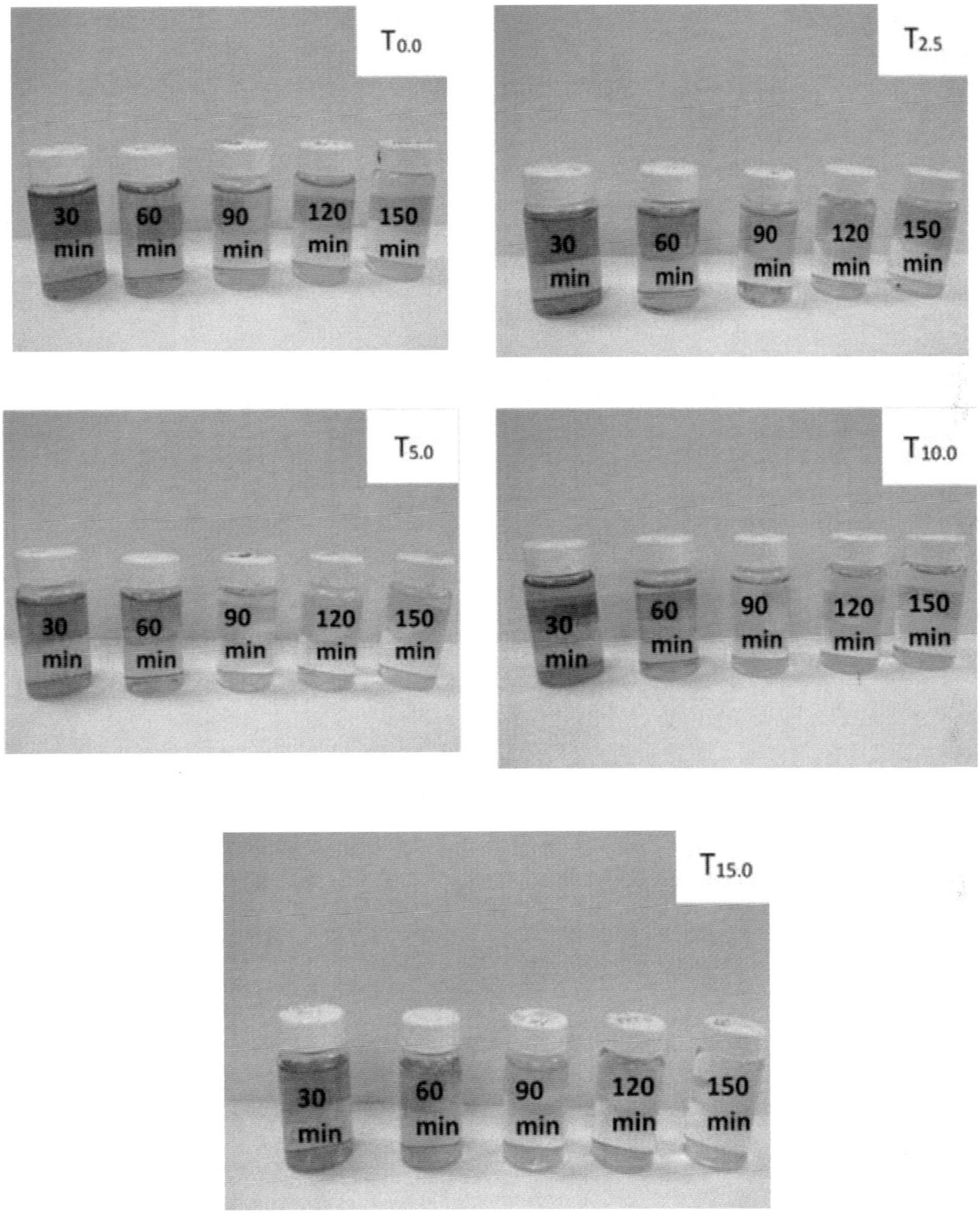

Fig. 2. Degradation MB solutions under sunlight as a function time of TiO_2-geopolymer paste

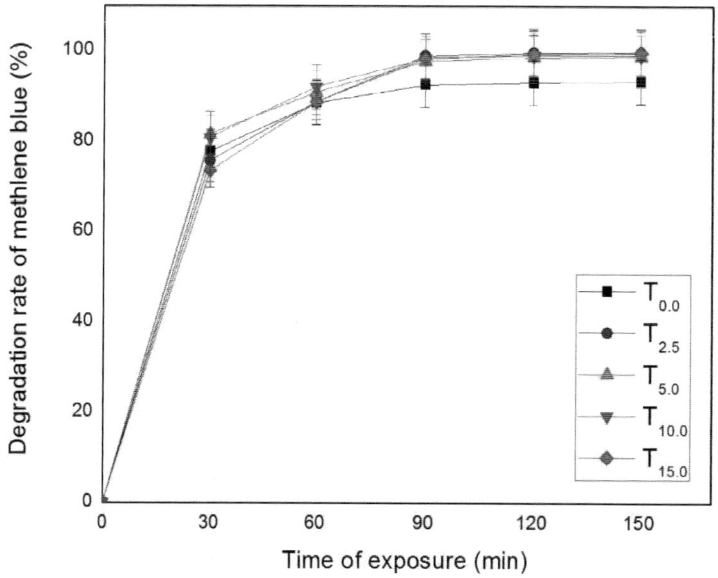

Fig. 3. The relationship between degradation rate of MB and time of exposure with different amount of nano-TiO_2 under sunlight.

The analysis of ultraviolet spectrophotometer was done for all specimen in order to confirm the photocatalytic behaviour of TiO_2-geopolymer paste. Fig. 3 shows the relationship between degradation rate of MB as a function of time with different amount nano-TiO_2 under sunlight. It is observed that there is a similar degradation trend for all sample which indicated that the degradation ratio is proportionally with the exposure time under sunlight. From Fig. 3, sample T_{15} exhibit the highest photocatalytic performance with degradation ratio of MB dye of 0.99 % after 150 minutes. This is due to the nano-TiO_2 that was activated by sunlight and play its role as photocatalyst [5-6].

It also being observed from Fig. 3, after 150 min, sample T_0 which is geopolymer paste without addition photocatalyst nano-TiO_2 also showed the discoloration trend of MB solution. This is due to contents of fly ash, which consists of different amount of mineral oxides such as iron oxide (Fe_2O_3), zinc oxide (ZnO) and TiO_2, which is sensitizers to light [7-9]. In addition, the changes colour of MB was believed not only due to the photocatalytic activity but also due to the fly ash that contain in the geopolymer matrix, which act as an adsorbent for the dye [1,9].

3.2 Surface degradation of TiO2-geopolymer paste under sunlight

The decomposition of organic compounds was examined through surface degradation test. The surface degradation testing was carried out for all specimen under direct exposing to the sunlight. Fig. 4 shows surface degradation of organic dye MB directly applied on the surface of TiO_2-geopolymer paste which exposed under sunlight for 150 min. It can be observed that colour degradation of MB was increased as the amount of nano-TiO_2 increased. This demonstrate that nano-TiO_2 nanoparticles act as catalyst on photocatalytic degradation reaction on the MB as an organic compound [10]. Hence, it can be concluded that nano-TiO_2 nanoparticles have the potential as a photocatalyst to be applied in cementitious materials.

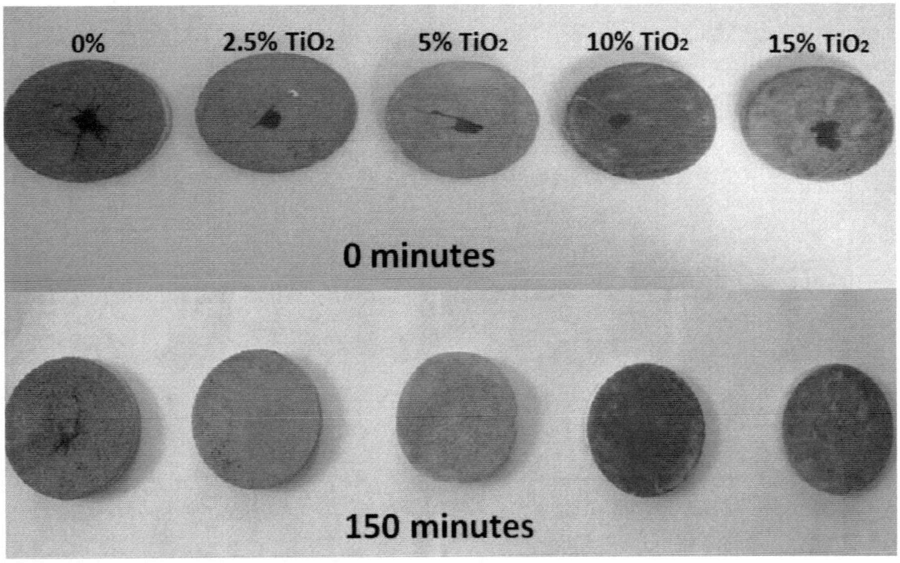

Fig. 4. Surface degradation of methylene blue applied on surface of TiO2-geopolymer paste under sunlight.

4 Conclusion

The following conclusions were reached based on the result analysis as presented in this paper; the addition of TiO_2 nanoparticles as photocatalyst to geopolymer paste enhance the self-cleaning performance. The photocatalytic activity of geopolymer paste with different amount TiO_2 nanoparticles is effective in degrading methylene blue under the sunlight. By increasing the amount of addition TiO_2 nanoparticles, the degradation rate also increased. The TiO_2-geopolymer paste showed a photocatalytic performance compared to the sample without TiO_2 nanoparticles. However, without addition of nano-TiO_2, the geopolymer paste had indicated MB photocatalytic degradation due to some mineral oxide presence in the raw material when exposed directly to UV rays. Thus, TiO_2 nanoparticles have the potential as photocatalyst in order to be applied in building as surface finishing materials.

The authors acknowledge the Center of Excellence Geopolymer & Green Technology (CeGeoGTech), Universiti Malaysia Perlis (UniMAP) for providing the support with all the necessary facilities. Special thanks to those who contributed to this project directly or indirectly.

References

1. J. Kumar, A. Srivastava, A. Bansal, Int. J. Innov. Res. Sci. Eng. Technol. **2(7)**, 2688–2693 (2013)
2. M.Z.N. Khan, Y. Hao, H. Hao, F.U.A. Shaikh, Cem. Concr. Compos. **104**, 103343

(2019)
3. K. Carter, P. Ziehl, R. Anay, L. Assi, E. (Eddie) Deaver, J. Clean. Prod. **198**, 1641–1651 (2018)
4. A. Kitab, M. Alam., H. Riaz, S. Rauf, Int. J. Adv. Life Sci. Technol. **1(4)**, 47–53 (2014)
5. W. Shen, C. Zhang, Q. Li, W. Zhang, L. Cao, J. Ye, J. Clean. Prod. **87(C)**, 762–765 (2015)
6. R. Khataee, V. Heydari, L. Moradkhannejhad, M. Safarpour, S.W. Joo, J. Nanosci. Nanotechnol. **13(7)**, 5109–5114 (2013)
7. P. Krishnan, M.H. Zhang, L. Yu, H. Feng, Constr. Build. Mater. **44**, 309–316 (2013)
8. S.N. Zailan, N. Mahmed, M.M.A.B. Abdullah A. V. Sandu, IOP Conf. Ser. Mater. Sci. Eng. **133**, 1-7 (2016)
9. M.M.A.B. Abdullah, K. Hussin, M. Bnhussain, A.R. Rafiza, Y. Zarina, ACI Mater. J. **109(5)**, 503–508 (2012)
10. K. Loh, C.C. Gaylarde, M.A. Shirakawa, Constr. Build. Mater. **167**, 853–859 (2018)

IOP Conf. Series: Materials Science and Engineering **957** (2020) 012007 doi:10.1088/1757-899X/957/1/012007

The Effects of Epoxidised Natural Rubber (ENR-50) on Cure Characteristics and Tensile Properties of Recycled Natural Rubber Catheter Filled Ethylene Propylene Diene Monomer (EPDM) Vulcanizates

Muhammad Salihin Zakaria[1,2,*], *Nik Noriman* Zulkepli[1,3], *Razif* Muhammed Nordin[4,5], *Khairul Anwar* Abdul Halim[1,2] and *Hafizuddin* Hanafi[2]

[1]Center of Excellence Geopolymer and Green Technology (CEGeoGTech), School of Materials Engineering, Universiti Malaysia Perlis (UniMAP), 01007, P.O Box 77, D/A Pejabat Pos Besar, Kangar, Perlis, Malaysia
[2]School of Materials Engineering, Universiti Malaysia Perlis (UniMAP), 01007, P.O Box 77, D/A Pejabat Pos Besar, Kangar, Perlis, Malaysia
[3]Faculty of Engineering Technology (FETech), Universiti Malaysia Perlis (UniMAP), Level 1 Block S2, UniCITI Alam Campus, Sungai Chucuh, Padang Besar, 02100, Perlis, Malaysia
[4]Department of Chemistry, Faculty of Applied Sciences, Universiti Teknologi MARA, Perlis Branch, Arau Campus, 02600 Arau, Perlis, Malaysia
[5]Green and Functional Polymer Research Group, Universiti Teknologi MARA, 40450 Shah Alam, Selangor, Malaysia

Abstract. The effect of epoxidised natural rubber (ENR-50) on cure characteristics and tensile properties (tensile strength, modulus at 100% elongation and elongation at break) of recycled natural rubber catheter filled ethylene propylene diene monomer (EPDM) vulcanizates were studied. The scorch time, t_2 of EPDM/NRr blends with and without ENR-50 increased as NRr content increased while the cure time, t_{90} is decreased. However, with the presence of ENR-50 in EPDM/NRr blends, the t_2 was longer and t_{90} was shorter than the uncompatibilized EPDM/NRr blends. The minimum torque, M_L, of EPDM/NRr blend with and without ENR-50 relatively increased while maximum torque, M_H, decreased with increasing NRr content. With the presence of ENR-50, the compatibilized EPDM/NRr blends exhibit a lower value of M_L and M_H compared with uncompatibilized blends. The incorporation of ENR-50 in EPDM/NRr blends enhanced the tensile strength and tensile modulus of compatibilized EPDM/NRr blends compared to uncompatibilized EPDM/NRr blends. The elongation at break (E_b) of compatibilized EPDM/NRr blends is lower than the uncompatibilized

* Corresponding author: salihin@unimap.edu.my

EPDM/NRr blends at all blend ratios. This indicated that ENR 50 improved the curing characteristics and tensile properties of the compound.

1 Introduction

In many applications, such as automobiles, boots, toys, and others, the wide use of rubber materials has resulted in high post-consumer waste production. This rubber wastes significantly contribute to environmental pollution, due to the presence of three-dimensional networks in rubber goods, such rubber waste does not degrade rapidly. Poor disposal of rubber goods globally is a possible threat to human health or even to the living and decreases environmental quality [1].

The emergence of a higher percentage of rubber discarded in rubber industries is due to the unstable nature of the rubber compound and the strict quality requirements of the rubber products. Such discarded rubbers contain about 95% very high-quality rubber hydrocarbons, which are just slightly cross-linked [2]. These rejects, if not used accordingly, may cause severe environmental and ecological problems. An example of the rejected rubber product is the catheter which is mainly used in the medical and health care facilities. A catheter is a flexible tube inserted into some part of the body which provides a fluid passage channel or a medical device.

In this study, the effect of epoxidised natural rubber (ENR-50) of recycled natural rubber catheter filled ethylene propylene diene monomer (EPDM) vulcanizates on tensile properties such as tensile strength, modulus at 100% of elongation and elongation at break will be determined. Besides, the effect on the cure characteristics also will be studied.

2 Materials and Methods

2.1 Materials

Epoxidized natural rubber (ENR-50) was used as a compatibilizer in this study. Filler used were recycled natural rubber catheter (NRr) and ethylene propylene diene monomer (EPDM). Other additives used were Ncyclohexyl-2-benzothiazole sulfonamide (CBS), zinc oxide, sulphur and stearic acid. All materials are listed as shown in Table 1 within their respective source.

Table 1. List of material used.

Materials	Description	Source
EPDM	Elastomer	Malaysian Rubber Board
ENR-50	Compatibilizer	
Recycled Natural Rubber Catheter (NRr)	Filler	Teleflex Medical Sdn. Bhd.
CBS	Accelerator	Anchor Chemical Co (M) Ltd.
Zinc oxide	Activator	
Stearic acid		
Sulphur	Vulcanizing agent	

2.2 Compounding

The recycled natural rubber catheter (NRr) powder forming was done in a grinder machine (RT-34 Vertical High-Speed Pulverizing Machine) using 100 mesh (149 microns). ENR-50 was mixed using a two-roll mill for all the compounding ingredients. The rubber vulcanizates were prepared using different formulations, as shown in Table 2. The EPDM/NRr blends with ENR-50 were formulated with mixing ratios of 80/10/10, 70/20/10, 60/30/10, 50/40/10 and 40/50/10 respectively.

Table 2. The formulation for recycled natural rubber catheter filled EPDM vulcanizate.

Ingredients (phr)	Blend				
	R10	R20	R30	R40	R50
EPDM	80	70	60	50	40
NRr	10	20	30	40	50
ENR-50	0/10	0/10	0/10	0/10	0/10
Zinc oxide	5	5	5	5	5
Stearic acid	2	2	2	2	2
Sulfur	2	2	2	2	2
CBS	1	1	1	1	1

2.3 Measurements

2.3.1 Cure characteristics

The cure characteristics such as scorch time (t_2), cure time (t_{90}), minimum torque (M_L) and maximum torque (M_H) were measured using a rheometer model HT- M 2000 operated at 160°C according to ASTM D5289.

2.3.2 Mechanical properties

The dumbell-shaped samples were prepared according to ASTM D-412 for tensile testing [3]. The tensile test was performed at a crosshead speed of 500mm/min using Instron Machine. Tensile strength, elongation at break and modulus at 100% elongation (M_{100}) were measured.

3 Results and Discussion

3.1 Cure Characteristics

The results of scorch time (t_2) and cure time, t_{90} of EPDM/NRr blends with and without ENR-50 are displayed in Fig. 1. It was observed that the t_2 of EPDM/NRr blends with and without ENR-50 relatively increased as NRr content increased. This shows that the NRr has higher scorch times than the EPDM resulting in the increase in scorch time of the blends of both rubbers. The high value for R10 and the decrease for R50 with ENR-50 because the amount of mastication during mixing varies. The EPDM/NRr blends with ENR-50 exhibit longer scorch times than EPDM/NRr without ENR-50 which indicates improved processability after compatibilization of the blends [2].

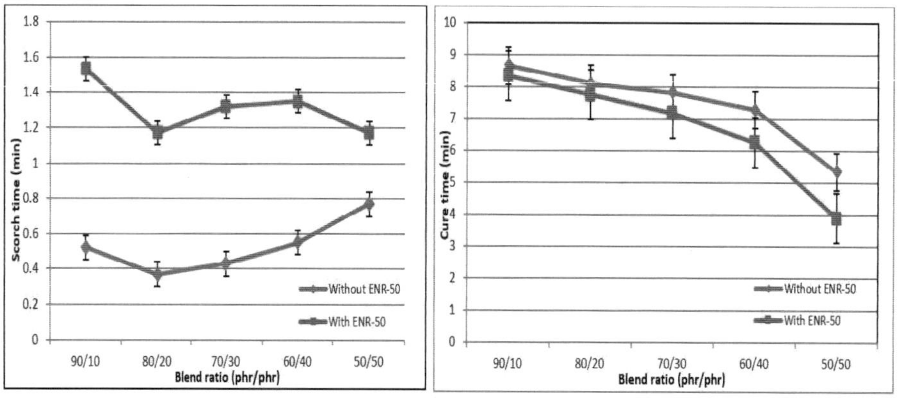

Fig. 1. The scorch time, t_2 and cure time, t_{90} of EPDM/NRr blends with and without ENR-50.

It also can be seen that cure time, t_{90} of EPDM/NRr blends with and without ENR-50 decreased as NRr content increased as shown in Fig. 1. As for the cure rate, NRr-rich blends displayed lower cure rates than the rich EPDM blends. The initial cure time value that is higher than NRr-rich blends can be due to the low efficiency of EPDM when vulcanized with the sulfur system, as well since EPDM has relatively low diene content [5]. It is also found that an increase in NRr content in rubber blends reduces cure time due to the presence of crosslinked precursors and unreacted curatives in recycled rubber, thus accelerating the vulcanization process [3].

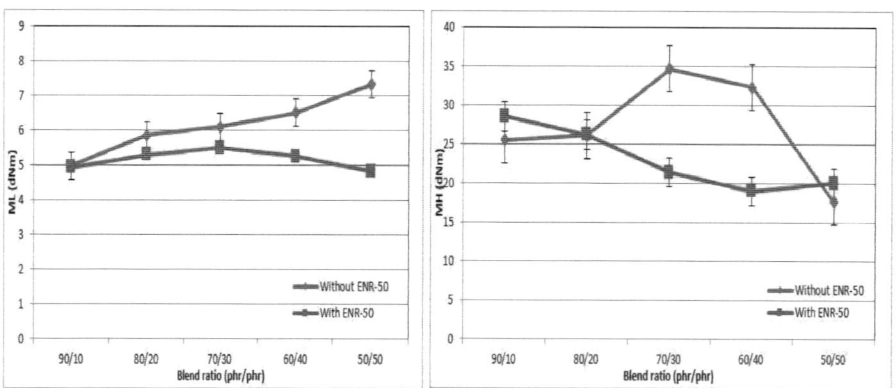

Fig. 2. The minimum torque, M_L and maximum torque, M_H of EPDM/NRr blends with and without ENR-50.

The minimum torque, M_L, of EPDM/NRr blends with and without ENR-50 is relatively increased with increasing NRr content as displayed in Fig. 2. This is possibly due to the presence of cross-linked NRr rubber and other additives. Once the content of NRr increases in the EPDM/NRr blends, the flowability of the blends is restricted and thus the M_L is increased. However, the EPDM/NRr blends with ENR-50 showed lower M_L especially in contrast to EPDM/NRr blends without ENR-50. This suggests improved processability after compatibilization of the blends [2].

Fig. 2 also shows the maximum torque, M_H, of EPDM/NRr blends without ENR-50 relatively decreased as NRr content increased. As the filler itself is rubber in this situation, the combined effect of cross-link density variation and the existence of cross-linked particles may result the variation. At 10 phr of NRr content, the presence of cross-linked particles

IOP Conf. Series: Materials Science and Engineering **957** (2020) 012007 doi:10.1088/1757-899X/957/1/012007

increases M_H, but sulfur migration is predominant at later stages of processing, leading to reduced cross-link densities [2]. This causes a reduction in M_H values. It is also can be seen that EPDM/NRr blends with ENR-50 relatively showed a lower value of M_H than EPDM/NRr blends without ENR-50. By reducing the stiffness of the compound, this proves that the ENR-50 improves the blends.

3.2 Tensile Properties

The tensile strength of EPDM/NRr blends with and without ENR-50 is shown in Fig. 3. Overall, both EPDM/NRr blends showed the same trend, with reduced tensile strength as NRr content equivalent to the EPDM content. These findings are due primarily to the blends being incompatible.

NRr content in the blend mainly attributed to the tensile strength of these binary blends. Binary NR and EPDM blends were incompatible and demonstrated weak interfacial adhesion, leading to poor strength properties. As the filler content increases, the tensile strength increases significantly and reaches a maximum value at 30 phr content. This is primarily due to the presence of the crosslinked structure in the recycled natural rubber [5]. The tensile strength values show a decrease or leveling-off behavior beyond the 30 phr material. There will be a strong tendency for agglomeration at higher filler loading, which can diminish the interfacial bonds [2]. Therefore, tensile strength drops suddenly.

Nevertheless, the tensile strengths of the EPDM / NRr blends have been enhanced with the inclusion of ENR-50 in EPDM / NRr blends. This can be clarified by a better distribution of NRr in the EPDM matrix with the existence of ENR-50, thus enhancing the interfacial adhesion of EPDM/NRr blends by decreasing the interfacial energy between phases [2]. A contrast of the 70/30 EPDM/NRr blend with and without compatibilizer presents that the tensile strength of blend with ENR-50 is higher than the blend without compatibilizer. The increase in tensile strength increases gradually up to 30 phr of NRr in the EPDM/NRr blend. However, with the amount of NRr increased to 40 phr, the tensile strength decreases again. This is because there is no co-curing between the ENR-50 and the other rubber forms, as it has a fully saturated structure. The depletion of co-cure has a detrimental effect on the final properties [6]. It is evident that 10 phr of ENR-50 is inadequate to provide enhanced tensile properties for the blend.

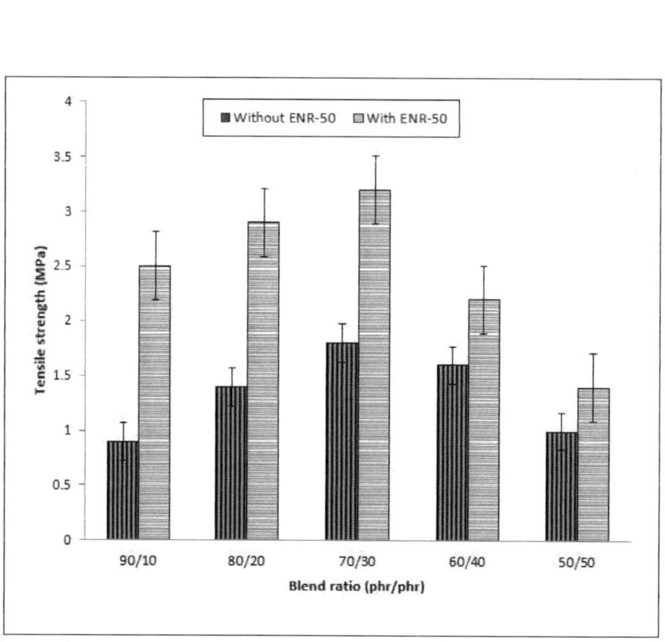

Fig. 3. The tensile strength of EPDM/NRr blends with and without ENR-50.

The effect of ENR-50 on the tensile modulus (M_{100}) of EPDM/NRr blends is shown in Fig. 4. It does seem that for both EPDM/NRr blends with and without ENR-50, it increased gradually as NRr content is increased, the emergence of NRr into the rubber matrix has increased the vulcanizates' stiffness. The increase in M_{100} of compatibilized EPDM/NRr blends was attributed to an enhancement in cross-linking density.

With the introduction of ENR-50 as a compatibilizer in melt compounds of natural rubber composites, the tensile modulus of the natural rubber composites has been reported to increase due to an increase in crosslink density [3]. With the inclusion of ENR-50 in EPDM /NRr blends, partial compatibility may occur through hydrogen bonding due to the interaction between NRr and the ENR-50 epoxy group. The isolated double bond in ENR-50 deterred the formation of intra-molecular sulfide links, which increased the rate of crosslinking, leading to an increase in the modulus [3].

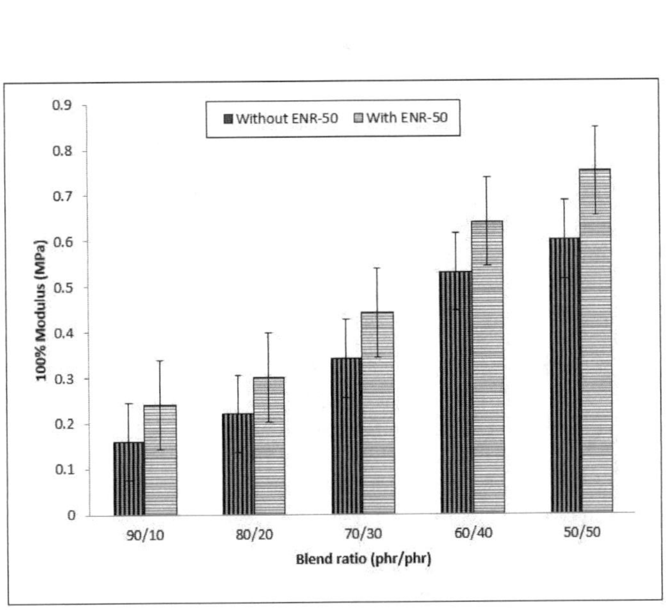

Fig. 4. The tensile modulus of EPDM/NRr blends with and without ENR-50.

Fig. 5 presents the effect of ENR-50 on elongation at break (E_b) of EPDM/NRr blends. The findings show that for EPDM/NRr blends with and without ENR-50 it decreased comparatively as NRr content increased. The increase in cross-linked density is stated to have led to a decrease in E_b value [3]. It would seem that the EPDM / NRr blends with ENR-50 were lower E_b than the EPDM / NRr blends without ENR-50. This finding is possibly due to the existence of various additives, crosslinked precursors and unreacted curative in the NRr itself, together with ENR-50, which impedes the flow and mobility of the EPDM/NRr blends.

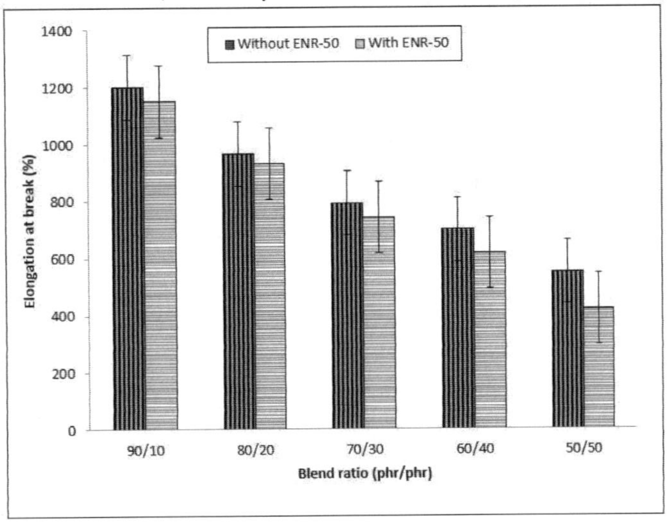

Fig. 5. The elongation at break (E_b) of EPDM/NRr blends with and without ENR-50.

4 Conclusions

The following conclusions can be drawn from this study;

1. The scorch time, t_2 of EPDM/NRr blends with and without ENR-50 increased as NRr content increased while the cure time, t_{90} is decreased. However, with the presence of ENR-50 in EPDM/NRr blends, the t_2 was longer and t_{90} was shorter than the EPDM/NRr blends without ENR-50.
2. The minimum torque, M_L, of EPDM/NRr blend with and without ENR-50 increased while the maximum torque, M_H, decreased with increasing NRr content. With the presence of ENR-50, the EPDM/NRr blends exhibited a lower value of M_L and M_H than EPDM/NRr blends without ENR-50.
3. The incorporation of ENR-50 in EPDM/NRr blends enhanced the tensile strength and tensile modulus of compatibilized EPDM/NRr blends compared to uncompatibilized EPDM/NRr blends. The elongation at break (E_b) of compatibilized EPDM/NRr blends is lower than the uncompatibilized EPDM/NRr blends at all blend ratios.
4. ENR-50 shows the effect of compatibilizer in EPDM/NRr blends by improving the curing properties and tensile properties of the compound.

The authors acknowledge School of Materials Engineering, Universiti Malaysia Perlis for the lab facilities. Special thanks to those who contributed to this project directly or indirectly.

References

1. M Nuzaimah et al, IOP Conf. Ser.: Mater. Sci. Eng. **368** 012016 (2018)
2. N.Z. Noriman, H. Ismail, A.A. Rashid, Polym. Test., **29**, 200-208 (2010)
3. H. Zhang, R. N. Datta, A. G. Talma, J. W. M. Noordermeer, Eur. Polym. J, **46**, 754–766 (2010)
4. H Ahmad, H Ismail, A Rashid, Sains Malays., **44**(6) 835–842 (2015)
5. Riyajan S A, Intharit I, and Tangboriboonrat P., Ind Crop Prod **36** 376-382 (2012)
6. H. Zhang, R. N. Datta, A. G. Talma, J. W. M. Noordermeer, Eur. Polym. J, 46, 754–766 (2010)

IOP Conf. Series: Materials Science and Engineering **957** (2020) 012008 doi:10.1088/1757-899X/957/1/012008

Structural, electronic and thermoelectric properties of SrTiO3 ceramic doped by lanthanum using first principles

Akeem Adekunle Adewale[1,3,4,*], *Abdullah* Chik[1,2,3], *and Ruhiyuddin* Mohd Zaki[2,3]

[1]Centre for Frontier Materials Research, Universiti Malaysia Perlis, 01000 Kangar, Perlis, Malaysia
[2]Center of Excellence Geopolymer and Green Technology (CEGeoGTech), School of Materials Engineering, Universiti Malaysia Perlis (UniMAP), 02600 Jejawi, Arau, Perlis, Malaysia
[3]School of Materials Engineering, Universiti Malaysia Perlis (UniMAP), Taman Muhibbah 02600, Jejawi, Arau, Perlis, Malaysia
[4]Departsment of Physics and Materials science, Kwara State University, Malete, Kwara State, Nigeria

Abstract. The $Sr_{0.92}La_{0.08}TiO_3$ model compound was studied by means of the first-principles method grounded on the full-potential linearized augmented plane wave with the local orbital method as implement in WIEN2k. The structures, electronic and thermoelectric properties of the model compound are examined. Geometry optimization of $Sr_{0.92}La_{0.08}TiO_3$ compound structure reveals the lattice constant of 3.9451 Å, which is the same as un-doped $SrTiO_3$. The direct bandgap of 2.48 eV (Γ-Γ) was measured from computed electronic band structure and its displayed density of states for $Sr_{0.92}La_{0.08}TiO_3$ sample. Transport properties were calculated using BoltzTraP code as implemented in the WIEN2k code. The results were analyzed as a function of the variable temperatures, the density of state and chemical potential. Temperature and density of state variation with transport properties are in the same trend. Chemical potential proves material as n-type via Seebeck coefficient. The thermoelectric performance ZT of 0.05 was obtained at room temperature, while maximum value of 3.79 was recorded at 1200 K.

1 Introduction

Thermoelectric (TE) devices convert waste thermal energy to electrical energy through the Seebeck effect principle and vice versa by the Peltier effect [1]. This conversion has been used for many applications in solid-state power generation and cooling [2-5]. The conversion efficiency of thermoelectric materials is related to dimensionless figure of merit (ZT) and it is expressed as [6]; $ZT = (S^2/\rho k)T$, where S is thermopower (Seebeck coefficient), ρ is electrical resistivity, k is thermal conductivity and T is temperature. The k is the summation of; lattice thermal conductivity k_{lat} and electrical thermal conductivity k_e. Strontium titanate ($SrTiO_3$) is a transition metal oxide that has attracted significant attention as alternative thermoelectric (TE) material due to its potentials such as less toxicity, high temperature

* Corresponding author: adewaleakeem12@gmail.com

resistance, etc. It is an n-type semiconductor with high Seebeck coefficient (~850μ VK^{-1}), low electrical conductivity and high thermal conductivity (6 Wm/K) at room temperature [7,8]. Hence, the thermoelectric performance of un-doped $SrTiO_3$ is low due to high thermal conductivity with a low electrical conductivity that emanates from the wide band-gap insulator. There are many approaches to improve the properties of $SrTiO_3$ such to doping, oxygen vacancy and many others.

Generally, the addition of impurity to thermoelectric materials has successfully enhanced the ZT value by reducing the thermal conductivity while conserving the electrical conductivity and thermopower [9-11]. The high power factor of 28 – 36 $\mu Wcm^{-1}K^{-2}$ had been obtained by the La-doped $SrTiO_3$ compound at room temperature [12] which is similar to state-of-art material Bi_2Te_3 compounds [3]. Likewise, a large Seebeck coefficient similar to that of Bi_2Te_3 was described by Okuda et al., 2001 for $Sr_{1-x}La_xTiO_3$ ($0 \leq x \leq 1$) [12]. The pentavalent impurity in $SrTiO_3$ with higher valence ions on Sr- or Ti-site and by oxygen vacancies [13-17], gives ZT values at range of 0.36 - 0.41 for single doping between temperature of 970 K and 1100 K[10,20,21]. In case of co-doping on A - and B - sites of ABO_3 perovskite compound using La- and Nb - atoms respectively, ZT value > 0.6 at a temperature above 1000 K was reported [16]. In this present study, first principle approach based on density functional theory within PBE-GGA approximation [20] was used as implemented in WIEN2k code. Structural and electronic properties of pure $SrTiO_3$ and La-doped $SrTiO_3$ were investigated. Thermoelectric transport properties of $Sr_{0.92}La_{0.08}TiO_3$ compound were calculated using BoltzTraP code [21]. The transport results were analyzed with their influence on temperature, the density of states and chemical potential.

2 Methods of calculation

The first-principles calculation was used to investigate the $Sr_{0.92}La_{0.08}TiO_3$ compound based on WIEN2k code [22]. Full potential linearized augmented plane wave plus local orbital method within density functional theory (DFT) calculation was performed including spin-orbit coupling. An exchange-correlation function Perdew-Burke Ernzerh (PBE) of generalized gradient approximation (GGA) [20] is adopted for calculation of total ground state energy as well for optimizing of the structural cells internal parameters. The kinetic energy cutoff K_{max} = 6/RMT was used for the plane-wave basis set during energy convergence. Muffin-tin radii (RMT) were set as 2.5 a.u for Sr and La atoms, 1.95 and 1.76 a.u for Ti and O atoms respectively. The maximum value for waves function within muffin-tin of the atomic sphere is l_{max} = 12, while the charge density is Fourier expanded up to G_{max} = 24 (a.u.). In self-consistent calculations, the total energy and charge convergence norms were set to 10^{-4} Ry and 0.001e, respectively. The core states and valence states separation energy was fixed to -6.0 Ry.

On-site electronic interaction Hubbard U of 0.8 eV was set for all calculations with the exchange parameter (J = 0). Based on the use of localized basis sets the U value was imposed on d state of Ti atom [23]. The Monkhorst-Pack's [24] k-points grid of 6×6×4 is used in the irreducible Brillouin zone (IBZ) for electronic structure calculations. Geometry optimization of the study compound was performed using the equation of state by Birch-Murnaghan's [25]. The thermoelectric transport properties were calculated using semi-classical Boltzmann transport equation as applied in BoltzTraP code [21]. A high dense k-point of 14×14×9 is used for transport calculation. Calculations are performed from temperature of 300 K to 1200 K. For a variation of the thermoelectric (TE) properties and temperature, TE results at the chemical potential of 0 K Fermi energy were used for plots. Likewise, TE values were plotted against the chemical potential of range -2 to 2 eV and as well as DOS (182 – 207 states/eV), in order to investigate the effects of DOS and chemical potential on thermoelectric properties.

3 Results and discussion

3.1 Structural properties

Structural optimization of the $Sr_{0.92}La_{0.08}TiO_3$ compound was carried out using the equation of state as earlier said in section 2. This is carried out by fitting the total energy of the unit cell against the volume. The optimized calculated structural parameters are in good agreement with other available computational works as displayed in Table 1. Distortion was not observed in the lattice constant as a result of the small percentage of La in the whole supercell. Figure 1 shows the variation in energy against change in volume during the structural optimized calculation. The bond lengths of computed compounds are measured using XCrysDen tool [26] and displayed in Table 2, the results are compared to available DFT work[27]. The calculated bond length is used to evaluate tolerance factors (t) that can be used to determine the stability of perovskite compound phase transition [28].

$$t = \frac{0.707(A-O)}{(B-O)} \quad (1)$$

where A-O and B-O are the compound bond length from A site to O atom and B site to O atom respectively. The $Sr_{0.92}La_{0.08}TiO_3$ compound is still maintaining cubic structure with space group number (Pm3m) since its 't' value is 1.00 which is in the range of 0.93-1.02 for cubic perovskite compound. In Figure 2, the crystal structure of the study compounds are shown. The La atom is substituted for one of 12 Sr - atoms in the supercell, surrounded by four TiO_6 octahedral. So, La/ Sr is coordinated by 12 O - atom. The Ti - atom is in six-fold coordinated by O – atom.

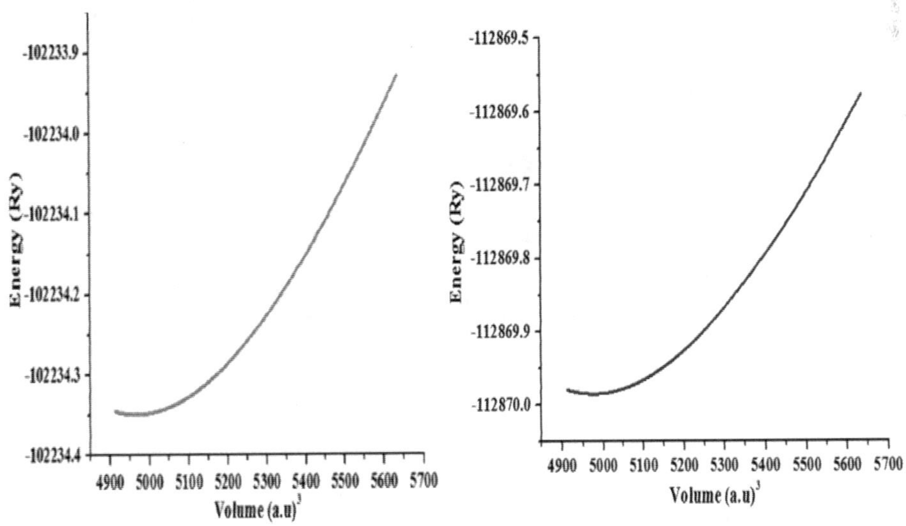

Fig. 1. Changes in energy against volume for $SrTiO_3$ and $Sr_{0.92}La_{0.08}TiO_3$ compounds.

Table 1. Optimized lattice parameter, volume, bulk modulus and bulk modulus derivative (B').

Samples	Lattice constant (Å)	Volume (Å^3)	Bulk modulus (GPa)	Derivative bulk modulus	Ref.
$SrTiO_3$	3.9451	735.84	189.17	6.62	Present work
$Sr_{0.92}La_{0.08}TiO_3$	3.9451	737.42	195.23	7.33	
$SrTiO_3$	3.9125		173.71		[29]
	3.9400		169.51		[27]

Table 2. Measured bond length for study compounds.

Samples	Bond length	Present work	Other computational work[27]	Tolerance factors (t)
$SrTiO_3$	A-O	2.7896	2.79	1.00
	A-B	3.4166	3.41	
	B-O	1.9726	1.97	
$Sr_{0.92}La_{0.08}TiO_3$	A-O	2.7896		1.00
	A-B	3.4166		
	B-O	1.9726		

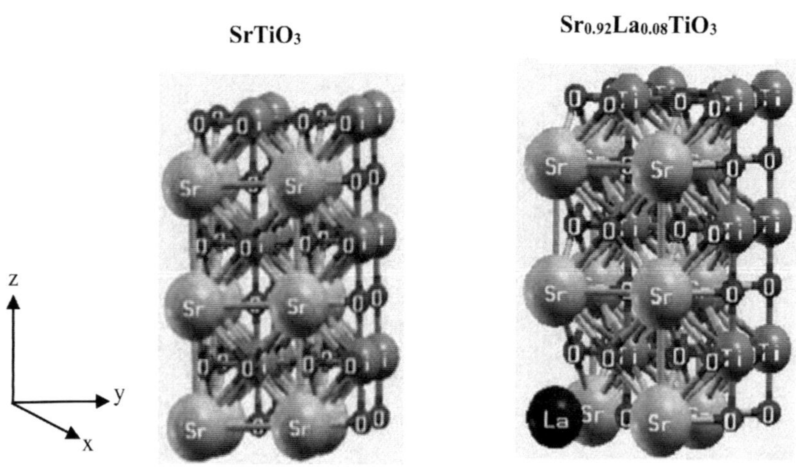

Fig. 2. Crystal structures of $SrTiO_3$ and $Sr_{0.92}La_{0.08}TiO_3$ compound, where the four kinds of atoms are distinguished by coloured spheres as Sr (cyan), La (blue), Ti (gray) and O (red).

3.2 Electronic properties

The computed band structure and density of the state of the study compound are shown in Figures 2 and 3. The band structures were plotted along symmetry points; R(0.5 0.5 0.5); Γ(0.0 0.0 0.0); X(0.0 0.5 0.0); M(0.5 0.5 0.0) and Γ(0.0 0.0 0.0) of the Brillouin zone as demonstrated in Figure 3. Possible electron transition can be investigated through the conduction and valence bands of the band structure. We obtained 3.26 eV bandgap in $SrTiO_3$ compound with the Hubbard U in DFT+U calculation. This is similar to the experimentally reported band gap of 3.25 eV[30]. $SrTiO_3$ band structure confirmed the semiconducting nature of the compound with the position of Fermi energy level (E_F) at the edge of the valence band (VB). The model compound $Sr_{0.92}La_{0.08}TiO_3$ displayed semi-metallic behavior as a result of overlapping E_F on the conduction band (CB). This caused the material to become an electron charge carrier (n-type). The direct bandgap (2.48 eV) was measured from CB minimum and VB maximum of the compound at Γ- Γ points of IBZ. The direct energy band is consistent with the previous study on $SrTiO_3$ band structure [31]. The presence of La doping shrinks the bandgap of $Sr_{0.92}La_{0.08}TiO_3$ compound [32,33].

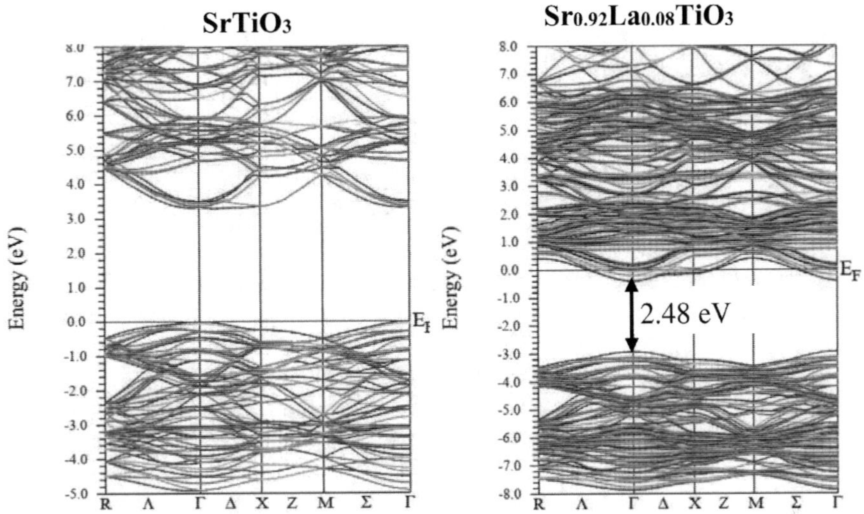

Fig. 3. Electron band structure for $SrTiO_3$ and $Sr_{0.92}La_{0.08}TiO_3$ compounds.

The contributions of each atom in band structure lines cluster were analyzed by computing total and partial density of state (TDOS and PDOS). TDOS and PDOS can be divided into two energy regions; conduction and valence. For $SrTiO_3$ compound, three prominent peaks of the electron were observed in TDOS; one from the conduction band and two from the valence band. PDOS of $SrTiO_3$ (Figure 4) shows that 2p state of O-atom and 3d state of Ti atom dominated at VB and CB respectively. This is similar to other simulation work on $SrTiO_3$[27,31]. A non-magnetic behavior was confirmed with identical spin up and spins down of DOS orbitals. In $Sr_{0.92}La_{0.08}TiO_3$ compounds, the conduction band composed of 3d state of Ti and Sr atoms with the 5d state of La. Each atom established its participation with a peak of an electron as shown in Figure 4. The VB is dominated by 2p state of O atom at both spin and the same time they are symmetry. In Figure 4, the semi-metallic behavior of the $Sr_{0.92}La_{0.08}TiO_3$ compound was confirmed by spin-up TDOS exhibiting metallic and spin-down displayed semiconducting nature. It is noticed that La substitution in $SrTiO_3$ leads to metallic nature [34] in the spin-up region.

IOP Conf. Series: Materials Science and Engineering **957** (2020) 012008 doi:10.1088/1757-899X/957/1/012008

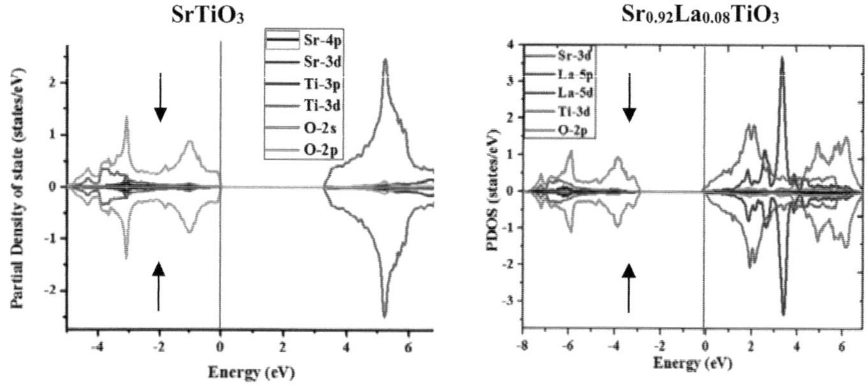

Fig. 4. Total and Partial density of state for $SrTiO_3$ and $Sr_{0.92}La_{0.08}TiO_3$ compounds.

3.3 Thermoelectric properties

Our transport properties calculation is based solely on compounds $Sr_{0.92}La_{0.08}TiO_3$. Thermopower, electrical conductivity, and electronic thermal conductivity are calculated in terms of temperature, the density of state, and chemical potential (μ) as shown in Figure 5-6. Figure 5 demonstrates the chemical potential dependence of parameters of transport at 1000 K. Contrary to the current potential repulsive force, the chemical potential indicates the energy need to insert an electron in the process. The sign of potential chemical value shows semiconducting regions of type n (negative) and type p (positive) [35].

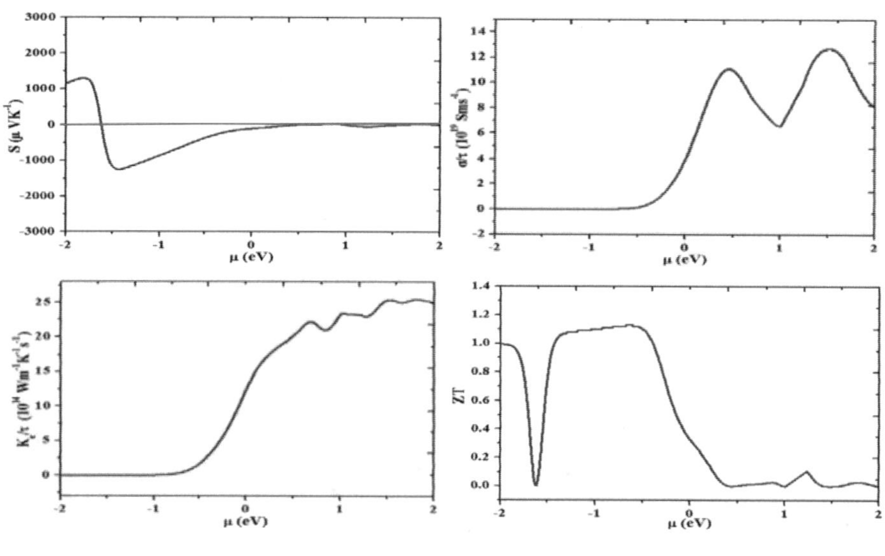

Fig. 5. Variation of chemical potential with thermoelectric properties for $Sr_{0.92}La_{0.08}TiO_3$ compounds.

IOP Conf. Series: Materials Science and Engineering **957** (2020) 012008 doi:10.1088/1757-899X/957/1/012008

A high Seebeck coefficient value of 947.72μVK^{-1} is recorded for μ of -1.69 eV, a negative value confirms the material's n-type existence as previously reported in the electronic properties analysis. A positive value of chemical potential, electrical conductivity and electronic thermal conductivity are increased. Electrical conductivity is directly proportional to chemical potential (μ) as calculated by relationship [5]; $\sigma = ne\mu$, where n is concentration of carrier and μ is mobility of carrier. Clearly, ZT of the study compound is record high value of ~1.25 at the chemical potential from -0.67 eV to -0.61 eV.

The temperature and DOS dependent on Seebeck coefficient were plotted in Figure 6 (a & b). The negative value of S indicates that major carriers are electrons. This confirmed the position of E_F in the band structure and density of state. The absolute S value increase in both cases. A negative value of S value can be attributed to La –atom impurity in the compound with shifting of E_F to CB region in the band structure. The S value at 1200 K is -107.59 μ VK^{-1}. A large value of DOS (206.6 eV) is responsible for such a high Seebeck coefficient. The electrical conductivity (s/t) increases with the temperature / DOS as shown in Fig. 6(c & d). This indicates the semiconducting nature of the compound; this confirms the DOS analysis (at spin down region) in electronic properties. Also with increasing temperature, the electrons get sufficient energy to cross the energy barrier at Fermi-level and therefore increasing electrical conductivity [36]. Since chemical potential is subjected to the temperature by the relation; $\mu \propto \sqrt{T^{-3}}$[37]. Electrical conductivity increases due to chemical potential increasing with temperature. The σ/τ increased from 3.52×1019 Sm.s^{-1} at room temperature to 4.23 ×1019 Sm.s^{-1} at 1200 K.

The thermal conductivity k consists of the electron part k_e and the lattice vibration (phonon) part k_l, as $k = k_e + k_l$. Only k_e/τ can be determined from BoltzTraP code [21,29]. The electronic thermal conductivity has been shown in Figure 6(e&f) to vary with temperature / DOS. With an increase in temperature / DOS, the measured k_e/τ increases monotonically. For temperature changing we recorded 2.75×10^{14} Wm/Ks of k_e/τ at 300 K and 15.49×10^{14} Wm/Ks at 1200 K. This can be due to the effect of electronic thermal conductivity, since with the increase in temperature, the charge carriers are heat excited, as a result, the k_e increasing [37].

In Figure 6(g & h), the dependent of the dimensionless figure of merit (ZT) on temperature/ DOS is present. The calculated ZT value increased with the increase in temperature which is the same trend with experimental work [38]. The ZT value can be attributed to the large value of S and higher k_e/τ at the same time. At the temperature of 300 K, ZT of 0.05 was recorded and at a higher temperature of 1200 K, it was 0.39. This result is similar to many experimental reports ~0.37 of ZT value at 973 K for 9.0 % La doping [38] and ZT of 0.28 at 773 K in $La_{0.12}Sr_{0.88}TiO_3$[39].

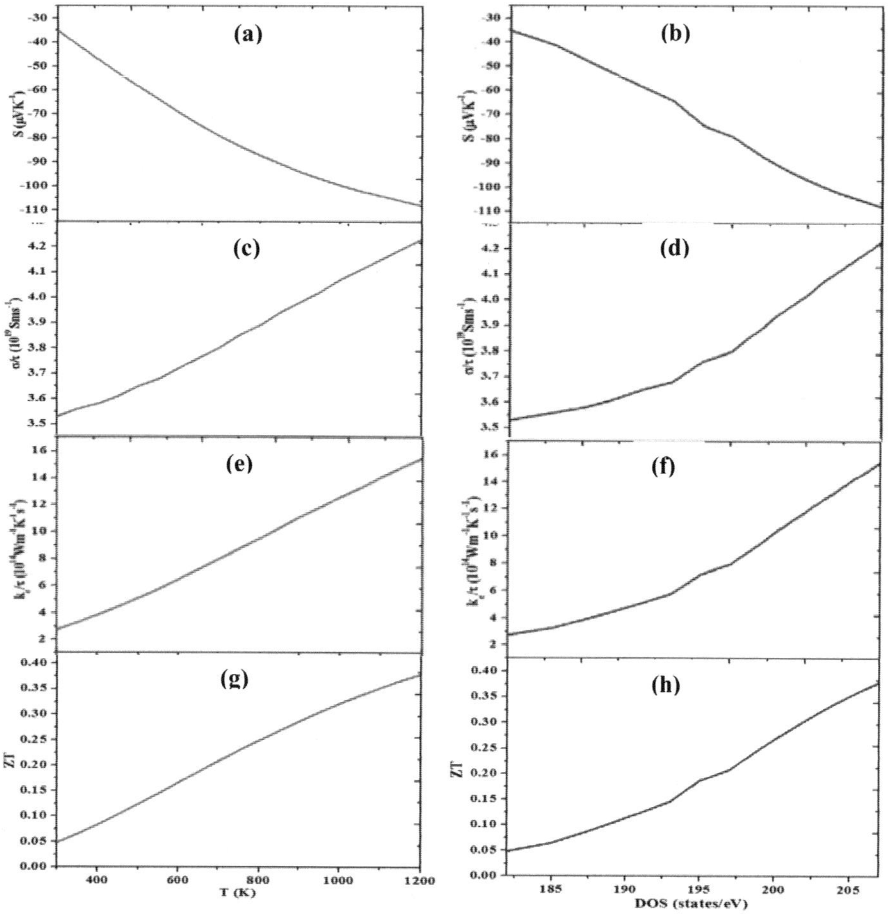

Fig. 6. Temperature / Density of state-dependent on thermoelectric transport properties of $Sr_{0.92}La_{0.08}TiO_3$ compounds.

4 Conclusion

Density functional theory calculation was implemented to study the structural, electronic and thermoelectric properties of $Sr_{0.92}La_{0.08}TiO_3$. Semi-classical Boltzmann transport equation was solved for thermoelectric transport properties using BoltzTraP code. The cubic structure of the doped compound, similar to un-doped, was obtained after structural optimization using the equation of state by Birch-Murnaghan's. During electronic properties analyses with band structure and density of states, the energy band gap of 2.48 eV was obtained which was attributed to La atom present in the compound. The spin-down of DOS reveals semiconducting nature and spin up shows the metallic behaviour. This was combining in the band structure plots as semi-metallic with the Fermi level partial crossing of the conduction bands. Thermoelectric properties were analyzed by plotting the variation of transport properties with chemical potential, temperature, and density of states. In chemical potential variation with transport properties, large negative Seebeck coefficient value was recorded show the n-type nature of compound by S. Temperature and DOS displayed the same trends with transport properties. A high ZT value of 0.39 was obtained at 1200 K. Our result is consistent with the past reported work on La atom doped $SrTiO_3$.

The authors would like to extend their appreciation to Centre for Frontier Materials Research, Center of Excellence Geopolymer & Green Technology (CEGeoGTech) and the School of Materials Engineering, Universiti Malaysia Perlis for their conducive environment provided for the research. Also, we thank the Malaysian Government for their financial support of this project through the Fundamental Research Grant Scheme (FRGS/1/2016/STG07/UNIMAP/02/3).

References

1. G. J. Snyder, E. S. Toberer, World Sci., 101-110 (2011)
2. P. V. Arribi, P. García-Fernández, J. Junquera, V. Pardo. Phys Rev B, **94**, 035124 (2016)
3. F. J DiSalvo. Sci., **285**, 703-706 (1999)
4. I. T. Witting, T. C. Chasapis, F. Ricci, M. Peters, N. A. Heinz, G. Hautier, G. J. Snyder, Adv. Electron. Mater., 1800904 (2019)
5. A. A. Adewale, A. Chik, R. M. Zaki, F. C. Pa, C. K. Yeoh, N. H. Jamil, *AIP Conference Proceedings*, **2030**, 020196 (2018)
6. V. K. Zaitsev, M. I. Fedorov, I. S. Eremin, E. A. Gurieva, D. M. Rowe, *Thermoelectrics handbook: macro to nano*, (2006)
7. H. Muta, K. Kurosaki and S. Yamanaka. J. Alloys Compd., **392**, 306-309 (2005)
8. D. Srivastava, C. Norman, F. Azough, M. C. Schäfer, E. Guilmeau, R. Freer, J. Alloys Compd., **731**, 723-730 (2018)
9. M. Ibáñez, D. Cadavid, R. Zamani, N. García-Castelló, V. Izquierdo-Roca, W. Li, A. Fairbrother, J. D. Prades, A. Shavel, J. Arbiol, A. Pérez-Rodríguez, Chem. Mater., **24**, 562-570 (2012)
10. K. Park, J. S. Son, S. I. Woo, K. Shin, M. W. Oh, S. D. Park, T. Hyeon, J. Mater. Chem. A, **2**, 4217-4224 (2014)
11. M. Saleemi, M. S. Toprak, S. Li, M. Johnsson, M. Muhammed, J. Mater. Chem., **22**, 725-730 (2012)
12. T. Okuda, K. Nakanishi, S. Miyasaka, Y. Tokura. Phys Rev B, **63**, 113104 (2001)
13. C. Chen, T. Zhang, R. Donelson, T. T. Tan, S. Li, J. Alloys and Compd., **629**, 49-54 (2015)
14. Z. Lu, H. Zhang, W. Lei, D. C. Sinclair, I. M. Reaney, Chem. Mater., **28**, 925-935 (2016)
15. A. A. Yaremchenko, S. Populoh, S. G. Patricio, J. Macías, P. Thiel, D. P. Fagg, A. Weidenkaff, J. R. Frade, A. V. Kovalevsky, Chem. Mater., **27**, 4995-5006 (2015)
16. J. Wang. Nano Energy, **35**, 387-395 (2017)
17. A. V. Kovalevsky, M. H. Aguirre, S. Populoh, S. G. Patrício, N. M. Ferreira, S. M. Mikhalev, D. P. Fagg, A. Weidenkaff, J. R. Frade, J. Mater. Chem. A, **5**, 3909-3922 (2017)
18. A. Tkach, J. Resende, K. V. Saravanan, M. E. Costa, P. Diaz-Chao, E. Guilmeau, O. Okhay, P. M. Vilarinho, ACS Sustain. Chem. Eng., **6**, 15988-15994 (2018)
19. B. Zhang, J. Wang, T. Zou, S. Zhang, X. Yaer, N. Ding, C. Liu, L. Miao, Y. Li, Y. Wu, Y, J. Mater. Chem. C, **3**, 11406-11411 (2015)
20. J. P. Perdew, K. Burke, M. Ernzerhof, Phys. Rev. Lett., **77**, 3865 (1996)

21. G. K. Madsen, D. J. Singh, Comput. Phys. Commun., **175**, 67-71 (2006)
22. P. Blaha, K. Schwarz, G. K. Madsen, D. Kvasnicka, J. Luitz, *An augmented plane wave+ local orbitals program for calculating crystal properties*, (2001).
23. M. Cococcioni. *From Models to Materials Modeling and Simulation*, (2012).
24. H. J. Monkhorst, J. D. Pack, Phys. Rev. B, **13**, 5188 (1976)
25. F. Birch, J. Appl. Phys., **9**, 279-288 (1938)
26. A. Kokalj, J. Mol. Graph Model., **17**, 176-179 (2000)
27. A. A. Mubarak, Int. J. Mod. Phys. B, **30**, 1650141 (2016)
28. Z. Ali, I. Ahmad, A. H. Reshak, Physica B: Condensed Matter, **410**, 217-221 (2013)
29. P. Sikam, P. Moontragoon, C. Sararat, A. Karaphun, E. Swatsitang, S. Pinitsoontorn, P. Thongbai, Appl. Surf. Sci., **446**, 92-113 (2018)
30. S. Challagulla, R. Nagarjuna, S. Roy, R. Ganesan, Chemistry Select, **2**, 4836–4842 (2017)
31. A. A. Adewale, A. Chik. IJNeaM, **12,** 11-18 (2019)
32. W. J. Xie, Y. G. Yan, S. Zhu, M. Zhou, S. Populoh, K. Gałązka, S. J. Poon, A. Weidenkaff, J. He, X. F. Tang, T. M. Tritt, Acta Mater., **61**, 2087-2094 (2013)
33. X. Yan, G. Joshi, W. Liu, Y. Lan, H. Wang, S. Lee, J. W. Simonson, S. J. Poon, T. M. Tritt, G. Chen, Z. F. Ren, Nano Lett., **11**, 556-560 (2010)
34. Y. Suzuki, H. Nakamura, Phys. Chem. Chem. Phys., **17**, 29647-29654 (2015)
35. T. Wei, B. Xu, X. Ji. Eur, Phys. J. B, **92**, 17 (2019)
36. M. J. Alrahamneh, A. A. Mousa, J. M. Khalifeh, Physica B Condens. Matter., **552**, 227-235 (2019)
37. K. Kaur, R. Kumar, Chin. Phys. B, **25**, 026402 (2016)
38. D. Liu, Y. Zhang, H. Kang, J. Li, Z. Chen, T. Wang, J. Eur. Ceram. Soc., **38**, 807-811 (2018)
39. P. P. Shang, B. P., Zhang, Y. Liu, J. F. Li, H. M. Zhu, J. Electron. Mater., **40**, 926-931 (2011)

IOP Conf. Series: Materials Science and Engineering **957** (2020) 012009 doi:10.1088/1757-899X/957/1/012009

Study of thermoelectric properties of Sr0.92A0.08TiO3 (A=Yb/ Tm) perovskite oxide using density functional theory method

Akeem Adekunle Adewale[1,3,4,*], *Abdullah* Chik[1,2,3] , and *Ruhiyuddin* Mohd Zaki[2,3]

[1]Centre for Frontier Materials Research, Universiti Malaysia Perlis, 01000 Kangar, Perlis, Malaysia
[2]Center of Excellence Geopolymer and Green Technology (CEGeoGTech), School of Materials Engineering, Universiti Malaysia Perlis (UniMAP), 02600 Jejawi, Arau, Perlis, Malaysia
[3]School of Materials Engineering, Universiti Malaysia Perlis (UniMAP), Taman Muhibbah 02600, Jejawi, Arau, Perlis, Malaysia
[4]Department of Physics and Materials science, Kwara State University, Malete, Kwara State, Nigeria

Abstract. The first principle methods have been employed to investigate electronic and thermoelectric properties of $Sr_{0.92}Yb_{0.08}TiO_3$ and $Sr_{0.92}Tm_{0.08}TiO_3$ perovskite-oxide based molded samples. Generalized gradient approximation (GGA) with Hubbard U parameter is used by WIEN2k code for the calculations. The straight band line was observed in the band structure of both studied samples. This was generated from 4f-orbitals as shown in partial density of state diagrams. It is also noticed that Yb and Tm doped in $SrTiO_3$ changed the perovskite-based oxide from a wideband insulator to metallic nature. A thermoelectric power factor of $Sr_{0.92}Tm_{0.08}TiO_3$ sample is higher than that of $Sr_{0.92}Yb_{0.08}TiO_3$, this is as a result of its huge electrical conductivity. The dependent of chemical potential to temperature was revealed in the study where high value of power factor was recorded for high temperature.

1 Introduction

Perovskite-based oxide compounds displayed similar crystal structure as the calcium titanate mineral ($CaTiO_3$). Ideally, cubic perovskite's oxide crystal structure can be described as ABO_3, where the A-site cation is 12-fold coordinate and the B site cation is 6-fold coordinate with oxygen which is anions, as present in Fig. 1[1]. Atom in A-site is typically alkaline or rare earth metal, and B-site is likely to be transition or non-transition metals [2–4]. These perovskite compounds have advantages of low toxicity and high abundance compared to the state-of-the-art materials; chalcogenides, telluride-base, etc. Nonetheless, the overall thermoelectric performance of oxide perovskites is relatively low as a result of its high thermal conductivity and low electrical conductivity. Various methods have been used to enhance oxide perovskite materials among them are doping, inducing oxygen vacancy, nanostructure, etc. These perovskite oxides have been used in many applications as well as in thermoelectric devices [5].

* Corresponding author: adewaleakeem12@gmail.com

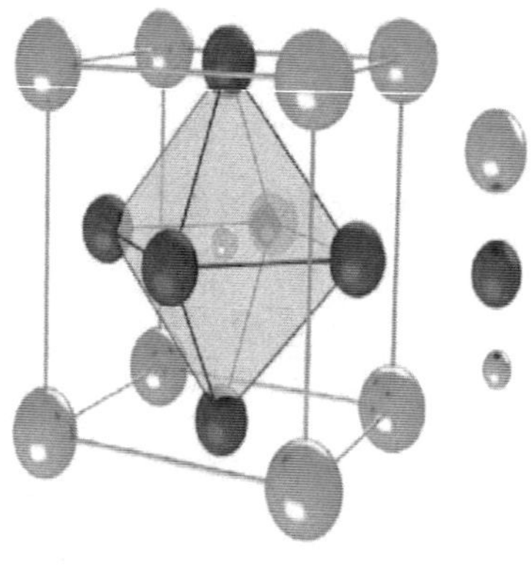

Fig. 1. ABO_3 perovskite-oxide structure [1].

Thermoelectric (TE) materials are capable of converting heat directly into electricity through the Seebeck effect [6-8]. The performance of thermoelectric material can be determined by the dimensionless Fig. of merit (ZT) expressed by;

$$ZT = \frac{Pf.T}{k} \quad (1)$$

where Pf is power factor, κ is thermal conductivity and T is temperature. The Pf is the product of electrical transport properties (Pf = $S^2\sigma$); Seebeck coefficient (S) and electrical conductivity(σ). The thermoelectric material's performance can be improved by enhancing Pf and reducing thermal conductivity [9]. $SrTiO_3$ is one of the most promising n-type materials because it has unusual combinations (for an oxide) of good electrical conductivity and a high Seebeck coefficient, resulting in a power factor of 3.5 mW m^{-1} K^{-2} in single crystals and epitaxial films. Ytterbium and thulium are rare earth elements, lanthanides with oxidation state of 3. Yb^{3+} and Tm^{3+} ion can be used as a dopant in perovskite oxide, especially at A-site which is A^{2+} ion. This will create the interaction between 4f orbitals in elements (Yb and Tm) and 3d-orbital in B-site [10]. Especially in $SrTiO_3$ when Yb^{3+} or Tm^{3+} is hosted into A-site, it can contribute an electron to Ti at its d state [11].

Hence an additional charge carrier would be induced and exchange interactions may cause improvement in electronic properties which will lead to enhancement of electronic transport properties. A current study exhibits the enhancement of thermoelectric properties in rare-earth-doped ABO_3 perovskite structured materials. Electronic and thermoelectric properties of $SrTiO_3$ doped by 8%Yb and 8%Tm were investigated by using first-principle methods. Band structure and density of state of $Sr_{0.92}Yb_{0.08}TiO_3$ samples are analysed for electronic properties. The thermoelectric parameter; Seebeck coefficient, electrical conductivity and power factor were studied with their dependence on temperature. Variation of power factor and chemical potential were also calculated for temperature of 300, 600, 900 and 1200 K.

2 Methods of calculation

In the present study solution of Kohn-Sham equation [12] was adopted in first-principles calculations based on the density functional theory in order to investigate the electronic and thermoelectric properties of $SrTiO_3$ doped by Yb/ Tm. This first principle was implemented by WIEN2k package [13] by means of full-potential linearized augmented plane wave (FP-LAPW) [14] basis set. The exchange-correlation functional Perdew, Burke, and Ernzerhof (PBE) [15] in generalized gradient approximation with Hubbard model (GGA+U) was used in the calculation. The Hubbard U value of 0.8 used was selected using semi-empirical method [16] based on the result of band structure of un-doped $SrTiO_3$ as implemented in WIEN2K.

The self-consistent calculation were performed using 6×6×4 mesh Monkhorst-Pack in k-points, where the angular momentum Lmax = 12, RKmax = 7 and Gmax = 24 $(a.u.)^{-1}$ were set. A mesh of 2000 k-points and the tetrahedral method [17] was used for the integration of the Brillouin zone. The thermoelectric properties are calculated within the basis of semi-classical Boltzmann theory using BoltzTraP software [18] and relaxation time $\tau = 5\times10^{-11}$s was adopted from Lemanov et al., (2004) work on dielectric relaxation in $SrTiO_3$: Mn[19].

3 Results and discussion

The electronic properties of un-doped $SrTiO_3$ were calculated in order to see the influence of dopant (Yb or Tm) on the $SrTiO_3$. These were analyzed via band structure and density of state results. The thermoelectric properties were determined through electrical transport properties; Seebeck coefficient and electrical conductivity. Variation of these transport properties and power factors with temperature at range of 300 to 1200 K were calculated.

3.1 Electronic structure properties

In electronic band structure, the possible energy ranges available for electrons to occupy and those regions energy which cannot be occupied; the two regions are named as energy band and forbidden or bandgap respectively. These regions are so important that the electronic properties of material can be analysed through them. They are used to illustrate the possible electronic transitions from valence band maximum (VBM) to conduction band minimum (CBM). In Fig. 2, the band structures of $SrTiO_3$, $Sr_{0.92}Tm_{0.08}TiO_3$, and $Sr_{0.92}Yb_{0.08}TiO_3$ samples were presented. The band energy was plotted against Brillouin zone; R, Γ, X, M, and Γ as shown in Fig. 2.

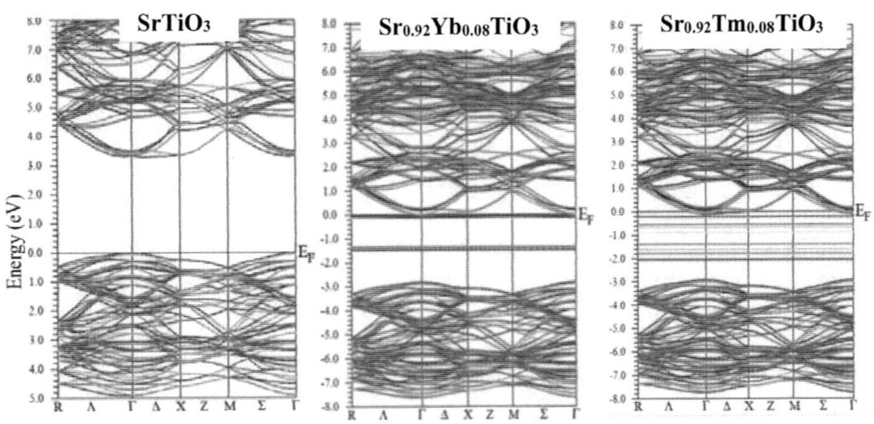

Fig. 2. Electronic Band structure for $SrTiO_3$, $Sr_{0.92}Yb_{0.08}TiO_3$ and $Sr_{0.92}Tm_{0.08}TiO_3$ samples.

In the process of determining the energy band gap of $SrTiO_3$ closed to experimental value, several Hubbard U are used [20] and at the end 0.8 of U give band energy that is in good agreement. The direct bandgap of 3.26 eV was recorded at Γ - Γ point for $SrTiO_3$ sample (Fig. 2) similar to experimental report of indirect 3.25 eV [21]. Metallic nature in doping samples $Sr_{0.92}Yb_{0.08}TiO_3$ and $Sr_{0.92}Tm_{0.08}TiO_3$ were observed by the straight band line on the Fermi level (E_F), that originated from doped element Yb and Tm, as compared to un-doped $SrTiO_3$ in Fig. 2.

In order to visualize the orbital presence in the band structure, density of state calculation were performed. In Fig. 3 the partial density of state for $Sr_{0.92}Yb_{0.08}TiO_3$ and $Sr_{0.92}Tm_{0.08}TiO_3$ were presented. Fig. 3a and 3b are for $Sr_{0.92}Yb_{0.08}TiO_3$ sample while Fig. 3c is for $Sr_{0.92}Tm_{0.08}TiO_3$. It is clearly shown that 4f-orbital from dopant element (Yb and Tm) with 3d-orbital from Ti caused $SrTiO_3$ to become metallic in nature for both studied samples. Likewise, it is confirmed that this 4f-orbital formed a straight band line in the band structure diagram. Fig. 3b is displayed from scale of 10 to -10 DOS in order to see the participation of low DOS orbital in $Sr_{0.92}Yb_{0.08}TiO_3$ sample. The conduction band is composed of 3d-orbital of Ti, 4d-orbital of Yb/Tm, and 3d-orbital of Sr. The 4d and 3d from Yb/Tm and Sr atoms have little or contribution in the electronic influence of the material since the position is far from E_F. The valence band composed of 4f-orbital of Yb/Tm and 2p-orbital of O atom are as shown in Fig. 3b and Fig. 3c.

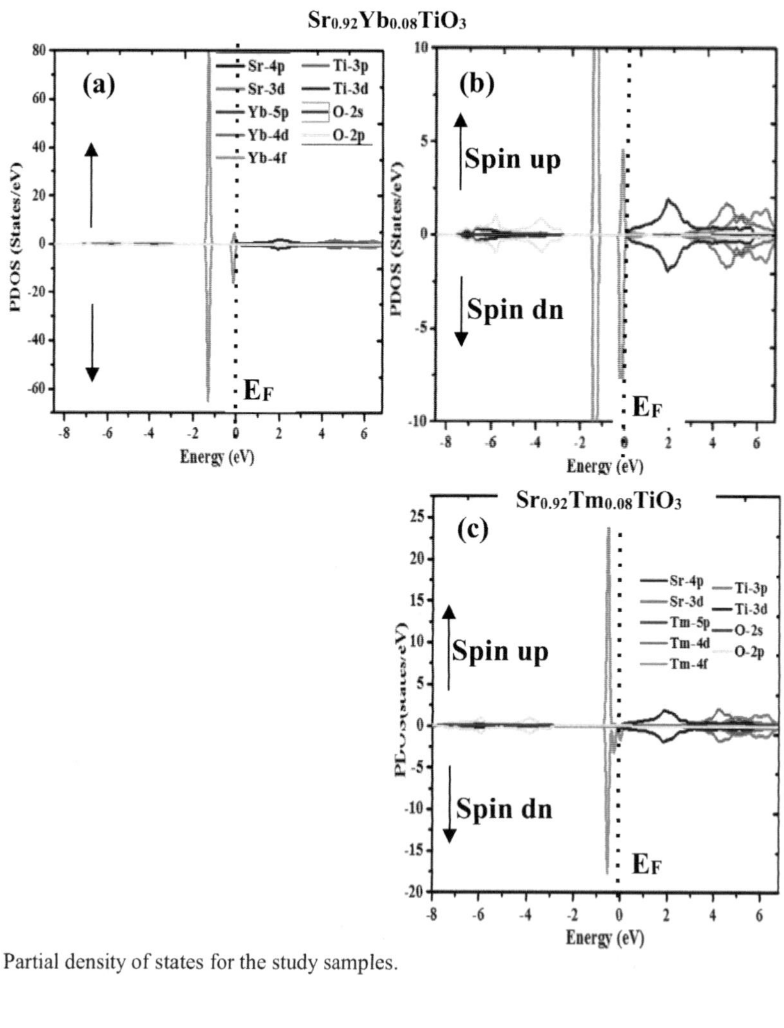

Fig. 3. Partial density of states for the study samples.

3.2 Thermoelectric properties

The thermoelectric transport properties are calculated via; Seebeck coefficient(S), electrical conductivity and power factor at the temperature of 300 to 1200 K as presented in Fig. 4. The Seebeck coefficients for both doped samples are negative; which indicate the n-type nature of the $SrTiO_3$ at these Yb and Tm doping levels. Absolute S value in Yb doped of $SrTiO_3$ is higher than that of Tm and this can be attributed to high DOS at the Fermi level in partial density of state diagram [22]. The electrical conductivity of $Sr_{0.92}Yb_{0.08}TiO_3$ and $Sr_{0.92}Tm_{0.08}TiO_3$ model sample increases with an increase in temperature which is general behaviour of semiconductors, despite the presence of doped in sample. The Tm doped model have higher electrical conductivity than the Yb doped model, as shown by the higher number of flat bands close to Fermi level in Fig. 2. It is clearly shows that the high power factor recorded for $Sr_{0.92}Tm_{0.08}TiO_3$ sample is due to large electrical conductivity compared to the $Sr_{0.92}Yb_{0.08}TiO_3$ sample. Performance of $SrTiO_3$ doped by Yb and Tm were plotted using power factor against chemical potential at range of 2 and -2 eV are displayed at constant temperature of 300 K, 600 K, 900 K and 1200 K in Fig. 5. In this Fig. 5, dependent on power factor on temperature are revealed, and both samples show high power factor at similar chemical potential value.

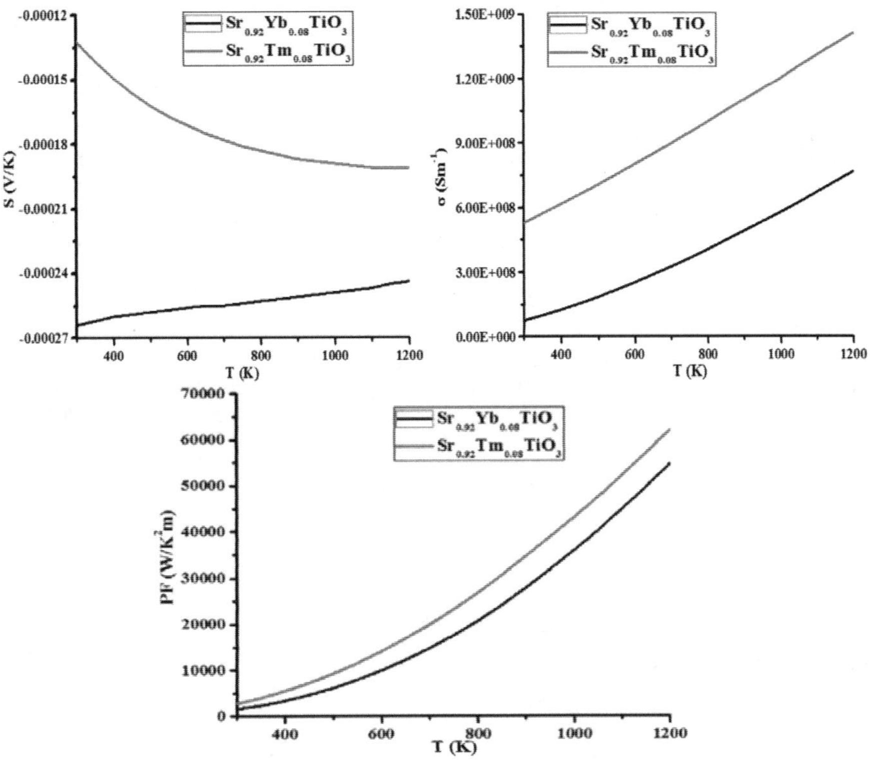

Fig. 4. Temperature-dependent of Seebeck coefficient, electrical conductivity and power factors for $Sr_{0.92}Yb_{0.08}TiO_3$ and $Sr_{0.92}Tm_{0.08}TiO_3$ samples.

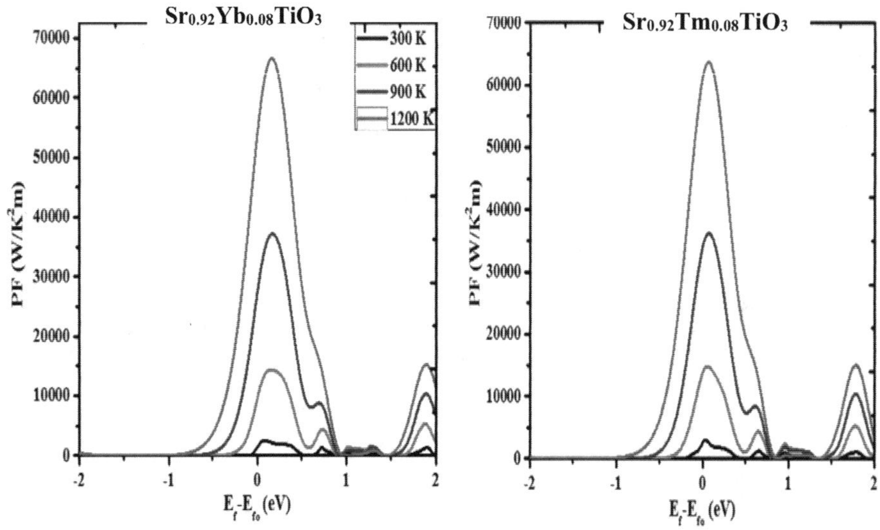

Fig. 5. Changing in power factor of $Sr_{0.92}Yb_{0.08}TiO_3$ and $Sr_{0.92}Tm_{0.08}TiO_3$ against chemical potential.

4 Conclusion

Ab initio method in density functional theory is employed to investigate the electronic and thermoelectric properties of $SrTiO_3$ doped by 8% of Yb/ Tm. It was noticed that present of Yb/Tm doped transforms $SrTiO_3$ from insulating oxide to metallic. A high density of state in doped sample is created by 4f-orbital from dopant Yb/Tm. By comparison, Yb doped sample has dense 4-f orbital at Fermi level than Tm doped sample, which leads to high Seebeck coefficient value record by $Sr_{0.92}Yb_{0.08}TiO_3$ sample. Electrical conductivity in $SrTiO_3$ doped by Tm is higher than Yb counterpart and this is a result of large number of flat band lines around Fermi level. The power factor value of $Sr_{0.92}Tm_{0.08}TiO_3$ is high than that of $Sr_{0.92}Yb_{0.08}TiO_3$, and that was attributed to its large electrical conductivity. $Sr_{0.92}Yb_{0.08}TiO_3$ and $Sr_{0.92}Yb_{0.08}TiO_3$ are excellent candidates for high temperature n-type thermoelectric devices have been recorded in variation of power factor and chemical potential.

The authors would like to extend their appreciation to Centre for Frontier Materials Research, Center of Excellence Geopolymer & Green Technology (CEGeoGTech) and the School of Materials Engineering, Universiti Malaysia Perlis for their conducive environment provided for the research. Also, we thank the Malaysian Government for their financial support of this project through the Fundamental Research Grant Scheme (FRGS/1/2016/STG07/UNIMAP/02/3).

References

1. Y. Chen, L. Zhang, Y. Zhang, H. Gao, H. Yan. RSC Adv, **8**, 10489-10508 (2018)
2. S. A. Khandy, D. C. Gupta. RSC Adv, **6**, 97641-97649, (2016)
3. O. Fursenko, J. Bauer, G. Lupina, P. Dudek, M. Lukosius, C. Wenger, P. Zaumseil, Thin Solid Films, **520**, 4532-4535 (2012)
4. G. Lupina, G, Kozłowski, J. Dabrowski, C. Wenger, P. Dudek, P. Zaumseil, G. Lippert, C. Walczyk, H. J. Müssig, Appl Phys Lett., **92**, 062906 (2008)

5. K. A. Parrey, N. Devi, R. Khenata, S. A. Khandy. Comput. Condens. Matter, **17**, 00340 (2018)
6. T. M. Tritt, Annu. Rev. Mater. Res., **41**, 433-448 (2011)
7. A. Shakouri Annu. Rev. Mater. Res., **41**, (2011).
8. A. A. Adewale, A. Chik, R. M. Zaki, F. Che Pa, Y. C. Keat, N. H. Jamil, Solid State Phenom, **280**, 3-8 (2018)
9. P. V. Arribi, P. García-Fernández, J. Junquera, V. Pardo. Phys Rev B, **94**, 035124 (2016)
10. J. Inaba, T. Katsufuji. Phys Rev B, **72,** 052408 , (2005)
11. M. Muralidharan, V. Anbarasu, A. E. Perumal, K. Sivakumar. J. Mater. Sci. Mater. Electron, **25**, 4078-4087 (2014)
12. W. Kohn and L. J. Sham. Phys Rev, **140**, A1133 (1965)
13. P. Blaha, K. Schwarz, G. K. Madsen, D. Kvasnicka, J. Luitz. *An augmented plane wave+ local orbitals program for calculating crystal properties*, (2001).
14. O. K. Andersen, PHYS REV B, Vol. **12**, 3060 (1975)
15. J. P. Perdew, K. Burke, M. Ernzerhof. Phys. Rev. Lett, **78**, 1396 (1996)
16. M. Cococcioni. *From Models to Materials Modeling and Simulation*, (2012).
17. S. A. Dar, V. Srivastava, U. K. Sakalle, V. Parey, G. Pagare. Mat Sci Eng B -Adv, **236**, 217-224 (2018)
18. G. K. Madsen, D. J. Singh. Comput. Phys. Commun, **175**, 67-71 (2006)
19. V. V. Lemanov, E. P. Smirnova, A. V. Sotnikov, M. Weihnacht, Phys. Solid State, **46**, 1442-1448 (2004)
20. A. A. Adewale, A. Chik, IJNeaM, **12**, 11-18 (2019
21. S. Challagulla, R. Nagarjuna, S. Roy, R. Ganesan, Chemistry Select, **2,** 4836-4842 (2017)
22. M. J. Alrahamneh, A. A. Mousa, J. M. Khalifeh, Physica B Condens. Matter, **552**, 227-235 (2019)

IOP Conf. Series: Materials Science and Engineering **957** (2020) 012010 doi:10.1088/1757-899X/957/1/012010

The Effect Of Filler Content and Chemical Modification On The Properties Of Poly(Lactic Acid)/Pandanus Amaryllifolius Biocomposite Films

Mohd Rasidi, Mohd Syahmie[1,2,*], *Chew*, Kang Wei[1], *Mohd Nasib*, Amira[3]

[1]School of Material Engineering, Universiti Malaysia Perlis, 02600 Arau, Perlis, Malaysia
[2]Centre of Excellence Geopolymer and Green Technology (CEGeoGTech), Universiti Malaysia Perlis (UniMAP), Komplek Pusat Pengajian Jejawi 2, 02600 Arau, Perlis, Malaysia.
[3]Chemical Engineering Technology, Faculty of Engineering Technology, Universiti Malaysia Perlis, Kampus UniCITI Alam, Sungai Chuchuh, 02100 Padang Besar, Perlis, Malaysia.

Abstract. The biocomposite films of Polylactic acid (PLA) and Pandanus Amaryllifolius filler (PAF) were prepared using solvent casting method. The effect of PAF content and modification of PAF by silane coupling agent on mechanical, morphological and enzymatic biodegradation of PLA/PAF biocomposite films were determined. PAF content in PLA/PAF biocomposite films caused the tensile strength and elongation at break to decrease, meanwhile Young's modulus of biocomposite films increased. The SEM micrographs showed poor filler-matrix interaction between PAF and PLA matrix by showing rough surface morphology. The chemical modification of PAF using 3-APE silane coupling agent improved Young's modulus and tensile strength in biocomposite films however elongation at break has exhibited the trend of decrease. The morphological studied showed that modified PAF had better dispersion and interaction with PLA matrix by showing smoother surface morphology. The changes of functional group of unmodified and modified biocomposite films were proven by Fourier Transform Infrared study.

1 Introduction

PLA is one of the biodegradable and biocompatible thermoplastics that could be formed by fermentation from sustainable and renewable resources [1]. Nowadays, PLA has become an alternative to petrochemical polymer for various applications. Main application of PLA today is in packaging field due to its rigidity, clarity and ease of processing [2]. Resistance of PLA towards aliphatic molecules also one of the properties attributes to food packaging field. PLA can easily process into desired configuration such as moulded part and films by using conventional thermoplastic processing techniques. However, low impact resistance and brittle feature has restricted the use for application which require higher stress level [3]. To

* Corresponding author: syahmie@unimap.edu.my

overcome these limitations, reinforcing filler can be added to PLA to enhance its mechanical properties.

Nowadays, natural filler is getting more focused due to its biodegradability, light weight, less abrasiveness and its high specific strength comparing with synthetic filler [4]. The properties of natural fillers depend on several factor such as growing condition, region of origin chemical composition, structural and other characteristics [5]. Thus, there is a need to master more knowledge and have a well usage of them because these will affect the desired properties of biocomposites [6]. Jute [7], kenaf [8], abaca [9], bamboo [10], flax and Cordenka [11] have been extensively investigated by material scientists around the world and used as filler for biocomposites variety of applications.

Pandanus amaryllifolius filler (PAF) or commonly known as Pandan leaves is widely distributed in parts of south-eastern Asia like Malaysia, Thailand and Indonesia [12]. PAF was applied due to its high fibre contents, low prices and high strength. Besides that, PAF has been reported to be one of the best 2-acetyl-1pyrrolin (2-AP) sources [13]. The colour and unique aroma from 2-AP contribute to a creative composite packaging with a unique Pandan aroma.

However, natural fillers are hydrophilic, which is incompatible with hydrophobic nature of PLA. This will result in poor interfacial adhesion between filler and matrix. Poor interfacial adhesion of filler-matrix will affect the physical and mechanical properties of biocomposites [14]. Furthermore, the hydroxyl groups which are rich in the surface of natural filler will also cause moisture uptake from the environment too. Therefore, chemical modification has been to done on natural fillers to increase the compatibility and interaction with the matrix.

This work was focused on the effect of PAF content and chemical modification of PAF on the mechanical properties and morphological properties of biocomposite films. Biodegradability of biocomposite films were tested by enzymatic biodegradation test. FTIR was utilized to determine the change in functional group for PLA /PAF biocomposite films.

2 Methodology

2.1 Materials

Polylactic acid was supplied from TT Biotechnologies Sdn Bhd., Penang. Whereas chloroform, ethanol 95% were purchased from HmbG chemicals, Syarikat Saintifik Jaya, Malaysia. 3-Aminopropylethoxysilane (3-APE) was supplied by Fkuka, Penang. Pandanus Amaryllifolius filler (PAF) was obtained around Perlis. PAF was first cleaned and dried in oven at 80 °C for 24 hours. The dried PAF were then grounded to fine powder, sieved by scientific siever to obtain 53μm particle size.

2.2 Methods

2.2.1 Chemical Modification of Pandanus

3-APE was dissolved in ethanol 3:1 v/v at ambient temperature. After that, PAF were added into the silane solution and stirred for 1 hour by mechanical stirrer. Mixture was left overnight. The PAF was filtered by Whatman filter paper and dried in the oven at 80 °C overnight.

2.2.2 Preparation of PLA/Pandanus biocomposites films

Pellets of PLA were dissolved in 100ml chloroform and stirrer in water bath at 45°C. The mixture were stirred at 1 hour at 350rpm to make sure PLA pellets completely dissolved. Once PLA was completely dissolved, Pandanus were loaded constantly. The mixture was placed in water bath at 45°C and stirred for another 1 hour by mechanical stirrer. PLA/PAF solution was casted into a clean glass mould and left for 2 days at ambient temperature. The dried films was gently been removed from the glass mould. Same procedure was repeated for chemical modification of PLA/PAF biocomposite films. The biocomposites were compounded followed the formulation as shown in Table 1.

Table 1. Formulation of PLA/PAF biocomposite films.

Materials	Unmodified PLA/PAF biocomposite films	Modified PLA/PAF biocomposite films
PLA (php)	100	100
PAF (php)	0,5,10,15,20,25	5,10,15,20,25
3-APE (%)	-	3

2.3 Testing and Characterization

2.3.1 Tensile Test

Tensile test of biocomposite films were carried out by using Instron Machine 5960 Series followed ASTM D 882. The sample was been prepared in rectangular shape. Crosshead speed set was 20mm/min. Tensile strength, elongation at break and Young's modulus were obtained for each tested specimen.

2.3.2 Morphological Analysis

Scanning Electron Microscopy (SEM) with the model of JEOL-6400-LA was applied to study the tensile fracture surface and morphology of biocomposite films. Tensile fracture end of specimens was mounted onto aluminium stubs and a thin layer of palladium was sputter-coated to prevent from electrostatic charging during observation. The SEM micrographs were obtained at 10Kv.

2.3.3 Fourier Transform Infrared Spectroscopy (FTIR) Analysis

FTIR was used to characterize the presence of functional groups for unmodified PAF and modified PAF. The analysis was carried out by using the FTIR spectra resulted from the Perkin Elmer, Model L 128044. The range of scanned was set between 400 and 4000cm^{-1} with a resolution of 4cm^{-1}.

3 Result & Discussion

3.1 Tensile Properties

Fig. 1 demonstrated the effect of PAF content on tensile strength of unmodified and modified PLA/PAF biocomposite films. It can be seen that tensile strength of unmodified and modified biocomposite films decreased as PAF content increased. However, biocomposite films which undergo silanization demonstrated a higher tensile strength in comparison with unmodified biocomposite films. The increased in tensile strength was resulted from enhancement of interfacial adhesion between PLA matrix and PAF. Proper wetting between filler and matrix lead to formation of better filler-matrix interface and van der Waals forces between them [16]. Jandas et al. (2011) also reported that chemical treatment of banana fibre (BF) will result in an improvement in interfacial bonding of the fibres with the PLA matrix. This is predominantly due to formation of a bridge between fibre and matrix and shows the improvement of the interaction for filler and matrix. Hence, it showed the trend of improvement in the tensile strength treated biocomposites [17].

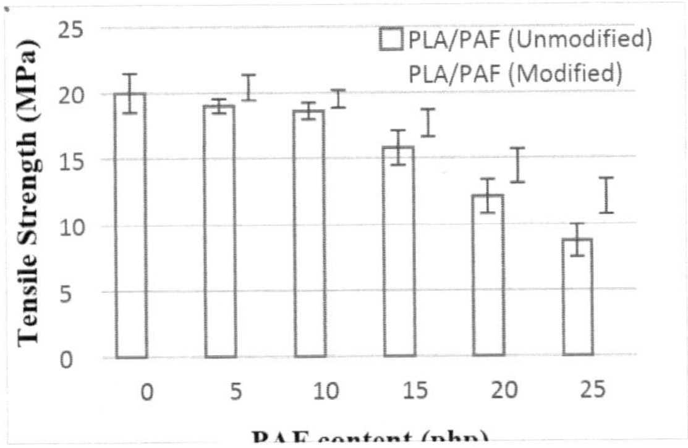

Fig. 1. Effect of PAF content on tensile strength of unmodified and modified PLA/PAF biocomposite films with 3-APE.

Fig. 2 illustrated the effect of 3-APE on elongation at break of unmodified and modified PLA/PAF biocomposite films. It had found that the increased in PAF content had decreased elongation at break for unmodified and modified PLA/PAF biocomposite films. This is because PLA matrix and PAF experienced a decreased in deformability of rigid interface. The movement of PLA chains aligned during elongation had been restricted. At similar PAF content, the modified biocomposites had shown lower elongation at break in comparison with unmodified biocomposites. After silane treatment, tensile strength of biocomposites had an improvement in rigidity and decrement in ductility, which consequently lower elongation at break of PLA/PAF biocomposite films. Similar observation was reported by Syahmie et al. (2015) which was focused on silane treatment of Nypa Fruticans husk filler and the elongation at break of biocomposites was reduce due to the decrement of ductility [18]. Hence, it had shown that chemical modification of filler will enhance interfacial attraction and driven to lower mobility of the polymer chain.

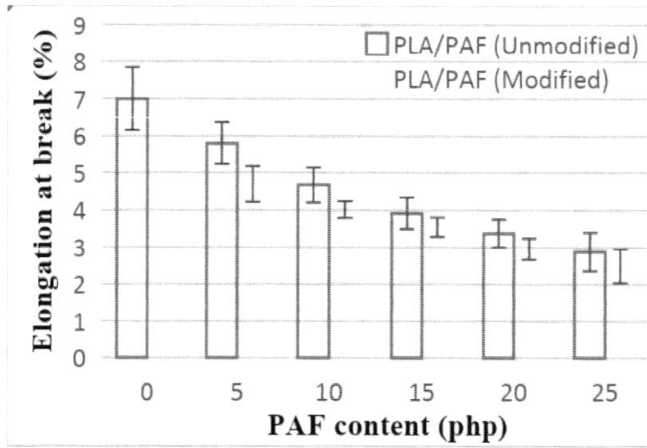

Fig. 2. Effect of PAF content on elongation at break of unmodified and modified PLA/PAF biocomposite films with 3-APE.

Fig. 3 shows the effect of PAF content on Young's modulus of unmodified and modified PLA/PAF biocomposite films with silane coupling agent. Unmodified and modified PLA/PAF films exhibited an increase of Young's modulus in comparison with neat PLA. However, modified PLA/PAF biocomposite films exhibited a higher increment than unmodified PLA/PAF biocomposite films. 3-APE treatment on PAF had increased the Young's modulus of the PLA/PAF biocomposite films. This indicated that the chemical modification of PAF has led to a better and higher efficiency of load transfer which provide for better continuity and a stronger interface between the matrix and natural filler [19]. 3-APE has the ability to interact with polylactide chain which was crucial to make it compatible with PAF. Silane can contribute to the interface with hydrophilic properties, the molecule of silane had bifunctional groups which make them capable to form a bridge with the hydrophilic and hydrophobic phase. During the modification, 3-APE was hydrolysed to silanol, which can form bonding effectively with carboxylic groups of PAF through ester linkage. Amino group which contain in 3-APE can also form hydrogen bonds with C=O sites on the PLA backbone. Hence, this led to compatible of PAF with the PLA matrix [17]. Same result in PLA/durian ring cellulose were found by P Penjumeas et al. (2016). Previously worked by Syahmie et al. (2015), it can be seen that Young's modulus of treated PLA/rLDPE/NFF biocomposites were increased slightly by silane treatment compared to untreated PLA/CSP biocomposites. It shows interfacial adhesion between NFF filler and PLA inherent in the mobility of the polymer chain [18].

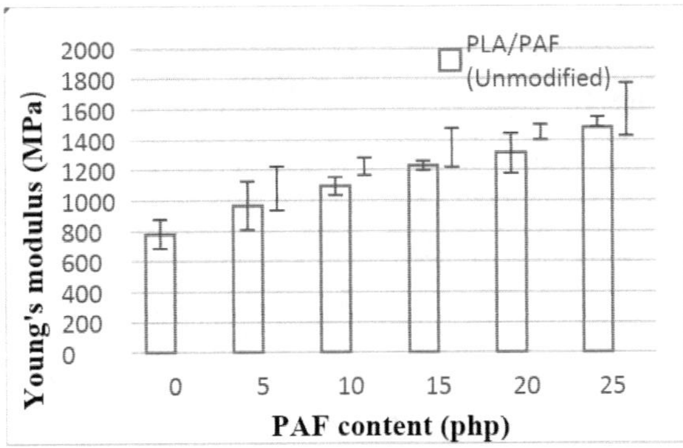

Fig. 3. Effect of PAF content on Young's modulus of unmodified and modified PLA/PAF biocomposite films with 3-APE.

3.2 Morphological Analysis

Fig. 4(a) illustrates the micrograph of tensile fracture surface of PLA/PAF biocomposite films at 10 php of PAF while Fig. 4(b) shows the micrograph of tensile fracture of 20php of PAF. From the Fig. 4(b) the micrograph shows rougher microstructure compare to the microstructure in Fig. 4(a). Notable gaps had been displayed between PLA matrix and filler. The presence of gaps attributed to the debonding during tensile test or poor dispersion during biocomposite preparations which indicated a weak filler-matrix interaction. The poor interfacial interaction between PLA matrix and PAF filler also resulted in filler pulled out during tensile test. In addition, filler agglomeration and poor dispersion has shown in Fig. 4(b). This is because higher PAF content led to tendency of filler-filler interaction to form agglomerate and poor dispersion. The micrographs of tensile fracture surface of modified PLA/PAF biocomposite films with 3-APE silane coupling agent were shown in Fig. 5(a) and (b). The micrograph of modified PLA/PAF biocomposite films exhibited a smooth surface and significantly voids had been reduced compared to unmodified PLA/PAF biocomposite films. Moreover, at higher filler content as illustrated in Fig. 5(b), PAF well embedded and created interlocking mechanism with the surface of PLA matrix. This is because modified PLA/PAF biocomposite fims had good interfacial interaction between matrix of PLA and PAF. Silanization of filler modified its surface properties and lead to improvement the filler-matrix interaction rather than filler-filler interaction. The improvement in interfacial adhesion led to better stress transfer from the PLA matrix to PAF which result in better tensile performance compared to unmodified PLA/PAF biocomposite.

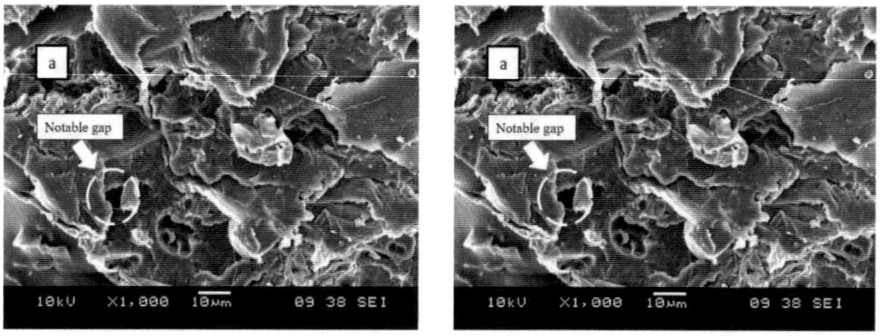

Fig. 4. Scanning electron micrograph of unmodified PLA/PAF biocomposite films (a) 10 php of PAF and (b) 20php.of PAF.

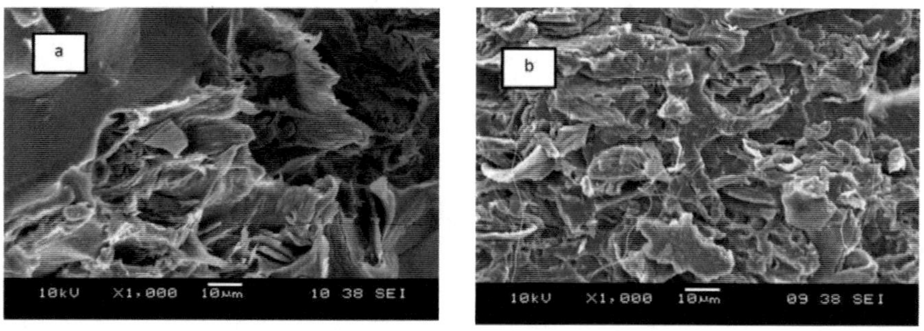

Fig. 5. Scanning electron micrograph of modified PLA/PAF biocomposite films
(a) 10 php of PAF and (b) 20 php of PAF.

3.3 Fourier Transform Infrared Spectroscopy (FTIR) Analysis

FTIR spectra of unmodified PAF and modified PAF/PLA biocomposite films was shown in Fig. 6. The spectra of biocomposite specimen is assumed to include a combination of PLA and filler transmitance peak. From the spectra of unmodified PLA/PAF, the peaks at the 1752.14 cm^{-1} in the C=O stretching region, peak at 1191.30 cm^{-1} in C-H bending region and the peak at 1085.60 cm^{-1} in C-O stretching region indicated the peak for between PLA. Meanwhile, for modified PLA/PAF spectra, peak at 1752.27 cm-1 in C=O stretching region, 1190.94 cm^{-1} in C-H bending region and 1085.73 cm^{-1} in C-O stretching region represented the peak for PLA. It can be seen on the IR spectrum of unmodified PLA/PAF biocomposite at peak 3302.70 cm-1 assigned to hydroxyl group (O-H) belonging to PAF. The peak at 2994.35 cm-1 assigned to C-H stretching vibration. Besides, carbonyl group (C=O) from hemicellulose was shown at 1752.14 cm^{-1}. While 1450.80 cm^{-1} peak indicated C-H deformation from lignin. Cellulose and lignin attributed to peak at 1369.4 cm^{-1}. Furthermore, the peaks in the range of 1000-1150 cm-1 were C-O-C and C-O groups from main carbohydrates of cellulose and lignin. The peaks at 700-900 cm-1 were attributed to C-H vibration. The modified PAF found that the IR spectrum has a significant decreased from 3302.70 cm^{-1} to 3249.84 cm^{-1}. This is well agreed for the formation of hydrogen bond from the interaction of hydroxyl group in PAF and polar group of 3-APE. Peak at 692.33 cm^{-1} corresponding to Si-O-C bending in 3-APE.

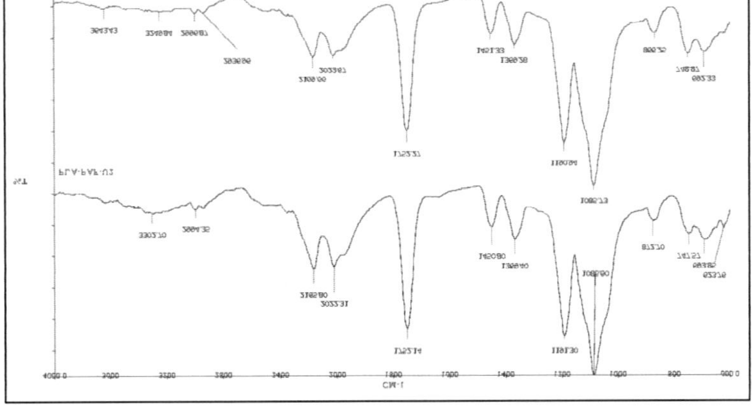

Fig. 6. FTIR spectra of modified (a) and unmodified (b) PLA/PAF biocomposite films

4 Conclusion

Tensile strength and elongation at break of PLA/PAF biocomposite films decreased with increased in PAF content. However, there was an increase in Young's modulus with increasing in PAF content. The morphological studied of tensile fracture surface showed poor interfacial interaction between hydrophobic PLA matrix and hydrophilic PAF. The chemical modification using 3-APE silane coupling agent, modified PLA/PAF biocomposite films exhibited higher tensile properties in terms of tensile strength and Young's modulus. However, elongation at break demonstrated lower value due to restriction in mobility of PLA chains. The SEM micrograph of modified PAF showed good dispersion and high compatible of PAF in the PLA matrix. Furthermore, FTIR anlaysis observed hydroxyl groups on PAF had decreased due to reaction of 3-APE with PAF.

References

1. Y. Byun, & Y.T. Kim, Acad PR (2014).
2. S. Farah, D.G., Anderson & R. Langer, Adv Drug Deliver Rev **107**, 367 (2016).
3. M. A. Elsawy, K. H. Kim, J. W. Park, & A. Deep, A., Adv Mater Res-Switz **79, 1346** (2017).
4. S. Roy, S. Bhowmik, J. P. Davim, & K. Kumar, In Green Approaches to Biocomposite Materials Science and Engineering, 96 (2016).
5. O. Faruk, A. K. Bledzki, H. P. Fink, & M. Sain Prog Polym **37**, 1552 (2012).
6. R. Hoffmann, D. D. S. Morais, C. J. F. Braz, K. Haag, R. M. R. Wellen, E. L. Canedo, L. H. de Carvalho & K. Koschek Compos Part A- Appl S **124**, 105472 (2019).
7. M. H., Hasan, M. S. Mollik, & M. M. Rashid, Int J Adv Manuf Tech **94**, 1863 (2018).
8. N. A. M. Radzuan, A. B. Sulong, M. R. Mamat, I. Tharazi, D. Tholibon, R. Dweiri, & M. S. Hammadi, Int J Integr Eng **10,** 15 (2018).
9. Y. Liu, Y. Ma, J. Yu, J. Zhuang, S. Wu, & J. Tong, Compos Interface **26**, 67 (2019).

10. R. S. M. Ghani, R. Wahab, N. M. C. Musthafa, N. Mokhtar, M. S. Sulaiman, M. S., & M. D. Lee, J Eng Sci Res **2**, 6 (2018).
11. S. Frackowiak, J. Ludwiczak, & K. Leluk, K. J Polym Environ **26**, 4360 (2018).
12. L. S. M. Ooi, E. Y. L. Wong, S. S. M. Sun, and V. E. C. Ooi, Peptides **27**, 626 (2006).
13. N. Laohakunjit & O. Kerdchoechuen, Food Chem. **101**, 339 (2006).
14. Y. Xie, C. A. S. Hill, Z. Xiao, H. Militz, & C. Mai, Compos Part A Appl Sci Manuf **41**, 806 (2010).
15. B. S. Yoon, M. H. Suh, S. H. Cheong, J. E. Yie, S. H. Yoon, & S. H. Lee, J Appl Polym Sci **60**, (1996)
16. M. A. Kashfipour, N. Mehra, & J. Zhu, Adv Compos & Hyb Mater **1**, 415 (2018).
17. P. J. Jandas, S. Mohanty, S. K. Nayak, & H. Srivastaca, Soc Plast Eng **116**, 1689 (2011).
18. M. S. Rasidi, H. Salmah, H., & P. L. Teh, P. L. Polym Eng & Sci **55**, 1733 (2015).
19. P. Penjumras, R. Abdulrahman, R. A. Talib, & K. Abdan, IOP Conf. Ser. Mater. Sci. Eng **137**, 1 (2016).

IOP Conf. Series: Materials Science and Engineering **957** (2020) 012011 doi:10.1088/1757-899X/957/1/012011

Effect of Surface Treatment on Mechanical Properties of Rice Husk Reinforced Recycled High Density Polyethylene (rHDPE) Composites.

Mohd Nazry Salleh[1,*], *Roslaili* Abdul Aziz[2] , *Mohd Fairul Sharin* Abdul Razak[1], *Luqman* Musa[1] and *Hong* Li Ying[1].

[1]School of Materials Engineering, Universiti Malaysia Perlis, 02600 Arau, Perlis
[2]Institute of Engineering Mathematics, Universiti Malaysia Perlis, 02600 Pauh Putra, Perlis

Abstract. The aim of this work was to investigate the effect of surface treatment on rice husk reinforced recycled high density polyethylene (rHDPE) composites. Three types of surface treatment on rice husk were carried out which were maleated treatment, alkaline treatment and acrylic acid treatment. The characteristics and mechanical properties of the composites were analysed. The results for all treatments showed that tensile strength and break elongation of composites decreased significantly when the rice husk filler loading increased. Further findings showed that maleated treatment on rice husk can further enhance the mechanical properties due to the present of ester bond formed in FTIR analysis.

1 Introduction

Recently, natural fiber reinforcements had been used in polymeric green composites to decrease environmental problems. Composite with natural reinforcements can decrease the environment burden, makes composite light in weight, economic and easy to process with comparable mechanical and thermal properties [1]. In early of the 1960s, the demand for high strength, tougher and low density materials for use in construction, transportation and aerospace industries becomes crucial. Due to the high demand on performance of products, extensive research and development had been carried out in the field of composite materials [2].

Nowadays, there has been increasing awareness of more sustainable and environmental friendly development, which has grown the interest of using agriculture waste as reinforcement in polymer composite [2]. Rice husk is one of the natural waste product which is widely available. It consists of 35% of cellulose, 25% of hemicellulose, 20% of lignin and 17% of ash content by weight. The rice husk ash with higher content of silica encourage the use of rice husk ash as inorganic reinforcement [3]. The rice husk is separated from the rice grain in the milling process. The function of rice husk is to provide the protection layer to the rice grain. It is the outermost which is cover the rice grain. Thus, many researchers are

* Corresponding author: nazry@unimap.edu.my

currently conducted on the studies of rice husk reinforced polymer composite to produce high performance natural fiber-reinforced composites.

In this research, the major concern is the poor compatibility between the hydrophobic thermoplastic matrix and hydrophilic fibers. This may cause to the poor dispersion of fibers into the matrix. Due to the hydrophilic nature of rice husk fiber, its highly moisture sensitivity causing swelling and delamination in composites. This is because there is a strong hydrogen bond exist between the fibers which is to hold them together. Weak interface will be resulted and cannot transfer stress from matrix to the fiber [4]. Therefore, chemical treatment and coupling agent is required to improve the adhesion of fiber and matrix interface. Surface modification is needed to modify the surface of rice husk in order to render the surface of rice husk to become more hydrophobic and more compatible to polymer matrix [5]. Chemical treatment is one of the mostly used for surface modification of natural fibers. Acrylic acid treatment and alkaline treatment are the common chemical treatment methods. This can help to enhance the adhesion properties between the matrix and fibers [6]. Maleic anhydride polyethylene (MAPE) is used as coupling agent in this polymer composite. The purpose of using coupling agents is to enhance the interaction between natural fibers and hydrophobic thermoplastic matrices. Hence, this research is intended to study the effect of chemical treatment on rice husk reinforced recycled HDPE and rice husk composites.

2 Materials and Methods

2.1 Raw Materials

The density of rHDPE is 923 kg/m^3, melt flow index is 0.72 g/10min at 190 ^{0}C and melting point is 135 ^{0}C. The rHDPE was obtained from Nippon Pigmen (M) Sdn. Bhd with. While rice husk was sieved using 1mm of sieve which is 0.0394 inches. The rice husk was sieved to obtain an average size of rice husk and to remove the cadelle beetle in order to enhance the processing. Rice husk was oven dried at 80 °C for 24 h before use in order to reduce the moisture content.

2.2 Chemicals

MAPE was used as maleated agent to produce composite with a melting temperature of 135.2 ^{0}C. MAPE used is supplied by ALDRICH Chemistry. Its density is 0.92 g/mL at 25 ^{0}C. Acrylic acid was used to carry out the acid treatment on the rice husk. The molar weight of acrylic acid is 72.063 g/mol with a clear and colorless liquid. The melting and boiling point of acrylic acid is 14^{0}C and 141^{0}C respectively. Sodium hydroxide (NaOH) was used in the process of alkaline treatment on the rice husk, supplied by HmbG Chemicals Company. It may be corrosive to metals. The molar mass of NaOH is 40.00 g/mol.

2.3 Procedures

2.3.1 Acrylic Acid Treatment of Rice Husk

Acrylic acid solution was used in acid treatment of rice husk. Acrylic acid solution was prepared by mixing acrylic acid (10 vol. %) with 800ml of water in a 1000ml glass beaker. 150g of rice husk were stirred completely with acrylic acid in the beaker for 10 minutes. Soaking of rice husk was carried out for 4 hours. The treated rice husk was then washed by tap water to completely remove the acid over the surface of rice husk until it reached pH 7. The wet rice husk was spread and lay over a tray and dried in an oven overnight at 60 ^{0}C. The dried clumps was break using a roller. The acrylic treated rice husk then were kept in a moisture free sealed polybag.

2.3.2 Alkaline Treatment of Rice Husk

NaOH solution was used in alkaline treatment of rice husk. 80g of NaOH was mixed with 800ml of distilled water in a 1000ml glass beaker to produce aqueous solution of NaOH. 150gms of rice husk were mixed and stirred with NaOH solution for 10 minutes. The treatment were carried out at room temperatures for 4 hours. The treated rice husk was then washed with tap water, dried and rolled prior to store in a sealed polybag.

2.3.3 Maleated Treatment on Rice Husk

The weight of MAPE, rHDPE and rice husk were prepared by following the composites formulation as shown in Table 1. MAPE, rHDPE and rice husk were mix homogeneously before compounded by using twin screw extruder.

2.3.4 Composite Compounding

The rHDPE and rice husk were compounded by using a co-rotating twin screw extruder at 70 rpm speed. The temperature from the feeding zone to die zone were set at 185 ^{0}C, 215 ^{0}C, 195 ^{0}C, and 185 ^{0}C. Then the rHDPE and rice husk were compounded into long strand through the twin screw extruder. The polymer composite strand was later pelletized using a crusher. Th e pelletized compound was dried in the oven to eliminate the moisture before molded using a compression machine. The temperature of the hot press were set at 180 ^{0}C upper dan lower platen, respectively to produce the polymer composites. The period of preheating, venting and full pressing were set at 3, 2 and 10 minutes, respectively and cold press was set to 5 min to cool the samples. The pressure used to press the samples was set at 10 MPa for both hot and cool press.

Table 1. Composites formulation.

Surface Treatment	Sample Notation	rHDPE (wt%)	Rice Husk (wt%)	MAPE (php)
Maleated Treatment	M-10	90	10	3
	M-20	80	20	3
	M-30	70	30	3
Acrylic Acid Treatment	AC-10	90	10	0
	AC-20	80	20	0
	AC-30	70	30	0
Alkaline Treatment	AK-10	100	10	0
	AK-20	80	20	0
	AK-30	70	30	0

2.4 Testing

2.4.1 Fourier Transform Infrared Spectroscopy (FTIR)

Fourier transform infrared spectroscopy (FTIR) was used to analyze the IR spectra of the samples formed. In this study, The Perkin Elmer Spectrometer 2000 was used to obtain the result. An infrared spectra represent the functional group of a sample with absorption peaks. 32 times of scan were carried out to obtain a better resolution on the reading. The transmittance spectra were obtained in the range of 4000 and 6000 cm^{-1} with a resolution of 4 cm^{-1}.

2.4.2 Tensile Test

Tensile test is one of the testing to determine the mechanical properties of composites. All samples were tested using the Instron tensile test machine. The standard method of tension test for plastic properties is ASTM D638. The testing was carried out at room temperature with a crosshead speed of 10mm/min. All the reported results of tensile tests are the average of five replicated for each formulation.

3 Results and Discussion

3.1 Fourier Transform Infrared (FTIR)

Fig. 1 shows the structural analysis and structural comparison of rice husk reinforced rHDPE composite. The FTIR spectra showed the composite of M-20, AK-20 and AC-20. From Figure 1, there are a few of same peaks in these three spectra. The absorption band at the range of 3350 cm^{-1} to 3430 cm^{-1} attributed to carboxylic group in the rice husk filler and rHDPE. While the absorption at around 2905 cm^{-1} and 2860 cm^{-1} were indicated to the vibration of carbon-hydrogen (CH) bonds, which is present in three types of surface treatments. Besides, at the absorption ranges of between 2024 cm^{-1} to 2026 cm^{-1} showed the stretching of aromatic carbon upon C=C. Another carbonyl vibration was found at 1660 cm^{-1}, which could be attributed to the stretching vibration of carboxylic groups or the interaction of cell window. Another peak which showed at 1452 cm^{-1} attributed to the vibration of carbon-hydrogen bonds [7].

For the M-20 which is shownin red line, there is one weak absorption band at 1031 cm^{-1} appeared in the spectrum which is absent in alkaline treatment and acrylic acid treatment. This peak showed the carbonyl stretching vibrations of the ester, ether and aldehyde groups in the hemicellulose, lignin and natural fats and wax. Esterification reaction occurred in maleated treatment due to the ester bond form between the hydroxyl groups on the surface of rice husk and the ahydride group of coupling agent [8]. This can enhance the interfacial adhesion of the composite. Esterification reaction happened due to the presence of MAPE, therefore when after the alkali treatment and acrylic acid treatment, this peak was disappeared. The function of alkaline treament and acrylic acid treatment are to remove the hemicellulose, lignin and wax. Furthermore, the absorption band showed at 1660 cm^{-1} which indicated the presence of grafted maleic anhydride group in the polymer matrix due to the C=O stretching.

Next, for the AK-20 which is showed in blue line, the FTIR spectrum showed one extra absorption band at 3630 cm^{-1} which is –OH vibration. This peak occurred after the alkaline treatment on rice husk due to the removal of surface impurities from the rice husk filler. Therefore, the –OH groups are increasingly exposed on the rice husk filler surfaces.

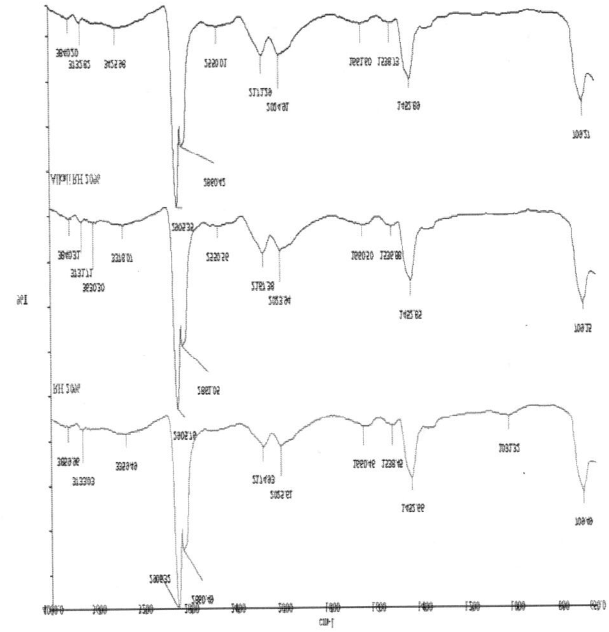

Fig. 1. Structural comparison of M-20, AK-20 and AC-30 of rice husk loading.

3.2 Tensile Test

3.2.1 Tensile Strength

Fig. 2 shows the result of tensile strength of different loading of rice husk with different surface treatment on rice husk. The tensile strength of rice husk reinforced rHDPE composite are influenced by the content of rice husk loading and the surface treatment on rice husk. Based on the Figure 2, 10% loading of rice husk in this composite showed the highest tensile strength value among these three content of rice husk. The content of rice husk resulted in significant effect on the tensile strength.

For the maleated treatment on rice husk, 10% loading of rice husk (M-10) had the highest tensile strength which is 16.544 MPa. This is because the rice husk fillers were effectively transfer the stress applied due to the better interfacial bonding between the hydrophilic group in fillers and the hydrophobic group in matrix modified with MAPE [9]. The higher loading of rice husk content attributed to the lower the tensile strength of the rice husk reinforced rHDPE composite. The value of tensile strength of M-20 reduced to 13.902 MPa and tensile strength of M-30 reduced to 13.545 MPa.

While for the alkaline treatment and acrylic acid treatment, there was a slightly difference with maleated treatment which is there is a slightly decrease in tensile strength of the composite when the rice husk loading increase from 10% to 20%. The decrease in tensile strength with the rice husk fillers loading may be due to the increase in weak interfacial area between the rice husk fillers and rHDPE matrix [10]. This also might be due to too much loading of rice husk loading and causing insufficient wetting of rice husk by the polymer matrix. This will lead to the filler agglomeration and hence affecting the stress transfer.

Maleated treatment of rice husk reinforced rHDPE shows the highest tensile strength compared to alkaline treatment and acrylic acid treatment. Maleated coupling agent is widely used to enhance the strength of composites which is consists of fillers and polymer matrix. Maleated treatment is different with other chemical treatment. This is because the maleic anhydride is not only modify the surface of rice husk but also the PE matrix in order to obtain better interfacial bonding in composites [11]. Therefore, maleated coupler forms carbon-carbon bond to the polymer chain with the matrix. This covalent bonding between the hydroxyl groups of the fibre and the anhydride groups of the maleic anhydride make the bridge interface for efficient interlocking [12]. This had proved by the result of FTIR before which is showed in Figure 1. The absorbance band which show at 1660 cm^{-1} proved the presence of grafted maleic anhydride group in the composite, therefore, maleated treatment had the highest tensile strength.

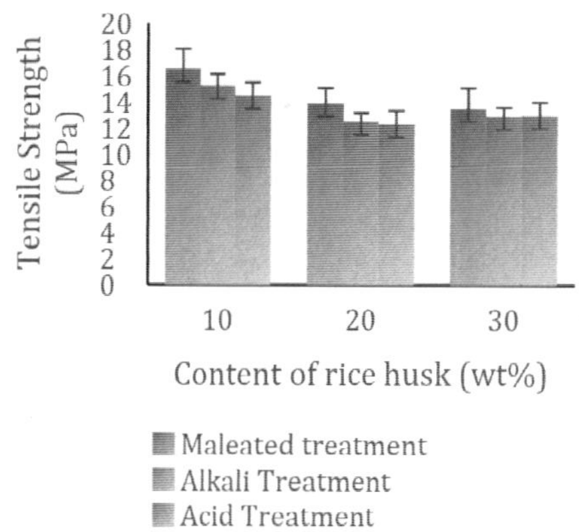

Fig. 2. Tensile strength biocomposite with different rice husk loading and surface treatments.

3.2.2 Elongation at Break

Figure 3 shows the elongation at break of the composite with different rice husk loading and surface treatments. The elongation at break tend to reduce when the loading of rice husk increases. AK-30 and AC-30 had the lowest value of elongation at break which is 3.9%. This happened due to the brittle properties of rice husk in the composite and this can reduce the ductility of the polymer matrix [13].

Besides, based on the Figure 3, maleated treatment composite has the highest elongation at break compared to alkali treatment and acrylic acid treatment. M-10 of rice husk loading shows the highest value which is 9.5%, while M-20 loading of rice husk is 6.6% and M-30 is 4.8%. For alkaline treatment on rice husk, AK-10 of rice husk loading shows 9.3% of break elongation while AK-20 rice loading is 4.5% and AK-30 of rice husk loading is 3.9%. Acrylic acid treatment has a lower break elongation compared to maleated treatment and alkaline treatment. AC-10 of rice husk loading with acrylic acid treatment shows 7.9% of break elongation while AC-20 is 5.2% and AC-30 is 3.9%.

According to Jia Ying Tong et al., maleic anhydride grafted matrices used as coupling agent are able to improve the elongation at break of the composite. This is because the formation of ester bonds between anhydride groups of coupling agents entered into an esterification reaction with the surface hydroxyl groups of rice husk. This can be proved by the result in FTIR. The peak of 1031 cm^{-1} was detected and this showed the formation of ester bond. The hydrophilic group in rice husk is chemically bonded by coupling agent. Coupling agent was blended by wetting in the polymer chain. Therefore, maleated treatment composite had the highest elongation at break among the three surface treatment [8].

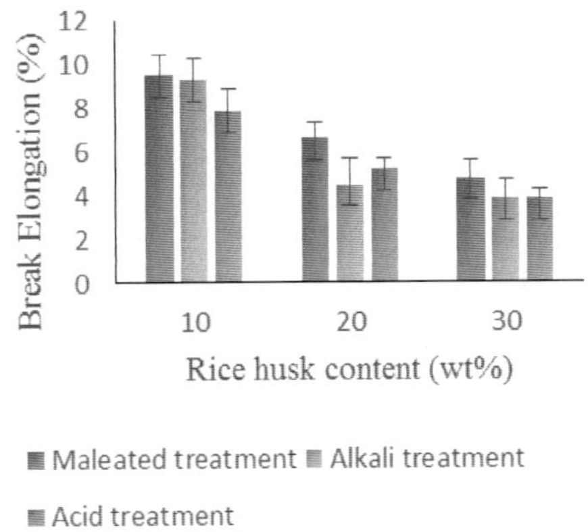

Fig. 3. Elongation at break of biocomposite with different rice husk loading and surface treatment

4 Conclusion

In a conclusion, the FTIR shows the spectrum of M-20, AK-20 and AC-20. The peak of carbonyl stretching vibrations of ester, ether and aldehyde groups appeared in spectrum of maleated treatment. The peak was disappeared after the alkaline and acrylic acid treatments. Due to the ester bond formed in the maleated treatment composites, the reading of tensile strength and break elongation were the highest among the three surface treatments. The spectrum of alkaline treatment shows one extra peak which is –OH vibration. This is resulted by increasingly exposure of –OH groups on the rice husk surface after the alkali treatment which removed the surface impurities. In addition, the higher the loading of rice husk filler, the weaker the tensile properties of the rice husk reinforced rHDPE composites. 10% of rice husk loading shows the optimum tensile strength and break elongation due to the effective transfer of applied stress. The higher the loading of rice husk, the lower the tensile strength and break elongation achieved by the composite. This is because of the insufficient wetting of rice husk by the polymer matrix.

The author would like to acknowledge the support from the Fundamental Research Grant Scheme-RACER under a grant number of RACER/1/2019/TK05/UNIMAP/2

References

1. Chanda, A. K., Hazra, A., Praveen Kumar, M., Neogi, S., and Neogi, S. (2015). Chemical treatments of rice husk filler and jute fiber for the use in green composites, Fibers and Polymers, 16(4), pp. 902–910.
2. Kenechi, N.-O., Linus, C., and Kayode, A. (2016). Utilization of rice husk as reinforcement in plastic composites fabrication, 1(3), pp. 32.
3. Panthapulakkal, S., Law, S., and Sain, M. (2005). Enhancement of processability of rice husk filled high-density polyethylene composite profiles, Journal of Thermoplastic Composite Materials, 18(5), pp. 445–458.

4. Razavi-nouri, M., Jafarzadeh-dogouri, F., Oromiehie, A., and Langroudi, A. E. (2006). Mechanical properties and water absorption behaviour of chopped rice husk filled polypropylene composites, Iranian Polymer Journal, 15(9), pp. 757–766.
5. Hafizuddin, M., Ghani, A., Royan, N., Royan, R., Kang, S. W., Sulong, A. B., Husk, R. (2015). Effect of alkaline treated rice husk on the mechanical and morphological properties of recycled hdpe / rh composite, 10(5), pp. 138–144.
6. Emdadi, Z., Asim, N., Yarmo, M. A., and Sopian, K. (2015). Effect of chemical treatments on rice husk (rh) water absorption property, International Journal of Chemical Engineering and Applications, 6(4), pp. 273–276.
7. Chen, R. S., Ab Ghani, M. H., Salleh, M. N., Ahmad, S., and Tarawneh, M. A. (2015). Mechanical, water absorption, and morphology of recycled polymer blend rice husk flour biocomposites, Journal of Applied Polymer Science, 132(8), pp. 1–12.
8. Tong, J. Y., Royan, N. R., Ng, Y., Ghani, M. A., and Ahmad, S. (2014). Study of the mechanical and morphology properties of recycled hdpe composite using rice husk filler, Advances in Materials Science and Engineering, pp. 1–10.
9. Rahman, M. R., Islam, M. N., Huque, M. M., Hamdan, S., and Ahmed, A. S. (2010). Effect of chemical treatment on rice husk (rh) reinforced polyethylene (pe) composites, BioResources, 5(2), pp. 854–869.
10. R. Chen, S. Ahmad, M. Ghani. (2014). Optimization of high filler loading on tensile properties of recycled HDPE/PET blends filled with rice husk, 46, pp. 46-51.
11. Li, X., Tabil, L. G., & Panigrahi, S. (2007). Chemical treatments of natural fiber for use in natural fiber-reinforced composites: a review. Journal of Polymers and the Environment, 15(1), pp. 25–33.
12. M. Kabir, H.Wang, K. Lau. (2012). Chemicsl treatments on plant-based natural fibre reinforced polymer composites: an overview. Journal of Composites Part B: Engineering, 43(7), pp. 2883-2892.
13. K. Hardinnawirda and I. S.Aisha. (2012). Effect of rice husks as fillers in polymer matrix composites," *J. Mech. Eng. Sci.*, vol. 2, pp. 181-186.

Corrosion Investigation of Sn-0.7Cu Pb-Free Solder in Open-Circuit and Polarized Conditions

Mohd Zohirul Haziq Aziz[1], *Nasihah* Zainon[1], *Ahmad Azmin* Mohamad[2] and *Muhammad Firdaus Mohd* Nazeri[1,*]

[1]Center of Excellence Geopolymer and Green Technology (CEGeoGTech), School of Materials Engineering, Universiti Malaysia Perlis, 02600 Jejawi, Kangar, Perlis, Malaysia
[2]School of Materials and Mineral Resources Engineering, Universiti Sains Malaysia, 14300, Nibong Tebal, Penang, Malaysia

Abstract. The corrosion performance of Sn-0.7Cu solder alloy was investigated in 1 M hydrochloride acid by means of open-circuit potential and potentiodynamic polarization techniques. In open-circuit potential measurement, the potential showed almost constant reading after achieving potential -550 mV, consistent with single dissolution peak observed in potentiodynamic polarization. This revealed single element dissolution process of Sn took place. Supporting morphological and phase analyses were done by using scanning electron microscope and X-ray diffraction. Phase analysis confirmed the corrosion product are SnO and SnO_2. Morphological analysis showed two type of oxide layer formed as corrosion product, which are compact oxide layer and loose rod-like oxide layer after polarization, while rod-like corrosion product was diminished in open-circuit condition. This indicate that complete protection from further corrosion was impossible due to the incomplete passivation layer formation in potentiodynamic polarization.

1 Introduction

Tin-lead (Sn-Pb) solders are commonly used in the electronic industry due to advantages such as low melting temperature, good mechanical properties and cheap. However, this alloy has been banned for use in many applications attributed by its toxic nature [1]. Hence, Pb-free solders emerge as an alternative.

In the last decade, Pb-free solders from Sn-based alloys of tin-copper (Sn-Cu), tin-bismuth (Sn-Bi), tin-silver (Sn-Ag) and tin-zinc (Sn-Zn) have been developed [2]. However, the best Pb-free solder alloys is yet to be found as each alloy and composition which has its advantages and limitations.

The Sn-0.7Cu is one of the outstanding candidate replacements for Sn-Pb solders as this alloy offers good electrical conductivity and much lower price compared with the other candidates. In term of the melting temperature, Sn-0.7Cu melts about 227 °C of which comparable with the Sn-Pb. According to Farina [3] one of advantages of Sn-0.7Cu is the

* Corresponding author: firdausnazeri@unimap.edu.my

excellent resistant to corrosion as similar as traditional Sn-Pb solder in sodium chloride (NaCl) solution.

Best of our knowledge, there are many literature studies focused on the correlations between the microstructure and the corrosion properties of solder alloys in saline solution, especially in 3.5 wt. % NaCl. However, there are apparent lack of studies on electrochemical corrosion behaviour of Sn–0.7Cu solder alloy in other common corrosive solution for example in acidic medium of hydrochloric acid (HCl). A solder alloy requires good corrosion behaviour as part of reliability issue, especially when the solder joints are directly exposed to aggressive medium in an industrial environment [4-6]. In this work, corrosion performance of Sn–0.7Cu solder alloy will be investigated in 1 M HCl in open circuit and potentiodynamic polarization condition. Subsequent characterization techniques of XRD and SEM will be implemented to support the electrochemical findings.

2 Methodology

The Sn-0.7Cu solder (5 mm diameter and 3 mm thickness) pellets was individually attached to copper wire using soldering process to allow electrical conduction. The solder was then cold mounted with epoxy resin (Fig. 1). Each mounted sample were ground and polished to produce a smooth and mirror-like surface. For OCP measurement, the sample and reference electrode of saturated calomel electrode (SCE) were attached to a voltmeter equipped with data recording software. A potentiostat (Autolab PGSTAT 30) was used for potentiodynamic polarization of Sn-0.7Cu in 1 M HCl. The scan potential range used are from -2 V_{SCE} to 1 V_{SCE} at the scan rate of 1mV/s. The sample served as the working electrode, while SCE and platinum were used as reference and counter electrodes, respectively. For supporting characterizations, X-ray Diffraction (XRD) model (Shimadzu XRD 6000) was used for phase analysis of pristine Sn-0.7Cu and after corrosion. The analysis is performed at 2Θ of 20-90° with the scan rate of 5°/min. The phase peak was matched using Xpert Highscore Plus software. Scanning Electron Microscope (SEM), JEOL JSM-6460-LA was used for morphological analysis of sample before and after electrochemical studies.

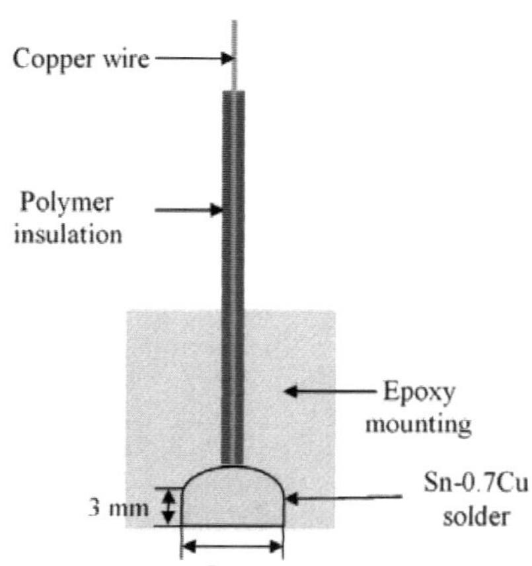

Fig. 1. Schematic diagram of mounted Sn-0.7Cu

3 Results and Discussion

3.1 Corrosion Analysis

The initial potential drop from -530 to -550 mV_{SCE} in the first 3000 seconds (stage I) after immersion in HCl solution (Fig. 2). Potential drop from positive to negative value indicates that the system becomes more electrochemically active. This reveals that the solder surface is corroding at this stage. Dissolution of Sn is expected as the least noble element in this solder. The potential become more stable with small fluctuations ranging -550 ± 5 mV_{SCE} after 10000 seconds to 80000 seconds (stage II).

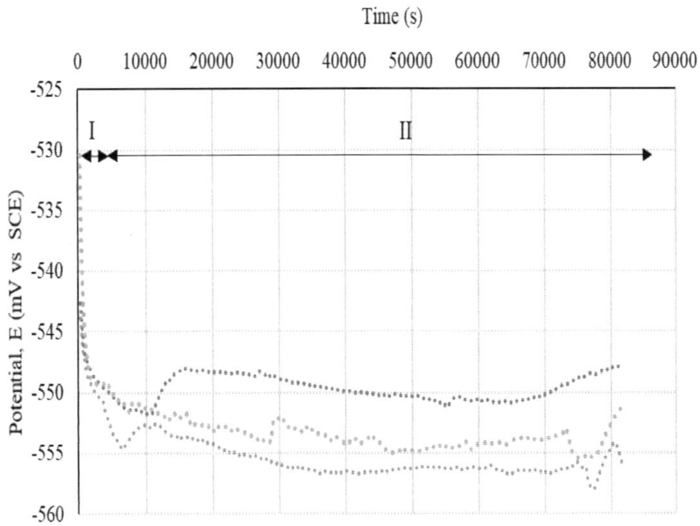

Fig. 2. Open circuit potential plot of Sn-0.7Cu in 1 M HCl

According to Wang et. al [7], OCP value influenced by the metal dissolution and formation of corrosion products. Metal dissolution drives the OCP value to shift to a negative value while positive value indicates the protection of corrosion products. The OCP value become constant with time when the surface of the solder reached an equilibrium between metal dissolution and formation of corrosion products.

Fig. 3 shows the potentiodynamic polarization scan of a Sn-0.7Cu in 1 M HCl at scan rate of 1.0 mVs^{-1}. The polarization can be divided into two regions which are anodic and cathodic regions. The scan started in the cathodic region at potential of -2.0 V_{SCE} (point A). The current rapidly decreased when the applied potential increased. The current reduction stopped when the potential reached -0.55 V_{SCE} (point B). This indicates the initiation of anodic region. The current rapidly increased ahead from this potential producing linear region known as a primary activation region (BC).

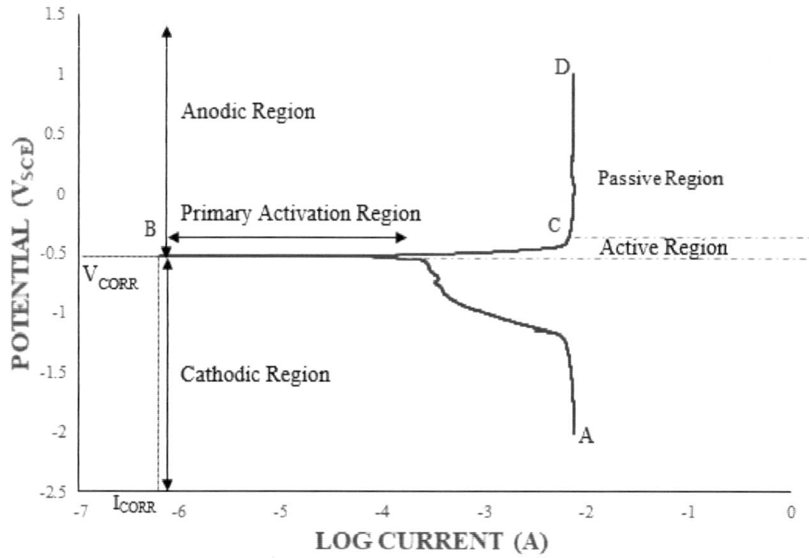

Fig. 3. Potentiodynamic polarization graph of Sn-0.7Cu

This linear region illustrate the point where dissolution of active material began. Rasid et. al [8] reported that Sn was first dissolve from Sn-0.7Cu solder. Single dissolution peak seen at -0.55 V_{SCE} in accordance with the dissolution potential seen in OCP analysis. Dissolution of Sn can be represented by Eq. 1 [9]:

$$Sn \rightarrow Sn^{2+} + 2e^{-} \quad (1)$$

The current decreases slightly at point C due to passivation of the solder. Sn^{2+} deposited on the surface of the solder and formed passive layer. Oxides of SnO and SnO_2 are produced according to the reaction [9]:

$$O_2 + 4e^{-} + 4H^{+} \rightarrow 2H_2O \quad (2)$$

$$2Sn^{2+} + 2OH^{-} \rightarrow 2SnO + 2H^{+} \quad (3)$$

$$Sn + 2H^{+} \rightarrow Sn^{2+} + H_2 \quad (4)$$

$$2Sn^{2+} + O_2 + 6H_2O \rightarrow 2Sn(OH)_4 + 4H^{+} \quad (5)$$

$$2Sn(OH)_4 \rightarrow 2SnO_2 + 4H_2O \quad (6)$$

The slopes interception of linear part of anodic and cathodic regions represent the corresponding equilibrium corrosion potential (E_{corr}) and corrosion current (i_{corr}) for Sn-0.7Cu solder. The E_{corr} and i_{corr} values in 1 M HCl obtained from tafel extrapolation are -0.55 V_{SCE} and 1.854 μA, respectively.

3.2 Phase analysis

Fig. 4 shows the XRD pattern of the Sn-0.7Cu solder prior corrosion analysis. The phases detected are tetragonal β-Sn (ICDD 00-089-2958) and monoclinic Cu_6Sn_5 (ICDD 03-065-2303).

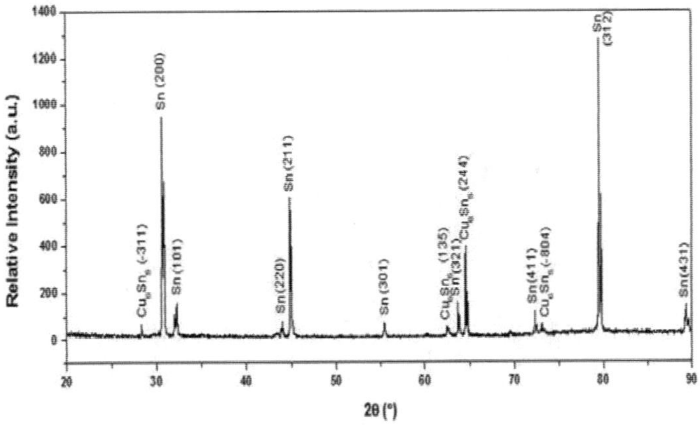

Fig. 4. X-Ray diffraction pattern of pristine Sn-0.7Cu solder alloy

Fig. 5a shows the XRD pattern for the Sn-0.7Cu alloys after immersion in 1 M HCl in open-circuit condition. Significant phase changes were observed after the immersion where the intensities of the Sn planes at 2Θ ranging from 30° to 80° were significantly reduced. This result proves that Sn reacted during immersion to produce tin oxides. Similar finding was also previously reported [10]. Identical observation was also seen after potentiodynamic polarization (Fig. 5b).

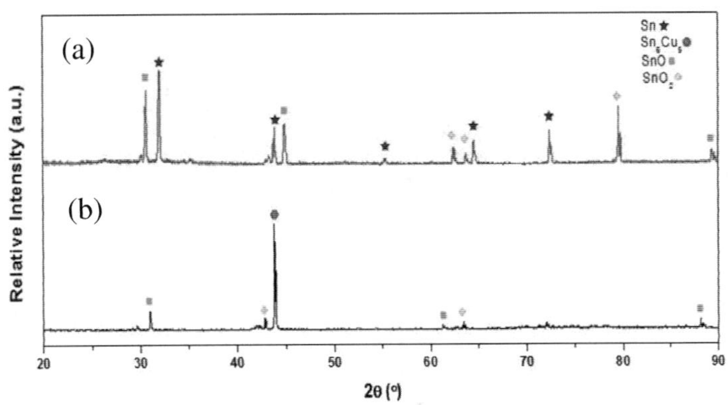

Fig. 5. XRD pattern of the Sn-0.7Cu solder after (a) open-circuit and (b) potentiodynamic polarization analyses

In both conditions, two new phases were observed after corrosion process took place. The phases are SnO (ICDD: 00-050-1429) and SnO_2 (ICDD: 00-006-0395). This has proved that the dissolution of Sn is occurred and further react with OH^- to form SnO and SnO_2. The obtained result is same with the report of Huang et al. [11] that reported that both phase were the passivation barrier that prevent the solder from further corrosion.

3.3 Morphological Analysis

The microstructure of Sn-0.7Cu solder after open-circuit immersion reveals that the corrosion products covered then entire surface, with number of porous spots and small cracks (Fig. 6). This explains the stable potential region seen produced by Sn-0.7Cu in the OCP profile at the later stage of immersion.

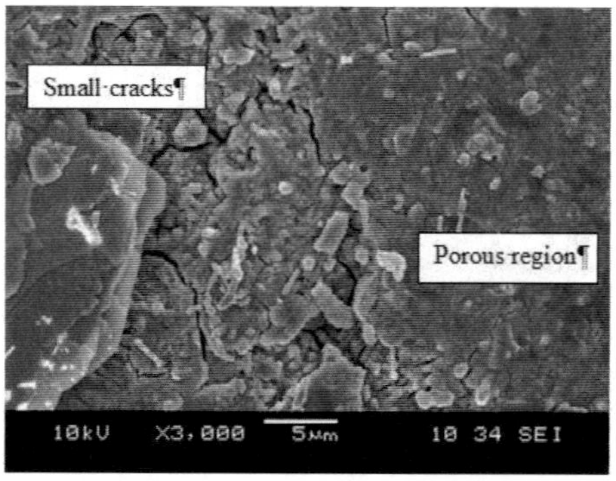

Fig. 6. SEM photographs of Sn-0.7Cu solder alloy immersion in 1 M HCl for OCP polarization at magnification X3000

After polarization, it was found that the solder surface was completely covered by the rod-like structure and sheet-like corroded products (Fig. 7). The presence of pores between the lamellar corrosion products indicated that further corrosion penetration is possible. This lamellar structure may have formed because of the higher Sn removal rate at a scan rate of 1.0 mVs^{-1}. It can be said that higher removal rate in polarized condition altered the morphology of the passive film formed. The lack of morphology uniformity was introduced in polarized condition as the deposition process of oxides was forced by the external power source applied [12].

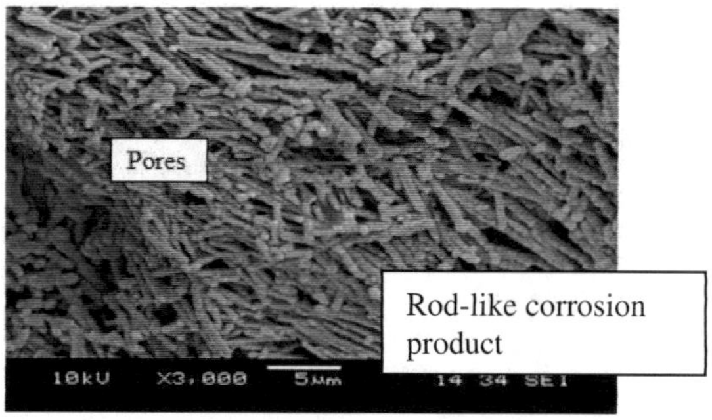

Fig. 7. Scanning Electron Micrograph of Sn-0.7Cu after potentiodynamic polarization at magnification X3000

4 Conclusion

The electrochemical corrosion behavior of Sn-0.7Cu has been investigated through open-circuit and polarized conditions. Single element dissolution process of Sn was seen in both conditions. XRD analysis confirmed the existence of passive oxide film which formed by corrosion by-product of SnO and SnO_2 that act as a barrier to increase the corrosion resistance in both conditions. However, morphological analysis had found some that the passivation layer contained porous region made of loose-oxide formation. This was identified as weak spot for full corrosion protection.

The authors would like to acknowledge the support from the Fundamental Research Grant Scheme (FRGS) under a grant number of FRGS/1/2019/TK05/UNIMAP/02/5 from the Ministry of Education Malaysia.

References

1. Abtew, M. and G. Selvaduray, Mater. Sci. Eng. R Rep. **27**, 95-141 (2000)
2. Efzan, E. and A. Marini, Int. J. Eng. Sci. **1**, 2305-8269 (2012)
3. Farina, S. and C. Morando, J. Mater. Sci. : Mater. Electron. **26(1)**, 464-471 (2014)
4. Garcia, L.R., L.C. Peixoto, W.R. Osório, and A. Garcia, Mater. Lett **63**(15), 1314-1316 (2009)
5. Osório, W.R., E.S. Freitas, J.E. Spinelli, and A. Garcia, Corros. Sci. **80**, 71-81 (2014)
6. Wang, M., J. Wang, H. Feng, and W. Ke, Corros. Sci. **63**, 20-28 (2012)
7. Zhenghong Wang, C.C., Jinting Jiu, Shijo Nagao, Masaya Nogi, Hirotako Koga, Hao Zhang, Gong Zhang, Katsuaki Suganuma, J. Alloys Compd **716**, 231-239 (2017)
8. Mohd Rasid, Z.A., M.F. Omar, and M.F. Mohd Nazeri, Mater. Sci. Forum **888**, 394-399 (2017)
9. Nordarina, J., H. Mohd, A. Ahmad, and F. Muhammad. IOP Conf. Ser. Mater. Sci. Eng (2018)
10. Liew, M.C., I. Ahmad, L.M. Lee, M.F.M. Nazeri, H. Haliman, and A.A. Mohamad, Metall Mater Trans A 43(10), 3742-3747 (2012)
11. Huang, H.-z., D. Lu, G.-w. Shuai, and X.-q. Wei, T Indian I Metals **69**(8), 1537-1543 (2016)
12. Nazeri, M.F.M., A.B. Ismail, and A.A. Mohamad, J. Alloys Compd **606**, 278-287 (2014)

IOP Conf. Series: Materials Science and Engineering **957** (2020) 012013 doi:10.1088/1757-899X/957/1/012013

High Temperature Cyclic Oxidation of Ni-based 800H Superalloy at 700℃ in Air

Noraziana Parimin[1,*], and *Esah* Hamzah[2]

[1]School of Materials Engineering, Universiti Malaysia Perlis, 02600 Arau, Perlis Malaysia
[2]Faculty of Mechanical Engineering, Universiti Teknologi Malaysia, 81310 Skudai, Johor, Malaysia

Abstract. The high temperature cyclic oxidation has been investigated on 800H superalloy, composed of Fe-33Ni-19Cr alloy at 700°C for 150 cycles in laboratory air. Two types of samples has been used, where the Fe-33Ni-19Cr alloy has been experienced a different heat treatment temperatures, which are 1000°C and 1100°C, which has differ grain size. The alloy was subjected to a short oxidation period at 700°C for one hours, followed by cooling for 20 minutes for each cycle. The oxide phase analysis has been analyzed using x-ray diffraction (XRD) technique. The cross sectional line scan analysis has been examined using field emission scanning electron microscope (FESEM) equipped with energy dispersive x-ray (EDX) spectrometer. The phase analysis recorded four types of structure has formed on the oxidized sample, composes of austenite phase represent the base metal, corundum oxides, spinel oxides and fluorite oxides structure. The cross sectional analysis indicate that the oxidized samples were formed a several oxide layer, mainly compose of Cr-Mn and Cr-Ti rich oxide with detection of internal oxide precipitates compose of Al-oxide and Si-oxide precipitates.

1 Introduction

Ni-based superalloy are commonly used in high temperature application due to the excellent performance such as high temperature strength, toughness, oxidation resistance, creep resistance and corrosion resistance [1-4]. Ni-based superalloy are commonly used in aircraft engine, turbine blade, turbine disks, power plants and other application [1-2]. In these aforementioned applications, this superalloy were encounter a repeated thermal cycling which can affect material degradation. Ni-based 800H superalloy consists of Fe-33Ni-19Cr alloy has been considered as a suitable materials used at high temperature condition because of their ability to maintain their mechanical properties at increased temperature coupled with their superior oxidation resistance [4-6].

The superalloy have been developed to achieve oxidation resistance by utilizing the selective oxidation concept [2, 7]. These concepts are affected by many factors such as alloy composition, surface conditions, oxidizing environment, and oxide scale behaviour [2]. When a metal is exposed at high temperature conditions to an oxygen rich environment, oxide nuclei will formed and start to impinge on each other, to developed a

* Corresponding author: noraziana@unimap.edu.my

IOP Conf. Series: Materials Science and Engineering **957** (2020) 012013 doi:10.1088/1757-899X/957/1/012013

thermodinamycally stable and slow growing oxide scales, hence encapsulate the alloy surface to formed a protective thin oxide layer. At the initial stage, the rate of oxidation is rapid and all element in the alloy is oxidized to formed various oxide phases. The high amount of Cr as the oxide forming element will form a slow growing surface oxide scale of Cr_2O_3 [7-9]. In addition, the amount of various oxides is approximately depend on the alloying element in the alloy such as Mn, Ti, Al and Si, to formed external oxide scale and internal oxide precipitates. Mn and Ti may incorporate into the scale, forming a stable outer oxide layer [1, 3, 6-7, 9-11]. Whereas, Al and Si will precipitate as internal oxides [1, 7, 10, 12-13].

Since the physical properties of oxides are different from alloy, it will cause a stress generation and relief in oxide or metal during cooling and heating subjected to cyclic condition. Two stress sources are growth stress and thermal stress. Most superalloy undergo thermal cycling during service. The temperature during cyclic condition will cause loss of protective scale by spallation effect. Normally, spallation and crack formation are driven by oxide growth stress and thermal stress. Oxide stress occur during oxide thickening, whereas thermal stress formed through heating and cooling, developed a coefficient of thermal expansion mismatch at metal-scale interface [6]. In order to avoid spallation, it is important to improve metal-scale adherence. There are several methods that have been used to improve scale adherence such as surface treatment, aging treatment, thermomechanical treatment and heat treatment that focused on the alteration of alloy grain structure [3, 14-15].

Therefore, in this study, the effect of different grain size of alloy obtained from different heat treatment temperature exposure will be investigated in high temperature cyclic oxidation condition, to examined the effect of grain size alteration on the oxide scale growth of Ni-based 800H superalloy.

2 Methodology

Ni-based 800H superalloy was used in this study, which is Fe-33Ni-19Cr alloy (ASTM-B-409: ASME-SB-409, UNS N08810). The chemical composition of this alloy has been analyzed using optical emission spectrometer (OES), presented in Table 1.

Table 1. Chemical composition of Fe-33Ni-19Cr alloy in weight percent (wt. %) with balance Fe.

Alloy / Element	Ni	Cr	C	Al	Ti	Si	Mn	P	Cu	N
Fe-33Ni-19Cr	33	19	0.078	0.534	0.489	0.315	0.556	0.007	0.082	0.02

The Fe-33Ni-19Cr alloy has experienced a heat treatment process at two different temperature in order to varies the grain size of the alloy. The heat treatment process were done at 1000°C and 1100°C for 3 hours soaking time, followed by water quench. These samples are denoted as H10 and H11 for samples heat-treated at 1000°C and 1100°C, respectively. H10 produced an average grain size of 56.72μm, which is fine grain size, while H11 produced an average grain size of 64.77μm, which is coarse grain size. Both heat-treated samples were undergo a cyclic oxidation test at 700°C for 150 cycles in laboratory air using CARBOLITE tube furnace type 3HA 12/300 system. Both samples were subjected to a short oxidation period at 700°C for one hour, followed by cooling process near a room temperature for 20 minutes for each cycle. This process was repeated until 150 cycles. The oxide phase analysis was examined using x-ray diffraction (XRD) technique model SIEMENS machined coupled with BRUKER D5000 XRD instrument

IOP Conf. Series: Materials Science and Engineering **957** (2020) 012013 doi:10.1088/1757-899X/957/1/012013

software. The phase analysis was accomplished with the aid of DiffracPlus Evaluation Software and PDF-2 search/match applications to identify the phases of the oxide scales. The oxidized alloy was examined in cross sectional view using field emission scanning electron microscope (FESEM) attached with energy dispersive x-ray (EDX) analyzer model CARL ZEISS 35 VP.

3 Results and Discussion

3.1 Phase Analysis

The XRD spectrum of both oxidized Fe-33Ni-19Cr alloy of H10 and H11 samples after cyclic oxidation exposure for 150 cycles was depicted in Fig. 1. The four major phases was identified, with corresponding the crystal structure, which are austenite, corundum, fluorite and spinel structure. The austenite peaks corresponding to the based alloy were recorded at five different diffraction angle 2θ.

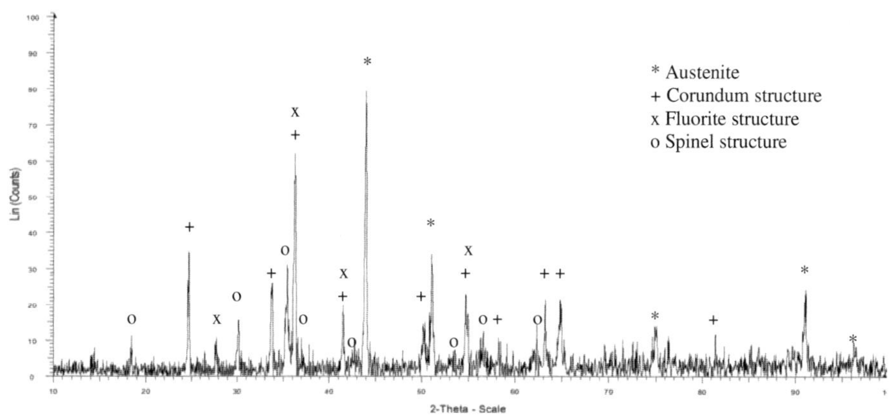

(a) H10

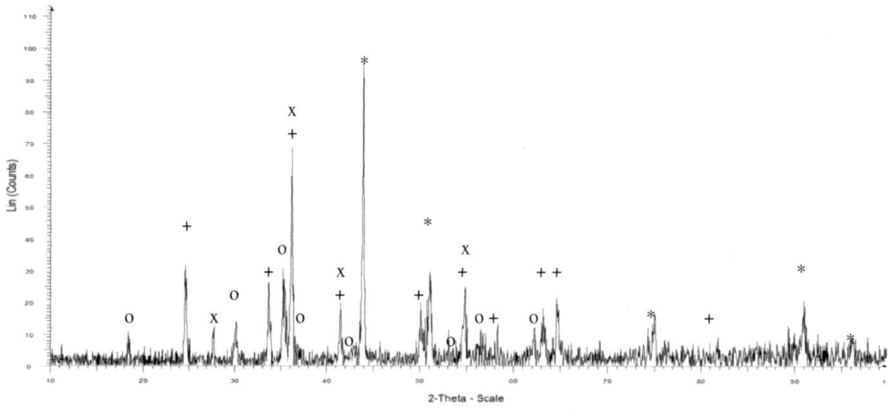

(b) H11

Fig. 1: XRD pattern of oxidized Fe-33NI-19Cr alloy after 150 cycles: (a) H10 and (b) H11.

The oxide phases with corundum structure was detected which are Cr-rich oxides composed of Cr_2O_3 and $(Cr_{0.88}Ti_{0.12})_2O_3$ and also hematite Fe_2O_3 phase. The Cr-rich oxide and hematite peaks were overlaps on each other due to similar crystal structure of the oxide. The formation of Cr-Ti rich oxide was said to mitigate the Cr evaporation effect due to the lower Cr vapour pressure of $(Cr,Ti)_2O_3$ compared to Cr_2O_3 [6]. In addition, the flourite oxide structure of Ti-rich oxides were detected consists of TiO_2 and $(Ti_{0.97}Cr_{0.03})O_2$. The presence of Ti-rich oxides was confirmed by the peak at 2θ 27.5 (110). The formation of Ti-rich oxides in Fe-Ni-Cr alloy system was also reported by other researchers [1, 10]. On the other hand, the spinel oxides structure compose of $MnCr_2O_4$, $NiCr_2O_4$, $NiFe_2O_4$, $FeCr_2O_4$ and magnetite Fe_3O_4 phases were detected. The overlap of spinel phases was also reported by other researcher [13], hence cannot be clearly distinguish in between each other. Other researcher also reported the formation of $NiCr_2O_4$ oxide [1-2, 14] $NiFe_2O_4$ oxide [13] and $FeCr_2O_4$ oxide [16, 18] on oxidized of Ni-basad alloy [19-23].

The formation of chromia oxides composes of Cr_2O_3 and $MnCr_2O_4$ has been reported by several researcher [3, 9, 11]. These oxide act as a barrier against further oxidation due to its very low diffusion coefficient for oxygen and metal, and then offer a high oxidation resistance. According to [11], the addition of a small amount of manganese will reduced the oxidation rate at high temperature exposure by forming a continuous $MnCr_2O_4$ spinel layer. Mn is found to be diffuse into the scale, developing an outer layer of Cr-Mn spinel, which can reduces the creation of volatile Cr-species [7, 10]. Mn may diffuse into the outer oxide layer with Cr due to its large solubility in Cr.

Moreover, the formation of protective oxide layer of Cr_2O_3 oxide, offer a resistance to high temperature oxidation due to the continuous and slow growing scale behaviour. The formation of protective Cr_2O_3 oxide on oxidized Ni-based alloy was also reported by numerous researchers [1-2, 4-5, 8, 16-17]. On the other hand, internal oxidation is a common phenomenon observed in metal alloys at high temperatures, usually greater than 600°C [18]. The formation of Al-rich and Si-rich oxide precipitates on oxidized Ni-based alloy has been reported by several rechearcers [1, 7, 10, 12-13]. However, in this study, the detection of Al and Si precipitates through XRD cannot be seen due to thick oxide layer formed on the sample surface. However, further analysis using FESEM-EDX technique has identify the formation of these internal oxide. Al_2O_3 and SiO_2 oxide precipitates usually formed beneath the surface oxides, which both oxides contribute to the protective consequence at temperature condition. Similar finding has been reported [7, 10].

3.2 Cross Sectional Analysis

Fig. 2 (a) shows a FESEM image with EDX line scan analysis of cross sectional fine grain H10 sample after cyclic oxidation for 150 cycles. Four oxide layers was produced on the oxide scale, compose of top outer layer (area 1) enriches in elements O, Cr, Mn and Ti, proposing the formation of Cr-rich oxide containing Mn and Ti, such as Cr_2O_3, $MnCr_2O_4$, $(Cr_{0.88}Ti_{0.12})_2O_3$ and/or $(Ti_{0.97}Cr_{0.03})O_2$. According to other researchers [6-7, 10], Mn and Ti may incorporate into the scale, forming a stable outer oxide layer, thus reduces the Cr volatility of Cr containing oxides. The inner oxide layer of area 2 revealed the enrichment of elements Fe and Ni with trace amount of Cr suggesting the formation of spinel oxides phases, which are $FeCr_2O_4$, $NiCr_2O_4$ and/or $NiFe_2O_4$. Layer 3 which is close to the scale-metal interface revealed the high amount of element Si, while, layer 4 indicated the presence of Al as the internal oxide. Similar observation was reported by other researcher, which is Al and Si normally formed an internal oxides on Fe-Ni-Cr alloy system [7, 10]. In addition, the EDX line scan analysis also revealed a slightly high amount of Al at area 5, suggesting the formation of internal precipitate Al oxide.

Fig. 2 (b) shows a FESEM image with EDX line scan analysis of cross sectional coarse grain H11 sample after cyclic oxidation for 150 cycles which formed three oxide layers. The outer oxide layer (area 1) consist of elements O, Cr and Ti, while, layer 2 which close to scale-metal interface enriched in elements O and Si. The internal oxide layer (area 3) consists of elements O and Al. These results suggesting that outer oxide layer composed of Cr-rich containing Ti oxide, while Si oxide and Al oxide precipitates as internal oxide. No evidence of element Mn detected on the oxide layer, indicating that the outer oxide layer of H11 sample did not formed a Cr-Mn rich oxide. The probability of oxide formed on the outer layer are Cr-Ti rich oxide and Cr-rich oxide. This observation shows that the coarse grain H11 sample only formed Cr-rich and Cr-Ti rich oxides, where these oxide has a beneficial effect to lessen the Cr evaporation effect.

Whereas, H10 sample was developed a Cr-rich, Cr-Ti rich and Cr-Mn rich oxides. The formation of Cr-Ti and Cr-Mn rich can reduces the creation of volatile Cr-species, hence producing good protective oxide layer on the alloy surface. The chromia oxides consists of Cr_2O_3 and $MnCr_2O_4$ offering a high oxidation resistance. The results indicate that fine grain H10 sample favour the outward ion diffusion of Mn element to incorporated with Cr in the outer oxide layer by the rapid diffusion through grain boundary area. Fine grain sample has higher grain boundary area compared to coarse grain sample. Even though the XRD analysis of H11 sample has detected the formation of corundum and spinel oxide structure, but the oxide peaks was overlap, hence reducing the availability to recognized each of oxide phases formed on the alloy surface.

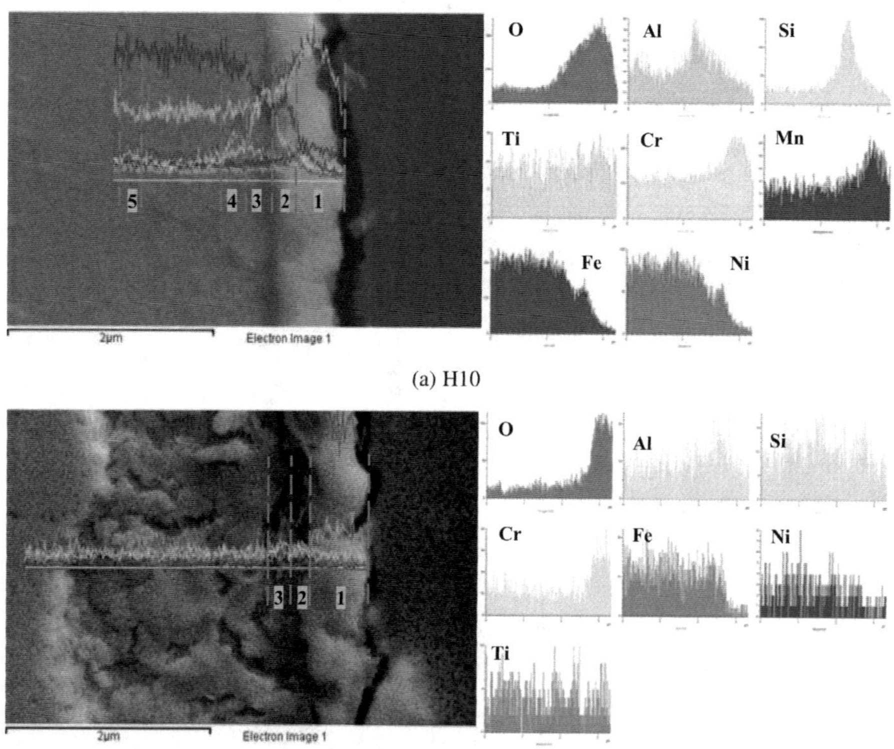

Fig. 2: Cross sectional FESEM image with corresponding EDX line scan analysis of oxidized Fe-33Ni-19Cr alloy after 150 cycles: (a) H10 and (b) H11.

4 Summary

The cyclic oxidation behaviour of Ni-based 800H superalloy, which is Fe-33Ni-19Cr alloy was studied in air at 700°C up to 150 cycles. Based on the oxidation test, the formation of oxide phases compose of corundum, fluorite and spinel oxide structure is responsible for imparting resistance against high temperature oxidation of the studied alloy. The formation of protective Cr-rich oxide scales composed of Cr-Mn and Cr-Ti rich oxides were formed on the outer oxide layer. Whereas, Al-rich and Si-rich oxide precipitates formed as internal oxide beneath the scale-metal interface. Both oxide precipitates contribute to the protective consequence at temperature conditions.

The authors would like to thank the Ministry of Education Malaysia for the research fund under the Fundamental Research Grant Scheme (FRGS) (Project Code. FRGS/1/2016/TK05/UNIMAP/02/4).

References

1. J. Jiang, G. Xiao, Y. Wang, Y. Liu, J. Alloys & Comp. **784** (2019)
2. D. Saber, I.S. Emam, R. Abdel-Karim, J. Alloys & Comp. **719** (2017)
3. X. Wang, J.A. Szpunar, J. of Alloys and Comp. **752** (2018)
4. S. Bsat, B. Xiao, X. Huang, S. Penttila, Oxid. Met. **89** (2018)
5. B. Adam, L. Teeter, J. Mahaffey, M. Anderson, L. Arnadottir, J.D. Tucker, Oxid. Met. **90** (2018)
6. L. Tan, X. Ren, K. Sridharan, T.R, Allen, Corros. Sci. **50** (2008)
7. J. Zurek, D.J. Young, E. Essuman, M. Hänsel, H.J. Penkalla, L. Niewolak, W.J. Quadakkers, Mater. Sci. & Eng.: A **477** (2008)
8. S. Liu, J. Shen, X.H. Guo, L.L. Liu, Y. Niu, Corros. Sci. **135** (2018)
9. A. Col., V. Parry, C. Pascal, Corros. Sci. **114** (2017)
10. D.J. Young, J. Zurek, L. Singheiser, W.J. Quadakkers, Corros. Sci. **53** (2011)
11. T.D. Nguyen, J. Zhang, D.J. Young, Corros. Sci. **112** (2016)
12. C. Pascal, V. Parry, E. Fedorov, M. Braccini, P. Chemelle, N. Meyer, D. Oquab, D. Monceau, Y. Wouters, M. Mantel, Corros. Sci. **93** (2015)
13. T.J. Pan, Y.S. Li, Q. Yang, R.F. Feng, A. Hirose, Corros. Sci. **53** (2011)
14. L. Ni, Z. Wu, C. Zhou, Prog. In Nat. Sci.: Mater. Int. **21** (2011)
15. I.A. Sustaita-Torres, S.H. Rodríguez, M.P. Guerrero-Mata, M. Garza, E. Valdés, F. Deschaux-Beaume, R. Colás, Mater. Chem. and Physics **133** (2012)
16. M.S. Elbakhshwan, S.K. Gill, A.K. Rumaiz, J. Bai, S. Ghose, R.B. Rebak, L.E. Ecker, Appl. Surf. Sci. **426** (2017)
17. Y.X. Xu, J.T. Lu, X.W. Yang, J.B. Yan, W.Y. Li, Corros. Sci. **127** (2017)
18. B. Langelier, S. Y. Persaud, R. C. Newman, G. A. Botton, Acta Materialia **109** (2016)
19. N. Parimin, E. Hamzah, IOP Conf. Ser.: Mater. Sci. Eng. **701** (2019) 012022
20. N. Parimin, Singh, IOP Conf. Ser.: Mater. Sci. Eng. **701** (2019) 012063
21. N. Parimin, Jemahri, IOP Conf. Ser.: Mater. Sci. Eng. **701** (2019) 012062
22. N. Parimin, Z. Zulnuraini, S.A.C. Sakdun, Solid State Phenomena **280** (2018)
23. N. Parimin, E. Hamzah, A. Amrin, Solid State Phenomena **280** (2018)

IOP Conf. Series: Materials Science and Engineering **957** (2020) 012014 doi:10.1088/1757-899X/957/1/012014

Effect of Dynamic Vulcanization on Properties of Linear Low Density Polyethylene/Natural Rubber Filled with Calcium Carbonate

Loh Annie[1] , *Mohamad Kahar* Ab Wahab[1,*] , *Elammaran* Jayamani[2]

[1]School of Materials Engineering, University Malaysia Perlis, 02600, Arau, Perlis, Malaysia;
[2]Discipline Leader Mechanical Engineering, Swinburne University of Technology, Kuching, Sarawak, Malaysia

Abstract The effect of dynamic vulcanization with different loading of calcium carbonate ($CaCO_3$) on linear low density polyethylene/natural rubber using N,N'-m-phenylene bismaleimide (HVA-2) vulcanized and dicumyl peroxide (DCP) vulcanized as a cross-linker were investigated. Optimum ratio of LLDPE/NR blends was chosen is 70/30 in this study. Calcium carbonate loading was varied at 0wt%, 5wt%, 10wt% and 20wt% as filler in LLDPE/NR blends. The study showed that the tensile properties of the un-vulcanized LLDPE/NR blends were deteriorated with increasing in $CaCO_3$ content. The dynamic vulcanization approach by using the HVA-2 and DCP dynamic vulcanization agents was applied on the LLDPE/NR blends in order to compensate the deterioration properties caused by the incorporation of $CaCO_3$. According to the results, the addition of HVA-2 and DCP showed the composites were better in water resistance and tensile properties as compared to the standardized counterpart. The morphological analysis by using the scanning electron microscopy (SEM) found out that the interfacial adhesion and dispersion of $CaCO_3$ was improved in both vulcanization systems.

1 Introduction

Thermoplastic elastomers (TPE) are polymers that exhibit the property of cross-linked elastomer at ambient temperature and plastic properties at elevated temperature [1]. This effect is associated with certain inter-chain secondary valence forces of attraction, which have the effect of typical ordinal covalent cross-links, but at elevated temperature, the secondary bond dissociates, and the polymer possess thermoplastic behavior [2]. Thermoplastic elastomers are composed of two phase that is a soft phase (amorphous domains) and hard phase (crystalline domains), which provides elastic properties and thermoplastic segment respectively [3]. Thermoplastic elastomers provide both processing ease and full recyclability to the material.TPEs can be produced by a unique skill known as dynamic vulcanization. Dynamic vulcanization (DV) involves a choosing cross-linking of rubbers component and dispersing finely in molten thermoplastic continuous phase through intensive mixing to produce thermoplastic vulcanizate [4]. Compatibilization between the matrix and

* Corresponding author: kaharwahab@unimap.edu.my

the modifier occurs in the extruder allowing a finer dispersion of the latter [5]. This results in greater adhesion between the two phases. Simultaneously, another reaction stabilizes the morphology of the impact modifier by cross-linking. DV can promote characteristics such as tensile characteristics, heat resistance and fluid resistance [6].The use of filler in the TPE was due to the advantage that can be obtained, such as improvement in process-ability, cost reduction of the product and achievement of specific [7-12]. Generally, fillers become the minor dispersed phase in a polymer matrix. Various fillers have been employed in polymer blend such as kaolin and calcium carbonate. Mineral filler is used to produce lower cost and promote the mechanical properties of polymer blend [13].

2 Experimental

2.1 Materials

Linear low density polyethylene (LLDPE) was purchased from Lotte Chemical Titan Malaysia with density 0.920 g/cm^3. Standard Malaysian Rubber - 10 (SMR - 10) was purchased from the Kilang Hevea Bukit Perak (KHBP) in Perak with T_g -69°C. Calcium Carbonate ($CaCO_3$) in odorless and fine white powder was purchased from Sigma Aldrich Chemical Company, St Louis, USA with density 2.83g/cm^3. N, N-m-phenylenebismaleimide (HVA-2) and dicumyl peroxide (DCP) were supplied from Aldrich Chemical Company, St. Louis, USA).

2.2 Preparation of the sample

2.2.1 Preparation of LLDPE/NR/$CaCO_3$

The LLDPE was added into the heated two roll mill until all the LLDPE pellets were molten. The NR was then added to blend together with the molten LLDPE until homogeneous. The $CaCO_3$ was incorporated into the LLDPE/NR blend and processed until the $CaCO_3$ was homogeneous and well dispersed into the matrix. The compounding recipes for the HVA-2 dynamic vulcanized system and DCP dynamic vulcanized system were as shown in Table 1. The compounding process was processed using heated two roll mill with speed 15rpm at 140°C for 15 minutes. The compounding was then molded into a 1mm sheet by using compression molding machine at 140°C for testing and characterizations purpose.

Table 1. Compounding recipe for each of the vulcanization system

VULCANIZATION SYSTEM	INGREDIENTS	PARTS PER HUNDRED (PHR)
Un-vulcanized system	LLDPE/NR/$CaCO_3$	100
HVA-2 dynamic-vulcanized system	LLDPE/NR/$CaCO_3$	100
	HVA-2	2
DCP dynamic-vulcanized system	LLDPE/NR/$CaCO_3$	100
	DCP	2

IOP Conf. Series: Materials Science and Engineering **957** (2020) 012014 doi:10.1088/1757-899X/957/1/012014

3 Characterization and Testings

Water absorption test was performed by following ASTM D570 standard. The samples were dried and each sample was weighed by using electronic balance. The samples were then submerged in distilled water at room temperature and the samples were weighed at 1, 2, 3, 5, 10, 15, 20, 25, 30 days. For each composition ratio, an average out of five samples were prepared for measurement. Instron tensile test machine 3366 was used to analyze the mechanical properties of the blends. The specimens were cut into a rectangular shape, with the dimension of total length 100mm, 50mm gauge length, 30mm width and 1mm thickness for the tensile test under the ISO 527-3. The cross-head speed was set at 50mm/min under an ambient temperature. An average value was taken out of 5 samples in order to get an accurate value of tensile strength, modulus of elasticity and elongation at break. The morphological studies was carried out on the tensile fracture surface by using the Analytical Scanning Electron Microscopy (SEM) (Model JEOL JSM-6460LA) with the supply of accelerating voltage at 10kV. The tensile fracture sample was coated with palladium layer at about 1.5-3.0nm thickness with aid of utilizing the Auto Fine Coater (Model JEOL JFC 1600) in order to prevent the charging effect on the surface morphology. Fourier transform spectroscopy (FTIR) (Model Perkin Elmer Spectrometer 2000) was used to inspect structural changes at a resolution of 4cm^{-1} and in the range of 4000-650cm^{-1}. The samples were scan for 32 times for a good resolution on the reading.

4 Discussion

4.1 Water Absorption Test

The water absorption test is conducted for 30 days, where reading of LLDPE/NR/$CaCO_3$ composites is weighted at 1, 2, 3, 4, 5, 10, 15, 20, 25, 30 days. The outcome of water absorption for un-vulcanized LLDPE/NR/$CaCO_3$ composites was shown in Figure 1(a). The curve depicts classic sorption performance with a quickly water absorption for the first few days of submerged for all the samples. This is mostly assigns to the storage of water absorption at the LLDPE/NR/$CaCO_3$ composites interphase [14]. Lack of wetting between hydrophilic natures of calcium carbonate and hydrophobic LLDPE/NR blend phase produces more voids that cause water molecules to store and flow in the polymer sample [15]. Excessive water take up described as the driving force between the dissimilar in water content and concentration point in the samples [16]. Higher water take up rate at 30% filler content can be described with the large aggregate of calcium carbonate increase the affinity of the sample with water molecules. The outcome of water take up for DCP and HVA-2 vulcanized systems are shown in figure 1 (b) and (c), respectively. From the figures, it is clearly showed that the vulcanized system shows better water resistance if compared with un-vulcanized (controlled) system. This is because capacity of DCP and HVA-2 to attach itself at the LLDPE/NR/$CaCO_3$ composite interphase and improved the interaction between $CaCO_3$ and LLDPE/NR matrix. Well embedded and dispersed calcium carbonate in the vulcanized LLDPE/NR matrix reduces the particle expansion and restriced water accumulation within the calcium carbonate phase [17]. HVA-2 vulcanized system shows better water resistance when compared to DCP vulcanized system. This is because DCP vulcanized system tended to undergo decomposition upon processing at high temperature and so the polymer chains would undergo chain scission into shorter chains and this made the DCP vulcanized system to be inferior to that of HVA-2 vulcanized system [18].

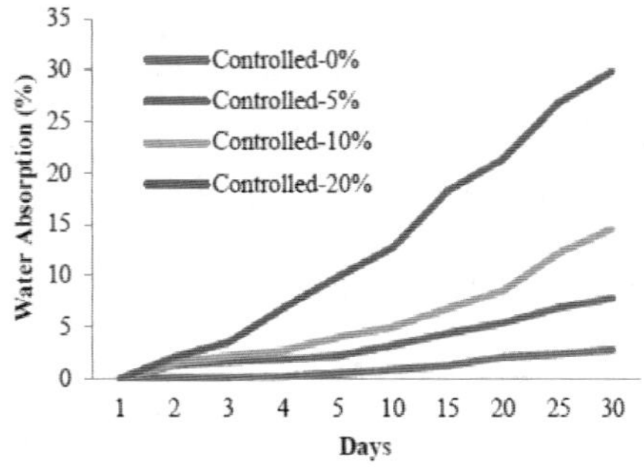

(a)

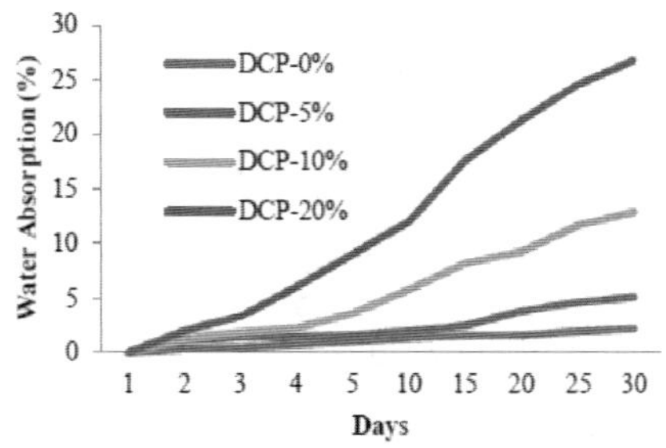

(b)

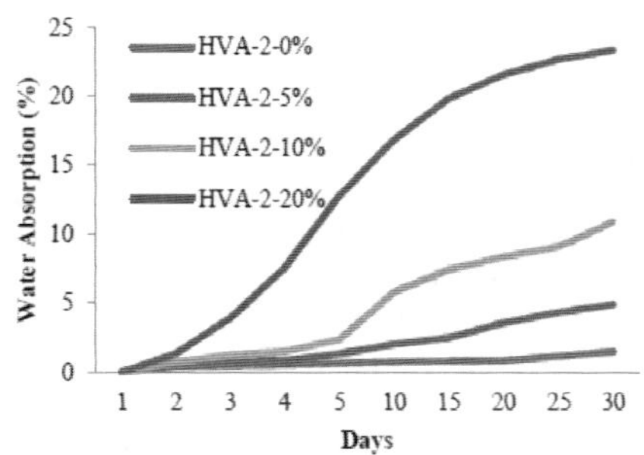

(c)

Fig. 1. Water absorption percentage for (a) un-vulcanized, (b) DCP vulcanized, and (c) HVA-2 dynamic vulcanized LLDPE/NR/$CaCO_3$ composites.

4.2 Tensile properties

Figure 2 depicts the tensile strength of LLDPE/NR/$CaCO_3$ composites for un-vulcanized, HVA-2 vulcanized and DCP vulcanized systems. The graph showed that the increasing in calcium carbonate content will decrease tensile strength of the LLDPE/NR/$CaCO_3$ composites for all cases. This was due to the incorporation of calcium carbonate content leaded to poor adhesion with LLDPE/NR blends. The formation of agglomeration of calcium carbonate acted as the stress concentration point and caused the LLDPE/NR blend broke easily when the load was applied [19]. Its due to the composite was unable to bear the high stress and transfer the stress effectively between the LLDPE/NR/$CaCO_3$ phases [20]. The distribution and dispersion of the calcium carbonate particles were the main reason of adhesion and tensile properties of the blend. Thus, the higher the calcium carbonate content, the higher was the tensile properties deterioration and these resulted in the sample being broken easily when the external force was applied. The same trend of deterioration of mechanical properties could be observed when the HVA-2 vulcanized system and DCP vulcanized system were used when calcium carbonate content increase from 0wt% to 20wt%. However, there was an interesting found-out that the tensile strength of both HVA-2 dynamic vulcanized system and DCP dynamic vulcanized systems was higher than that of un-vulcanized system. The improvement in tensile strength was as outcome of the better adhesion between the LLDPE/NR phases by the formation of cross-linking in the NR phase [21]. The reduction of tensile strength as a result of increasing the calcium carbonate content was also less significant in vulcanized system [22]. This is because the calcium carbonate content were becoming smaller and distributing better well in the LLDPE/NR matrix, thus, the formation of weak concentration points would not be crucial [23]. Furthermore, the vulcanized rubber phase could prevent the chain slippage and provided better dimensional stability to the blend. With respect to the comparison between DCP dynamic vulcanized system and HVA-2 dynamic vulcanized systems, the tensile strength of the HVA-2 dynamic vulcanized systems was better than DCP dynamic vulcanized system. The highest tensile strength of HVA-2-0% was 14.42MPa while for DCP-0% was merely 8.76MPa as shown in Figure 2. The better in strength was assign to the substantial cross-links formation in the HVA-2 vulcanized blends [24]. This indicated that the HVA-2 dynamic vulcanized system was more efficient cross-linkers to be used in this vulcanization system as compared to using DCP dynamic vulcanized system [22].

Similar trend happened to the modulus of elasticity of the LLDPE/NR/$CaCO_3$ blends as shown in the graph, whereby the increase in calcium carbonate content would reduce the modulus of elasticity of the blends. The modulus of elasticity of controlled-0% un-vulcanized system, DCP-0% and HVA-2-0% dynamic vulcanized system without calcium carbonate content was 77.50MPa, 82.00MPa and 93.00MPa and reduced to 29.33MPa, 43.00MPa and 52.33MPa respectively when the calcium carbonate content was increased up to 20wt%. The incorporation of calcium carbonate would interfere the LLDPE/NR chains allignment and give the the softening effect to the blends. However, the application of dynamic vulcanization approach compensated the loss in the modulus of elasticity values as a result from the calcium carbonate incorporation. The vulcanized rubber phase could prevent the chain slippage and provided better dimensional stability to the blends. The graph also shows the effect of calcium carbonate loading on the elongation at break for the un-vulcanized, DCP and HVA-2 dynamic vulcanized systems in LLDPE/NR blends. The percentage of elongation at break would reduce with increasing the calcium carbonate content. The major reason to the reduction in elongation at break value was due to the weak dispersion and distribution of the calcium carbonate within LLDPE/NR blends [24]. The higher the amount of calcium carbonate content would lead to more weak stress concentration points, thus it broke easier when the tensile force was being applied. It could further be based on the data obtained from the elongation at break for controlled-0%, DCP-0% and HVA-2-0% vulcanized blends which were 459.62%, 463.84% and 473.34% and reduced to as low as 339.75%, 359.34% and 437.87% respectively when the calcium carbonate content was increased up to 20wt%. However, there were some improvements whereby all the dynamic vulcanized blends showed better elongation at break than un-vulcanized blends. The contribution was as an outcome of the cross-linked NR in the rubber phase thereby improving the flexibility of the blends and restricted the deformability of the LLDPE/NR phase.

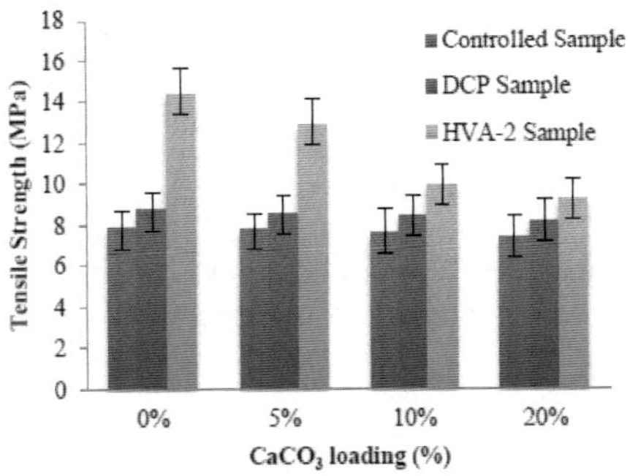

(a)

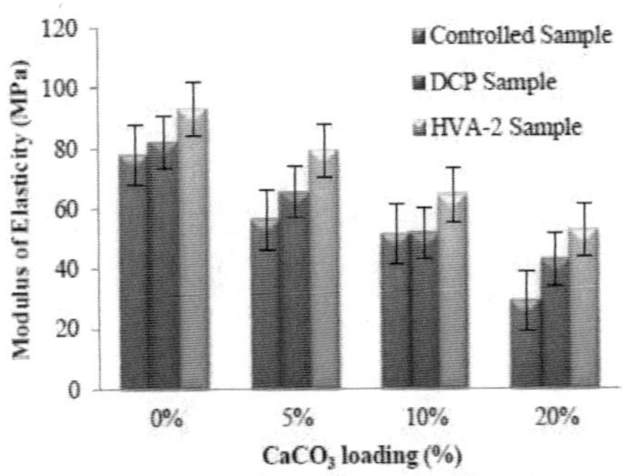

(b)

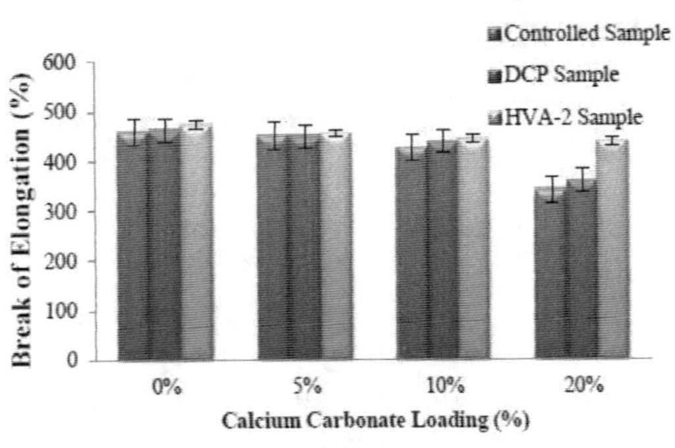

(c)

Fig.2 Tensile properties of different vulcanization system of LLDPE/NR/$CaCO_3$ composites.

4.3. Morphological Study

The surface morphology of the polymer sample was conducted by Scanning Electron Microscopy (SEM) in order to analyze the interfacial interaction and phase separation of the dispersed phase. The SEM micrographs with x500 magnification of fractured surfaces of LLDPE/NR blends were shown in Figure 3. All the LLDPE/NR blends showed phase separated morphology where the rubber phase occurred as a small droplets in LLDPE matrix. The phase morphology evidently showed poor interfacial adhesion evidences by the voids on the surface. Due to the incompatibility between the LLDPE and NR, this would create some voids and holes on the blends surface. Holes formation could also be as a result of rubber pull out during the tensile testing. In Figure 3(a), there was an incorporation of 20wt% of $CaCO_3$ into the LLDPE/NR blend. The incompatibility of $CaCO_3$ with the LLDPE/NR blend would create more and larger void formations and this explained why there was a reduction in the mechanical properties when the $CaCO_3$ content was increased [25]. On the other hand, it could be observed that the HVA-2-0% had a smoother surface than that of HVA-2-20% as shown in Figure 3 (b). The HVA-2-0% had lesser void formation and better interfacial interaction as a result from the cross-linking formation in NR phase and made the blend being more compact. It could be observed that the DCP-0% had also a smoother and compact surface than un-vulcanized system (Figure 3 (c)). As the calcium carbonate content was higher, there would be more incompatible in the phase structure due to the poor mixing creating more void formations and agglomeration [21]. It could also be observed that the controlled-20% had a larger void formation than HVA-2-20% and DCP-20%. Without the addition of curing agents, the interfacial bonding between the LLDPE/NR/$CaCO_3$ composites was weak [26]. With the incorporation of curing agent, the elastomer polymer chains will undergo cross-linking which control the chain mobility and increasing in viscosity in the rubber phase resulting in high shear which lead to the better dispersion and distribution of calcium carbonate [27].

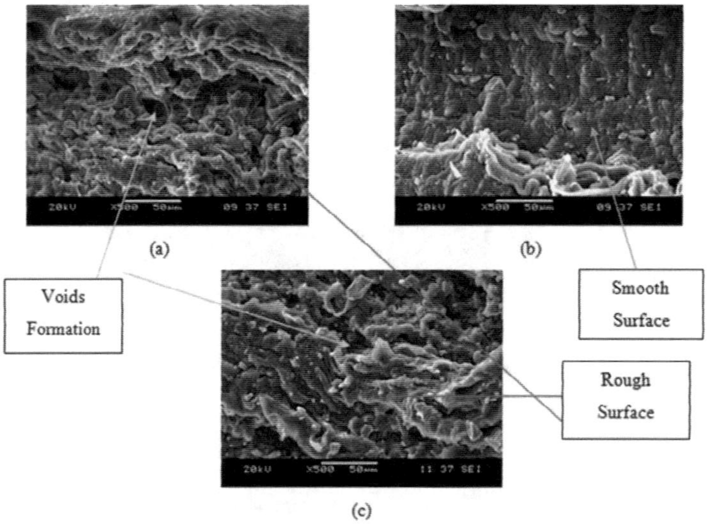

Fig. 3. SEM micrographs of un-vulcanized and vulcanized blends at x500 magnification with 20% $CaCO_3$ content for (a) un-vulcanized controlled system (b) HVA-2 vulcanized system (c) DCP vulcanized system

IOP Conf. Series: Materials Science and Engineering **957** (2020) 012014 doi:10.1088/1757-899X/957/1/012014

4.4 Structural Analysis

The infra-red spectra of LLDPE/NR blends were analyzed in order to investigate the structural changes after incorporation of $CaCO_3$. The FTIR test was carried out onto un-vulcanized LLDPE/NR blend with 0wt% and 20wt% of $CaCO_3$ is shown in Figure 4 (a). The most prominent peak to differentiate the controlled-0% and controlled-20% would be the band range occurring at 1650.81cm^{-1} representing stretching vibrations of C=O groups. The controlled-0% and controlled-20% had the almost similar characteristics for the rest of the peaks. The band range occurring at 2900 cm^{-1} correlating to the C-H streacth and symmetry stretching vibrations of CH_2 group. The chemical structure of the LLDPE and NR constituted a lot of $-CH_2$ substituent leaving a lower %T. The absorption peak at 1451.58cm^{-1} and 1446.44cm^{-1} would be due to the methyl group $-CH_3$ bending vibration for controlled-0% and controlled-20% respectively. The absorption peak at 1451.58cm^{-1} and 1446.44cm^{-1} were attributed to the C=C single bond. Furthermore, the absorption peak at 710.86cm^{-1} represent $-CH_2-CH_2-$ (cis) and this constituted from natural rubber (NR) [21]. Another analysis was done to study the effect of vulcanization systems by using different vulcanizing agents of HVA-2 and DCP by comparing between controlled-20%, HVA-2-20% and DCP-20% with 20wt% of $CaCO_3$ as shown in Figure 4 (b). Both the HVA-2 vulcanized and DCP vulcanized samples showed a variation of spectra from un-vulcanized blends. It was found that the HVA-2 had a peak at 2906.32cm^{-1} with highest percentage of transmittance representing C-H stretch and symmetry stretching vibrations of CH_2 groups and while the reduction of percentage of transmittance in DCP was due to the chain scission occurring into the blend causing the NR to be in a shorter chain, so more detections at this peak were observed. Besides that, after subjecting to dynamic vulcanization, the both systems were almost giving the similar absorption peak and the structural changes were varying slightly while comparing the un-vulcanized blends and vulcanized blends.

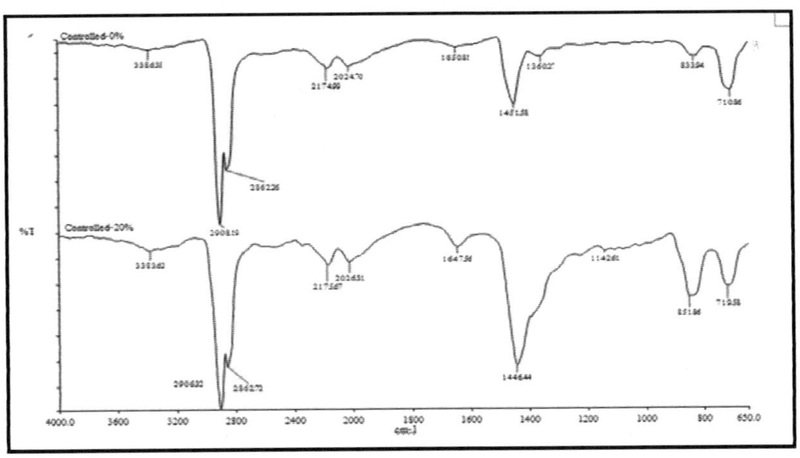

Fig. 4(a). Infrared spectra of un-vulcanized LLDPE/NR and LLDPE/NR/$CaCO_3$-20% composites.

IOP Conf. Series: Materials Science and Engineering **957** (2020) 012014 doi:10.1088/1757-899X/957/1/012014

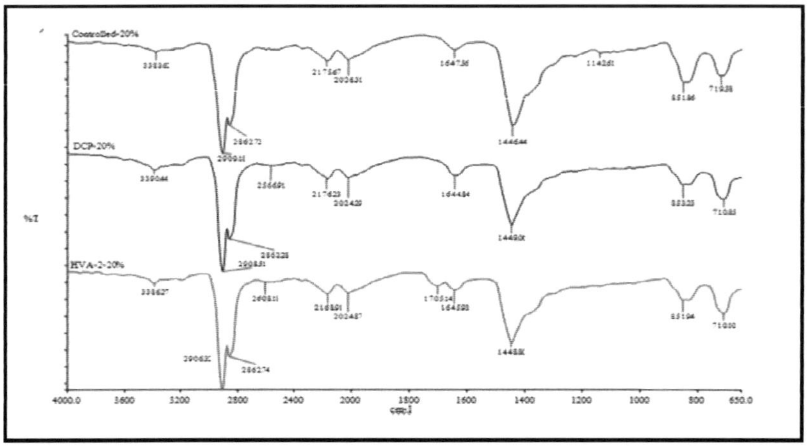

Fig. 4(b). Infrared spectra of un-vulcanized, HVA-2 vulcanized and DCP vulcanized systems for LLDPE/NR/$CaCO_3$-20% composites.

5 Conclusions

In conclusion, the incorporation of $CaCO_3$ had influence on the reduction in tensile properties of the LLDPE/NR blend. By using the SEM, it could be observed that the decrease in tensile properties was due to the incompatibility and week interfacial interaction of $CaCO_3$ with the LLDPE/NR blend which would create more and larger void formations as more $CaCO_3$ content was increased. The interaction between the $CaCO_3$ and the LLDPE/NR matrix was also studied by using the FTIR testing. The incompatibility was shown whereby no new formation or interaction between them in the un-vulcanized blend. The dynamic vulcanization was done to crosslink the rubber phase and the mechanical properties promoted greatly as differentiate to the un-vulcanized counterpart. The improvement in the tensile properties was mainly because of the promoted in crosslink density in the polymer. With respect to different vulcanization agents being used, it was found out that the HVA-2 dynamic vulcanized blends were greater than DCP dynamic vulcanized blends in the aspect of both properties. This was because HVA-2 was an effective cross-linking agent which was able to vulcanize NR even without the presence of free radical initiator at a sufficiently high temperature if compared with DCP. With the incorporation of curing agent, the elastomer polymer chains would undergo cross-linking which restricted the chain mobility and increasing in viscosity in the rubber phase resulting in high shear and therefore improved the dispersion and distribution of $CaCO_3$.

Acknowledgement

The authors wish to gratefully acknowledge the School of Materials Engineering, as well as Research and Innovation Department under Universiti Malaysia Perlis and Kementerian Pengajian Tinggi Malaysia (Ref: FRGS/1/2016/TK05/UNIMAP/02/9) for supporting this research project.

References

1. A. W. M. Kahar, H. Ismail, A. Abdul Hamid, A J. Therm. Anal. Calorim., **123(1)**, 301-308, (2016).

2. E. Prut, T. Medintseva, O. Kochanova, N. Erina, L. Zhorina, V. Kuleznev, J. Thermoplast. Compos. Mater., **28(8)**, 1202-1216, (2013).
3. S. Sanchez-Valdes, L.F. Ramo-De Valle, O. Manero, Handbook of Polymer Synthesis, Characterization, and Processing, (eds Saldívar-Guerra, E., Eduardo Vivaldo‐Lima, E.), John Wiley & Sons, Inc, 505-517, (2013).
4. A.W.M. Kahar, H. Ismail., N. Othman, J. Vinyl Add. Technol. **18(3)**, 192-197, (2012)
5. Y. Huang, C. Zhang, Y. Pan, W. Wang, L. Jiang, Y. Dan, J. Polym. Environ., **21**, 375-387, (2013).
6. B. Eleonora, G. Vadim, M. Tatiana, P. Aleksandra P., Eduard, S. Victor, Mater. Sci. Forum, **666**, 84-84, (2011).
7. A.W.M. Kahar, M. Lingeswarran, M.Z. Amirah Hulwani, H. Ismail, Polym. Bull., **76(2)**, 747-762, (2019).
8. M. Darinya, M., Natthapong, P. Nipon, Aca. J. Base, **6(2)**, 249-254, (2012).
9. A. Khalil, S.N. Shaikh, Z.R. Nudrat, S. Khaula, Adv. Mater. Phy. Chem., **2**, 90-97, (2012).
10. R. J. Walster, R. Rahman, A. W. M. Kahar. IOP Conf. Series: Mater. Sci. Eng., **429,** 012043, (2018).
11. A.W.M. Kahar, N. Sarifuddin, H. Ismail, H., Ira. Polym. J., **26 (2)**, 149-159, (2017).
12. A.A. Azrem, N.Z. Noriman, M.N. Rafiz, Key Eng. Mater., 867-871, (2014).
13. C. Thongpin, S. Klatsuwan, P. Borkchaiyapoom, S. Thongkamwong, J. Met., Mater. and Miner., **23(1)**, 53-59, (2013).
14. N.F. Alias, H. Ismail, A.W.M. Kahar,. Bio-res, **12(4)**, 9103-9117, (2017).
15. Ahmad, M., 2005. The Effect of Dynamic Vulcanization on the Mechanical Properties of EPDM/PP Thermoplastic Elastomer. Inter. J. Polym. Mater., **54**, 619-631
16. A.H. Mohd Zain, A.W.M. Kahar, I. Hanafi, J. Polym. Environ, **26(2),** 691–700, (2018).
17. K. Das, D. Ray, N.R. Bandyopadhyay, A. Gupta, S. Sengupta, S. Sahoo, M. Misra, Ind. Eng. Chem. Resear., **49(5)**, 2176-2185,(2010).
18. A.W.M. Kahar, H. Ismail., N. Othman, J. Appl. Polym. Sci., **128(4)**, 2479–2488, (2013).
19. A.G. Mustafa, H. Azman, Y. Abdirahman, J. App. Polym. Sci.,**121**, 2413 – 2421, (2011).
20. A.W.M. Kahar, H. Ismail. J. Vinyl Add. Technol. **22(3)**, 191-199, (2016).
21. G. Jayalatha, K.N.K. Sunil, J. Elast. Plast., **47(2)**, 153-169, (2013).
22. O.S. Dahham, N.Z. Noriman, A.W.M. Kahar, H. Ismail, S.T. Sam, J. Appl. Sci. & Agric., **10(5)**, 33-39, (2015).
23. A.H. Mohd Zain, A.W.M. Kahar, N.Z. Noriman, Proced. Chemist. **19**, 638-645, (2016)
24. K. Ji-Hoo, K. Gue-Hyun, Inter. J. Eng.Technol. Inno., **7(1)**, 39–47, (2017).
25. N. Ibrahim, A.W.M. Kahar, D.L Ngoc, H. Ismail, H. Bio-res., **12(2),** 3076-3087, (2017).
26. F. Qinghong, S. Bo, T. Tiam-Ting, T.S. Lee, H. David, B. Soo-Tueen, Composite, Part B, **60**, 561-567, (2014).
27. H. Salmah, H., Ismail, A.A. Bakar, J. Appl. Polym. Sci., **107(4)**, 2266-2273, (2007).

Preparation and characterization of LDPE/TPS blend filled with calcium carbonate

Anis Adila Ab Halim[1] , and *Mohamad Kahar* Ab Wahab[1,*]

[1]School of Materials Engineering, University Malaysia Perlis, 02600, Arau, Perlis, Malaysia

Abstract The measurement of low-density polyethylene / thermoplastic starch (LDPE/TPS) composite with different loading of calcium carbonate ($CaCO_3$) as filler was conducted under several testing such as the tensile test and scanning electron microscopy. The preparation of LDPE/TPS/$CaCO_3$ composite was divided into three stages. The first stage involved the preparation of TPS with and without modification with carboxylic acid. The second stage, LDPE was blended with TPS to produce LDPE/TPS blend. LDPE/TPS blends with blend ratio of 100:0, 90:10, 80:20, 70:30, 60:40 and 50:50 was prepared. Each composition will undergo tensile testing to analyze their mechanical properties. The optimum LDPE/TPS blend ratio was chosen for subsequent study. Third stage involved the compounding of the chosen LDPE/TPS blend ratio with $CaCO_3$. The result indicate that LDPE/TPS with incorporation of $CaCO_3$ gave higher value of tensile properties as compared to LDPE/TPS blend. Based on the SEM micrograph, treated TPS with citric acid and ascorbic acid performed better interfacial phase as compared to untreated TPS, with no agglomeration and voids formation.

1 Introduction

Starch has widespread use since it is low cost material, and production from renewable resources. Botanical sources that contain starch such as wheat, corn, potato and rice. Biodegradable starches can be processed by using injection molding, blow molding, film blowing, foaming, thermoforming and extrusion [1-5]. Starch consists of granule and two major parts of granules are amylose and amylopectin. In addition, higher proportion of short chains gives low gelatinization temperature in amylopectin.

Starch is a hydrophilic material, brittleness and has poor mechanical strength. However, the properties of starch can be enhanced by blended with other biodegradable polymer such as polyvynil alcohol, polylactic acid and poly (ε-caprolactone) [6-8]. The addition of plasticizer in starch film will affect the tensile strength, flexibility, elongation at break, glass transition temperature and hardness [9-12]. Commonly, plasticizers are added into polymer to reduce brittleness, reduce inter-molecular forces and increasing flexibility. Thermoplastic starch (TPS) was produced from native starch which it capable to flow and mixed with other

* Corresponding author: kaharwahab@unimap.edu.my

synthetic polymer such as low density polyethylene. TPS is purchased from plasticized starch when the starch is mixed with plasticizer such as glycerol and conducted under mechanical process. Blending TPS with other synthetic polymers gives high ductility, high modulus with toughness, low cost, good mechanical properties, and improve biodegradability. On the other hand, crosslinking agent such as citric acid and ascorbic acid are used to form bonds between starch polymer chains in order to strengthen the material. Besides that, crosslinking agent also used to reduce the water sensitivity of starch and to improve the barrier properties of TPS [13-15]. In order to enhance the properties of product, the incorporation of filler is required. Fillers usually used to improve the performance of rubbers and other polymeric materials. Clays, talc and calcium carbonate are the types of fillers. In this study, calcium carbonate was used as filler for blending with PE/TPS blends. Calcium carbonate is used to improve mechanical properties such as strength, modulus and toughness.

2 Experimental

2.1 Materials

LDPE is a thermoplastic resin that purchased from Titan Chemical, Malaysia was used for this study. Density of LDPE is 0.91 g/cm3 and the melting point is between the ranges 105°C to 115°C. Potato starch was manufactured by ThyeHuat Chan Sdn. Bhd. Penang. The moisture content of potato starch is 20%. Potato starch is a white powder in appearance and thedensity is 1.5 g/cm3. Calcium carbonate was used as reinforcing filler to increase the mechanical properties of the compounding. Glycerol is made by HmBG® Chemicals and was used as a plasticizer in preparation of TPS. The density of glycerol is 1.261 g/cm3 and the melting point is 17.8°C. Citric acid is provided by SIGMA-ALDRICH and used as a crosslinking agent to reduce the water sensitivity of starch and improving the mechanical properties of TPS. Ascorbic acid is purchased from SIGMA-ALDRICH and used as a crosslinking agent to improve the properties of TPS. The melting point of ascorbic acid is between the ranges 190-194°C.

2.2 Preparation of the sample

2.2.1Modification of TPS with citric acid and ascorbic acid

Modification of TPS was done by mixing 65% of potato starch and 35% of glycerol in the presence of 3% of citric acid or ascorbic acid. The mixtures were left at room temperature for 24 hours after preparation. Then, the mixtures were mixed by using heated two roll mill machine at temperature 150°C for 10 minutes.

2.2.2 Preparation LDPE/TPS/$CaCO_3$ composite

LDPE was blended with TPS by using heated two roll mill with the different blends ratio. The blend proportion of LDPE/TPS blend were100:0, 90:10, 80:20, 70:30, 60:40 and 50:50.Then, LDPE/TPS blend was molded by using hot press with 1mm dimension of mould at temperature 150°C for 6 minutes. The mechanical properties of LDPE/TPS blend were investigated under tensile testing. The best optimum measured data was used to ratio for subsequent study.The choosen LDPE/TPS blend ratio was further blended with calcium carbonate. Calcium carbonate were vary from 5% to 20% relative to the LDPE/TPS blends. The compounding process was prepared by using heated two roll mill at temperature 150°C for 10 minutes. Next, the compounding was molded by using hot press with 1mm dimension of mould at temperature 150°C for 6 minutes.

3 Characterization and Testings

The tensile properties are determined by using Instron Machine 5569 Universal Tester. The specimens were cut from compression moulded sheets according to the ASTM D-882 standard. The samples were cut into dumbell shape with average thickness is approximately 1mm ± 0.2mm. The crosshead speed was set 5mm/min and the gauge length was set 50mm. Three samples were taken from each of the formulation to get the average of each parameter.The morphology of the potato starch was observed under scanning electron microscopy model JEOL JSM 6460LA. SEM is used to determine the phase separation between blend components and the interaction occurred at the interphase.

4 DISCUSSION

4.1 Tensile properties

Effect of thermoplastic starch content towards the properties of LDPE can be determined by blend both materials with different TPS content. TPS has low tensile properties while LDPE pose high tensile properties. Blending of TPS with LDPE will affect the properties of LDPE which it tends to reduce the mechanical properties.The effect of TPS content on tensile strength of LDPE/TPS blends is shown in Fig. 1(a). As shown from the graph, the tensile strength of LDPE which is 9.1 MPa decrease as the TPS content increase until 4.7 MPa at 50 wt%. This results shows that increasing in TPS content can decrease the tensile strength of LDPE/TPS blends. This is due to the TPS cannot transfer stress or force applied from the outside and well distribute throughout the sample. Besides, the formation of defects such as low dispersion of thermoplastic starch will become stress concentration point and cause thermoplastic easily breaks when force is applied.The incorporation of TPS content also leads to poor adhesion with LDPE/TPS phase. The effects of these problems will cause the blends inability to support the high stress and unable to transfer the stress efficiently between phase in LDPE/TPS blends. The dispersion and distribution of TPS particles is the major factor of adhesion and tensile properties of the blends [16-17]. Therefore, as the TPS increased, the deterioration of tensile properties also increased and resulting the blends easily breaks when external force applied. Fig. 1(c) shows the effect of TPS content on elongation at break of LDPE/TPS blend. According to Kahar *et.al* [18], the percentage of elongation at break decreases with increasing of TPS content. The major cause to the decreases of break elongation values is due to the poor dispersion and distribution of TPS within LDPE/TPS blends. As shown in the graph, the elongation at break of LDPE which is 121.6 % decrease until 12.3 % as the TPS content increase. This is due to the weak interaction at the interphase of LDPE and TPS which leads to phase separation. Therefore, LDPE/TPS blends will break easier when the force is applied to the samples.The modulus elasticity of LDPE/TPS blends was found to be decreased as TPS content is increased. The results found that for LDPE/TPS blends at 10% TPS content is 125.8 MPa, but on 50% TPS content the value become 66.9 MPa Fig.1 (c)). The incorporation of TPS will soften the LDPE/TPS blends due to the presence of glycerol during TPS preparation. In the other words, the stiffness and rigidity of the blends will decrease as the TPS content increases [19]. At the lower TPS content, the modulus elasticity in LDPE/TPS blends is higher and possess better stiffness. From tensile properties, LDPE/TPS with blend ratio of (80:20) was chosen for subsequent study due to acceptable tensile properties. The tensile strength of LDPE/TPS (80:20) is 7.0 MPa with 44% of elongation at break which higher than LDPE/TPS (70:30), LDPE/TPS (60:40) and LDPE/TPS (50:50).

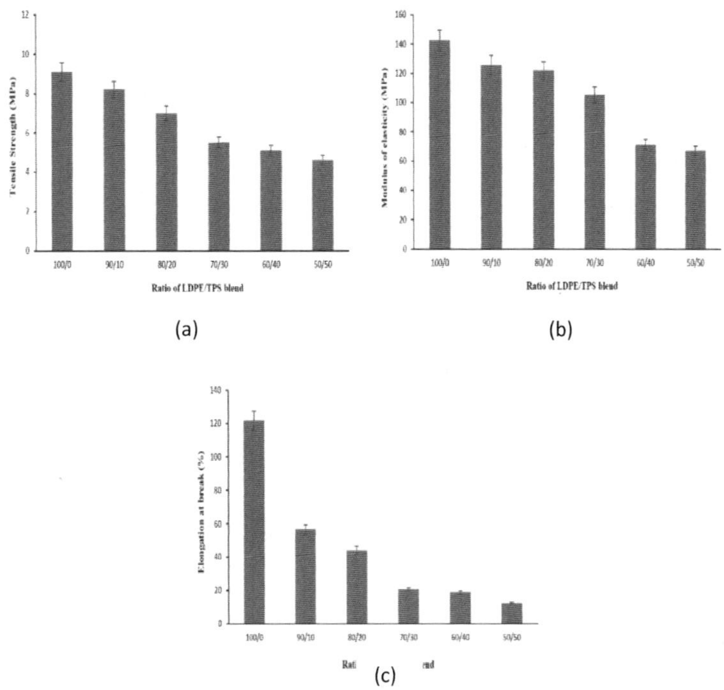

Figs 1 (a) – (c). Effect of TPS content on elongation at break of LDPE/TPS blend

From the preliminary result, LDPE/TPS blends with 80/20 blend ratio was selected due to the acceptable mechanical with optimum TPS content. For the subsequent study, LDPE/TPS blends was prepared by treated TPS with citric acid and ascorbic acid. Citric acid and ascorbic acid was added in LDPE/TPS blends to acts as a crosslink agent in order to improve the compatibility between LDPE/TPS interphase. Based on the Fig. 2, the tensile strength of LDPE/TPS, LDPE/TPSCA and LDE/TPSAA improves steadily as calcium carbonate is added. The tensile strength is improved which increased from 6.9 MPa to 7.5 MPa for LDPE/TPS, 7.4 MPa to 9.3 MPa for LDPE/TPSCA and 7.1 MPa to 8.9 MPa for LDPE/TPSAA. The initial increase in strength of the composite is due to good filler-matrix interaction. Thus, it enables more stress to be transferred from the matrix to the filler during external loading. However, further increase in calcium carbonate loading was decreased in the tensile strength of LDPE/TPS, LDPE/TPSCA and LDPE/TPSAA. The tensile properties decrease along with further addition of calcium carbonate. The increasing filler loading will attributed to the formation of filler aggregates, which are made up of particles held together by weak forces [20-21]. The aggregates are easily debond from the matrix and resulted in decrease the tensile strength of the composite. The increase in strength of LDPE/TPS composite at lower calcium carbonate loading are attributed by the homogeneity dispersed of calcium carbonate particles. The well dispersed particles made the crack propagation path longer, absorb a portion of energy and enhanced the plastic deformation [22]. However, with the increased of calcium carbonate loading, the detachment of the filler particle from the LDPE/TPS matrix caused voids which increased the size to form crack. In addition, the agglomeration of calcium carbonate results decrease in mechanical strength due to low strength of agglomerates itself. On the other hand, the tensile properties of LDPE/TPS blends that treated with citric acid and ascorbic acid increased as compared to untreated LDPE/TPS blends. As can be seen from the graph, the addition of citric acid and ascorbic acid in TPS was improved the tensile strength of the LDPE/TPS blends. Citric acid and ascorbic acid was improved the interaction of the starch molecule as the citric acid and ascorbic acid increased the slippage of starch molecule. The acidity of citric acid and ascorbic acid was induced to fragmentation of starch and rupture of starch granules that weaken the chain entanglements.

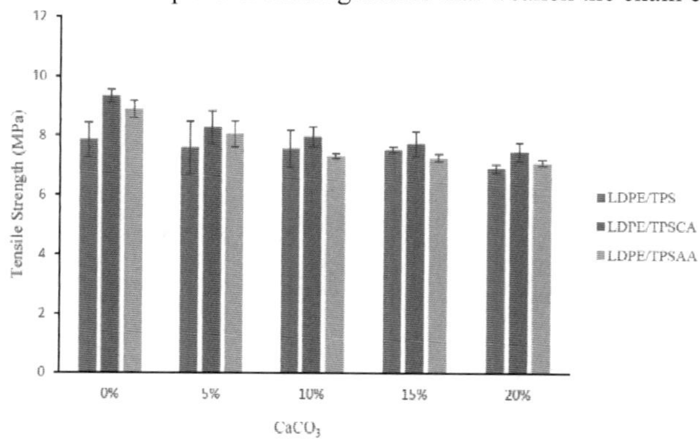

Fig. 2. Effect of filler loading on tensile strength of LDPE/TPS, LDPE/TPSCA and LDPE/TPSAA.

4.2 Morphology analysis

Fig. 3 (a) – (c) show the SEM images of LDPE/TPS, LDPE/TPSCA and LDPE/TPSAA blends. The images from both 100x and 300x magnification shows that there are significant voids between the interphase of LDPE and TPS. The voids in LDPE/TPS blends indicating poor degree of interfacial interaction between LDPE and TPS. This is due to compatibility problem between hydrophilic TPS and hydrophobic LDPE. TPS particle does not tend to disperse homogenously in LDPE. In the Fig. 3 (a), thermoplastic starch particles were found around 100 µm – 150 µm which indicating poor dispersion and higher degree of phase separation. On top of that, a large particle size will weaken the adhesion bonding between LDPE/TPS blends [23-24]. On the other hand, as can be seen from Fig. 3(c) and (d), the SEM micrographs for LDPE/TPSCA and LDPE/TPSAA blends shows good interfacial phase as compared to LDPE/TPS blends. The interfacial phase and adhesion was found to be improved by well dispersed of TPS particles that was treated with citric acid and ascorbic acid in LDPE/TPS blends. The size range of thermoplastic starch particle that treated with citric acid and ascorbic acid were found around 30 µm – 50 µm. As can be seen from the Figs. 3 (b) and (c), the smaller TPS particles that treated with citric acid and ascorbic acid are homogeneously dispersed throughout the LDPE/TPS matrix. Besides, the voids formation between polymer phase also reduced and no agglomeration was found in LDPE/TPSCA and LDPE/TPSAA blends. Continuity of a phase was favoured by citric acid and ascorbic acid may attributed to low interfacial tension.

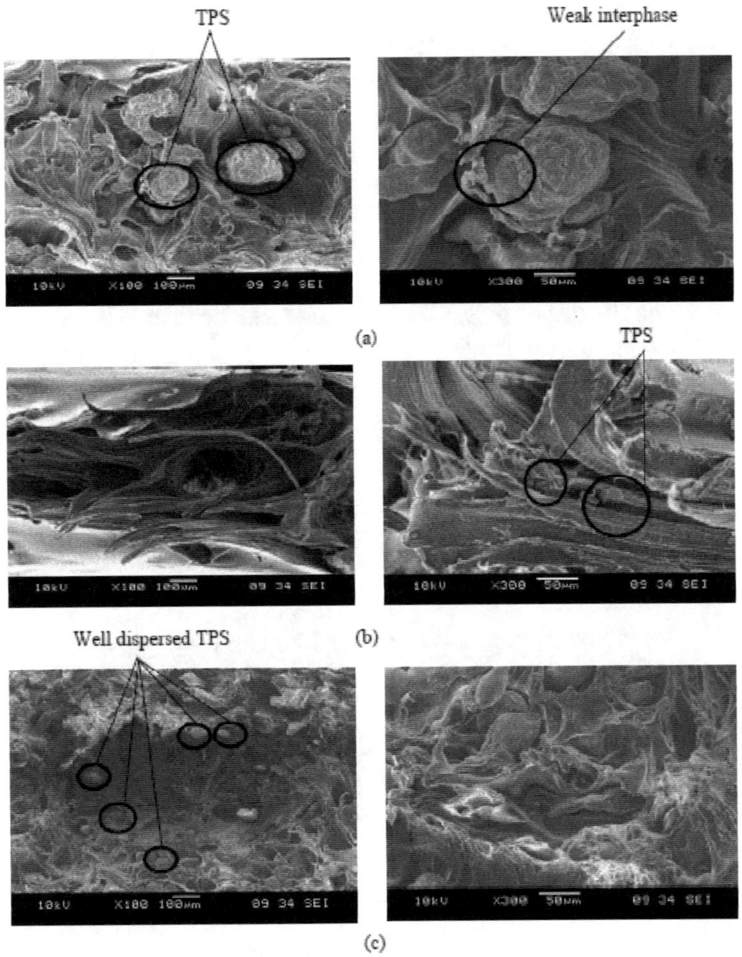

Fig.3. SEM micrographs of (a) LDPE/TPS, (b) LDPE/TPSCA and (c) LDPE/TPSAA blends at 100x and 300x magnification

The thermoplastic starch that treated with citric acid and ascorbic acid was dispersed and dissolves in LDPE phase, thus reducing the voids on the blends surface. These TPSCA and TPSAA well dispersed in LDPE in smaller size of particles due to LDPE phase was adhered well with TPS particles. The agglomeration of TPS in LDPE also found to be reduce and approved that citric acid and ascorbic acid has been improved the adhesion of LDPE/TPS blends. The breakdown of TPS particles due to citric acid and ascorbic acid contributed to homogeneous dispersion and well embedded in LDPE/TPS blends. In order to study the interaction between matrix and dispersed phase of LDPE/ TPS/$CaCO_3$, SEM micrographs of fracture surface was observed under 100x and 300x magnification. Fig. 4 (a) – (c) show the SEM micrograph of LDPE/TPS, LDPE/TPSCA and LDPE/TPSAA blends with 20% of calcium carbonate. The images from both 100x and 300x magnification shows that there are clear voids between the interphase of CaCO3 and LDPE/TPS matrix blend as compared to $CaCO_3$ filled LDPE/TPSCA matrix and LDPE/TPSAA matrix. The voids in LDPE/TPS/$CaCO_3$ composite indicates poor degree of interfacial interaction between LDPE and TPS as well as between $CaCO_3$ and LDPE/TPS matrix. This is due to the compatibility problem between hydrophilic TPS and hydrophobic LDPE.

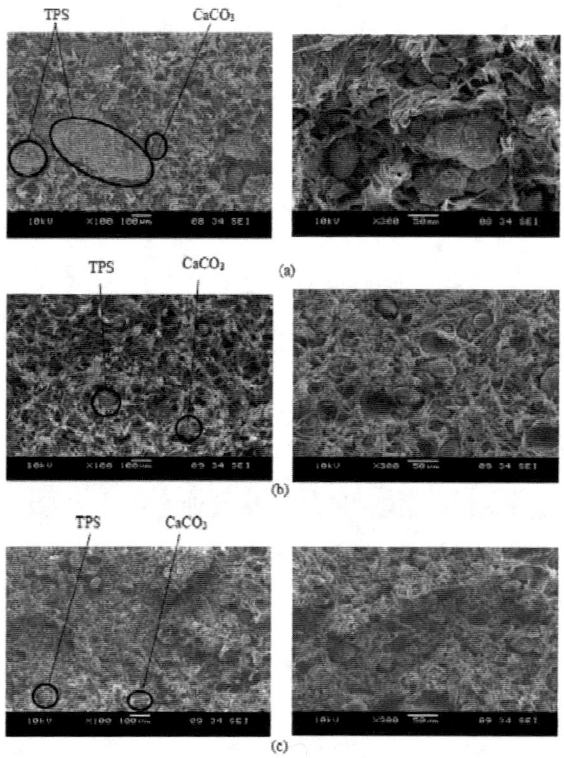

Fig. 4: SEM micrographs of (a) LDPE/TPS, (b) LDPE/TPSCA and (c) LDPE/TPSAA blends with calcium carbonate at 100x and 300x magnification

In addition, a large particle size will weaken the adhesion bonding between LDPE/TPS/$CaCO_3$ composites. The size range of calcium carbonate particle were found around 10 μm – 20 μm and the size range of thermoplastic starch particle that were found around 100 μm – 150 μm. Besides, as can be seen from Fig. 4 (b), the SEM micrographs for LDPE/TPSCA/$CaCO_3$ composites shows better interfacial phase as compared to LDPE/TPS/$CaCO_3$ composites. The interfacial phase and adhesion has been improved by well dispersed of TPSCA particles and calcium carbonate particles in LDPE phase. The size range of thermoplastic starch particle that treated with citric acid is smaller compared to the size of untreated thermoplastic starch. The size of TPSCA particles were found in SEM micrograph is around 20 μm – 50 μm. Based on the Figs. above, the smaller TPS particles that treated with citric acid is homogeneously dispersed throughout the LDPE/TPSCA/$CaCO_3$ matrix. For the LDPE/TPSAA/$CaCO_3$ composites, SEM micrograph exhibit good interfacial phase as compared to LDPE/TPS/$CaCO_3$ composites. The adhesion phase of LDPE/TPSAA/$CaCO_3$ composites has been improved by addition of ascorbic acid in TPS particles. TPS particle well embedded in LDPE phase. Therefore, the micrograph found there is no agglomeration and voids formation occurred in LDPE/TPSAA/$CaCO_3$ composites. The size of TPSCA particles were found in SEM micrograph is around 20 μm – 50 μm and the size of $CaCO_3$ particle were found around 10 μm – 20 μm. On the other hand,

IOP Conf. Series: Materials Science and Engineering **957** (2020) 012015 doi:10.1088/1757-899X/957/1/012015

the presence of e $CaCO_3$ also can be observed in SEM micrograph. Increasing of $CaCO_3$ content in LDPE/TPS matrix causes voids which increase in size to form a crack. Agglomerate of $CaCO_3$ form by the interaction between the filler particle. The particle tends to attract each other cause it grouping together forming the agglomerations. In addition, the agglomeration of $CaCO_3$ particle results in a decrease in tensile properties of LDPE/TPS composite compared to LDPE/TPSCA and LDPE/TPSAA composite due to low strength of the agglomerates itself.

5 Conclusions

TPS had been blended with LDPE to produce a partial biodegradable polymer. According to the study, several composition between LDPE and TPS are prepared with different LDPE/TPS blend ratio. Each composition provided different characteristics based on the amount of TPS content. Tensile test demonstrated that high TPS content will give low mechanical properties of LDPE/TPS blends. From tensile properties, LDPE/TPS with blend ratio of 80:20 was chosen for subsequent study due to the acceptable tensile properties.In order to achieve better properties of LDPE/TPS blend, TPS was treated with citric acid and ascorbic acid to improve the mechanical properties. Calcium carbonate filler was added in LDPE/TPS, LDPE/TPSCA and LDPE/TPSAA at different weight percentage which are 5%, 10%, 15% and 20% in order to performed better properties. Based on the SEM micrograph, treated TPS performed better mechanical and thermal properties as compared to untreated TPS. This is due to the TPS particle well embedded in LDPE phase. On the other hand, the micrograph was found there is no agglomeration and voids formation occurred in LDPE/TPSCA/$CaCO_3$ composites and LDPE/TPSAA/$CaCO_3$ composites.

The authors wish to gratefully acknowledge the School of Materials Engineering, as well as Research and Innovation Department under Universiti Malaysia Perlis and Kementerian Pengajian Tinggi Malaysia (Ref: FRGS/1/2016/TK05/UNIMAP/02/9) for supporting this research project.

References

1. L. Yu, K. Dean, L. Li, Prog. Polym. Sci. **31,** 576–602, (2006).
2. P.L. Tran, D.H.D. Nguyen, V.H. Do, Y.L. Kim, S. Park, S.H. Yoo, Food Sci. Technol., **62**, 1091-1098, (2015)
3. N.F. Alias, H. Ismail, A.W.M. Kahar,. Bio-res, **12(4)**, 9103-9117, (2017)
4. A.W.M. Kahar, H. Ismail., N. Othman, J. Appl. Polym. Sci., **128(4),**2479–2488, (2013).
5. D. Lu, C.M. Xiao, S.J.Xu, eXPRESSPolyem. Lett.,**6**, 366–375, (2009).
6. N. Ibrahim, A.W.M. Kahar, D.L Ngoc, H. Ismail, H. Bio-res., **12(2),** 3076-3087, (2017).
7. H. Chi, K. Xu, D. Xue, C. Song, W. Zhang, P. Wang, P. Food Resear. Intern.,**40**, 232-238, (2007).
8. A.H. Mohd Zain, A.W.M. Kahar, I. Hanafi, J. Polym. Environ, **26(2),** 691–700, (2018).
9. M.A. Huneault, H. Li, Polymer, **48**, 270-280, (2007).
10. Y. Yang, Z. Tang, Z.Xiong, J. Zhu, Intern. J. Biol. Macro., 77, 273-279, (2015).
11. A.W.M. Kahar, N. Sarifuddin, H. Ismail, H., Ira. Polym. J., **26** (**2**), 149-159, (2017).
12. G.L. KatarzynaLeja, Polish J. Environ. Stud**. 19,** 255-266, (2010).
13. N.N.S. Thomas, S. Mohan, E. Francis, Euro. Polym. J. **65,**224-335, (2012).

14. A.W.M. Kahar, M. Lingeswarran, M.Z. Amirah Hulwani, H. Ismail, Polym. Bull., **76(2),** 747-762, (2019)
15. A.W.M. Kahar, H. Ismail., N. Othman, J. Vinyl Add. Technol.**18(3),** 192-197, (2012)
16. H. Liu, F. Xie, L. Yu, L. Chen, L., Li, Prog.Polym. Sci., **34(12),** 1348-1368, (2009).
17. A.H. Mohd Zain, A.W.M. Kahar, N.Z. Noriman, Procedia Chem., **19,** 638-645, (2016)
18. A.W.M. Kahar, H. Ismail. J. Vinyl Add. Technol. **22(3),** 191-199, (2016).
19. Y. Lua, M.C.Cueva, L.C. Edgar, S.Ozcan, Carbohyd. Polym.,**131**, 208-217, (2015).
20. O.S. Dahham, N.Z. Noriman, A.W.M. Kahar, H. Ismail, S.T. Sam, J. Appl. Sci. & Agric., **10(5),** 33-39, (2015).
21. C. Wang, X.W. He, Q. Huang, F.X.Luo, X. Fu, Food Chemist., **141**, 3381-3385, (2013).
22. Z. Xiong, Y. Yang, J. Feng, X. Zhang, C. Zhang, Z. Tang, Carbohyd. Polym.,**92**, 810-816, (2013).
23. R. J. Walster, R. Rahman, A. W. M. Kahar. IOP Conf. Series: Mater. Sci. Eng., **429,** 012043, (2018).
24. A. W. M. Kahar, H. Ismail, A. Abdul Hamid, A J. Therm. Anal. Calorim.,**123(1),** 301-308, (2016).
25. T. Mekonnen, P. Mussone, D. Bressler, H. Khalil, J. Mater. Chemist.,**1,** 13379-13398, (2013).

IOP Conf. Series: Materials Science and Engineering **957** (2020) 012016 doi:10.1088/1757-899X/957/1/012016

Structural and dimensional analysis of LDPE/TPS blend filled with calcium carbonate

Anis Adila Ab Halim[1]. *Mohamad Kahar* Ab Wahab[1,*]

[1]School of Materials Engineering, University Malaysia Perlis, 02600, Arau, Perlis, Malaysia

Abstract This study aims to investigates the changes in structural, thermal stability and water absorption properties of thermoplastic starch (TPS) before and after modification with citric acid and ascorbic acid. The first stage of study involved the preparation of TPS from native potato starch. Starch will be converted into thermoplastic like in the presence of glycerol as plasticizer. For the second stage, LDPE was blended with TPS to produce LDPE/TPS blend with and without modification. Then, the influence of calcium carbonate as particulate filler on the properties of LDPE/TPS blends was also investigated. Structural findings based on the FTIR analysis showed the formation of new ester bond on the TPS backbone after the modification, which correspond to the formation of crosslinking network in TPS phase. Water absorption test was used to determine the effectiveness of TPS modification in term of physical interaction between the phases.

1 Introduction

There is increasing developments in the use of partial biodegradable polymer in daily applications resulting in highly demand of natural sources such as starch [1-3]. Since plastics are very important in daily life, production of biodegradable plastics to make plastics more compatible with the environment is necessary [4-5]. According to Katarzyna et.al, [6], primary degradation mechanism of biodegradable polymer involves the action of microorganism such as bacteria, fungi and algae. Biodegradable materials degrade into biomass, carbon dioxide and methane under aerobic or anaerobic conditions. In addition, cellulose, chitin, starch, polyhydroxyalkanoates, polylactide, polycaprolactone, and collagen are one of the types of biodegradable polymers [7].

Most researcher use biodegradable polymer in their studies as its exhibit more benefits which are sustainable, low cost and environmental. Thermoplastic starch is processing from native starch which it capable to flow and mixed with other synthetic polymer such as low density polyethylene Blending TPS with other synthetic polymers gives high ductility, high modulus with toughness, low cost, good mechanical properties, and improve biodegradability. On the other hand, crosslinking agent such as citric acid and ascorbic acid

* Corresponding author: kaharwahab@unimap.edu.my

are used to form bonds between starch polymer chains in order to strengthen the material. Besides that, crosslinking agent also used to reduce the water sensitivity of starch and to improve the barrier properties of TPS [8-11].

In order to enhance the properties of product, the incorporation of filler is required. Fillers usually used to improve the performance of rubbers and other polymeric materials. Clays, talc and calcium carbonate are the types of fillers. In this case, native potato starch undergoes the plasticization process with glycerol and treated with carboxylic acid, namely; citric acid and ascorbic acid. Calcium carbonate was used as a particulate filler in order to improve the dimension and thermal stability of LDPE/TPS blends, as well as their resistance to the moisture absorption.

2 Experimental

2.1 Materials

LDPE is a thermoplastic resin that purchased from Titan Chemical, Malaysia was used for this study. Density of LDPE is 0.91 g/cm3 and the melting point is between the ranges 105°C to 115°C. Potato starch was manufactured by Thye Huat Chan Sdn. Bhd. Penang and was chosen in this study. The moisture content of potato starch is 20%. Potato starch is a white powder in appearance and the density is 1.5 g/cm3. Calcium carbonate was used as reinforcing filler. $CaCO_3$ was used to increase the mechanical properties of the compounding. Glycerol is made by HmBG® Chemicals and was used as a plasticizer in preparation of TPS. The density of glycerol is 1.261 g/cm3 and the melting point is 17.8°C. Citric acid is provided by SIGMA-ALDRICH and used as a crosslinking agent to reduce the water sensitivity of starch and improving the mechanical properties of TPS. Ascorbic acid is purchased from Sigma-Aldrich and used as a crosslinking agent to improve the properties of TPS. The melting point of ascorbic acid is between the ranges 190-194°C.

2.2 Preparation of the sample

2.2.1 Modification of TPS with citric acid and ascorbic acid

Modification of TPS was done by mixing 65% of potato starch and 35% of glycerol in the presence of 3% of citric acid or ascorbic acid. The mixtures were left at room temperature for 24 hours after preparation. Then, the mixtures were mixed by using heated two roll mill machine at temperature 150°C for 10 minutes.

2.2.2 Preparation LDPE/TPS/$CaCO_3$ composite

LDPE was blended with TPS by using heated two roll mill with the different blends ratio. The blend proportion of LDPE/TPS blend were100:0, 90:10, 80:20, 70:30, 60:40 and 50:50. Then, LDPE/TPS blend was moulded by using hot press with 1mm dimension of mould at temperature 150°C for 6 minutes. The mechanical properties of LDPE/TPS blend were investigated under tensile testing. The best optimum measured data was used to ratio for subsequent study. The chosen LDPE/TPS blend ratio was further blended with calcium carbonate. Calcium carbonate were vary from 5% to 20% relative to the LDPE/TPS blends. The compounding process was prepared by using heated two roll mill at temperature 150°C for 10 minutes. Next, the compounding was moulded by using hot press with 1mm dimension of mould at temperature 150°C for 6 minutes.

3 Characterization and Testings

Thermogravimetric analysis (TGA) is used to determine the thermal behaviour of the thin film. TGA is used to measure the weight loss of the sample as a function of temperature and time. The experiment was conducted according to standard ASTM D258. 10mg of specimen was put in the platinum pan. The samples were heated at 20°C /min from 30°C to 600°C. The data were collected using Perkin-Elmer TGA 7. The decomposition of composite and its thermal stability was observed based on the curves produced. The structure of polymer blends was investigated by using FTIR Spectroscopy. Water absorption test was conducted according to standard ASTM D570. The samples were cut in dimension 30mm × 30mm. Then, the samples were dried in the oven at a temperature 50°C for 24 hours. After that, the samples were allowed to cool in room temperature and each of the samples was weighted using electronic balance. The samples were immersed in distilled water at room temperature for 30 days. The water absorption measurements were taken for every 3 days. Three samples were prepared for each of the formulation.

4 DISCUSSION

4.1 Structural analysis

Fig. 1 shows the TPS, TPSCA and TPSAA spectra in range of 4,000-650 cm−1 region. The IR spectrum of The peaks at 3,485 cm^{-1} and 2,928 cm^{-1} in potato starch was attributed to O–H and C–H bond stretching [12] while the peaks at 1,406 cm^{-1} was attribute to the bending modes of H–C–H and C– H [13]. The characteristic peaks 1078 cm^{-1} and 1150 cm^{-1} in TPS indicate the C-O bond stretching of C-O-H group [14]. Another two peaks at 1650 cm^{-1} was attributed to C=O bond [15]. Besides, the spectrum also shows the effect of citric acid and ascorbic acid modification on TPS. The existence of new shoulder peak at 1709 cm^{-1} in TPSCA and 1729 cm^{-1} in TPSAA illustrates that the ester bond existed [16]. Peak at 1729 cm^{-1} reflect the possible of chemical reaction between TPS with citric acid and ascorbic acid. It is noted that, the citric acid and ascorbic acid is capable to form covalent bond with TPS by detecting the appearance of new ester peak. Similar results were reported by Bower and Maddams [17] which states that the C=O stretching was resulted from the formation of ester groups between TPS and carboxylic acid. The functional groups of LDPE/TPS/$CaCO_3$, LDPE/TPSCA/$CaCO_3$ and LDPE/TPSAA/$CaCO_3$ composite were presented in Fig. 1(b). The presence of calcium carbonate was identified by assigned of O-C-O bond at absorption peak of 925 cm^{-1} [18]. No additional peak was observed after the addition of $CaCO_3$ which confirmed the only physical interaction occurred between $CaCO_3$ and LDPE/TPS matrix.

IOP Conf. Series: Materials Science and Engineering **957** (2020) 012016 doi:10.1088/1757-899X/957/1/012016

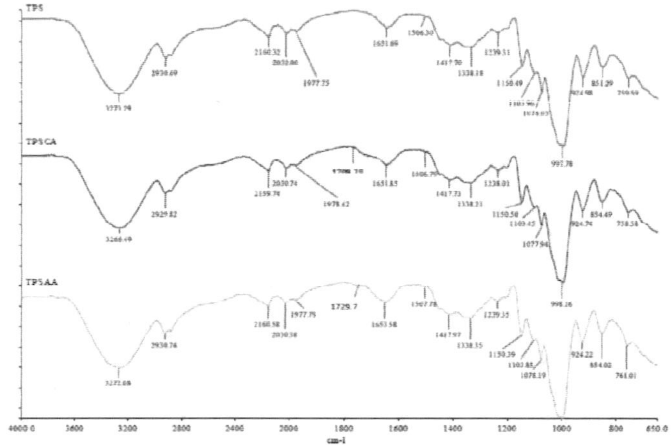

Fig. 1 (a). FTIR spectrum for TPS, TPSCA and TPSAA.

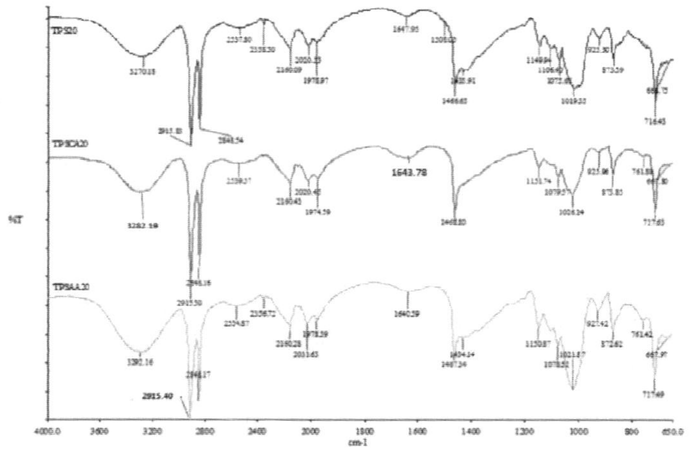

Fig. 1(b). FTIR spectrum for LDPE/TPS/CaCO3, LDPE/TPSCA/$CaCO_3$ and LDPE/TPSAA/$CaCO_3$ composite.

4.2 Water absorption

Fig. 2(a) shows the percentage of water uptake or water absorption of LDPE/TPS, LDPE/TPSCA and LDPE/TPSAA blends. As can be seen from the graph, the water absorption for the first 3 days was increased rapidly for all the series. This phenomenon was due to the water molecule start to penetrate inside the voids exist in LDPE/TPS blends. Then, the graph show steady increase after water molecule start to interact with hydroxyl groups from TPS (Stage 2) before its become saturated with surrounding (Stage 3) . On te other hand, the graph for LDPE/TPS blend shows higher water absorption when compared to LDPE/TPSCA and LDPE/TPSAA blends. There was significant voids between the LDPE and TPS interphase that allows the water molecule to penetrate and accupied inside. The formation of voids between LDPE and TPS molecules indicating to poor interfacial interaction between LDPE and TPS. In addition, high content of hydrophilic nature in TPS will affected the quantity of water absorption. Potato starch are hydrophilic and having OH group that can interconnect with the water molecule through the hydrogen bonding [19]. On contrary, the weight of water absorption for LDPE/TPSCA and LDPE/TPSAA blends were lower than LDPE/TPS blends. The modified TPS with citric acid and ascorbic acid were able to improve the adhesion at the interphase and thus lower the possibility of voids formation. Besides, citric acid and ascorbic acid also react as a crosslink agent [20] that can prevent the swelling of starch as well as restricts the penetration of water molecules inside the starch.

Fig. 2(b) shows the effect of water absorption on LDPE/TPS/$CaCO_3$, LDPE/TPSCA/$CaCO_3$ and LDPE/TPSAA/$CaCO_3$ composites. The water absorption of LDPE/TPS/$CaCO_3$ composite, LDPE/TPSCA/$CaCO_3$ composite and LDPE/TPSAA/$CaCO_3$ composite stage 1 was show similar trend with the blends without $CaCO_3$ due to the of water molecule occupied the voids at the interphase LDPE/TPS matrix and CaCO3. The diffusion of water molecules into the miniaturized voids and micro gaps in the polymer matrix, was due to the lack interfacial interaction between phases [21]. In stage 2, the water uptake slowly increase due to the hydrophilicity of TPS could form strong interaction with water molecule. In the third stage, the graph become saturated and reach the equilibrium with the surrounding. In all series, the water absorption for LDPE/TPS/$CaCO_3$ composite was found to have significant water uptake as compared to another systems. The formation of voids between LDPE and TPS molecules indicating to poor interfacial interaction between $CaCO_3$ with LDPE/TPS matrix. On the other hand, there is less apparent voids formation between the LDPE/TPSCA and LDPE/TPSAA with $CaCO_3$ which reflects lower water absorption as compared to LDPE/TPS system.

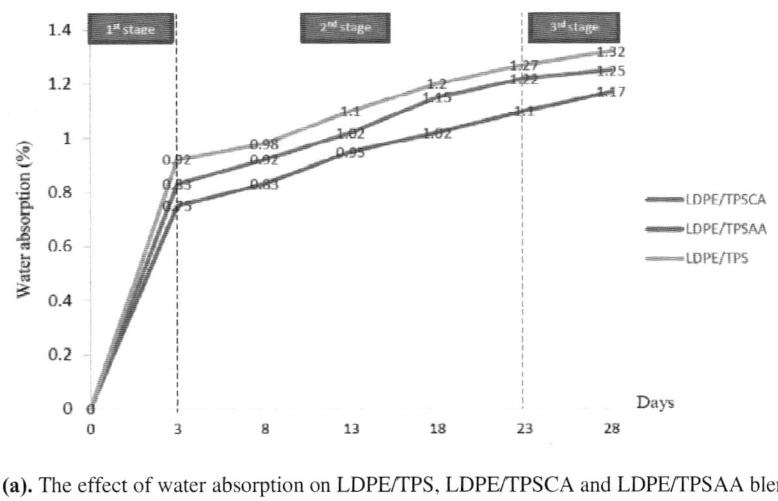

Fig. 2 (a). The effect of water absorption on LDPE/TPS, LDPE/TPSCA and LDPE/TPSAA blends.

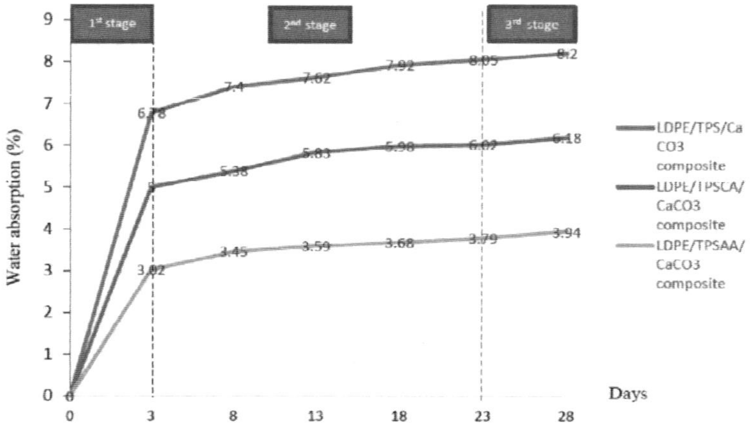

Fig. 2(b). The effect of water absorption on LDPE/TPS, LDPE/TPSCA and LDPE/TPSAA with 20% calcium carbonate.

4.3 Thermal stability

The TGA thermograms observed in Fig. 3(a) shows the thermal decomposition for LDPE/TPS blends and LDPE/TPSAA blends. There were three decomposition stages found for LDPE/TPS blends and LDPE/TPSAA blends which are corresponding to their compound's degradation. The degradation temperature observed at 38.61ºC to 110ºC was correspond to volatilization of moisture and another degradation occurred at the range of 110ºC to 250ºC due to the of glycerol that used in preparation of TPS [22]. The first major weight loss is referring to the degradation of thermoplastic starch at about 250ºC to 350ºC [23]. Another major weight loss for LDPE/TPS blends and LDPE/TPSAA blends occurred at 331.5ºC and 298.9ºC respectively, which attributed to the decomposition temperature for LDPE phase.

Based on the degradation temperature curves, LDPE/TPS blends shows higher thermal stability as compared to LDPE/TPSAA blends. This due to the presence of ascorbic acid in TPS will partially hydrolyse the TPSAA chains and thus lowering its thermal stability. Fig. 3(b) shows the TGA thermogram for LDPE/TPS/CaCO3 composite and LDPE/TPSAA/CaCO3 composite at 20% $CaCO_3$ loading. The thermograms graph for both composites exhibit almost similar characterization. First stage decomposition start at temperature 150°C–250°C which related the evaporation of glycerol in TPS phase. For the second stage, at temperature 250°C–350°C was attributed to decomposition of TPS in the composite. At temperature 350°C–470°C was attributed to LDPE decomposition [24]. Although, TPS and LDPE was completely decomposed at 500°C, the remaining residue above that temperature was attributed to calcium carbonate. The decomposition temperature for calcium carbonate was reported at 850°C–900°C [25].

By comparing the thermal decomposition of both series, the weight loss of LDPE/TPSAA/$CaCO_3$ composite is lower than LDPE/TPS/$CaCO_3$ composites. LDPE/TPS/$CaCO_3$ composite indicate poor degree of interfacial interaction between LDPE and TPS as well as between calcium carbonate and LDPE/TPS matrix and thus lowering their thermal stability. Better interfacial interaction between LDPE/TPSAA matrix with $CaCO_3$ was believed to contribute better thermal stability for LDPE/TPSAA/$CaCO_3$ composite. Good interaction at the interphase allows $CaCO_3$ to prevent LDPE/TPS chain movement and therefore give resistance to the thermal decomposition.

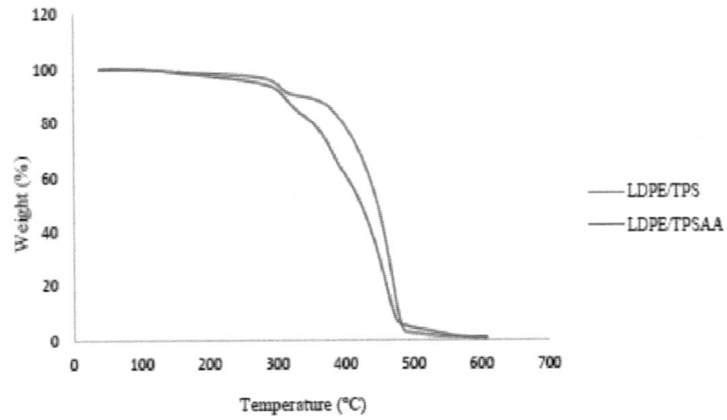

Fig. 3 (a). TGA curves for LDPE/TPS blends and LDPE/TPSAA blends.

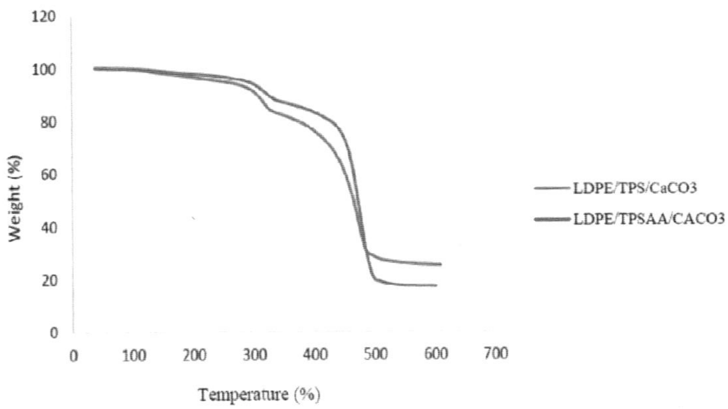

Fig. 3(b). TGA curves for LDPE/TPS/$CaCO_3$ composite and LDPE/TPSAA/$CaCO_3$ composite.

5 Conclusions

In order to achieve better thermal and water resistance properties, TPS were treated with citric acid and ascorbic acid. Treatment with citric acid and ascorbic acid creating new fuctional groups as confirmed with FTIR analysis. Ester bonding that appeared after the treatment could initiate the crosslink network within the TPS phase. This would lead to the better thermal stability and better resistance to the water absorption. Moreever, $CaCO_3$ as a particulate filler was added in LDPE/TPS, LDPE/TPSCA and LDPE/TPSAA in order to improved the targeted properties. Treated TPS with citric acid and ascorbic acid were found to have better compatibility with $CaCO_3$ as its reflected to their thermal and water resistance properties.

The authors wish to gratefully acknowledge the School of Materials Engineering, as well as Research and Innovation Department under Universiti Malaysia Perlis and Kementerian Pengajian Tinggi Malaysia (Ref: FRGS/1/2016/TK05/UNIMAP/02/9) for supporting this research project.

References

1. P.L. Tran, D.H.D. Nguyen, V.H. Do, Y.L. Kim, S. Park, S.H. Yoo, Food Sci. Technol., **62**, 1091-1098, (2015).
2. S. Berkesch, Biodegrad. Polym. **1**, 1-14, (2005).
3. N.F. Alias, H. Ismail, A.W.M. Kahar,. Bio-res, **12(4)**, 9103-9117, (2017).
4. A.H. Mohd Zain, A.W.M. Kahar, N.Z. Noriman, Procedia Chem., **19**, 638-645, (2016)
5. Z. Xiong, Y. Yang, J. Feng, X. Zhang, C. Zhang, Z. Tang, Carbohyd. Polym., **92**, 810-816, (2013).
6. G.L. Katarzyna Leja, Polish J. Environ. Stud. **19**, 255-266, (2010).
7. P. Jariyasakoolroj, S. Chirachanchai, Carbohyd. Polym., **106**, 255-263, (2014).
8. N. Ibrahim, A.W.M. Kahar, D.L Ngoc, H. Ismail, H. Bio-res., **12(2),** 3076-3087, (2017).
9. N.N.S. Thomas, S. Mohan, E. Francis, Euro. Polym. J. **65,** 224-335, (2012).
10. A.W.M. Kahar, N. Sarifuddin, H. Ismail, H., Ira. Polym. J., **26 (2**), 149-159, (2017).
11. A.H. Mohd Zain, A.W.M. Kahar, I. Hanafi, J. Polym. Environ, **26(2),** 691–700, (2018).
12. Y. Yang, Z. Tang, Z. Xiong, J. Zhu, Intern. J. Biol. Macro., 77, 273-279, (2015).
13. J. Ren, H. Fu, T. Ren, W. Wuan, Carbohyd. Polym., **77(3),** 576-582, (2009).
14. C.K. Hong, I. Hwang, N. Kim, D.H. Park, B.S. Hwang, C. Nah, Ind. Eng. Chemist., **14**, 71-76, (2008).
15. A.W.M. Kahar, H. Ismail., N. Othman, J. Vinyl Add. Technol. **18(3),** 192-197, (2012)
16. A.W.M. Kahar, H. Ismail. J. Vinyl Add. Technol. **22(3),** 191-199, (2016).
17. D. I. Bower, W. F. Maddams, "*The Vibrational Spectroscopy of Polymers*," Cambridge University Press, New York, N.Y., 154, (1989).
18. H.A. Al-Hosney, S. Carlos-Cuellar, J. Baltrusaitis, V.H. Grassian, Phys. Chemist. Chem. Phy., **7(20),** 3587-3595, (2005).
19. J. Aburto, I. Alric, S. Thiebaud, E. Borredon, D. Bikiaris, J. Prinos, C. Panayiotou, J. Appl. Polym. Sci.,**74(6),** 1440-1451, (1999).
20. Noorizzah, A.W.M. Kahar, D.N. Uylan, J. Polym. Mater., **33 (1),** 201, (2016).
21. A.W.M. Kahar, H. Ismail., N. Othman, J. Appl. Polym. Sci., **128(4),** 2479–2488, (2013).
22. R. J. Walster, R. Rahman, A. W. M. Kahar. IOP Conf. Series: Mater. Sci. Eng., **429,** 012043, (2018).
23. T. Mekonnen, P. Mussone, D. Bressler, H. Khalil, J. Mater. Chemist., **1**, 13379-13398, (2013).

A. W. M. Kahar, H. Ismail, A. Abdul Hamid, A J. Therm. Anal. Calorim., **123(1),** 301-308, (2016).

A. Awal, M. Rana, M. Sain, Mech. Mater., **80**, 87-95, (2015).

The Effect of Modified Jackfruit (Artocarpus Heterophyllus Lam.) Seed Starch on the Properties of Poly Lactic Acid

Nur Amira Sahirah Abdullah [1], and *Mohamad Kahar* Ab Wahab[1,*]

[1]School of Materials Engineering, University Malaysia Perlis, 02600, Arau, Perlis, Malaysia

Abstract. This work investigates the effect of the modified jackfruit seed starch with silane coupling agent on the properties of polylactic acid (PLA). Jackruit seed starch will converted into thermoplastic like in the presence of glycerol as plasticizer. The physicochemical properties and thermal stability of the native and modified jackfruit seed starch blends with PLA were comparatively investigated. The structural changes after modification was analyzed by using Fourier Transmission Infra-red technique. Its confirmed the successful of the silane treatment from the newly peak observed at 1020 cm^{-1} and 1105 cm^{-1} for Si-O-Si and starch-O-Si bonds, respectively. For tensile strength, the value shows slightly decrese with the increasing of jackfruit seeds starch content. However the tensile properties for silane treatment shows the better improvement as compared to untreated counterpart. In the case of plasticized jackfruit seed starch, the elongation at break value increase with filler loading due to the ductile behavior of the blends system. The modification of the starch with silane coupling agent gives better interfacial adhesion between starch and PLA matrix while the glycerol gives the ductility and hydrophobicity to the starch and gives better elongation at break for the PLA/starch blends. The thermal stability of the silane treated

[*] Corresponding author: kaharwahab@unimap.edu.my

PLA/starch blends was also improved as compared to the native PLA/starch blends.

1 Introduction

Jackfruits (Artocarpus heterophyllus Lam.) is widely found in the tropical countries such as Malaysia, Thailand and Vietnam. The jackfruit has yellow flesh with bell shape and smooth surface seeds. Jackfruit seeds have high content of carbohydrate, protein, and starch [1-3]. The jackfruit seed starch is completely biodegradable and highly hydrophilic. Jackfruit seed starches can be modified physically or chemically to improve the physicochemical properties of the starch films. The starch can be converted into the thermoplastic starch (TPS) by addition of plasticizer. The commonly used of plasticizer in the synthesis of TPS are water, glycerol, urea, and citric acid. Other than that, the chemical modification of the jackfruit seed starch can be done by binding the hydroxyl group on the starch with other functional group to improve the hydrophobicity of the starch. The examples of the chemical that can be used in this modification are dodecenyl succinic anhydride (DSA), octenyl succinic anhydride (OSA), maleic anhydride (MA), and 3-chloropropyl trimethoxysilane [4,5]. Blending TPS with PLA is one of the approach to produce biodegradable plastics. PLA have high modulus, high strength, excellent clarity, good biocompatibility and biodegradability. The PLA have widely used in the biomedical application, and industrial packaging. The limitations of the PLA are highly in cost and brittleness [6-8]. Thus, the TPS was used in PLA matrix in order to obtain fully biodegradable composites with cost effective and improved impact resistance. However, PLA and starch are thermodynamically immiscible due to the different in hygrascopicnature. The interfacial adhesion between PLA and starch will be poor if they are blended directly. There are some modifications to improve the interfacial adhesion and mechanical properties between PLA and starch blend such as by addition of coupling agent or reactive compatibilizer. Besides, the composites are usually brittle if PLA is blended with granule starch especially at high starch conent. The plasticized starch or thermoplastic starch can be as alternative blend with PLA. In this research, the modifications of the jackfruit starch with plasticizer and silane coupling agent was focused and studied in details.

2 Experimental

2.1 Materials

Jackfruit seeds was obtained from a local market in Perlis, Malaysia and have bell shape and smooth surface. The poly (lactic acid) (PLA) was supplied by Titan Chemical, Malaysia in granular form. Analytical grade glycerol and 3-(trimethoxysilyl)propyl methacrylate (MPS) was supplied by Sigma Aldrich, Germany and used as received.

2.2 Preparation of the sample

2.2.1 Isolation of Starch (JF) from Jackfruit Seed

The jackfruit seeds were cleaned manually and the white aril was peeled off and removed. Then, the seeds will soak into the 5% concentration of NaOH to remove the brown spermoderm cover for two hours at room temperature. After two hours, the NaOH solution was removed and the seeds were rinsed with distilled water [3]. The seeds were scrubbed by using a soft brush in the tap water to remove the brown spermoderm to obtain the cotyledons of the seeds. The seeds were dried in the laboratory oven at temperature of 50°C for 9 hours. Then, the dried seeds were sliced into small pieces and undergo grinding process by using ring mill to obtain the jackfruit seed starch in powder form. The starch powder is then dried again in the laboratory oven at temperature of 80°C for 6 hours to remove residual water content. Lastly, the dried starch powder was sieved at 63 μm by using laboratory sieve.

2.2.2 Jackfruit Seed Starch Modification with Silane (JFS)

The jackfruit seed starch modification with silane was followed the method proposed by Kahar et al. [11].The 600 ml mixture of the ethanol/water was prepared with ratio of 80/20. The 150g of jackfruit was suspended in the mixture of ethanol and water. The 18 ml of 3 w/w% 3 -(trimethoxysilyl)propyl methacrylate (MPS) was added into the suspension of the starch and the pH was adjusted to 4 by addition of acetic acid. The mixture was stirred at room temperature for 2 hours. The MPS-modified starch was filtered and washed with ethanol to remove excess silane. The MPS-modified starch was dried in the laboratory oven at 50°C for overnight. The MPS-modified starch was grinded and sieved at 63 μm.

2.2.3 Preparation of Modified Jackfruit Thermoplastic Starch (JFT)

Glycerol was used as a plasticizer in order to convert the granule starch into thermoplastic starch. The 97.5 g of granule starch (65 w/w%) was dry mixed with 52.5 g of glycerol (35 w/w%) at room temperature and let to overnight [9]. Then, the mixture was compounded by using heated two roll mills at temperature of 120°C for 10 minutes.

2.2.4 Preparation of PLA/jackfruit Seed Starch Blend.

PLA/jacruit seed starch blends were prepared by using heated two roll mill at 170°C with rotation rate of 10 rpm. Firstly, the PLA was added to the nip of the heated two roll mill and processed until it melts. Then, the JF was added and mixed with PLA until homogeneous. Then, the PLA/JF compound was undergoing hot press molding to produce a 1 mm PLA/JF sheet. The hot press molding process was take place at 170°C for 6 minutes.

3 Characterization and Testings

The tensile testing was carried out by using an Instron machine with model 5569. The standard used for tensile testing was ASTM D638 for plastic. The cross head speed was set at 50 mm/min and the gap between upper jaw and lower jaw was set at 50mm. The tensile strength and elongation at break were measured by calculating the average values of five identical samples for each formulation. The Fourier Transform Infrared Spectroscopy (FTIR) was used to identify the functional group present of the PLA/JF, PLA/JFT, and PLA/JFS compounds. The model of the FTIR used was Perkin-Elmer FTIR200. The sample was scanned with wavelength between 4000cm^{-1} to 40000cm^{-1} 00 The total scans used were 32 scans. The Scanning Electron Microscope (SEM), model JEOL JSM-6460LA was used to analyze the morphology of the PLA/JF, PLA/JFT, and PLA/JFS blends by observing the tensile fracture surface of the samples with the supply of accelerating voltage at 10kV. The samples were coated with platinum at about 1.5 – 3.0 nm thickness in order to prevent the electrostatic charge during the experiment. The Thermogravimetric Analysis (TGA) model Pyris Diamond, Perkin-Elmer was used to measured the weight changes as a function of temperature and time in a controlled atmosphere. The TGA and DTA curves were obtained by heating samples from the 20°C to 700°C at 10°C/min under the nitrogen atmosphere.

4 Discussion

4.1 Structural analysis

For the native jackfruit seed starch spectra shown in Fig. 1, the band appear at 763 cm^{-1} indicating the C-C stretching while the C-H bending appear at 860 cm^{-1} which attributed to the vibration of starch glucosidic ring. The peak at 1020 cm^{-1} and 1079 cm^{-1} shows the C-O-H and C-O-C anhydroglucose unit while at 932 cm^{-1} represents the C-O stretching in a C-O-C bond [12]. The band at 1647 cm^{-1} originated from tightly bounded water to starch particle and those band was observed at 2931 cm^{-1} and 3394 cm^{-1} of the C-H and O-H bonds respectively. For the thermoplastic starch spectra, the peak at 2930cm^{-1} was abserved and represent the intermolecular bonds between starch and glycerol hydroxyl group [13]. This show that the glycerol was tightly bounded and trapped in the starch network during the plasticization process. The shifted this peaks to the lower wavenumber indicated the hydrogen bonding was formed between glycerol and C-OH bond in starch. The changes in starch structure after the modification with silane can be determine by formation of the Si-O-Si bond and starch-O-Si bond at the fingerprint region due to silane linkage with jackfruit seed starch. By referring to Fig. 2, the JFS reveals an absorption peak at 1019 cm^{-1} that represented to polysiloxane Si-O –Si bridge. The shoulder peak at 1105 cm^{-1} was observed as Si-O-C peak which due to the covalent bond between starch and silane. This low intensity peak also related to the residual unhydrolyzed Si–O–CH_3 groups where most of the silane had been hydrolyzed [14].

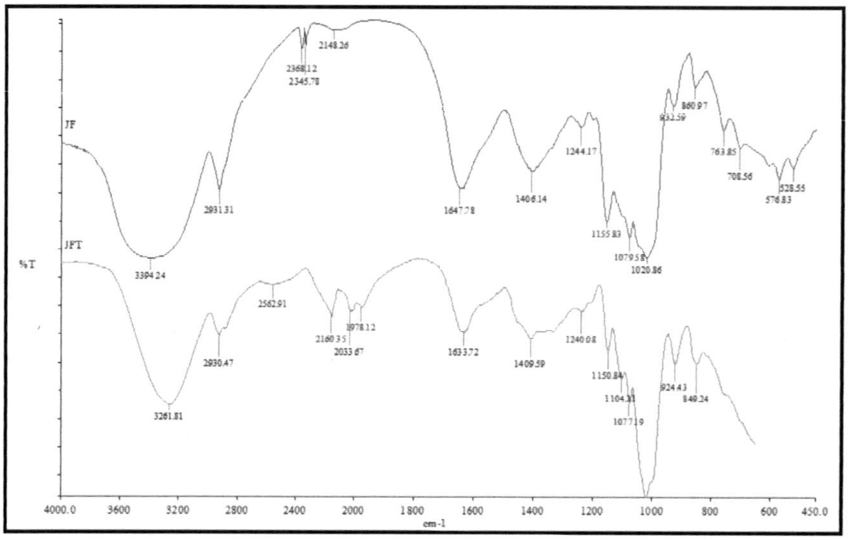

Fig. 1. FTIR spectra of JF and JFT.

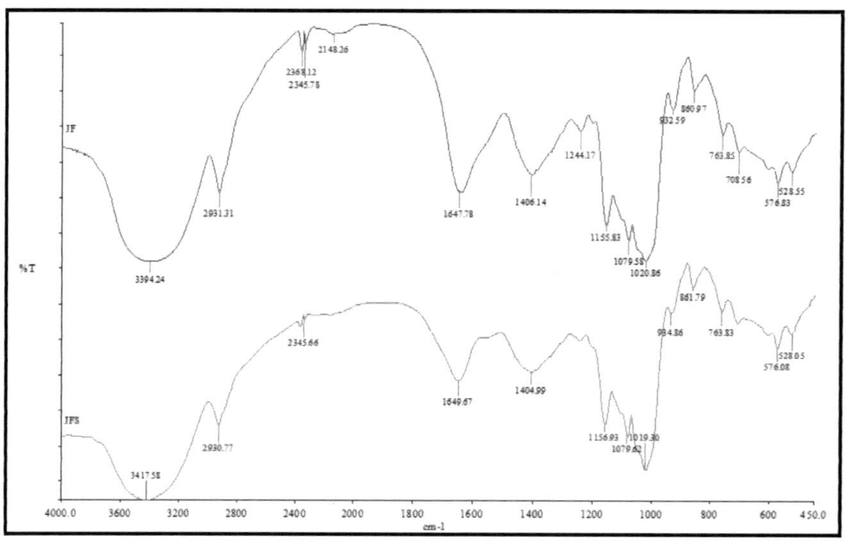

Fig. 2. FTIR spectra of JF and JFS.

4.2 Tensile properties

Fig. 3 shows the tensile strength at diferent PLA/JF, PLA/FS and PLA/JT blends ratio. For all the blend series, at 10% of jackfruit seed starch loading shows the highest tensile strength value. The increament in tensile strength was due to reinforcement effect on the PLA matrix through the mechanical interlock,which prevent the movement of PLA chains. At further jackfruit seed starch loading, the alignment of the PLA chains will be disrupt by the agglomeration of starch [15-16]. This phenomenon will lead to the tensile strength decrease as the starch content increase. The highest tensile strength recorded for PLA/JF blends at 10% jackfruit seed starch loading is 45.2 MPa and the value is decreases until 35.2 MPa for 30% of starch loading. The similar trend was observed for PLA blend with silanized jackfruit seeds starch (PLA/JFS). However, as compare to PLA/JF blend, silanized starch (PLA/JFS) shows better improvement in tensile strength. At 10% silanized jackfruit seed starch loading the tensile strength of PLA/JFS blends is 47.7 MPa. Modification with silane based, 3-(trimethoxysilyl)propyl methacrylate coupling agent can improve the interfacial adhesion between PLA matrix and starch. Good intherphases interaction allow the stress to transfer from PLA matrix into starch and prevent the crack formation at the interphase. The silanized jackfruit seed starch also improve the compatibility with PLA by attaching the hydrophobic character to starch backbone [17-18]. For PLA/JFT blends, as the JFT loading increase, the tensile strength was slightly decreases.. The highest tensile strength recorded at 5% of the JFT loading which is 39.6 MPa. In this blend system, the particulate starch was converted into the thermoplastic starch by addition of the glycerol as plasticizer. In the present of glycerol, the dispersion between JFT in PLA matrix was improved and provide some degree of reinforcement in the JFT matrix. However, the present of glyceron in JFT will soften the PLA phase and hence the tensile strength reduces as the JFT increase [19-23].

The Young's modulus of the silanized starch modification shows the highest value across all the series as shown in Fig. 3. As the starch content increases up to 10%, the Young's modulus increase due to reinforcement from incorporation of jackfruit seed starch. At further jackfruit seed starch loading, Young's modulus decreases due to the ductility of the PLA/starch blends increases. In the case of PLA/JFS blends, the silanized modification of starch shows better improvement in Young's modulus value. It's due to the silanized starch improved the interfacial adhesion between the phases and prevent the distortion of the PLA matrix. Contrary with PLA/JFT blend, the present of the glycerol as plasticizer will soften the blend and effect Young's modulus value. However, the incoprpotation of thermoplastic jackfruit seed starch in the PLA matrix, makes the compound more ductile as compared to PLA/JF and PLA/JFS. The dispersion of starch within the matrix was better due to the plasticity and the ability of JFT to flow during melt processing. The elongation at break of the PLA/JFT blends was the highst as compared to other series, which due to the flexible characteristic of the thermoplastic jackfruit seed starch.

IOP Conf. Series: Materials Science and Engineering **957** (2020) 012017 doi:10.1088/1757-899X/957/1/012017

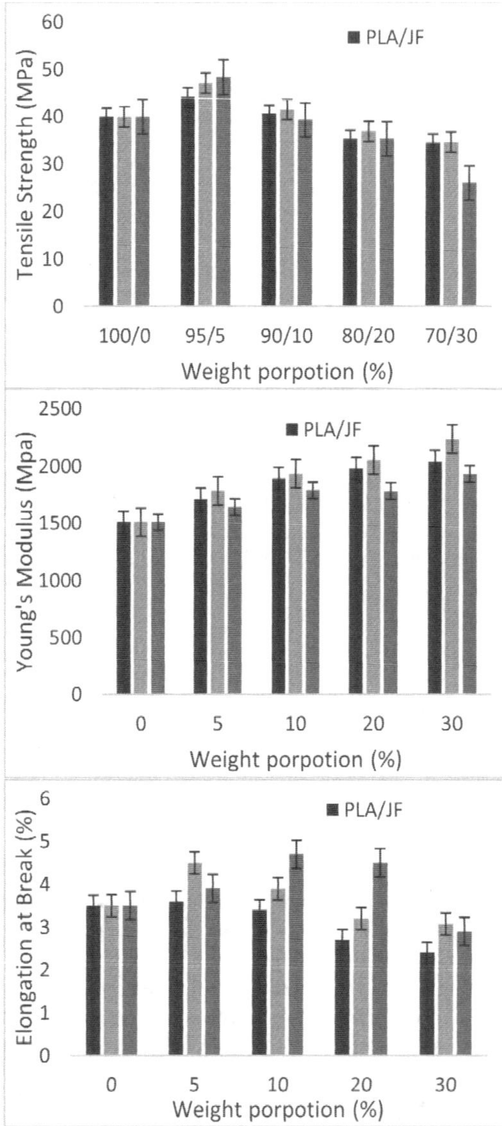

Fig.3. Tensile properties of PLA/JF, PLA/JFS and PLA/JFT blends

4.3 Morphological Study

The SEM image for the PLA/JF blend at 10% of jackfruit seed starch loading is shown in Fig. 4 (a). The surface of JF ganules are quite smooth with no PLA matrix adhering to the starch particle. The size of the starch granule used was around 60 µm and remain unchanged with visible phase separation. The PLA/JF blend shows that the blend system is immiscible and incompatible due to the appearance of holes separating the starch granules and PLA matrix. As refer to the Fig. 4 (a), there is a formation of a voids along the PLA matrix that due to weak interfacial interaction between phases. Fig. 4 (b) shows the SEM image of the PLA blend with silanized jackfruit seed starch at 10% starch loading. The interfacial adhesion phases was improved and the starch particles is embedded into PLA matrix. The voids formation was reduces due to the hydrophobicity characteristic of the silanized starch. Fig. 4 (c) shows the morphology of the PLA blend with thermoplastic jackfruit seed starch at 10% of JFT loading. The JFT was observed as co-continuos phase in a PLA matrix. The disrupted the crystalline structure of starch granules allow JFT to well dispersed and existed as co-continuos phase in PLA matrix. Characteristic of plasticized starch allows JFT to elongate better with PLA matrix, which can be correlated to the better elongation at break values.

IOP Conf. Series: Materials Science and Engineering **957** (2020) 012017 doi:10.1088/1757-899X/957/1/012017

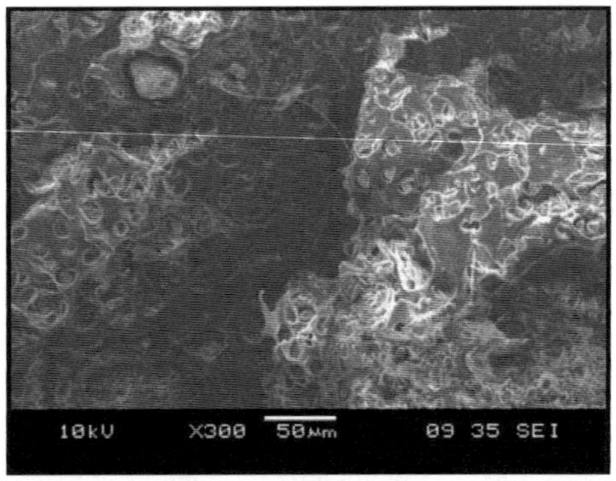

(a)

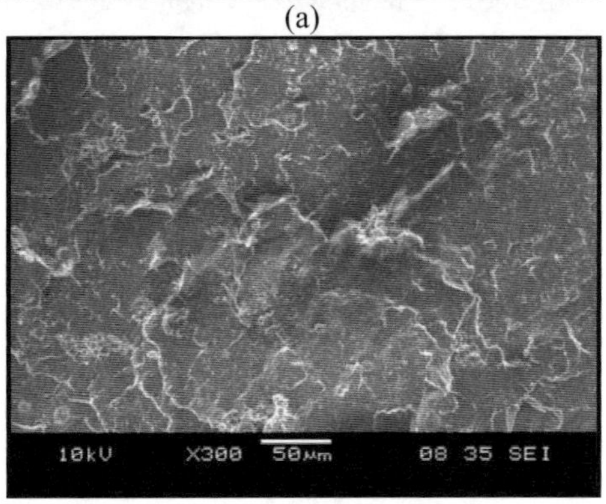

(b)

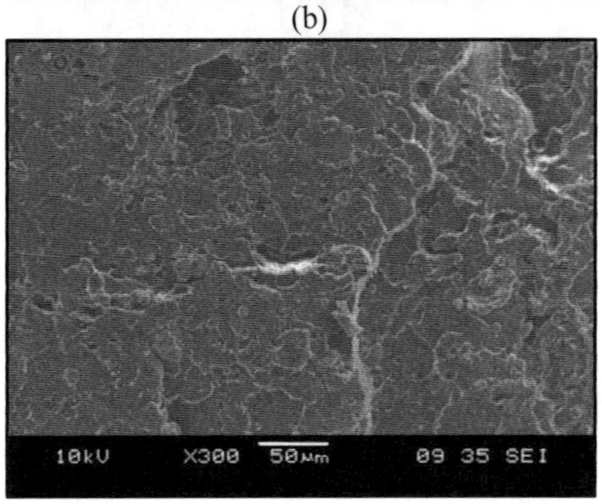

(c)

Fig. 4. Scanning electron micrograph of (a) PLA/JF blend (b) PLA/JFS blend, and (c) PLA/JFT blend at 300 X magnification.

4.4 Thermogravimetric Analysis (TGA)

The thermogravimetric analysis was used to study the thermal stability and the amount of the volatiles compounds present in the blend system. In this section, the weight loss curves and its derivatives of the blend system were investigated. The Fig. 5 (a) and (b) show the TGA curves and DTG curves of neat PLA an different PLA/jackfruit seed starch (90/10) blends system. From the Fig. 7 (a), the significant weight loss occurs at temperature range of ~50°C-100°C is due to the loss of moisture, at ~250°C-320°C is related to depolymerization of amylose and amylopectin in starch [24], and at ~320°C-410°C is a degradation of more thermally stable PLA phase [25]. For neat PLA, 5% weight loss appears at about 335.2°C while the PLA/JF was occurred at about 278.8°C. The incorporation of starch will interfere the alignment of PLA chains and therefore reduce thermal stability of the blends. For another blends series, 5% weight loss for PLA/JFS and PLA/JFT were occurred at 302.3°C and 277.2°C respectively. At degradation temperature of 300°C, the PLA/JFS blend show the highest thermal stability among the blends series, which the weight loss was 4.5%, while for PLA/JF and PLA/JFT were 24.5% and 21.1%, respectively. PLA/JFS blend after silane modification show good thermal stability due to the better interfacial improvement as confirmed by SEM morphology. As refer to the Fig. 8, the maximum thermal degradation temperature for PLA/JFS blends was 347.7 °C, while for PLA/JF blends and PLA/JFT blends were slightly shift to lower temperature of 311.4°C and 3.18.8°C respectively. The modification of the starch with silane was improved the thermal stability of the PLA blend systems and makes heat well distributed upon thermal introduction.

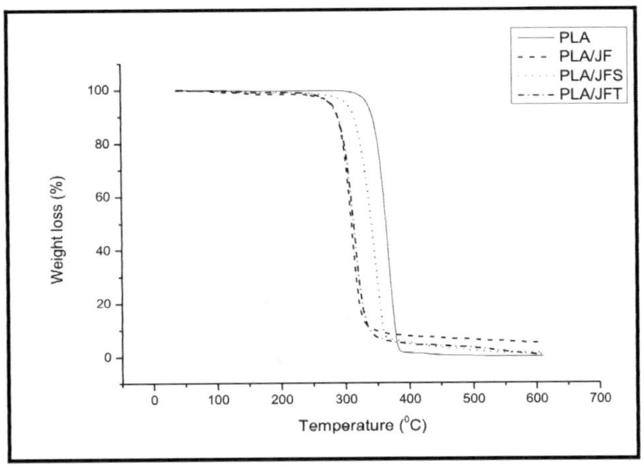

Fig. 5(a). The TGA curve of neat PLA and PLA blends of ratio 90/10.

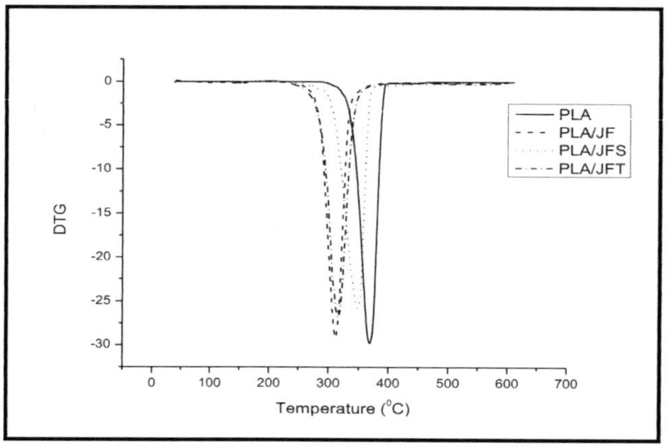

Fig. 5(b). The DTG curve of neat PLA and PLA blends of ratio 90/10.

5 Conclusions

The silane coupling agent and glycerol as plasticizer can be employed to obtain the modified jackfruit seed starch with improvement in tensile properties, interfacial interaction and thermal stability. The JFS based blend system show the highest tensile strength and Young's modulus values as compared to other series counterparts. The silanized starch also improve the compatibility and hydrophobicity with PLA matrix and resulting less void formation as proved from the SEM micrographs. The thermal stability of the blend system was also improved after jackfruit seed starch was treated with silane coupling agent. On the other hand, JFT blend systems gives the ductile behavior to the compound. The presence of the glycerol in JFT triggered the hydrogen bonding between the hydroxyl group of starch and glycerol and converting the particulate starch into a thermoplastic like. The JFT was observed as embedded phase in a PLA matrix and appeared as a co-continuous phase system. Thus, it gives homogenous morphology on the PLA/JFT blends and exhibited better elongation at break value.

The authors wish to gratefully acknowledge the School of Materials Engineering, as well as Research and Innovation Department under Universiti Malaysia Perlis and Kementerian Pengajian Tinggi Malaysia (Ref: FRGS/1/2016/TK05/UNIMAP/02/9) for supporting this research project.

References

1. O.S. Kittipongpatana, N. Kittipongpatana, Food Sci. Technol., **44**, 1766-1773 (2011)
2. A.W.M. Kahar, M. Lingeswarran, M.Z. Amirah Hulwani, H. Ismail, Polym. Bull., **76(2)**, 747-762 (2019)
3. P.L. Tran, D.H.D. Nguyen, V.H. Do, Y.L. Kim, S. Park, S.H. Yoo, Food Sci. Technol., **62**, 1091-1098 (2015)
4. N.F. Alias, H. Ismail, A.W.M. Kahar,. Bio-res, **12(4)**, 9103-9117 (2017)
5. A.W.M. Kahar, H. Ismail., N. Othman, J. Appl. Polym. Sci., **128(4),** 2479–2488 (2013)
6. H. Chi, K. Xu, D. Xue, C. Song, W. Zhang, P. Wang, P. Food Resear. Intern., **40**, 232-238 (2007)
7. P. Jariyasakoolroj, S. Chirachanchai, Carbohyd. Polym., **106**, 255-263 (2014)
8. N. Ibrahim, A.W.M. Kahar, D.L Ngoc, H. Ismail, H. Bio-res., **12(2),** 3076-3087 (2017)
9. M.A. Huneault, H. Li, Polymer, **48**, 270-280 (2007)
10. Y. Yang, Z. Tang, Z. Xiong, J. Zhu, Intern. J. Biol. Macro., **77**, 273-279 (2015)
11. A.W.M. Kahar, N. Sarifuddin, H. Ismail, H., Ira. Polym. J., **26 (2)**, 149-159 (2017)
12. A.H. Mohd Zain, A.W.M. Kahar, I. Hanafi, J. Polym. Environ, **26(2),** 691–700 (2018)
13. J. Ren, H. Fu, T. Ren, W. Wuan, Carbohyd. Polym., **77(3),** 576-582 (2009)
14. C.K. Hong, I. Hwang, N. Kim, D.H. Park, B.S. Hwang, C. Nah, Ind. Eng. Chemist., **14**, 71-76 (2008)
15. A.W.M. Kahar, H. Ismail., N. Othman, J. Vinyl Add. Technol. **18(3),** 192-197 (2012)
16. H. Liu, F. Xie, L. Yu, L. Chen, L., Li, Prog. Polym. Sci., **34(12),** 1348-1368 (2009)
17. A.W.M. Kahar, H. Ismail. J. Vinyl Add. Technol. **22(3),** 191-199 (2016)
18. Y. Lua, M.C. Cueva, L.C. Edgar, S. Ozcan, Carbohyd. Polym., **131**, 208-217 (2015)
19. O.S. Dahham, N.Z. Noriman, A.W.M. Kahar, H. Ismail, S.T. Sam, J. Appl. Sci. & Agric., **10(5),** 33-39 (2015)
20. C. Wang, X.W. He, Q. Huang, F.X. Luo, X. Fu, Food Chemist., **141**, 3381-3385 (2013)
21. Z. Xiong, Y. Yang, J. Feng, X. Zhang, C. Zhang, Z. Tang, Carbohyd. Polym., **92**, 810-816 (2013)
22. R. J. Walster, R. Rahman, A. W. M. Kahar. IOP Conf. Series: Mater. Sci. Eng., **429,** 012043 (2018)
23. T. Mekonnen, P. Mussone, D. Bressler, H. Khalil, J. Mater. Chemist., **1**, 13379-13398 (2013)
24. A. W. M. Kahar, H. Ismail, A. Abdul Hamid, A J. Therm. Anal. Calorim., **123(1),** 301-308 (2016)
25. A. Awal, M. Rana, M. Sain, Mech. Mater., **80**, 87-95 (2015)

IOP Conf. Series: Materials Science and Engineering **957** (2020) 012018 doi:10.1088/1757-899X/957/1/012018

The effect of twin screw compounding parameters on the tensile properties of pineapple leaf/sea shell hybrid polymer composite using DOE approach

Ng Tian Ching[1], *Khairul Anwar* Abdul Halim[1,*], *Mohd Firdaus* Omar[1,2], *Azlin Fazlina* Osman[1,2] *and Muhammad Salihin* Zakaria[1,2]

[1] School of Materials Engineering, Universiti Malaysia Perlis (UniMAP), Kompleks Taman Muhibbah, Jejawi 2, Perlis, Malaysia

[2] Centre of Excellence Geopolymer and Green Technology (CEGeoGTech), Universiti Malaysia Perlis (UniMAP), Kompleks Taman Muhibbah, Jejawi 2, Perlis, Malaysia

Abstract. Pineapple leaf and sea shell were used as natural fillers in this research due to its biodegradable nature, wide availability, continuous resources and low cost. This research work was carried out to investigate the effect of compounding process using twin screw extruder on the tensile performance of pineapple leaf/ sea shell polymer composite using Design of Experiment (DOE) approach. A total of nine runs of were formulated and the resulitng hybrid composites were compounded using twin screw extruder. The short term tensile test was carried out to determine the tensile properties and the data were sunsequently analyse using DOE software. Pareto chart of the standardized effect and the main effect plot were employed to investigate the relationship between processing parameters and the tensile performance of the hybrid composite systems. Based on the initial DOE analysis, it is shown that compounding parameters had influenced the final mechanical behavior of the hybrid composites.

1 Introduction

Composite material comprises of two or more different materials that combined or mixed to form new material that is superior compare to its constituents. For instance, the addition of filler into polymer matrix is to improve the mechanical properties and increase the strength of the resulting composite materials [1]. In recent years, the use of natural fillers as reinforcement have gain attention due to low cost, lightweight, wide availability and biodegradable [2]. Natural fillers are referred to the filler which obtained from natural resources such as plant and animals. In this study, pineapple leaf and sea shell are used as the natural fillers to form polymer composite materials. Natural fibers such as pineapple leaf are

* Corresponding author: kanwar@email.unimap.edu.my

mainly constituted of lignin, cellulose and hemicelluloses whereas sea shell is contains of calcium carbonate, $CaCO_3$ or chitin.

Natural fibers are having lower hydrophilic natural fiber which gives the fibers high durability and resistance to water. The utilisation of natural fiber based composite, it is able to provide significance functional and economic growth and benefits due to its properties of high strength, durability features [3]. Natural fiber based composites are be chosen due to its properties of environmentally-friendly and abundant of supply, it is easy to process and cheap, and its biodegradable nature [1]. In this research, polypropylene, PP is used and it acts as matrix in the experiment whereas pineapple leaf fiber and sea shell are acts as the natural fillers to hybrid polymer composite materials.

Design of experiment (DOE) is a statistical method that uses optimization techniques to produce processing parameters. DOE is a statistical method that use as a control and also to optimize the parameter processing [4-5]. The main purpose of DOE in this research is to investigate the effect of twin screw compounding parameters on the tensile properties of pineaple leaf/seashell hybrid composites. The main parameters involve were such as filler loadings, barrel temperature and screw speed.

2 Materials and Method

Pineapple leaf and sea shell were utilised as natural fillers in this study. Pineapple leaf was collected from Bukit Tambun, Simpang Ampat, Pulau Pinang, Malaysia while sea shell was collected from the seaside of Persiaran Gurney, Pulau Pinang, Malaysia. Natural fillers was then compounded with polypropylene (PP) PROPELINAS G452 which was produced by Polypropylene Malaysia Sdn. Bhd.

2.1 Sample Preparation

The pineapple leaf and sea shell was washed and cleaned by tap water to remove the dirt particles that found on the surface. The pineapple leaf and sea shell was dried under sunlight for about 3 days. After the pineapple leaf and sea shell was fully dried, it was then undergoes alkaline treatment process by immersed pineapple leaf and sea shell into sodium hydroxide, NaOH solution for 24 hours for the purpose of softening and cleaning to remove the impurities that found on the raw materials. After immersion, pineapple leaf and sea shell were removed from sodium hydroxide, NaOH solution and put inside oven at 75 °C to remove moisture and water content of the pineapple leaf and sea shell. Pineapple leaf and sea shell were put inside oven for few days until it was completely dried and pineapple leaf become dry leaf and crispy. Next, when the pineapple leaf and sea shell were completely dried, pineapple leaf was blend by using blender whereas sea shell was grind by using grinding machine. After obtaining powder of two natural fillers, the powders were sieved by using 63 μm siever. After the two natural fillers powders were completely sieved by 63 μm siever, hybrid natural fillers were proceed to the next step which is compounding process. The compounding process was conducted by using twin screw extruder machine. Twin screw extruder was operated by according to the processing parameter based on DOE. After obtaining hybrid filler polymer composites pellets through compounding process, the next step was hot press. The hot press machine was set according to the desired temperature, pressure, time required and other related parameters.

2.2 Design of Experiment (DOE)

DOE was conducted for the compounding process using twin screw extruder where the factors of interest namely screw speed, barrel temperature and filler loading were selected and set out using the Minitab software as listed in Table 2.1.

Table 2.1: The DOE compounding parameters.

PARAMETER	LOW	MIDDLE	HIGH
Sea shell and pineapple leaf fiber loading (wt%)	5	15	25
Speed (rpm)	50	80	120
Max barrel temperature (°C)	180	185	190

A total of nine runs (including a middle run) with different hybrid filler loadings, speed of screw and maximum barrel temperature were formulated as listed in Table 2.2.

Table 2.2 Parameters and formulations of DOE (full fraction factorial design).

Run	Sea Shell and Pineapple Leaf Fiber loading (wt%)	Screw Speed (rpm)	Max Temperature (°C)	Code
1.	5	50	180	5% 50S 180T
2.	5	50	190	5% 50S 190T
3.	5	120	180	5% 120S 180T
4.	5	120	190	5% 120S 190T
5.	25	50	180	25% 50S 180T
6.	25	50	190	25% 50S 190T
7.	25	120	180	25% 120S 180T
8.	25	120	190	25% 120S 190T
9.	15	85	185	15% 85S 185T

2.3 Tensile Test

The tensile test was carried out to analyze the tensile strength, break elongation and young modulus of the polymer composite samples. Tensile test was conducted by using Universal Tensile Machine (UTM), Instron Model 5569, with 50 kN load cell, with a loading speed of 10 mm/ min, the average ambient temperature during the test is 25 ± 2 °C and referring to ASTM 638.

3 Results and Discussion

Based on the tensile data tabulated in Table 3.1, it can be clearly seen that for sample 15% 85S 185T has the highest tensile strength compared to the other runs which was

39.37 MPa. This also indicates that the optimum screw speed was 85 rpm where at this speed, it is believed that the hybrid fillers are well dispesed in the matrix phase. Further, it can be seen that at higher screw speed i.e 120 rpm, the the tensile strength were slightly decrease compared to lower screw speed of 85 rpm. This due to the hybrid fillers could not dispersed homogenously where agglomeration of fillers can be present.

Table 3.1: Tensile strength versus filler loading of PP/pineapple leaf and sea shell hybrid composites under various processing parameters.

No of Runs.	Samples	Tensile Strength (MPa)	Break Elongation (%)	Modulus of Elasticity (MPa)
1.	**5% 50S 180 T**	25.201	5.3	847.2
2.	**5% 50S 190 T**	34.202	9.7	860.2
3.	**5% 120S 180 T**	32.805	7.7	968.6
4.	**5% 120S 190 T**	31.463	7.4	1075.3
5.	**15% 85S 185 T**	39.365	8.7	1128.7
6.	**25% 50S 180 T**	28.383	5.5	835.7
7.	**25% 50S 190 T**	21.364	7.5	710.7
8.	**25% 120S 180 T**	30.758	7.9	930.6
9.	**25% 120S 190 T**	28.555	8.2	1050.2
*	**Virgin PP**	30.982	18.2	731.5

Nevertheless, the lowest tensile strength recorded was for run 25% 50S 190T. The tensile strength obtained was 21.36 MPa which is 7.9% lower than other runs. It can be observed that the tensile strength for samples with lower screw are decreased compared to virgin PP except for run 5% 50S 190T. The lower tensile strength observed is primarily due to the hybrid fillers could not effectively dispersed within the polymer matrix. The lower shear force at lower screw speed could not disperse the hybrid fillers effectively. Moreover, it is believe that this could lead to poor interfacial bonding between hybrid fillers and matrix. It can also be stipulaed that the fibres pullout from the surface and a gap existed between fibres and matrix as a result of poor wetting of fibre which can lead to decrease in tensile strength of the composite [6,7]. Furthermore, the presence of moisture due to poor drying of fillers and voids presence with the composite matrix can also lead to poor composite performance [8, 9]. Other researcher also reported that the polymer composites failed at the interface was due to weak interfacial bonding [10].

The tensile data was then analysed using statistical software for further analysis. The response recorded were tensile modulus, tensile strength and elongation of break where the factors investigated were filler loadings, screw speed and maximum barrel temperature. With the aid of DOE software tools, Pareto charts of effects and main effect plot were constructed. Pareto chart of effects plotted was to determine the magnitude and the importance of the effects. On the Pareto chart, bars that cross the reference line are statistically significant. Main effect plot was used to examine the differences between level means for one or more factors.

IOP Conf. Series: Materials Science and Engineering **957** (2020) 012018 doi:10.1088/1757-899X/957/1/012018

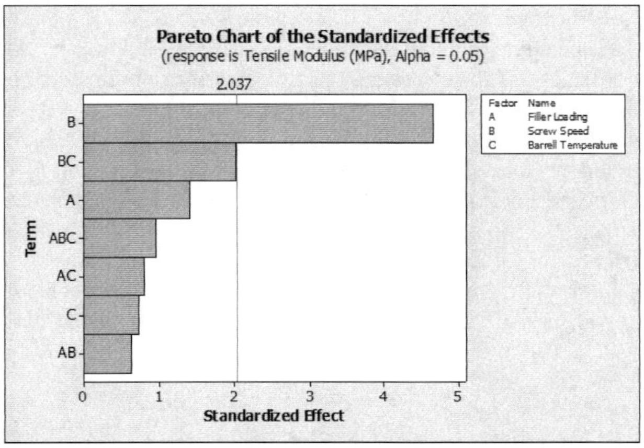

(a)

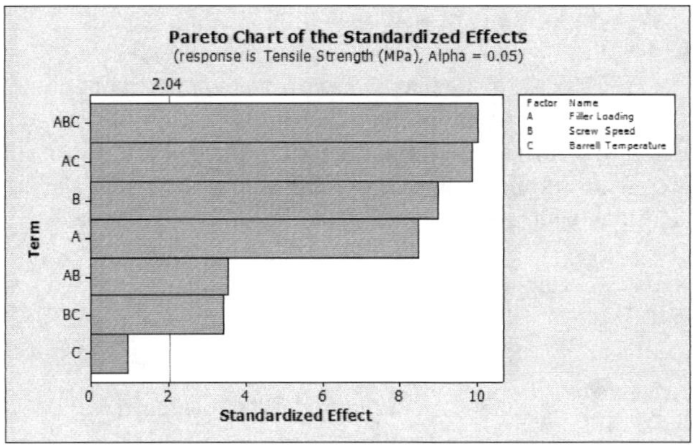

(b)

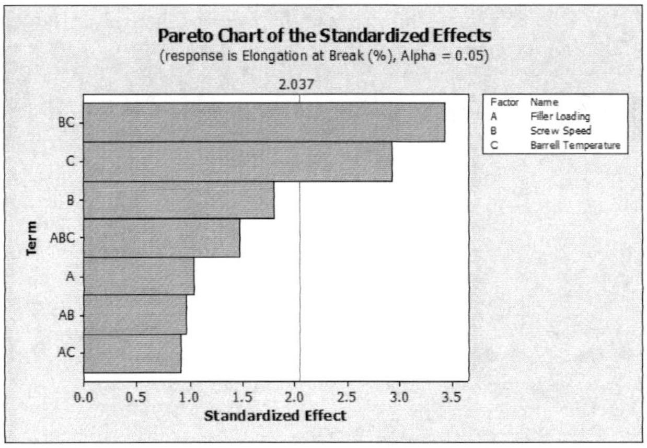

(c)

Figure 3.1: Pareto chart of effects for (a) tensile modulus (b) tensile strength (c) elongation at break of the hybrid polymer composites.

IOP Conf. Series: Materials Science and Engineering **957** (2020) 012018 doi:10.1088/1757-899X/957/1/012018

Figure 3.1 illustrates the samples from Pareto chart of the standardized effects for tensile modulus, tensile strength and elongation at break. On the Pareto chart, the reference line is drawn on the chart and the effects that passed over the reference line is considered as significant. Pareto chart is to analysed the result based on the factors (A, B, C). Based on the Pareto chart, it can be seen the significant of factors that includes two-way interaction and three-way interaction are located over the red line which means the factors or combination of factors are significant. For instance, as illustrated in Figure 3.1 (b), tensile strength is less affected by the barrel temperature if compared to filler loadings and screw speed. Nevertheless, combination of factors (ABC) exhibited as the highest and had significant effect on the tensile strength of polypropylene reinforced with pineapple leaf/seashell hybrid polymer composite materials. Then, followed by the two-way interaction of factors AC, AB and BC and also single factor of A and B. This indicates that the significance of factor A and B had on effect of tensile strength.

Figure 3.2 shows the main effect plot for tensile strength of hybrid polymer composites materials. Main effect plot was can effectively measure what a given variable, independent of the other process variables, has on the overall response being measured. In the main effect plot, the response mean for each of the factors level are connected by a line. The line used is for easier to study the plot diagram, make it to be more obvious and to show the comparison of the tensile strength based on each of the factors. For instance, as illustrated in Figure 3.2, it can be seen that screw speed has the larger main effect than filler loading and barrel temperature for all responses measured. The higher the screw speed, the higher the mean effect for tensile modulus, tensile strength and elongation at break :

The following are the example interpretation of main effect plot for tensile strength as illustrated in Figure 3.2 (b):

(a)	Filler loading	Increasing the hybrid filler loading resulted in decrease in tensile strength.
(b)	Screw Speed	Increasing the screw speed resulted in an increase in tensile strength.
(c)	Barrel temperature	Runs at higher barrel temperature had slight decrease in tensile strenght.

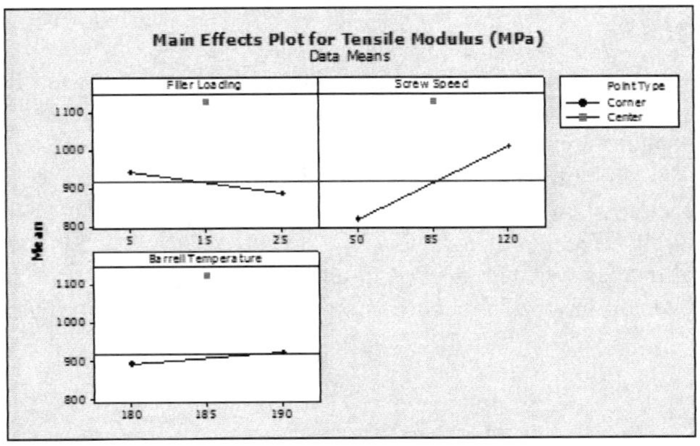

(a)

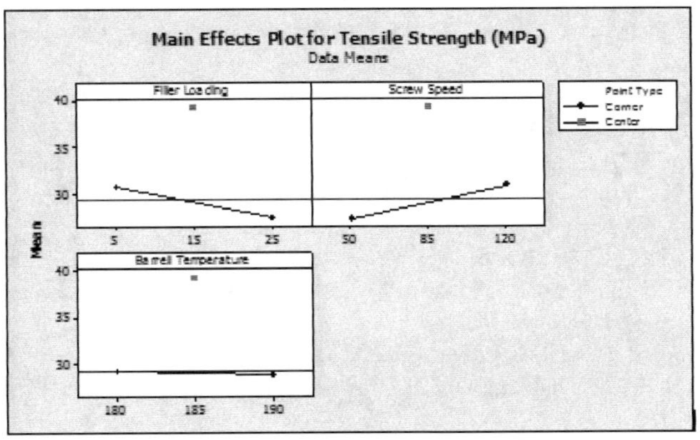

(b)

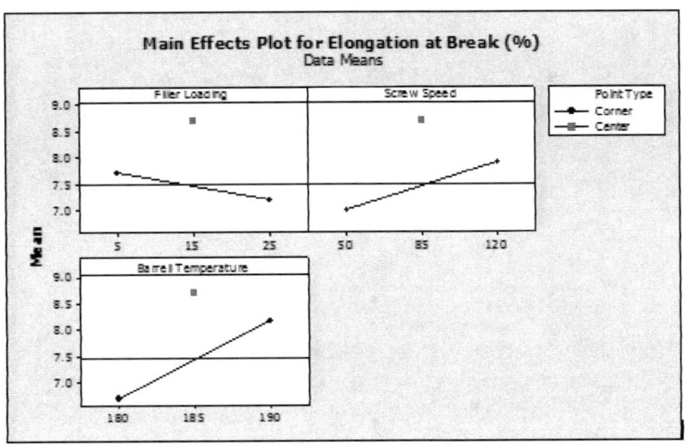

(c)

Figure 3.2: Main effect plot for for (a) tensile modulus (b) tensile strength (c) elongation at break of the hybrid polymer composites.

4 Conclusion

In this study, the effect of twin screw compounding parameters on the tensile properties of hybrid pineapple leaf/sea shell composites was investigated using DOE approach. It can be observed the optimum formulation achieved was for sample from run 15% 85S 185T where the highest tensile strength was recorded. Further, the pareto chart of effect showed the significance of factors where it was found that there were two-way and three-way interactions may exist. The main effect plots shows that by changing or increasing the screw speed and maximum barrel temperature had resulted in an increase in the responses recorded. It can observed that all factors have had influence on the tensile properties of the hybrid composites.

References

1. P. V. Ch. R. K. Santosha, V. Manikanth, Mater. Today, vol. **5**, pp. 5631-5635, (2018).
2. D.N. Saheb, J.P. Jog, Adv Polym Tech, vol. **18**, pp. 351-363, (1999).
3. M. Indra Reddy, P. V. Kumar Raju, Mater. Today, vol. **5**, pp. 5649- 5654, (2018).
4. Muhammad Azeem Munawar, Prog. Nat. Sci., vol. **28**, pp. 266- 274, (2018).
5. K. A. A. Halim, J. B. Farrell, and J. E. Kennedy, Mater. Chem. Phys., vol. **143**, pp. 336-348, (2013).
6. H. W. Chang, R. P. Smith, S. K. Li, A. W. Neumann, pp. 413-421, (1985).
7. IS Aji, ES Zainudin, K Abdan, SM Sapuan and MD Khairul, J. Compos. Mater., vol. **8**, pp. 981-989, (2015).
8. V. Manohara, C. G. Sreenivasa, K. N. Bharath, International Journal of Advanced Research in Mechanical and Production Engineering and Development, vol **1**, (2014).
9. M.R. Sanjay, P. Madhu, Mohammad Jawaid, P. Senthamaraikannan, S. Senthil, S. Pradeep, J. Clean. Prod., vol. **172,** pp. 566-581, (2018).
10. R.M.N. Arib, S.M. Sapuan, M.M.H.M. Ahmad, M.T. Paridah, H.M.D. Khairul Zaman, Mater. Des., vol. **27**, pp. 391-396, (2006).

IOP Conf. Series: Materials Science and Engineering **957** (2020) 012019 doi:10.1088/1757-899X/957/1/012019

The Effect of Polypropylene Grafted Maleic Anhydride on the Polypropylene / Natural Rubber Blends Filled with Cyperus Odoratus

Norfatulikmah Zainal Abidin[1] , and *Mohamad Kahar* Ab Wahab[1,1]

[1]School of Materials Engineering, University Malaysia Perlis, 02600, Arau, Perlis, Malaysia;

Abstract A new thermoplastic blend was developed by using polypropylene (PP) with natural rubber (NR) as the base material. The natural filler is added to enhance the properties of PP/NR blend. Two series of blends under varying conditions were prepared and investigated. The first series was to compare the properties of PP/NR blend with diferent filler loading. Second is to investigate the effect of polypropylene grafted maleic anhydride (PP-MAH) on the properties of PP/NR blend filled with *Cyperus Odoratus*. Process development, tensile properties, thermal analysis, FTIR analysis and morphological characterization of blends were investigated with the variation of the conditions of the blend. The increasing filler loading decrease the tensile properties of the PP/NR blend as support by morphology study of the PP/NR blends. Icorporation of PP-MAH was resulted in improved tensile and thermal properties compared to uncompatibilized counterparts. The increament in tensile strength was confirmed by the SEM study that showed better dispersion of NR in the PP matrix and improved adhesion between the two phases.

[1] Corresponding author: kaharwahab@unimap.edu.my

IOP Conf. Series: Materials Science and Engineering **957** (2020) 012019 doi:10.1088/1757-899X/957/1/012019

1 Introduction

Cyperus Odoratus can be considered as natural filler which contain lignocellulosic fibers. The major source of the lignocellulosic fibers or similar feedstock can come from recycling fiber based products, such as paper, waste wood and agricultural residues. The main advantages of using lignocellulosic fibers as an additive in thermoplastics is the low density, biodegradable, abundant, low cost, high specific stiffness and strength, desirable fiber aspect ratio, less abrasive nature to the equipment during process renewable, high filling levels possible, low energy consumption and wide variety of fibers available throughout the world [1-5]. Moreover, they can attract considerable interest as reinforcing fillers for thermoplastics especially those with a relatively lower melting point like polypropylene, high and low density polyethylene [6-8]. The properties deterioration of incorporation of *Cyperus Odoratus* into PP/NR blends is due to the difference between the surface properties of the natural filler and matrix. The natural filler is highly polar and hydrophilic while the matrix is generally non-polar and relatively hydrophobic [9-11]. Without compatibilizer, natural fillers embedded in a polymeric matrix generate unstable interfaces and the stress applied to the filler/polymer composite is not efficiently transferred from the matrix to the filler remains underexploited. The poor ability of the polymer to wet the filler hinders the homogenous dispersion of fillers within the polymeric matrix [12]. Therefore it is important to improve compatibility of the two polymeric phases of the blend to achieve better homogeneity and morphology for improved and desired mechanical properties [13-15]. Compatibilizing system that use polypropylene grafted maleic anhydride (PP-g-MAH) and carboxylated acrylonitrile butadiene rubber (NBR-RCOOH) as a compatibilizer was investigated by Soares and co workers [16] and found that there is an improvement in tensile and swelling properties of PP/NBR blends.

2 Experimental

2.1 Materials

The raw material that were used in mixing and compounding used in this study such as PP, NR, Cyperus Odoratus, ZnO, CBS, stearic acid, sulphur and PP-g-MAH are listed completely with the function and supplier in the Table 1.

Table 1: List of raw materials, their functions and suppliers

Raw materials	Function	Supplier
SMR L	Elastomer	RRIM Guthrie Group Sdn. Bhd.
Polypropylene	Plastic	Titan Polymer Sdn. Bhd
Cyperus Odoratus	Natural filler	Paddy field
Zinc oxide	Activator	Anchor Chemical Co (M) Ltd.
CBS	Accelerator	Anchor Chemical Co (M) Ltd.
Stearic acids	Activator	Anchor Chemical Co (M) Ltd.
Suphur	Vulcanizing agent	Anchor Chemical Co (M) Ltd.

2.2 Preparation of the sample

2.2.1 Preparation of PP/NR/ Cyperus Odoratus composites

Blends were prepared using heated two roll mill. Mixing was done at temperature 170°C and rotor speed 13rpm and 10 rpm for front and back mill for 12 min. The effect of compatibilizer of polypropylene grafted maleic anhydride (PP-g-MAH) on developed PP/NR composite. 3 wt% of the compatibilizer will be added. The mixing was carried out by using heated two roll mill with temperature 170ºC for 12 min. Samples of the blends were compression molded in an electrically heated hydraulic press. Procedure of moulding of all blends mentioned above except blends of PP/NR involved preheating of the sample for 3 min at 170°C, followed by compressing (pressure of 10 MPa) of the mould for 7 min at the same temperature and subsequently cooling under pressure for another 5 min.

3 Characterization and Testings

Tensile test were carried out according to ASTM D638 using the Instron tensile testing machine. A cross head speed of the tensile machine was maintained at 10mm/min and tests were performed at 25± 3°C. Stress at peak (MPa), Young's modulus (MPa) and elongation at break (%) were automatically calculated by the software. The average properties were calculated after taking the statistical average value of at least five measurements for each blends. Series of Fourier-transform infrared spectroscopy (FTIR) spectra were recorded with the IR spectrometer in the range of 550-4000cm^{-1} at resolution of 4 cm^{-1} using a Perkin-Elmer 2000 series instrument. All spectra were scanned for four times before the spectrum is confirmed. Thermogravimetric analysis (TGA) was carried out by using Perkin- Elmer Pyris 6 TGA analyzer with temperature range from 30°C to 600°C at a nitrogen flow of 50 ml/min and at a heating rate of 10 °C/min. Lost of weight in the sample was then recorded automatically from the analyzer. Morphology analysis for the blends tensile fracture surface was carried out using a scanning electron microsope (SEM) model JEOL JF6460LA. Samples that had already undergone tensile test were collected and subjected to the scanning electron microscope (SEM). All the test specimens were taken from tensile fracture with maximum care to avoid any damage and contamination that would affect the result. Fractured surfaces of the specimens were coated with a thin gold layer of about (1.5-3.0nm) thickness to avoid electrostatic charging during examination and poor image resolution. Differential scanning calorimetry or DSC is a thermoanalytical technique in which the difference in the amount of heat required to increase the temperature of a sample and reference are measured as a function of temperature. Both the sample and reference are maintained at nearly the same temperature throughout the experiment. DSC measurements were carried out under nitrogen on samples of 5-10 mg using a Perkin-Elmer DSC-7 apparatus. The samples were placed in standard aluminium pans with pierced lids. The scanning rate was of 10 °C/min. The first heating scan was carried out from 30-190°C and hold for 5 mins then 190-30 °C and hold for 5 mins then 30-230 °C for the determination of the glass transition temperature, $T_{g.}$

4 DISCUSSION

4.1 Structural analysis

Fourier transform infrared is used to determine the functional group in PP/NR blend with ratio (70:30). The FTIR spectra analysis of PP/NR blend is shown in Figure 1 below. From the figure, it can be seen several bands that indicates the functional group in the blend system. First, the shoulders observed in the region of 3100-3010cm^{-1} shows the =CH stretch and due to the presence of alkene groups in natural rubber. A strong absorption band appears at the range of 2950-2800cm^{-1} corresponds to the stretching vibration of C-H bonds that indicates

IOP Conf. Series: Materials Science and Engineering **957** (2020) 012019 doi:10.1088/1757-899X/957/1/012019

alkanes groups which is dominantly by polypropylene structure in the blend system [17]. Finally, the existence of strong peak at range of 1465-1537cm^{-1} represents the CH_2 bending. The different bands at 1712 and 1637 cm^{-1} which are associated with the stretching vibrations of the C=O and C=C groups of the acrylic, respectively [18]. The broad shape of the C=O peak suggests that this group is in interaction with the surface through hydrogen bonds with the hydroxyl groups [19]. The extent of carbonyl groups is observed in the region of 1712-1860 cm^{-1} with small peaks and shoulders detected in the spectra of the blend system [20]. Figure 1 shows the FTIR spectra of PP/NR-MAH with 10 wt% filler loading. The spectrum of grafted polypropylene (PP-g-MAH) compared to the one without compatibilizer, shows the appearance of an absorption band at 1790 cm^{-1}. This band which attributed to the symmetric and asymmetric stretching vibration of C=O group, proves that the grafting took place.

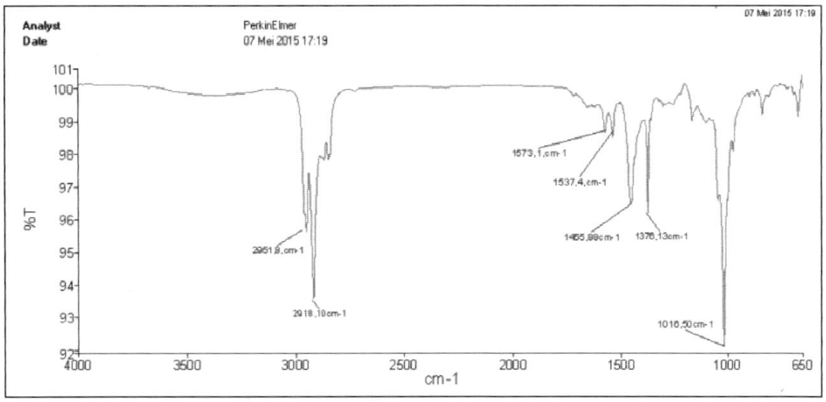

Fig. 1. FTIR spectra of PP/NR-MAH with 10 wt%

4.2 Tensile properties

The properties of thermoplastic rubber blends are dependent on the material properties of individual phases, blend ratio, the phase morphology and the interactions at the interface [21]. The tensile properties of immiscible polymer blends are affected by the addition of compatibilizers. The tensile properties namely tensile strength, Young's modulus and elongation at break of uncompatibilized blends of PP/NR filled with *Cyperus Odoratus* and compatibilized blends of PP/NR with maleic anhydride are demonstrated in Figures 2(a) (c) respectively. The tensile properties are found to be improved with addition of PP-g-MAH into the blends. The trend of the tensile strength when there is higher in filler loading is increasing from 14.99 MPa to 15.87 MPa. Then the tensile strength decreases as higher loading filler to 12.08 MPa and 11.5 MPa. The highest tensile strength is when 10 wt% of filler loading is incorporated into the blend system. The tensile strength of blend system that is added with compatibilizer is higher compared to PP/NR blend with filler loading. The increasing in tensile properties is due to the increased dipolar interaction between the PP-g-MAH and NR phase, which causes reduced in interfacial tension between the PP and NR phases [22-23]. However, the tensile strength is at highest value when 10 wt% of natural filler is added into the blend. This shows the highest value of natural filler into the blend system. As the filler loading increases, the tensile strength decreases because the filler doesn't interact with the matrix. The filler does not facilitate the stress transfer to the matrix. The effect of compatibilizer on the Young's modulus can be seen from Figure 2(b). From the graph, the highest Young's modulus is at 10 wt% filler loading with the presence of compatibilizer which is 596.3 MPa. The lowest is 400.8 MPa when 30 wt% filler loading. Studies by Pukansky [24] said that Young's modulus was measured during the process of elastic deformation in the composites where the movement of at the filler-matrix interfaces was very limited. When the polymer is filled with rigid particles, the polymer itself will contribute to the elasticity of the composites. The increase in the modulus of the composites due to the alignment of the polymer, bonding of the filler to the matrix and the fraction of filler packing. The effect of PP-g-MAH on elongation at break of PP/NR blend filled with Cyperus Odoratus is illustrated in Figure 4.16. Generally, adding a rigid filler into the PP/NR will decreased the elongation at break. From the Figure 2(c), that unfilled PP/NR exhibits ductile failure with the elongation at break 29.05%. However, with the incorporation of PP-g-MAH into filled PP/NR blend, the elongation at break has slightly improved compared to filled PP/NR without compatibilizers. The addition of PP-g-MAH into the same composition has increased the elongation at break from 17.76% to 26.5% and 27.4% respectively. Enhancement in ductility could be due to the presence of maleic anhydride in PP-g-MAH which acts as dispersing agents for Cyperus Odoratus into the matrix. Consequently, the ductility of Cyperus Odoratus filled PP/NR blend can be improved.

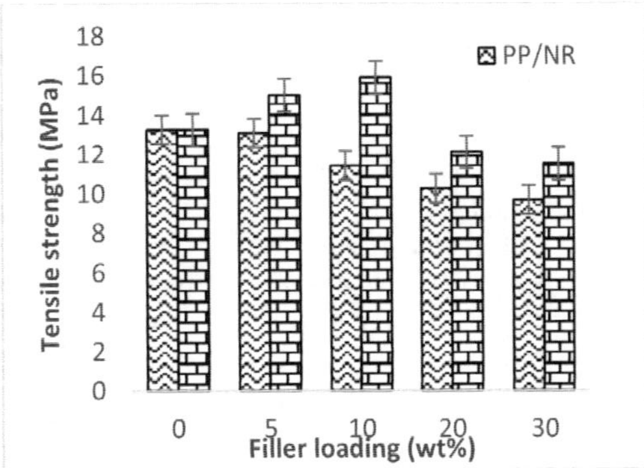

(a)

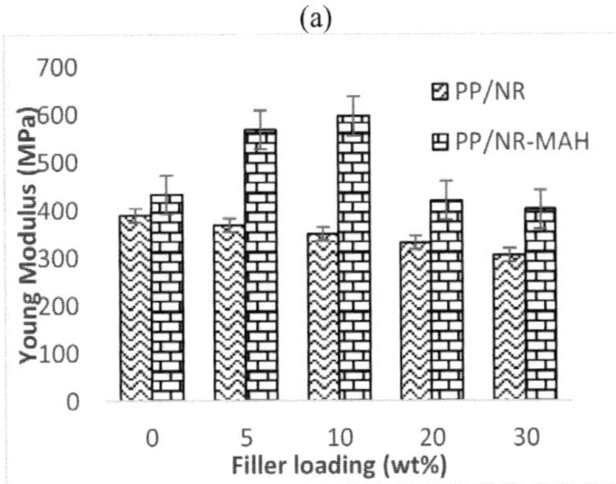

(b)

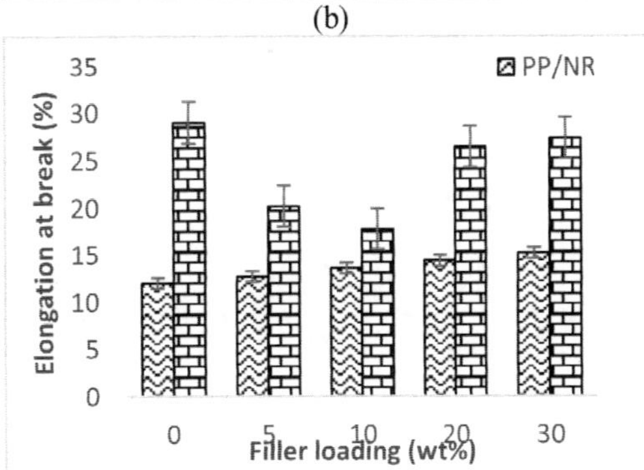

(c)

Fig. 2. Tensile properties of PP/NR and PP/NR-MAH composites.

4.3 Morphological Study

The effect of PP-g-MAH as a compatibilizer on the morphology of the PP/NR blends was investigated by scanning electron microscopy and the micrographs of tensile fractured surfaces of PP/NR with filler loading at 10% and 30% are illustrated in Figure 3 and 4. From the figure we can see that the blend system with PP-g-MAH as compatibilizer exhibit better dispersion of the fiber in the matrix. The fiber is well embedded in the continuous phase of the matrix. This shows a good adhesion between the natural filler and the matrix, PP. Generally polypropylene is categorized as a non-polar molecule while *Cyperus Odoratus* is polar molecule. PP becomes polarized by grafting with maleic anhydride. The addition of PP-g-MAH into the blend caused to increase dipolar interaction between the filler and matrix. Therefore, it could be achieved better adhesion and better dispersion of PP/NR and natural filler by compatibilizing the blends with PP-g-MAH.

IOP Conf. Series: Materials Science and Engineering **957** (2020) 012019 doi:10.1088/1757-899X/957/1/012019

(a)

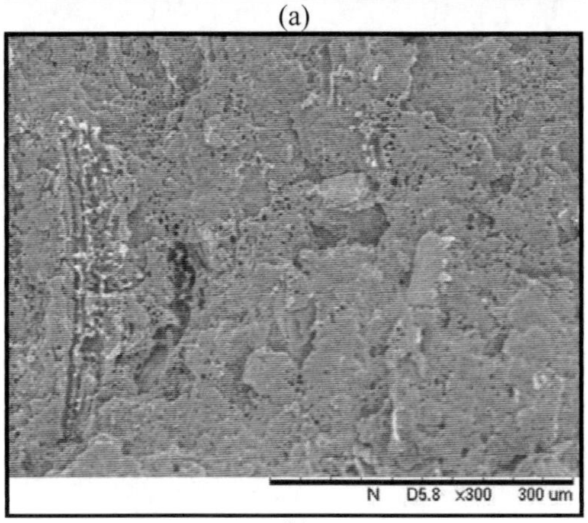

(b)

Fig. 3. Scanning electron micrograph of surface of PP/NR-MAH blend with 10 wt% filler loading at (a) x150 magnification (b) x300 filler magnification

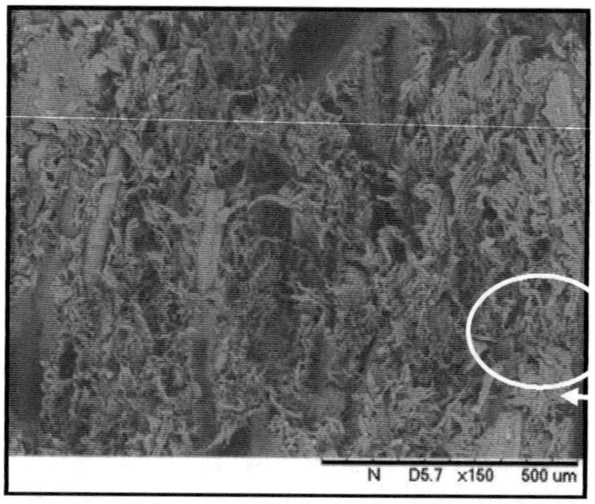

(a)

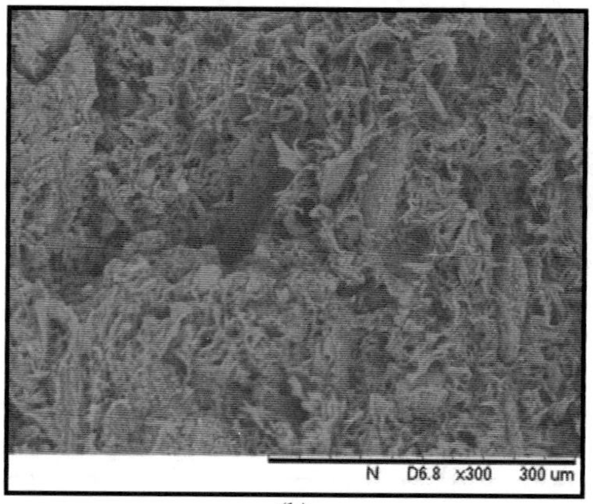

(b)

Fig. 4. Scanning electron micrograph of surface of PP/NR-MAH blend with 30 wt% filler loading at (a) x150 magnification (b) x300 magnification

4.4 Thermal stability

The Figure 5 (a) and (b) illustrated the TGA and DTG curves of PP/NR blend with filler loading and filled PP/NR blend with the presence of filler loading. The degradation process starts at 20 °C and complete at 580 °C. From the graphs, it can be seen that filled PP/NR with the presence of compatibilizer have more resistance against degradation and have a better

IOP Conf. Series: Materials Science and Engineering **957** (2020) 012019 doi:10.1088/1757-899X/957/1/012019

thermal stability compared to filled PP/NR blend. The control specimen has the highest weight loss followed by PP/NR blend with 10 wt% filler loading without the presence of compatibilizer. At the final decomposition point, total weight loss for of PP/NR blend with 30 wt% filler loading with the presence of PP-g-MAH lower compared to PP/NR blend with 10 wt% of filler loading. With the addition of compatibilisers, PP-g-MAH, the end degradation temperature were improved further. In this point of view, it can be concluded that the thermal stability of PP/NR blend filled with *Cyperus Odoratus* was improved with the addition of compatibilisers.

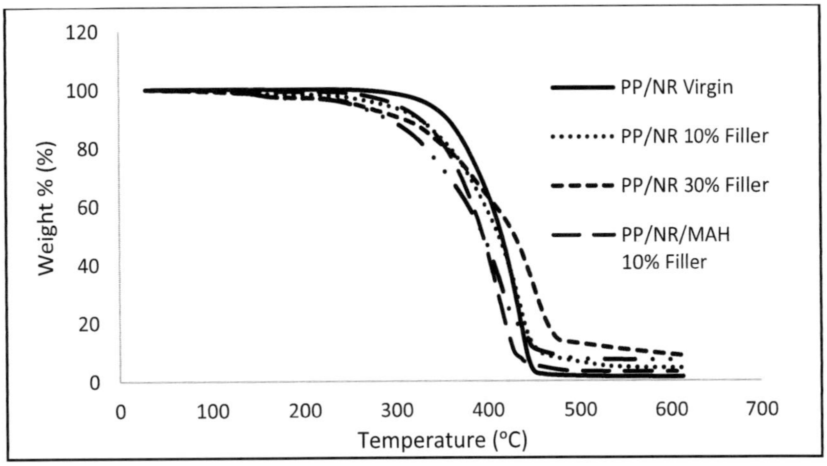

Fig. 5(a). The TGA curves of PP/NR blend with filler loading and PP/NR-MAH with filler loading

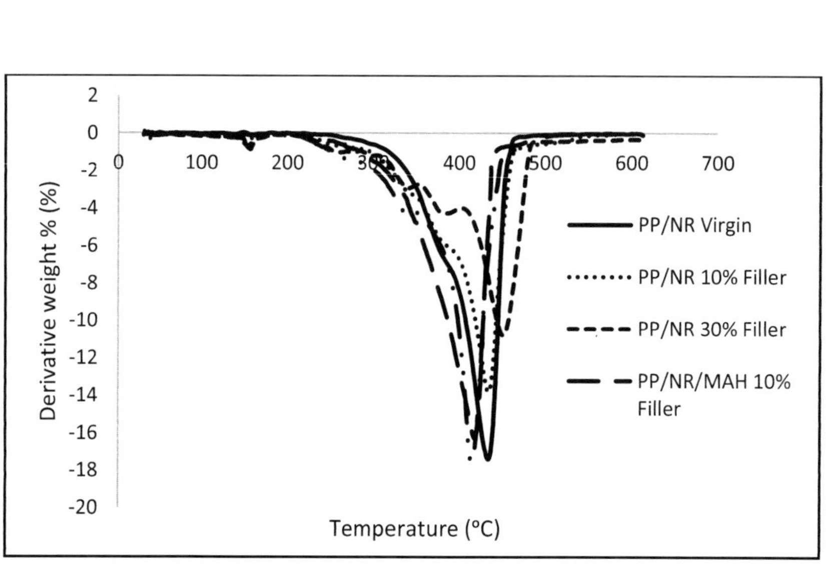

Fig. 5(b): The DTG curves of PP/NR blend with filler loading and PP/NR-MAH with filler loading

4.5 Differential Scanning Calorimetry (DSC)

The effect of compatibilisers (PP-g-MAH) on the blend can be seen as illustrated in Figure 6 (a) and (b) respectively. The effect of compatibilisers on thermal properties of PP/NR blend with filler loading been analysed in isothermal DSC experiments. In this study, details of themogram obtained for various compositions of filler loading with and without PP-g-MAH have been plotted. Compared to the virgin PP/NR blend the PP/NR blend with filler loading have lower crystallinity. Among the filled PP/NR blend composition the addition of PP-g-MAH has significant effect on increasing the crystallization and melting temperature, while the filled PP/NR blend without PP-g-MAH does not show significant effect on melting and crystallization temperatures of composites. Judging from the decreasing height of the peaks compatibilised filled PP/NR blend as compared to control, it appears that the incorporation of the compatibilisers reduce the level of crystallinity of PP/NR blend. Slower crytallisation rate of compatibilised filled PP/NR blend is observed by broadening of the crystallisation peak and reduction of the melting pont. Other reason for the reduction of the melting temperture of all compatibilised filled PP/NR blend due to the disturbance of crystallinity formation induced by the filler in the system during coolong process.

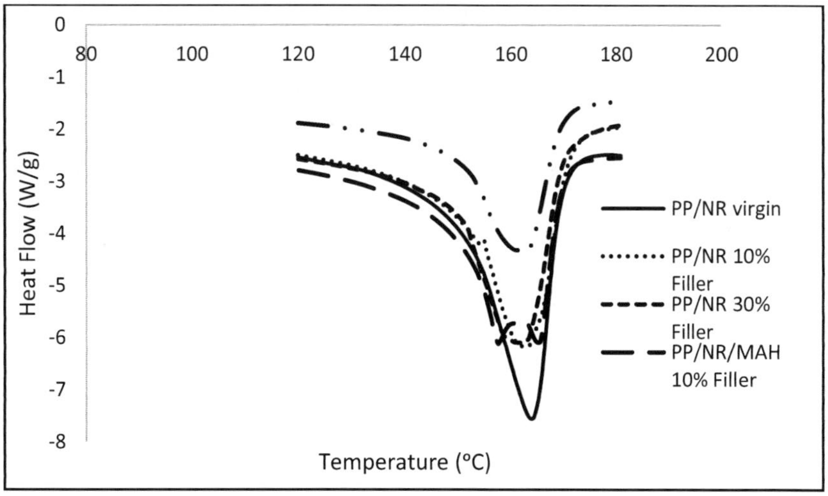

Fig. 6(a): The heating curve of PP/NR blend with filler loading and PP/NR-MAH with filler loading

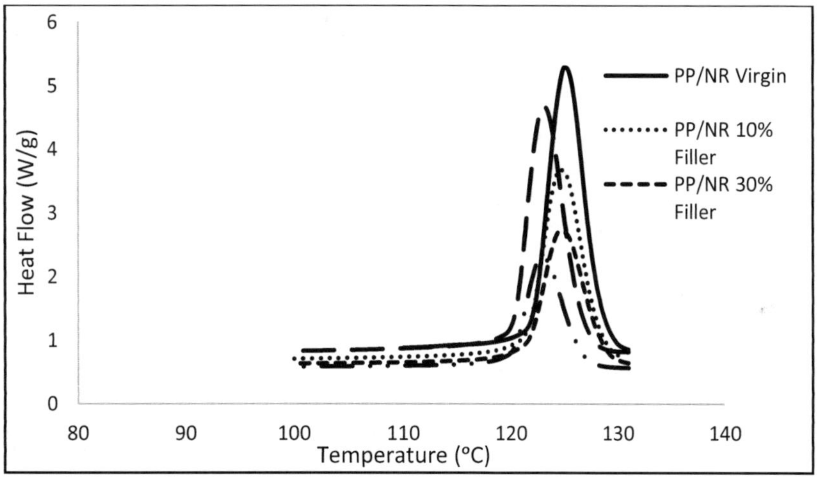

Fig.6 (b): The cooling curve of PP/NR blend with filler loading and PP/NR-MAH with filler loading

5 Conclusions

The properties analyzed from the tensile properties shows that the increasing of filler loading decreased the tensile strength. The Young's modulus also decreased with the increasing of filler loading however the elongation at break increases. Morphology of the blends indicates the filler is does not dispersed well in the PP matrix. Since PP/NR blends and natural filler are incompatible, an attempt was made to compatibilize in order to improve the morphology and the properties of the blends. Polypropylene graft maleic anhydride (PP-g-MAH) were used to compatibilize the blend. The compatibilizing of filled PP/NR with PP-g-MAH resulted in better tensile properties better than the uncompatibilized blends. A finer morphology with improved interfacial adhesion is obtained with all compatibilized blends.

The authors wish to gratefully acknowledge the School of Materials Engineering, as well as Research and Innovation Department under Universiti Malaysia Perlis and Kementerian Pengajian Tinggi Malaysia (Ref: FRGS/1/2016/TK05/UNIMAP/02/9) for supporting this research project.

References

1. X. Yuan, Y. Zhang, X.J. Zhang, J. Appl. Polym. Sci., 7, 333. (1999)
2. X. Colom, F. Carasco, P. Pages, J. Canavate, Compos. Sci. Technol., 63, 161. (2003).
3. A.W.M. Kahar, M. Lingeswarran, M.Z. Amirah Hulwani, H. Ismail, Polym. Bull., 76(2), 747-762, (2019)
4. A.W.M. Kahar, H. Ismail., N. Othman, J. Vinyl Add. Technol.18(3), 192-197, (2012)
5. D.E.E. Nashar, Abdul El-Messieh, A.H.J. Basta, J. Appl. Polym. Sci., 9, 3410. (2004)
6. J.J. Son, Thermoplast. Compos. 17, 509. (2004).
7. R. J. Walster, R. Rahman, A. W. M. Kahar. IOP Conf. Series: Mater. Sci. Eng., **429,** 012043, (2018).
8. H. Zaman, J. Song, L.S. Park, I.K. Kang, S.Y. Park, G. Kwak, K.B.Yoon. Polym. Bull. **67(1)**:187-198, (2011).
9. A. W. M. Kahar, H. Ismail, A. Abdul Hamid, A J. Therm. Anal. Calorim., **123(1),** 301-308, (2016).
10. N. Ibrahim, A.W.M. Kahar, D.L Ngoc, H. Ismail, H. Bio-res., **12(2),** 3076-3087, (2017).
11. A.H. Mohd Zain, A.W.M. Kahar, N.Z. Noriman, Procedia Chem., 19, 638-645, (2016)
12. F. Vilaplana, E. Strömberg, S. Karlsson, Polym. Degrad. Stabil., **95(11**), 2147–2161, (2010).
13. A. Joseph, S. George, K. Joseph, S. Thomas, J. Appl. Polym. Sci., 102, 2067-2080, (2006).
14. A.W.M. Kahar, H. Ismail. J. Vinyl Add. Technol. **22(3),** 191-199, (2016)
15. A.W.M. Kahar, N. Sarifuddin, H. Ismail, H., Ira. Polym. J., **26** (**2**), 149-159, (2017).

16. B.G. Soares, M.S.M. Almeida, M.V. Deepa Urs, G.N. Kumaraswamy, C. Rangathaiah, M.R. Siddaramaiah, J. Appl. Polym. Sci., 102, 4672-4681, (2006).
17. A.H. Mohd Zain, A.W.M. Kahar, I. Hanafi, J. Polym Environ, **26(2),** 691–700, (2018).
18. H. Salmah, H., Ismail, A. Bakar, J. Appl. Polym. Sci., **107(4),** 2266-2273. (2007).
19. M. Abdelmouleh, S. Boufi, M.N. Belgacem, A. Dufresne, Compos. Sci. Technol. 67, 1627-1639. (2007)
20. N.F. Alias, H. Ismail, A.W.M. Kahar,. Bio-res, **12**(4):9103-9117. (2017)
21. O.S. Dahham, N.Z. Noriman, A.W.M. Kahar, H. Ismail, S.T. Sam, J. Appl. Sci. & Agric **10** (5), 33-39
22. A.H. Mohd Zain, A.W.M. Kahar, H. Ismail, Mater. Resear. Exp., **6** (10), 105335, (2019).
23. H. Liu, R. Adhikari, Q. Guo, B. Adhikari, J. Food Eng., **116**(2), 588–597, (2013).
24. A.W.M. Kahar, H. Ismail., N. Othman, J. Appl. Polym. Sci., **128** (4), 2479–2488.

IOP Conf. Series: Materials Science and Engineering **957** (2020) 012020 doi:10.1088/1757-899X/957/1/012020

Dynamic Vulcanization of Polypropylene / Natural Rubber Blends Filled with Cyperus Odoratus

Norfatulikmah Zainal Abidin[1], and *Mohamad Kahar* Ab Wahab[1,*]

[1]School of Materials Engineering, University Malaysia Perlis, 02600, Arau, Perlis, Malaysia;

Abstract. A new thermoplastic blend was developed using polypropylene (PP) with natural rubber (NR) as the base material. The natural filler is added to enhance the properties of PP/NR blend. The first series was to compare the properties of PP/NR blend with different filler loading. Second series was to investigate the effect of dynamic vulcanization on the properties of PP/NR blend filled with *Cyperus Odoratus* with the intention of improving the blend properties. Process development, tensile properties, thermal analysis, FTIR analysis and morphological characterization of blends were investigated with the variation of the conditions of the blend. The increasing filler loading decrease the tensile properties of the PP/NR blend and it shows in the morphology study of the blend. Dynamic vulcanization of the blend resulted in improved tensile and thermal properties compared to unvulcanized counterpart. This result is confirmed by the SEM study that showed better dispersion of NR in the PP matrix and improved adhesion between the two phases.

1 Introduction

Thermoplastic elastomers (TPEs) are materials that are processed by the same method used for a rigid thermoplastic such as injection molding and extrusion, yet have the properties and performance normally describe to conventional thermoset rubber. The importance of TPEs has been gained due to their wide range of application in such as automotive parts, industrial supplies, electrical equipment, household appliances, food contact systems and medical apparatus [1-5]. It already started replacing many other conventional materials in various applications and have become a highly demanding class of polymeric materials. Some of the properties of TPEs which may not be found in any other materials are low cost and attractive properties such as light weight, superior mechanical strength, applicability at elevated

* Corresponding author: kaharwahab@unimap.edu.my

temperatures, corrosion resistance, ability to be tailored for specific engineering applications. The most important feature of TPEs is the repeated recyclability up to several times without significant loss of properties [6-8]. TPEs can be classified into two major groups which are block copolymers and rubber plastic blends. A class of TPEs based on rubber and thermoplastic compositions where the rubber phase is dynamically vulcanized is known as thermoplastic vulcanizates (TPVs) or dynamic vulcanizate. Dynamic vulcanization is the process of mixing a thermoplastic and rubber which is later cross-linked under dynamic conditions. In order to activate and to complete the vulcanization process, the process is performed at high shear rate above the melting temperature of the thermoplastics and also at sufficiently high temperature [9-10]. Due to the advantages of being low in both density and the cost, the product based on PP are very significantly commercial. Moreover, its crystalline structure and high melting point results in resistance to solvent and high temperature [11-13]. While natural rubber (NR) exhibits outstanding properties such as green strength and tensile strength because it can crystallize spontaneously when it is strained. However, for some specific applications, the properties of natural rubber such as modulus, hardness and abrasion resistance need to be improved [14-15]. The best choice of polyolefin to be blended with NR is polypropylene (PP) due to its high softening temperature 150°C and the low glass transition temperature of the blend [16].

2 Experimental

2.1 Materials

The raw materials that were used in mixing and compounding used in this study such as PP, NR, Cyperus Odoratus, Zinc oxide, CBS, stearic acid, sulphur and PP-g-MAH are listed completely with the function and supplier in the Table 1.

Table 1: List of raw materials, their functions and suppliers

Raw materials	Function	Supplier
SMR L	Elastomer	RRIM Guthrie Group Sdn. Bhd.
Polypropylene	Plastic	Titan Polymer Sdn. Bhd
Cyperus Odoratus	Natural filler	Paddy field
Zinc oxide	Activator	Anchor Chemical Co (M) Ltd.
CBS	Accelerator	Anchor Chemical Co (M) Ltd.
Stearic acids	Activator	Anchor Chemical Co (M) Ltd.
Suphur	Vulcanizing agent	Anchor Chemical Co (M) Ltd.

2.2 Preparation of the sample

2.2.1 Preparation of PP/NR blends via dynamic vulcanization approach

Blends were prepared using heated two roll mill. Mixing was done at temperature 170°C and rotor speed 13rpm and 10 rpm for front and back mill for 12 min. The effect of dynamic vulcanization (DV) on PP/NR blends filled with *Cyperus Odoratus* is determined. The curatives is added to the compounding based on the formulation given 1n Table 2 below. The mixing is still using heated two roll mill with temperature 170ºC for 12 min. Samples of the blends were compression molded in an electrically heated hydraulic press. Procedure of moulding of all blends mentioned above except blends of PP/NR involved preheating of the sample for 3 min at 170°C, followed by compressing (pressure of 10 MPa) of the mould for 7 min at the same temperature and subsequently cooling under pressure for another 5 min.

Table 2: Formulation table for dynamic vulcanization on PP/NR blends filled with *Cyperus Odoratus*

Ingredients			**Compositions (wt%)**		
PP/NR (70/30)	100	95	90	80	70
Cyperus Odoratus	0	5	10	20	30
Curatives					
Zinc oxide*			1.5		
Stearic acid*			0.6		
CBS*			0.75		
Sulphur			1.0		

*phr based on weight of NR

3 Characterization and Testings

Tensile test were carried out according to ASTM D638 using the Instron tensile testing machine. One millimeter thick dumbell tensile specimens were cut from the molded sheets with a Wallace die cutter. A cross head speed of the tensile machine was maintained at 10mm/min and tests were performed at 25± 3ºC. Stress at peak (MPa), Young's modulus (MPa) and elongation at break (%) were automatically calculated by the software. The average properties were calculated after taking the statistical average value of at least five measurements for each blend. Series of Fourier-transform infrared spectroscopy (FTIR) spectra were recorded with the IR spectrometer in the range of 550-4000cm^{-1} at resolution of 4 cm^{-1} using a Perkin-Elmer 2000 series instrument. All spectra were scanned for four times before the spectrum is confirmed. Thermogravimetric analysis (TGA) was carried out by using Perkin- Elmer Pyris 6 TGA analyzer with temperature range from 30°C to 600°C at a nitrogen flow of 50 ml/min and at a heating rate of 10 °C/min. Loss of weight in the sample was then recorded automatically from the analyzer. Morphology analysis for the blends tensile fracture surface was carried out using a scanning electron microsope (SEM) model JEOL JF6460LA. Samples that had already undergone tensile test were collected and subjected to the scanning electron microscope (SEM). All the test specimens were taken from tensile fracture with maximum care to avoid any damage and contamination that would affect the result. Fractured surfaces of the specimens were coated with a thin gold layer of about (1.5-3.0nm) thickness to avoid electrostatic charging during examination and poor image resolution. Differential scanning calorimetry or DSC is a thermoanalytical technique in which the difference in the amount of heat required to increase the temperature of a sample and reference are measured as a function of temperature. Both the sample and reference are maintained at nearly the same temperature throughout the experiment. DSC measurements were carried out under nitrogen on samples of 5-10 mg using a Perkin-Elmer DSC-7 apparatus. The samples were placed in standard aluminium pans with pierced lids. The scanning rate was of 10 °C/min. The first heating scan was carried out from 30-190°C and hold for 5 mins then 190-30°C and hold for 5 mins then 30-230°C for the determination of the glass transition temperature, T_g.

4 Discussion

4.1 Structural analysis

Fourier transform infrared (FTIR) is used to determine the functional group in PP/NR blend with ratio (70:30). The FTIR spectra analysis of PP/NR blend is shown in Figure 1 below. From the figure, it can be seen several bands that indicates the functional group in the blend system. First, the shoulders observed in the region of 3100-3010cm^{-1} shows the =CH stretch and due to the presence of alkene groups in natural rubber. A strong absorption band appears at the range of 2950-2800cm^{-1} corresponds to the stretching vibration of C-H bonds that indicates alkanes groups which is dominantly by polypropylene structure in the blend system [17]. Finally, the existence of strong peak at range of 1465-1537cm^{-1} represents the CH_2 bending [18]. From the Figure 2, it shows the FTIR spectrum from the PP/NR blend filled with 10wt % of filler loading which is *Cyperus Odoratus*. The different bands at 1712 and 1637 cm^{-1} which are associated with the stretching vibrations of the C=O and C=C groups of the acrylic, respectively [19]. The broad shape of the C=O peak suggests that this group is in interaction with the surface through hydrogen bonds with the hydroxyl groups [20]. The extent of carbonyl groups is observed in the region of 1712-1860 cm^{-1} with small peaks and shoulders detected in the spectra of the blend system [21]. The FTIR spectra analysis of and vulcanized PP/NR blend with 10 wt% filler loading can be seen from the Figure 3. From the figure we can see the stretching of CH3 at the region 2950-2800 cm-1. The peaks appeared in nearly 1715 cm-1 and 3345 cm-1 indicate the presence of carbonyl and hydroxyl/hydroperoxides group, respectively. However, it can be seen from the FTIR spectra between unvulcanised PP/NR with filler loading with PP/NR-DV with filler loading are almost identical indicating that there is no significant effect of vulcanization on the chemical changes during the processing. In the sulphur-cured system, S-S linkage are formed due to the presence of sulphur as vulcanizing agent [22].

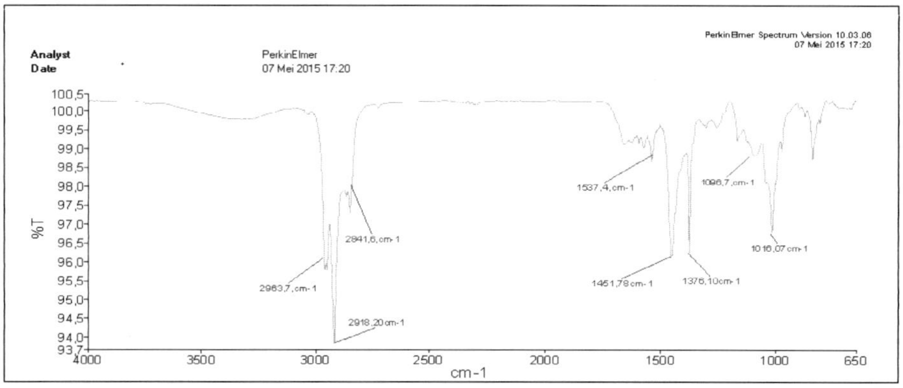

Fig. 1. FTIR spectra of virgin PP/NR (70:30) blend

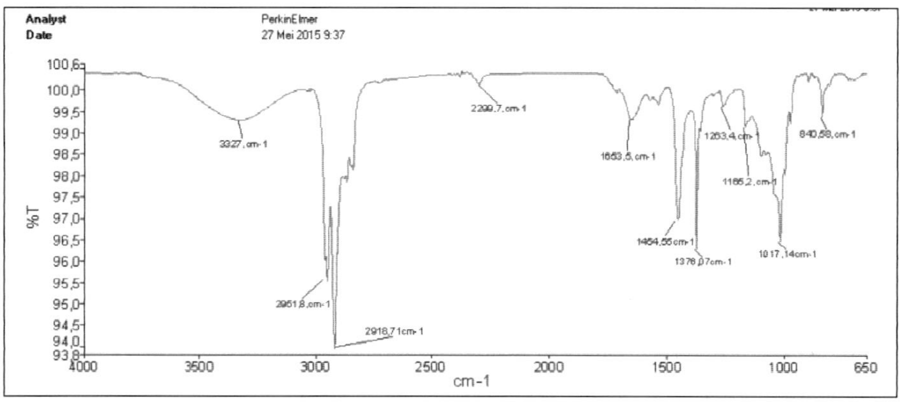

Fig. 2. FTIR spectra of PP/NR blend with 10 wt% of filler loading

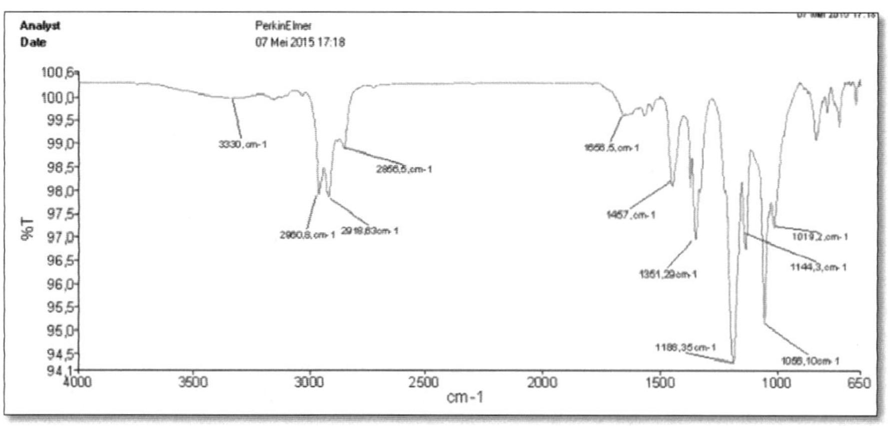

Figs 3. FTIR spectra of vulcanized PP/NR blend with 10 wt% of filler loading

4.2 Tensile properties

Figure 4 (a) shows comparison of tensile strength between unvulcanised PP/NR blend with filler loading and vulcanized PP/NR blend with filler loading. For PP/NR blend with 5 wt% filler loading, the tensile strength is 13.09 MPa and further decrease with filler loading 10 wt% that is 11.40 MPa. The lowest tensile strength is 9.66 MPa with incorporation of filler loading 30 wt%. From the graph, it can be seen that the vulcanized blend has higher tensile strength compared to unvulcanised blend system. The incorporation of 5 wt% of filler loading has shown slight increment compared to unvulcanised system which is 13.12 MPa. When the blend is vulcanized with 10 and 20 wt% filler loading the tensile strength decrease from 11.74 MPa to 11.38 MPa respectively. When dynamic vulcanization is performed during the mixing process, the viscosity of the mixture will increase. This will cause higher increment of friction during the mixing process so that particle will become smaller and the dispersion of NR phase in PP matrix are evenly distributed. Small size dispersion and the nature of cross-linked particles distribution facilitate the further initiation and movement of the matrix flow. Even though the tensile strength of vulcanized system is higher compared to unvulcanised system but the tensile strength for both systems decreased with the increasing amount of filler loading. This is due to the occurrence of filler agglomeration in the blend system. The second reason is the difficulty of achieving homogenous blend system by using roll-mill at higher filler content [23].

Similar trend was observed for the Young's modulus as shown in Figure 4 (b). It also decreases as the filler loading. The Young's modulus for vulcanized system is higher compared to the unvulcanised system. The Young's modulus for 5 wt% filler loading without vulcanization is 367.4 MPa while 385.1 MPa for unvulcanised system. The 30 wt% shows lowest Young's modulus with 305.6 MPa and 328.2 MPa for unvulcanised and vulcanized system respectively. This is due to the rubber-filler interaction decreases with increasing filler loading [24]. The dispersion of filler in the matrix poor in unvulcanised system hence the tensile strength and Young's modulus decreases as the filler loading increases. This will promote the discontinuity of the blends and at the end contributes to the reduction in tensile properties. The enhancement of tensile strength in vulcanised blend is due to the increased crosslink density in dynamically vulcanized PP/NR blend filled with *Cyperus Odoratus*. Formation of cross links in the elastomer phase facilitates better stress transfer which is the major contributing factor to such enhancements [25]. George et al. [26] reported that a large number of small, uniformly distributed cross-linked rubber particles can be observed in vulcanized samples and those entangled rubber molecules cannot slip past each other during loading and thereby resulting in higher tensile strength and Young's modulus. The effect of filler loading has increased the stiffness of the specimen and dynamic vulcanization has improved the stress transfer between the matrix and the filler hence increase the Young's modulus for the blend. Figure 4 (c) shows elongation at break for vulcanized PP/NR with filler loading compared to unvulcanised PP/NR blend with filler loading. The elongation at break correlates with the measures of the strain before it ruptures. It can be seen that the elongation at break is directly proportional to the Young's modulus of the blend system. The elongation at break for PP/NR-DV blend with 30 wt% shows the highest elongation at break which is at 16.4% due to the stress transfer between the filler and the matrix. The fine dispersion of NR in PP matrix and the cross-linked rubber particles helps the flow of the stress transfer and improved the interaction between filler-matrix.

IOP Conf. Series: Materials Science and Engineering **957** (2020) 012020 doi:10.1088/1757-899X/957/1/012020

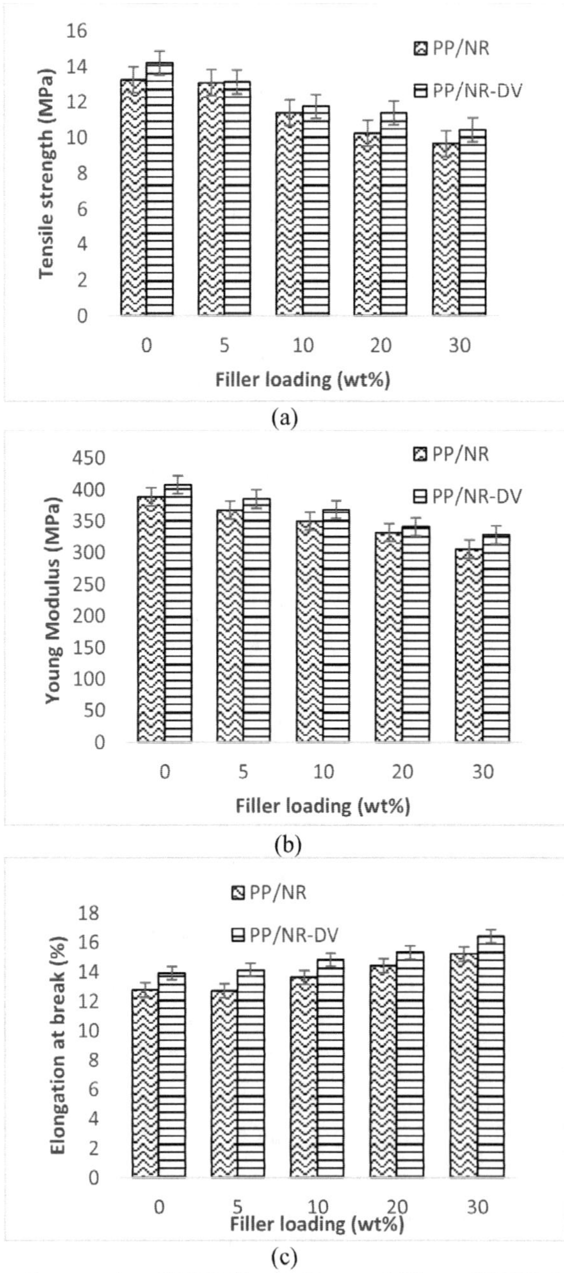

Figs 4 (a) – (c): Tensile properties of PP/NR blend with natural filler and PP/NR-DV with natural filler

4.3. Morphological Study

The phase morphology of a blend is determined by a several of factors including the relative concentrations of the polymeric components, the interfacial tension between them, the conditions used to process the mixture and the viscosity differences between the components [27-28]. The SEM pictures of tensile fractured surface of blends PP/NR (70:30) and PP/NR with 10% and 30% filler loading is shown in Figures 5(a) – (c). As shown in the figures, NR phase dispersed the continuous PP matrix. This is because the PP act as the domain to the system due to ratio 70:30. In Figure 5 (b), it can be seen that a number of voids and fiber pull outs on the tensile fractured surface of PP/NR blend with 10% filler loading. The presence of holes evidenced the retreat of particles, indicating low adhesion between the phases. The fiber is not wet by the PP matrix. In Figure 5 (c) shows the tensile fractured surface of PP/NR blend with 30% filler loading. There are more fiber pull out and voids that can be seen due to the incompatibility between the matrix and the fiber which is *Cyperus Odoratus*. This incompatibility is due to the difference between the surface properties of the natural filler that is highly polar and hydrophilic while matrix is generally polar and hydrophobic.

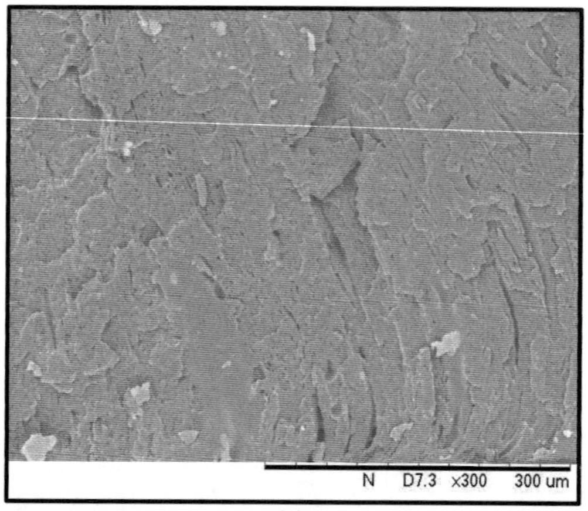

(a)

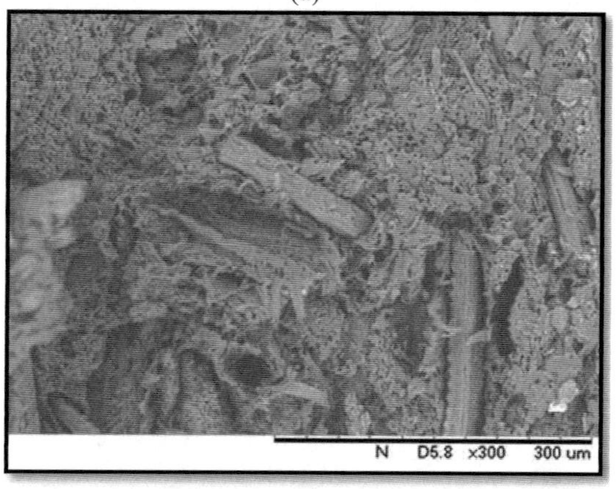

(b)

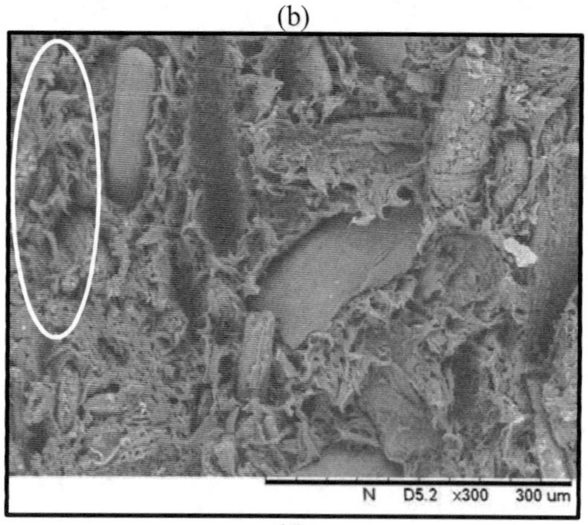

(c)

Figs 5: Scanning electron micrograph of surface of PP/NR blend (70:30) at (a) 0% filler loading, (b) 10% filler loading and (c) 30% filler loading at 300 X magnification.

Figure 6 (a) – (b) show the tensile fractured surface of PP/NR blend that undergoes dynamic vulcanization with 10% and 30% filler loading respectively. It can be observed in SEM pictures of tensile fractured surfaces of dynamically vulcanized blends has better dispersion of NR in the PP matrix and good adhesion between the natural fiber in the matrix compared to unvulcanised counterpart. The fiber shows good wetting and it is embedded in the matrix. The fiber is less agglomerate compared to the unvulcanised blend. This is a good indication to prove the improved tensile properties of vulcanized blends.

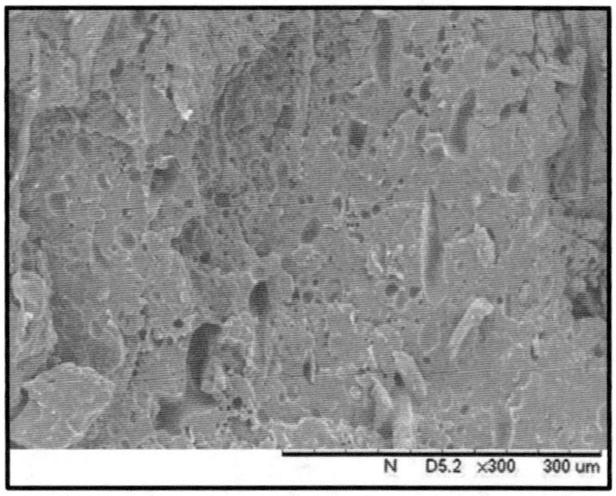

(a)

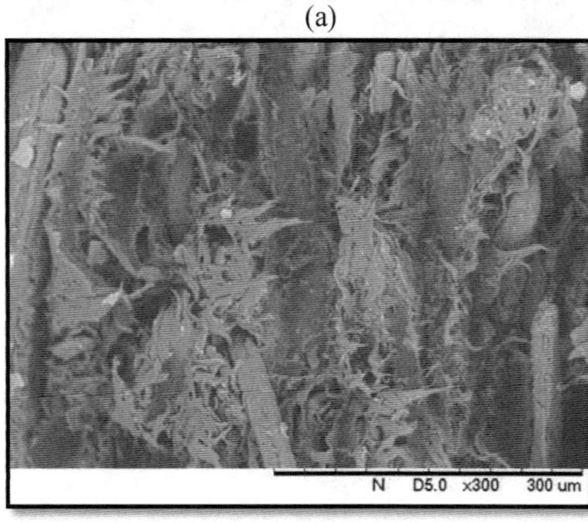

(b)

Figs 6: Scanning electron micrograph of surface of PP/NR-DV blends (70:30) at (a) 10% filler loading, and (b) 30% filler loading at 300 X magnification.

4.4 Thermogravimetric Analysis (TGA)

Thermogravimetric analysis or TGA is a standard procedure in which changes in the weight of a specimen is monitored as the specimen is progressively heated. The TGA and DTG curve of virgin PP/NR blend, PP/NR blend with 10 wt% and 30 wt% filler loading and vulcanized PP/NR blend with 10 wt% and 30 wt% filler loading is represented by the graph illustrated in Figure 7 (a) and (b) respectively. There are three stages of decomposition or weight loss happened on the system. The first stage is at range of 30-150 °C, the second stage is at the range of 150-300 °C and the third stage is at range of 300-490 °C. The initial weight loss between 30 °C and 150 °C is observed due to residual water. From the thermal degradation curves in the Figure 7 (a), the second stage of weight loss, it can be seen the initial degradation temperature for PP was 297.57 °C. Generally, the addition of filler has decreased the initial degradation temperature however when the PP/NR with filler loading is vulcanized the initial degradation were increased. This shows that the vulcanized PP/NR blend with filler loading have improved the thermal stability. The third stage weight loss can be seen is the end degradation temperature of PP was 457.25 °C. As the filler loading increase, the weight loss decreases as it needs more energy or high temperature to degrade the blends. From the Figure 7 (a) it can be seen that the weight loss of vulcanized PP/NR with 30 wt% filler loading is lower compared to unvulcanized PP/NR with 30 wt% filler loading. This shows that the dynamic vulcanization has increased the thermal stability of the blend system.

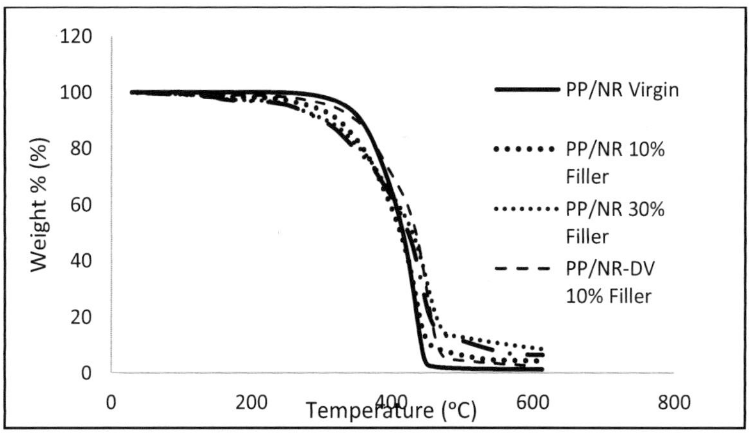

Fig. 7 (a): The TGA curves of PP/NR blend with filler loading and PP/NR-DV with filler loading

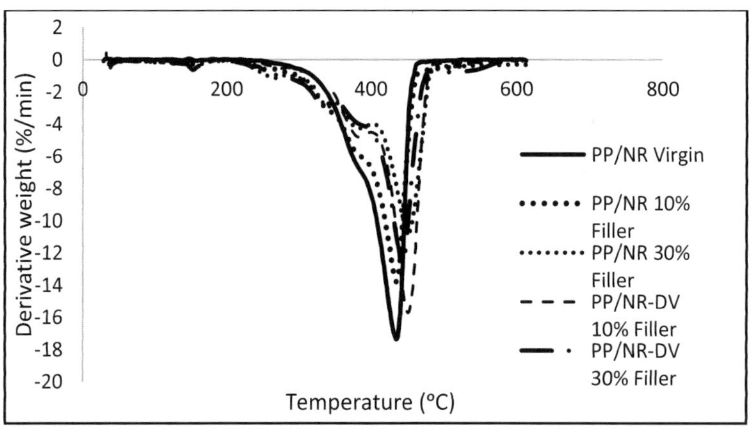

Fig. 7 (b): The DTG curves of PP/NR blend with filler loading and PP/NR-DV with filler loading

4.5 Differential Scanning Calorimetry (DSC)

The crystallization behaviour and melting characteristics of the composites were studied using differential scanning calorimetry (DSC). Figure 8 (a) and (b) shows the heating and cooling curve of PP/NR blend with filler loading and PP/NR-DV blend with filler loading. As shown in figure, the slight increased or remained essentially unchanged with increasing filler loading content. The melting point increased compared to unvulcanised system. It also can be observed that the use of lignocellulosic fiber can impart stability of the blend due to the presence of lignin as a natural oxidant. The increase in crystallinity is due to the arrangement of segmented molecules entangled molecules in the amorphous phase. When the dynamic vulcanization is done during the mixing process, the formation of carbonyl and hydroperoxides groups compete in the process of changing crystallinity. These chemical groups are attached to the molecular segments that cause and influx in disentangled chains, which it enabled the molecules to form a crystal structure, hence hindering the crystallization process.

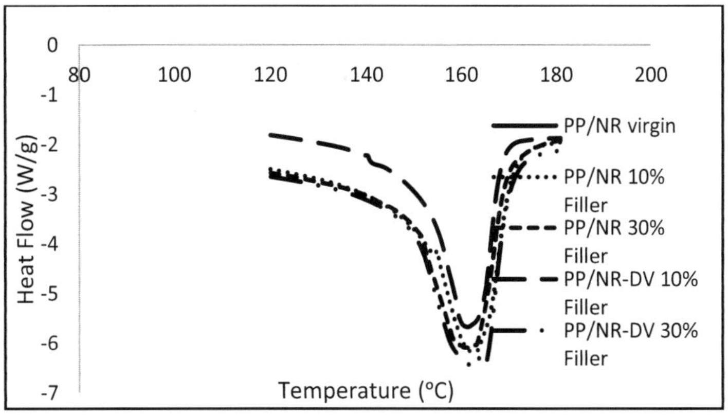

Fig. 8(a): The heating curve of PP/NR blend with filler loading and PP/NR-DV with filler loading

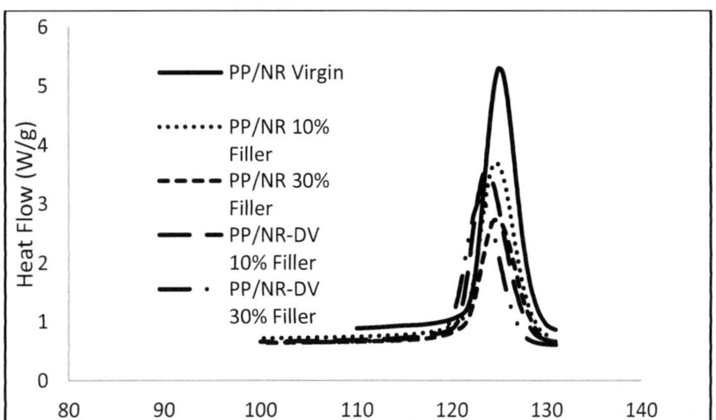

Fig. 8(b): The cooling curve of PP/NR blend with filler loading and PP/NR-DV with filler loading

5 Conclusions

The first series of the study focused on the comparison between the PP/NR blends and PP/NR blends filled with different composition of Cyperus Odoratus. The characterization and properties analyzed from the tensile properties shows that the increasing of filler loading decreased the tensile strength. The Young's modulus also decreased with the increasing of filler loading however the elongation at break increases. Morphology of the blends indicates the filler is does not dispersed well in the PP matrix. Since the PP act as dominant matrix and NR act the dispersed phase, dynamic vulcanization is done to crosslink the dispersed phase of NR. The tensile properties improved compared to the tensile properties of PP/NR blends filled with Cyperus Odoratus. The SEM studies revealed that better and uniform distribution of NR particles in the PP matrix when blends are subjected to dynamic vulcanization. The FTIR spectra analysis shows absorption bands that indicates that there is sulphur linkage due to the presence of crosslinking agent.

The authors wish to gratefully acknowledge the School of Materials Engineering, as well as Research and Innovation Department under Universiti Malaysia Perlis and Kementerian Pengajian Tinggi Malaysia (Ref: FRGS/1/2016/TK05/UNIMAP/02/9) for supporting this research project.

References

1. A.W.M. Kahar, M. Lingeswarran, M.Z. Amirah Hulwani, H. Ismail, Polym. Bull., 76(2), 747-762, (2019)
2. R. J. Walster, R. Rahman, A. W. M. Kahar. IOP Conf. Series: Mater. Sci. Eng., **429,** 012043, (2018).
3. A.W.M. Kahar, H. Ismail., N. Othman, J. Vinyl Add. Technol.**18**(3), 192-197, (2012)
4. H. Zaman, J. Song, L.S. Park, I.K. Kang, S.Y. Park, G. Kwak, K.B.Yoon. Polym. Bull. **67(1)**:187-198, (2011).
5. A. W. M. Kahar, H. Ismail, A. Abdul Hamid, A J. Therm. Anal. Calorim., **123(1),** 301-308, (2016).
6. G. Naderi, M.R. Nouri, M. Mehrabzadeh, G.R. Bakhshandeh, Ira. Polym. J., 8, 37-42, (1999)
7. N. Ibrahim, A.W.M. Kahar, D.L Ngoc, H. Ismail, H. Bio-res., **12(2),** 3076-3087, (2017).
8. F. Vilaplana, E. Strömberg, S. Karlsson, Polym. Degrad. Stabil., **95(11**), 2147–2161, (2010).
9. A.W.M. Kahar, H. Ismail. J. Vinyl Add. Technol. **22(3),** 191-199, (2016)
10. A.W.M. Kahar, N. Sarifuddin, H. Ismail, H., Ira. Polym. J., **26** (**2**), 149-159, (2017).
11. D. Trache, A. Donnot, K. Khimeche, R. Benelmir, N. Brosse, Carbohydr.Polym., **104**, 223-230, (2014)
12. S. Yasmeen, M. Kabiraz, B. Saha, M. Qadir, M., Gafur, S. Masum, Intern. Resear. J. Pure Appl. Chem., **10(4)**, 1–14, (2016).
13. A.H. Mohd Zain, A.W.M. Kahar, I. Hanafi, J Polym Environ, **26(2),** 691–700, (2018).
14. H. Salmah, H., Ismail, A. Bakar, J. Appl. Polym. Sci., **107(4),** 2266-2273. (2007).
15. N.F. Alias, H. Ismail, A.W.M. Kahar,. Bio-res, **12(4)**:9103-9117. (2017)
16. Z.Lin, C. Chen, Z. Guan, S. Tan, X. Zhang, J. Appl. Polym. Sci., **122(4)**, 2789-2797. (2011)

17. Z. Xiuju, S. Juncai, Y. Huajun, L. Zhidan, T. Shaozao, J. Thermo. Compos. Mater., **24(6),** 735-754. (2011).
18. A.H. Mohd Zain, A.W.M. Kahar, H. Ismail, Mater. Resear. Exp., **6** (10), 105335, (2019).
19. A.W.M. Kahar, H. Ismail., N. Othman, J. Appl. Polym. Sci., **128 (4)**, 2479–2488.
20. M. Abdelmouleh, S. Boufi, M.N. Belgacem, A. Dufresne, Compos. Sci. Technol. **67**, 1627-1639. (2007)
21. A.H. Mohd Zain, A.W.M. Kahar, N.Z. Noriman, Procedia Chem., **19**, 638-645, (2016)
22. C. Nakason, P. Wannavilai, A. Kaesaman, Polym. Test., **25**, 34-41, (2006).
23. A.A. Bahruddin, A. Prayitno, R. Satoto, Proced. Chemist, **4**, 146-153, (2012)
24. H. Ismail, H. Salmah, M. Nasir, M. Polym. Test., **20**, 819-823, (2001).
25. S. Siriwardena, H. Ismail, U.S. Ishiaku M.C.S. Perera, J. Appl. Polym. Sci., **85**, 438-452, (2002).
26. J.K.T. George, Varughese, S., Thomas, Polymer, **41**, 1507-1517, (2000).
27. J.D.V. Dyke, M. Gnatowski, A. Burczyk, Journal of Applied Polymer Science, **109**, 1535-1546, (2007).
28. J. Oderk, G. Groeninckx, Polymer, **43**, 2219-2228, (2002).

IOP Conf. Series: Materials Science and Engineering **957** (2020) 012021 doi:10.1088/1757-899X/957/1/012021

Tensile Properties of Polyethylene Composites Based Kaolin Geo-Filler

Loh Zhen Lun[2], *Yusrina* Mat Daud[1,2,*], *Farah Farhana* Zainal[1,2], *Mohammad Firdaus* Abu Hashim[1,2], *Hasri*[3], *Hartati*[3], *Tan Wei Shern*[2]

[1]Green Technology (CEGeoGTech), School of Materials Engineering, University Malaysia Perlis (UniMAP), P.O Box 77, d/a Pejabat Pos Besar, 01000 Kangar, Perlis, Malaysia
[2]School of Materials Engineering, Universiti Malaysia Perlis, 02600 Arau, Perlis, Malaysia
[3]Fakultas Matematika Dan Ilmu Pengetahuan Alam, Universitas Negeri Makassar, Kampus FMIPA UNM Parangtambung, Jl. Mallengkeri, Makassar, 90224, Indonesia

Abstract. The current work studies the tensile properties of polyethylene composites-based kaolin geo-filler. Polyethylene composites was prepared based on kaolin geo-filler at different loading content varies from 0,2,4,6,8 and 10 wt%. The optimum results were compared with polyethylene composites based on raw kaolin to study the effect both filler on tensile properties. Tensile test was conducted according to ASTM D638. Based on these research studies, the use of kaolin geo-filler is effectively improved the tensile properties of polyethylene as compared to the raw kaolin filler. As the result, 8% of kaolin geo-filler content demonstrate the optimum formulation to enhance the tensile properties of polyethylene composites.

1 Introduction

According to survey analysis from Plastics Europe Market Research Group (PEMRG), there are around 49.9 metric tons of plastic usage in 2016. Polyethylene (PE) shows the highest percentage of usage compared to other plastic. However, poor mechanical and physical properties of polyethylene can limit their large application. During the fabrication process of polyethylene industry, high operating speed can even more significantly degrade and leads to deterioration in its mechanical and physical properties [1]. This issue may influence the long-term use of polyethylene in the structural application. Therefore, polyethylene has been improved by addition of several fibrous and/or particulate filler to develop better properties [2-4]. Incorporation of kaolin as a filler is very common in polymer composites to any matrix of polymers since it has improved the mechanical properties of the polymer composites [5-6]. However, addition of kaolin based on geopolymerization process as a filler is rarely reported.

Kaolin geo-filler is an inorganic alumino-silicate structure which synthesized from predominantly silicon and aluminum material of geological in alkaline environment condition. Kaolin geo-filler is amorphous three-dimensional ceramic-like materials. It is

* Corresponding author: yusrina@unimap.edu.my

like natural zeolite except the amorphous microstructure. The alkaline environment allowed the geopolymerization to occur through poly-condensation process. Kaolin acts as main role in this process due to their highest SiO_2 and Al_2O_3 chemical composition. Under alkaline conditions, the geopolymerization occurs when the alkali hydroxide and silicate solution react to each other. It is essentially to prefer as low alkalinity system which high alkalinity may lead the common ions effect, and dissolution of C-S-H may be introduced [7-12]. Geopolymer based materials contribute to excellent mechanical properties, high early strength, abrasion resistance and thermal stability that lead to many application fields such ceramics, binder, matrices for hazardous waste stabilization, fire resistance and for high tech materials [13-16]. Therefore, in this study an attempt has been made to study the effect of geo-filler and polyethylene based on its tensile strength and modulus properties.

2 Experimental

2.1 Materials

Polyethylene pellets were sourced from Lotte Chemical Titan Holding Sdn. Bhd. with density 0.92 g/cm^3 and melting point 115 °C. Kaolin was supplied from Associated Kaolin Industries Sdn. Bhd. Malaysia. The chemical composition of kaolin from non-destructive analytical technique, X-ray Fluorescence shows as in Table 1.

Table 1: Chemical analysis of kaolin

Chemical Composition	SiO_2	Al_2O_3	K_2O	Fe_2O_3	TiO_2	ZrO_2	MnO_2	LOI
Mass (wt. %)	54.50	32.40	5.58	4.32	1.33	0.08	0.10	1.69

12 M of sodium hydroxide solution was prepared by using the sodium hydroxide pellets which supplied by HmbG Chemicals. The molar mass of sodium hydroxide is 40 g mol^{-1}. The density of sodium hydroxide is 2.13 g cm^{-3}. Sodium silicate pellets are used to prepare the sodium silicate solution with the molarity of 12 M. The sodium silicate pellets were sourced from the Sigma-Aldrich (M) Sdn. Bhd. The molar mass of sodium hydroxide is 122 g mol^{-1}. The density of sodium hydroxide is 2.4 g cm^{-3}.

2.2 Sample Preparation

Kaolin raw materials was placed in oven at 100 °C for 24 hours before mixed with the alkaline activator. Alkaline activator solution is prepared by using both sodium hydroxide solution and sodium monosilicate solution at 12 M with ratio 2:0. Kaolin was mixed with alkaline activator and the geopolymeric paste were placed in oven at temperature 80 °C for 24 hours. After 24 hours, the materials are crushed using ring mill and sieved by using siever size 150 μm. Polyethylene pellet and geo-filler were mixed according to Table 2. Preparation of composites was used twin screw extruder LabTech Scientific 20 mm co-rotating. The barrel, die temperature, feeder speed and screw speed are controlled according to Table 3.

Table 2: Formulations of polyethylene filled with kaolin geo-filler.

Sample	Polyethylene wt.%	Geo-filler wt.%
PE	100	0
PEKG-2 %	98	2
PEKG-4 %	96	4
PEKG-6 %	94	6
PEKG-8 %	92	8
PEKG-10 %	90	10

Table 3: Temperature and screw speed profiles of the extruder barrel

	Zone 1	Zone 2	Zone 3	Zone 4	Zone 5	Zone 6	Zone 7	Zone 8
Temperature	115 °C	120 °C	125 °C	130 °C	135 °C	140 °C	145 °C	150 °C
Screw speed	110 rpm							
Feeder speed	8 rpm							

2.2 Testing and Characterization

The composite sample of dumbell shape with thickness of 0.2 cm were fabricated for tensile testing using Instron Universal Testing Machine Model 5569 according to ASTM D638. The speed of testing used is 5 mm/min with the nominal strain rate of 1.5 mm/mm·minute.

3 Results and Discussion

The results of tensile strength and modulus for polyethylene composite (PEKG) with different geo-filler contents at varies loading from 0,2,4,6,8 and 10 wt. % were illustrated in Figure 1. Tensile modulus is directly proportional to tensile strength and it can be seen that tensile strength and modulus increased as geo-filler content increased up to 8 wt. % and reached the optimum tensile strength at 20.83 MPa and modulus at 476.67 MPa. However, the tensile strength and modulus has found to be decreased as the kaolin geo-filler reached 10 wt.%.

Kaolin geo-filler is an inorganic materials that acted as reinforcement in polyethylene composites and as geo-filler content increased, it will contributed to higher tensile strength. Inorganic reinforcing fillers are harder and stiffer than the matrix and deform less, causing an overall reduction in the matrix strain, especially in the vicinity of the particle as a result of the particle/matrix interface [17]. In addition, kaolin geo-filler is a concrete like materials and anticipated higher strength than matrix polymer. As the kaolin geo-filler incorporated in polyethylene matrix, these geo-filler tend to restrain movement of the matrix phase in the vicinity of each particles which bear a fraction of the load thus increase the tensile strength [18]. However, at higher geo-filler content the filler particles expected having poorly dispersed and acts as flaws in the composites structure. In addition, higher filler content will accumulate filler at the joint interface, and insufficient thermoplastic material is present to form a strong bond. So, the tensile strength will decreased at high percentage of filler content [19] .

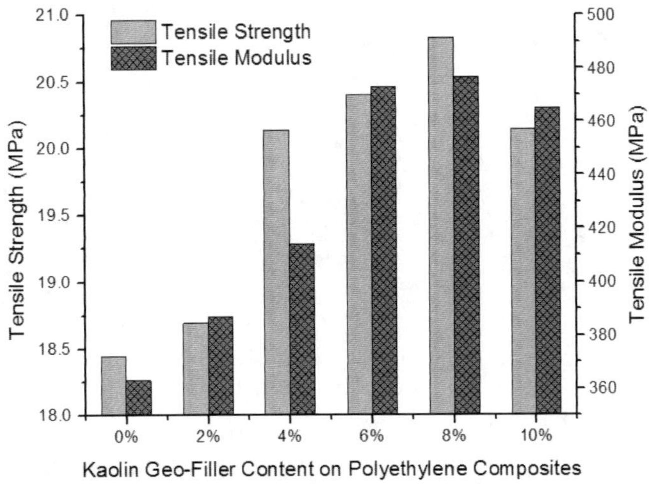

Fig.1: Tensile strength and modulus of polyethylene at different geo-filler content.

Tensile modulus is the ratio between tensile stress and tensile strain in the linear elasticity regime of a uniaxial deformation. From Figure 1, tensile modulus was increased from the origin polyethylene to addition of 2, 4, 6 and 8 wt.% of geo-filler content. Incorporation of geo-filler and polymer matrix clearly improve the stiffness of composites, which enhance the tensile modulus as the result. Higher filler content tend to lower the chain mobility. The limitation of chain mobility is the result of external load needed to strain the bonding of polymer [20]. However, tensile modulus was found decreased as the geo-filler increase to 10 wt.%. The reason is that the more shear force generated at higher filler concentration lead the filler agglomerates and decrease the free volume. It will restrict the mobility of chain movements and tends to lower the tensile modulus [21].

The optimum kaolin geo-filler content at 8 wt% in polyethylene composites was compared at even loading with raw kaolin in polyethylene composites (PEK) to study the efficacy of geo-filler in composites system. The results was illustrated in Figure 2. From the figure, addition of geo-filler kaolin obviously improved the tensile strength and modulus as geo-filler have cementing properties through geopolymerization process. Apart from that, PEKG also have rougher surface such in Figure 3. The rougher surface needs more energy to strain which support the high tensile strength and modulus [22].

IOP Conf. Series: Materials Science and Engineering **957** (2020) 012021 doi:10.1088/1757-899X/957/1/012021

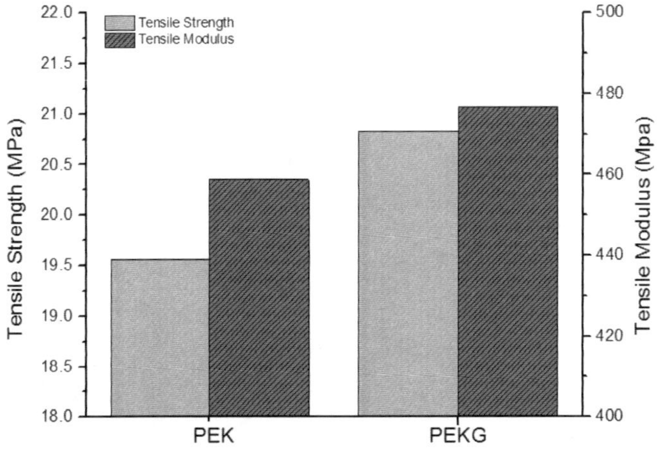

Fig.2: Tensile strength and modulus of polyethylene based kaolin filler (8 wt%) (PEK) and polyethylene based kaolin geofiller (8 wt%) PEKG

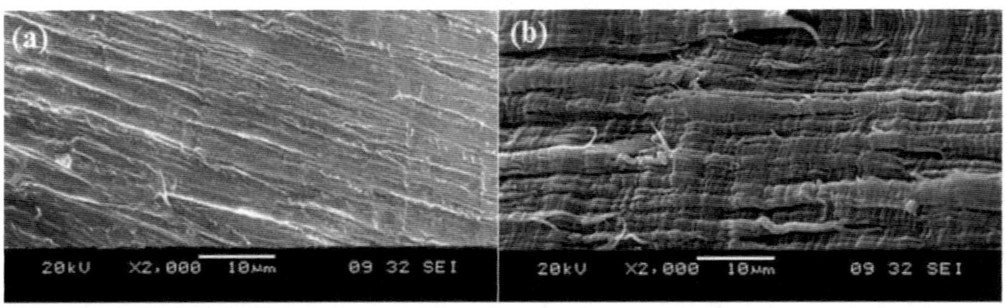

Fig.3: SEM morphology cross-sectional images: (a) polyethylene based kaolin filler (8 wt%) (PEK) and (b) polyethylene based kaolin geofiller (8 wt%) PEKG

4 Summary

In this study, kaolin geo-filler was produced from geopolymerization process using sodium hydroxide and sodium silicate was used as a filler in the polyethylene composites. Tensile properties of polyethylene composites has enhanced by addition of kaolin geo-filler with the optimum formulation of 8 wt.%.

The authors would like to thank the staffs of Center of Excellence Geopolymer & Green Technology (CEGeoGTech), School of Materials Engineering, Universiti Malaysia Perlis (UniMAP) for their involvement in the research. This work was supported and funded by the Universitas Negeri Makassar Kampus FMIPA UNM, Makassar, Indonesia.

References

1. Yahaya, S., Jikan, S. S., Badarulzaman, N. A., & Adamu, A. D. A., Path of Science, 3(10), 1001—1004 (2017)
2. H.D. Rozman, A.R.Rozyanty, L. Musa, G.S. Tay. J Wood Chem Technol **30**(2), 152-163 (2010)
3. Daud, Y.M., Yee, T.G., Adnan, S.A., Zaidi, N.H.A. AIP Conf. Proc 2030, 020048 (2018)
4. A.R. Rozyanty, H.D. Rozman, G.S. Tay. Adv. Mater. Res. **264-265**, 712-718 (2011)
5. H.D. Rozman, A.R. Rozyanty, G.S Tay, R.N. Kumar. J. Appl. Polym. Sci. **115**(5), 2677-2682 (2010).
6. Ibrahim, M., Jalil, J.A., Mahamud, S.N.S., Husseinsyah, S., Yahya, M.R. Key Eng. Mater. **594-595**, 775-779 (2014)
7. Zainal, F.F., Hussin, K., Rahmat, A., Abdullah, M.M.A.B., Shamsudin, S.R. AIP Conf. Proc. 040001 (2016)
8. Zainal, F.F., Fazill, M.F., Hussin, K., Rahmat, A., Abdullah, M.M.A.B., Wazien, W. Solid State Phenom **273**, 175-180 (2018).
9. Zainal, F.F., Amli, S.F.M., Hussin, K., Rahmat, A., Abdullah, M.M.A.B. 2017 Mater. Sci, **209** (1), art. no. 012026 .
10. Zainal, F.F., Md. Rahim, N.S., Wardan, R. Solid State Phenom, **28**, 226-230 (2018) .
11. Abu Hashim, M.F., Abdullah, M.M.A., Ghazali, C.M.R., Hussin, K., Binhussain, M AIP Conf. Proc **1835**, 020042 (2017).
12. Hashim, M.F.A., Abdullah, M.M.A.B., Sandu, A.V., Puskas, A., Daud, Y.M., Zainal, F.F., Faris, M.A., Hasri, Hartati. 2019, Mater. Sci. Eng. **572**, 012037.
13. Abu Hashim, M.F., Abdullah, M.M.A.B., Ghazali, C.M.R., Hussin, K., Binhussain, M., Omar, M.F. MATEC Web of Conferences, 78, 01066 (2016).
14. Daud, Y.M., Hussin, K., Osman, A.F., Ghazali, C.M.R., Al Bakri Abdullah, M.M., Sandu, A.V., Mater. Plast. **54**(3), 543-545 (2017)
15. Daud, Y.M., Hussin, K., Ghazali, C.M.R., Osman, A.F., Abdullah, M.M.A.B., Hussain, M.B. Key Eng. Mater. **673**,55-63 (2016)
16. Daud, Y.M., Kamarudin, H., Ruzaidi, C.M., Osman, A.F., Al-Bakri, M. Mater. Sci. Forum **803**, 58-62 (2015).
17. Yao, J., Zhu, H., Qi, Y., Guo, M., Hu, Q., & Gao, L. 2018 Mater. Sci. Eng. **359**, 012034 .
18. Yang, R., Yu, J., Liu, Y., & Wang, K. Polym. Degrad. Stab. **88**(2), 333-340 (2005)
19. Troughton, M. J. (2008). Handbook of plastics joining: a practical guide: William Andrew.
20. Manjhi, S., & Sarkhel, G. J. Appl. Polym. Sci. **119**(4), 2268-2274 (2011)
21. Salem, I.A.S., Rozyanty, A.R., Betar, B.O., Adam, T., Mohammed, M., Mohammed, A.M. J. Phys. **908** (1), 012002 (2017).
22. Kundie, F., Azhari, C. H., Muchtar, A., & Ahmad, Z. A., J. Phys. Sci. **29**(1), 141–165 (2018).

IOP Conf. Series: Materials Science and Engineering **957** (2020) 012022 doi:10.1088/1757-899X/957/1/012022

Failure Analysis on U-Bend Header Tube of Boiler Economizer

Nur Farhana Hayazi[1,*], *Shaiful Rizam* Shamsudin[1,2] and *Rajaselan* Wardan[1,2]

[1]School of Materials Engineering, Universiti Malaysia Perlis, Kompleks Pengajian Jejawi 2, Jejawi 02600 Arau Perlis.

[2]Center of Excellence Geopolymer & Green Technology (CEGEOGTECH), School of Materials Engineering, Universiti Malaysia Perlis, Kompleks Pengajian Jejawi 2, Jejawi 02600 Arau Perlis.

Abstract. A water tube boiler is a type of boiler in which water circulates in tubes heated externally by fire. It was found that the water tube economizers are often subject to the serious damage mainly because of the overheating and corrosion. Thus, the metallurgical investigation procedures starts with the visual inspection, stereomicroscopy, metallography and chemical analysis methods. It was identified that the failure of the u-bend tube was due to the cracking caused by bending forces during the procedures. This cracking causes a leak for the tube. The appearance of corrosion on the shell side is due to the presence of small content of sulphur under the boiler environment. Several welding imperfections were also detected, however, both corrosion and welding imperfections were not the primary caused of the tube failure. From the analysis, the use of SUS304 grade stainless steel was discovered to be the appropriate material selection for the tube boiler.

1 Introduction

In a boiler system, an economizer captures the heat of the flue gas and transfers it to the feed water [1]. As the flue gas exits the boiler, it enters an economizer with finned tubes. The finned tubes absorb the heat and raises the temperature of the boiler feed water, lowering the need energy input. The material for the economizer will depend on the fuel type and may be all steel, all cast iron or cast iron protected steel. All steel would be used for non-corrosive flue gases from burning natural gas, light oil and coal [2-4].

The u-bend header tube of boiler economizer found to be failed after 5-6 years in service (Figure 1(a)-(d)). A sudden pressure drop was recorded by the person in charge. From the observation, the boiler header was found leaked on the u-bend tube area. The materials of header tubes made of SUS 304 grade stainless steel. The objective of this analysis is to investigate the failure evidence presented by the u-bend tube and to examine the primary cause of the failure through the metallurgical analysis method.

* Corresponding author: farhanahayazi@unimap.edu.my

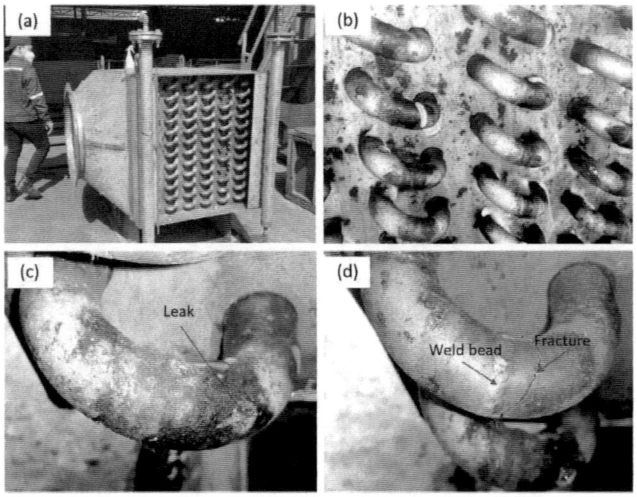

Figure 1:Camera photographs on boiler economizer; (a) header tube, (b) close-up views on defected u-bend tubes, (c) leaked area as indicated by the small white bubbles and (d) transverse crack adjacent to the weld bead.

2 Methodology

Visual inspection of the original condition on as-received u-bend tube was documented by digital DSLR camera (D3100, Nikon). Close-up viewing of crack area was done by stereomicroscope (SZX 10, Olympus). The u-bend tube was then sectioned using a diamond cutter at the slowest speed to ensure no external residual stress exerted onto the u-bend header. The sample once again been examined visually by using camera and stereomicroscope. The sample was cleaned by using stainless steel cleaner to remove all the dirt and rust inside (tube side) and outside (shell side) of the sample. The failed sample then grounded to 180, 240, 320, 440, 800 and 1200 SiC grit paper before polished using 6 and 1μ diamond solutions. The as-polished sample was inspected by metallurgical microscope (BX41, Olympus) to determine the presence of defects such as crack, inclusion, pin holes, void and/or microstructural anomalies. The as-polished samples were etched by 10 ml HNO_3, 10 ml CH_3COOH, 15 ml HCl and 5 ml glycerol to delineate specific microstructural features. Once etched, the microstructure of the material was viewed under metallurgical microscope (BX51M, Olympus). Surface morphology and micro-elemental analysis was conducted by scanning electron microscope, SEM (JSM-6490LA, JOEL) and energy dispersive spectroscopy, EDS (Oxford INCA Energy EDS system).

3 Results and Discussion

Figure 2 (a) - (d) show a camera photographs of as-received u-bend header tubes both on shell and tube sides. The surface of the header tube is found to be covered with brownish rust on both shell and tube side. The shell side surface looks more severe to corrode as compared to the tube side. In as-cleaned samples, there is a residual blackish oxide stain and slightly brownish rust (Figure 2(c)). While the tube side surface is seen to be shiny and rust-free (Figure 2(d)). These observations indicate that the rust is worse on the shell side rather than the tube side where the fluid flowing through in it.

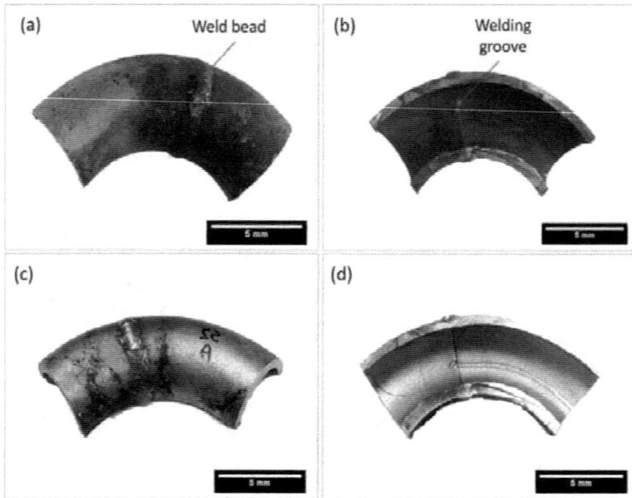

Figure 2:Camera photographs on (a) shell side, (b) tube side of as-received, and (c) shell side and (d) tube side images of as-cleaned of u-bend header tube.

The closed-up camera photographs further reveal the appearance of cracking on the u-bend header tube (Figure 3). At the shell side, there is a crack transversely propagates encircling the tube that occurs adjacent to the weld bead (Figure 3(a)). There are some blackish stains on the crack morphology which indicates that there is a corrosion reaction surrounding the crack area. Figure 3 (b) shows the cracks are found everywhere along the u-bent tube on the tube side. However, there were no signs of cracking on the weld bead. This proves that welding procedures are performed well without any defects. Additionally, no corrosion marks visually detected on the tube side.

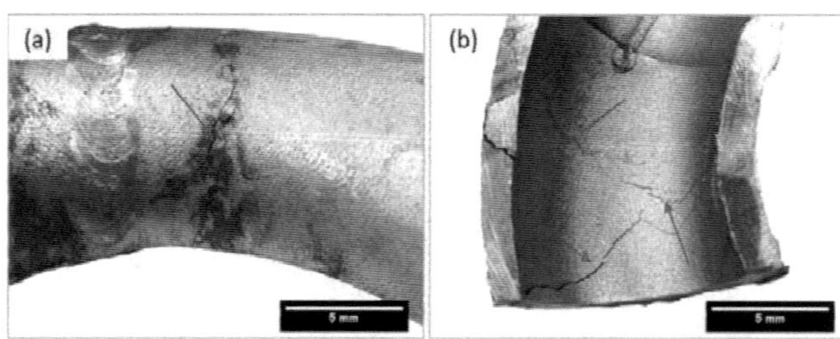

Figure 3:Closed-up camera photograph images on as-cleaned of typical (a) shell side and (b) tube side. Red arrows indicate the cracking morphology.

Observations by stereomicroscope around the weld bead both on shell and tube side found no signs of cracking and corrosion occurred (Figure 4(a) and (b)). Some welding imperfection such as linear misalignment and incomplete penetration (lack of root fusion) spotted on weld join. Generally, this imperfection act as an initiation site of the stress corrosion cracking (SCC) [3]. However, there is no SCC marks detected here. This further strengthens the fact that the welding works to join two semi-bent tubes do not have any problems at all, instead, it was carried out perfectly.

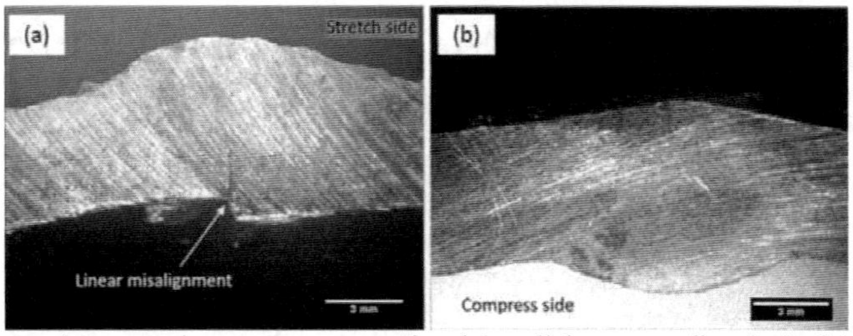

Figure 4: Stereomicroscope images at cross-sectional (parallel to stress direction) of the as-cut weld bead area on (a) stretched and (b) compressed side.

Etched metallographs images on the header tube metal are shown in Figure 5. There are single cracks together with hairline cracks morphology which dominated by transgranular cracking on the cross-sectional tube wall. This crack finally results in the enlargement of crack opening to be more severe. Both hairline and transgranular cracks indicate that the tube metal is under stress due to bending works that occur in a plastic manner beyond the materials yield point. In contrast, there is not a clear intergranular cracking that can be clearly detected. Therefore, the corrosion factor is not the main cause of the failure. Additionally, the absence of a slip band grains is believed due to the optimum heat treatment process performed on the u-bend tube works well and aidful [1-2].

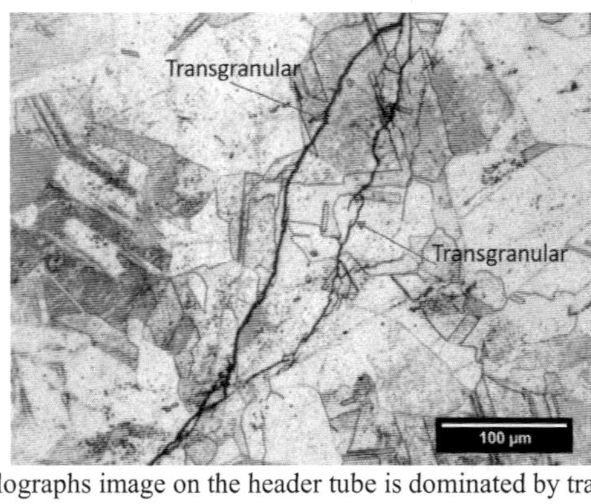

Figure 5: Etched metallographs image on the header tube is dominated by transgranular cracking.

Figure 6 shows the elemental EDS pattern on the shell side surface. The appearance of brownish corroded surface on the shell side was attributed to the oxidation and the presence of slightly sulphur elements which is believed as a result of sulphur content in the condensation of flue gas.

OES Results of the as-received u-bend header tube boiler is shown as in Table 1. This results proved that SUS304 grade stainless steel is the best material selection used for the tube boiler since it is not affecting the tube boiler failure. The material selection factor has been ruled out. The plant technical should perform a periodic NDT inspection for early detection and damage assessment to boiler tubes. Secondly, the application of u-bent header tubes by mechanical bending methods is indeed prone to crack failure, thus it is proposed to replace them with u-bend tubes fabricated by the casting method.

IOP Conf. Series: Materials Science and Engineering **957** (2020) 012022 doi:10.1088/1757-899X/957/1/012022

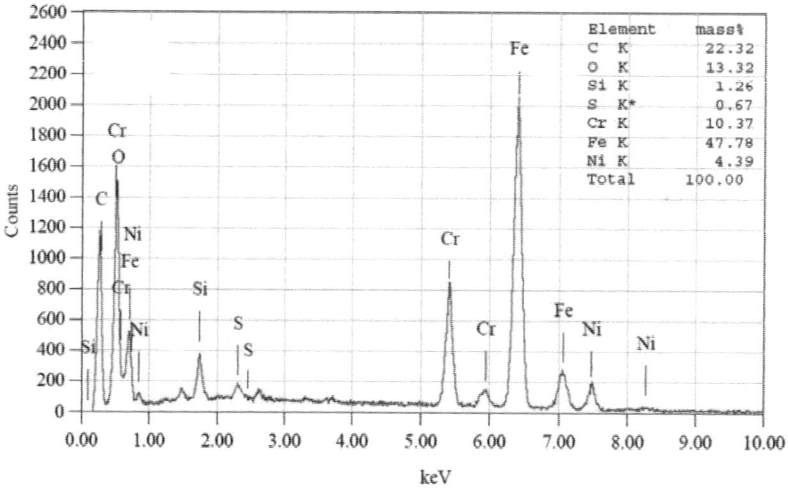

Figure 6: Elemental EDS pattern on u bend header tube at shell-side surface.

Table 1: OES Results of the as-received u-bend header tube boiler

Element	SUS304	Header Tube
C	0.08 max.	0.033
Si	1.00 max.	0.398
Mn	2.00 max.	1.76
P	0.045 max.	0.03
S	0.030 max.	0.008
Ni	8.00 – 10.50	8.499
Cr	18.00 – 20.00	18.586
Others	–	70.689

4 Conclusion

The leakage failure of the u-bend tube is due to a crack caused by stress due the bending procedure that may be performed by the hot or cold work methods. The lack of branching cracks and the absence of slip band grains indicate that the bending and heat treatment (or stress relief treatment) procedure works well, as the inadequacy signs of residual stress in the microstructural analysis observation. However, the cracks are believed to occur earlier that is during the bending procedure. The appearance of brownish rust on the tube surface due to the presence of small content of sulphur elements in the boiler environment. It does not penetrate inward and only affects the surface layer of the metal. However, at the opening cracks, this sulphur penetrates the metal tube through the crack narrow space which eventually speeds up the leakage of the tube. Corrosion is not the main root cause of header tube failure. As well as the welding imperfections. In conclusion, the material (SUS304 grade stainless steel) is found to be the appropriate materials selection for the header tube.

IOP Conf. Series: Materials Science and Engineering **957** (2020) 012022 doi:10.1088/1757-899X/957/1/012022

The author would like to acknowledge the support from the Prescott Engineering & Services Sdn Bhd, Pasir Gudang, Johor, Malaysia. Thanks to the School of Materials Engineering and Universiti Malaysia Perlis for providing a special financial support for enabling the publication of this manuscript.

References

1. A. Mecke, I. Lee, J.R. Baker jr., M.M. Banaszak Holl, B.G. Orr, Eur. Phys. J. E **14**, 7 (2004)

2. M. Ben Rabha, M.F. Boujmil, M. Saadoun, B. Bessaïs, Eur. Phys. J. Appl. Phys. (to be published)

3. F. De Lillo, F. Cecconi, G. Lacorata, A. Vulpiani, EPL, **84** (2008)

4. L. T. De Luca, *Propulsion physics* (EDP Sciences, Les Ulis, 2009)

IOP Conf. Series: Materials Science and Engineering **957** (2020) 012023 doi:10.1088/1757-899X/957/1/012023

The efficiency of Zero Charge Corrosion Protection on pure Mg by Planned Interval Test

Shaiful Rizam Shamsudin[1,*] , *Muhamad* Syazwan Sanusi[1], *Rajaselan* Wardan[1], *Azmi* Rahmat[1], and *Mohd* Nazree Derman[1]

[1]Center of Excellence Geopolymer & Green Technology, School of Materials Engineering, University Malaysia Perlis, Jejawi 02600 Arau, Perlis, Malaysia

Abstract. The concept of zero charge potential (E_{pzc}) is manipulated to eliminates the effect of the electric double layer in order to mitigate the corrosion reactions through the zero-charge corrosion protection (ZCCP) method. The efficiency of the ZCCP method was examined by planned intervals test on pure Mg metal in 3.5% NaCl solution. The test results showed that the tendency of both liquid corrosiveness and metal corrodibility are decreased from -1562.7mm/yr (E_{ocp}) to -98.2mm/yr (E_{pzc}). ZCCP technique was succeeded by corrosion protection efficiency of pure Mg which was recorded between 93 to 96% in 3.5% NaCl solution.

1 Introduction

A novel new technique in controlling and preventing corrosion of a metal by bringing the potential to a value known as zero-charge potential or potential of zero charges, E_{pzc} [1]. At E_{pzc}, the excessive charges on the metal surface are neutralized to zero charges. This condition works to turn-off the effect of the electric double layer, thus, weakening the charge adsorption force due to the electrostatic attraction of the electrode-electrolyte interface. This process is further hindered the charge transfer on the metal-solution interface, thus, prevent the corrosion reaction of the metal in aqueous solution.

The application of E_{pzc} concept is manipulated through a method called zero charge corrosion protection (ZCCP). ZCCP technique requires a power supply that potentiostatical capability of giving a negative current or voltage pulsed, hence bringing the potential of a metal to a net-potential value whereby no charges on the metal surface.[2]. This can be achieved by cyclic charging-discharging the metal surface between corrosion (E_{corr}) and polarized potential ($E_{impressed}$) repeatedly. The pulse charging-discharging of the metal to be protected can be restricted to a maximum of ±2 mV at about E_{pzc}. Since the E_{corr} is known, the corrosion current (i_{corr}) of metal also known, hence E_{pzc} can be determined by the application of pulse current/voltage equivalent to i_{corr} to the metal surface. This novel method of corrosion protection at E_{pzc} does not use the conventional concept of polarization of two electrodes (anodes & cathodes), it only needs a single electrode or metal to be protected.

* Corresponding author: rizam@unimap.edu.my

E_{pzc} can be applied to metal in various corroding environments such as acidic, neutral, and alkaline medium, static, and dynamic flow as well as any climate-changing environments [3].

The planned interval test (PIT) is a good technique for assessing the effect of time on metal corrosion and corrosion (solution) environment. Normal immersion methods can generally only measure normal corrosion rates during the exposure period in corrosive environments. However, this PIT can measure the variation of corrosion rates, both uniform and local corrosion with time throughout the test period. Measurement of corrosion rate through PIT is a study involving the formation or breakdown of protective film on metal surfaces during the test period. Variations in corrosion rates can also occur as a result of changes in the concentration of corrosion agents in an environment or solution medium.
The PIT procedure involves exposure of a metal sample for at least three periods. The exposure period selected is for initial, prolonged, and short periods of exposure after the long exposure period. Differences in corrosion rates between these three periods help to assess (i) differences in corrosion rates for short- and long-term exposure, and (ii) differences in corrosion rates for the same period at the beginning and end of the test period.

Magnesium (Mg) is a highly reactive metal and is easy to corrode when exposed to aqueous environments, especially in 3.5% NaCl solution. The reaction between the Mg surface with a 3.5% NaCl solution left a corrosion-resistant effect [4]. Mg alloys are often used as a sacrificial anode in underground environments due to their high reactivity properties. The reason for using pure Mg is because Mg is more reactive than steel. Therefore, if a corrosion protection method is effective at Mg, then it certainly works for steel [5]. The Mg reactivity factor and the precision of the weighing apparatus (± 0.5 mg) caused the measurement of weight loss to be detected within a short immersion period compared to the use of steel that took longer to detect weight loss [6]. The objective of the study is to measure the efficiency of corrosion protection with the zero charge techniques on pure Mg metal through a planned interval test.

2 Methodology

Pure Mg (99.8%) in the form of a solid rod was lathed to 3.3 cm in diameter. It was then cut into 2 cm high. The planned interval test (PIT) is performed per ASTM G31 without corrosion protection under open circuit potential (E_{ocp}) condition. All of these Mg samples were weighed by analytical balance (AL204, Mettler Toledo) to obtain the initial sample weight (W_i). Glass container containing 3.5% NaCl solution was prepared. Four Mg samples were used in this PIT, one for each interval. The planned interval test (PIT) was performed to assess the effect of immersion time and changes in the corrosive environment on metal corrosion. This test is to prove that corrosion protection at zero charge capacity, E_{pzc}, occurs on metal exposed to a corrosive solution. For comparison, immersion tests on pure Mg samples were performed on the open-circuit potential, E_{ocp}, and at zero charge potential, E_{pzc}. For this purpose, a pure Mg sample is used as a metal to be protected. Mg is known to be a more reactive metal than Fe where Mg reacts actively when exposed to 3.5% NaCl solution. The samples were inserted in different containers and each container was labeled with A_1, A_t, and A_{t+1}, which were marks to represent their respective immersion intervals. The immersion time interval is shown in the schematic diagram of Fig. 1.

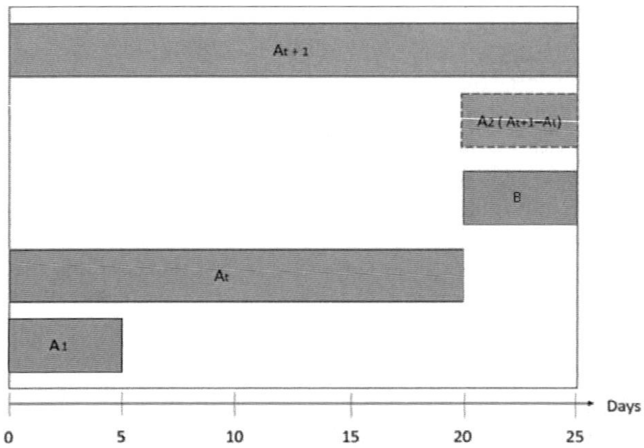

Fig. 1. Schematic diagram of the immersion time interval in the PIT.

After immersion for 5 days, sample in A_1 container was taken-out, for sample A_t, after 20 days of immersion it was removed and replaced by another sample and marked as B in the same container. Sample B was immersed in the same period as A_1 for 5 days. Finally, A_{t+1} samples were released after immersion for 25 days. This PIT step was repeated on the Mg sample subjected to the ZCCP method system.

ZCCP was applied by using a customized power supply capable of supplying current/ voltage pulses to steel samples. The ammeter with USB connection was connected in series in the circuit to measure electrical current consumption for both techniques at 1 reading/ hour interval. All corrosion protection tests were performed at room temperature with 1-ATM pressure.

Once the immersion period of PIT was completed, all samples were rinsed with distilled water. The samples were then soaked in 1000 ml of a solution containing 200 g of chromium trioxide (CrO_3), 10 g of argentum nitrate ($AgNO_3$), and 20 g of barium nitrate ($Ba(NO_3)_2$) to remove oxide and residue of chloride as suggested by ASTM G1. The cleaning was done for 5 minutes to ensure that all oxides were completely cleaned. The samples were weighed again to obtain the final weight (W_f). Weight loss (W) is calculated according to Eq. 1;

$$W = W_i - W_f \tag{1}$$

while corrosion rates are calculated from Eq. 2;

$$\text{Corrosion rate} = \frac{K \times W}{A \times T \times D} \tag{2}$$

where K is a constant (8.76×10^4), T is an immersion time (units of hours) with a precision of up to 0.01 hours, A is a surface area exposed in cm^2 units with a precision of up to 0.01 cm^2, W is a weight loss in gram (g) units with precision up to 1 mg and D is density in g/cm^3 units. Immersion time intervals with symbol A_2 are obtained by subtracting the weight loss obtained by A_{t+1} with the total weight loss A_t. The relationship between electrolyte corrosion and metal corrosion as a function of weight loss was analyzed and compared based on Table 1. The surface condition was then analyzed by stereomicroscope (SZX16, Olympus).

Table 1. Guidelines for corrosion analysis in planned interval test

Corrosion Factors	Observation of weight changes during immersion test	Criteria
Liquid corrosiveness	Unchanged	$A_1 = B$
	Decreased	$B < A_1$
	Increased	$A_1 < B$
Metal corrodibility	Unchanged	$A_2 = B$
	Decreased	$A_2 < B$
	Increased	$B < A_2$

3 Results and discussion

The results of these planned interval tests are shown in Tables 2 and 3. According to Table 2, the most significant weight loss occurred in the sample immersed under E_{ocp}. This indicates that the corrosion process occurs on Mg that immersed in a 3.5% NaCl solution without any protection procedure. On the other hand, the pure Mg metal immersed under E_{pzc} only recorded a very small decreased in weight loss, which actually due to non-protective procedure for 3 days to achieve stable and steady-state conditions before E_{pzc} pulse charged-discharged are performed. Mg samples in which immersed under E_{pzc} were found to have significantly lower corrosion rates than E_{ocp}. The ZCCP technique under E_{pzc} recorded the efficiency of corrosion protection as much as 93% to 96%. This percentage of efficiency is believed to increase further if it is practiced on steel.

Table 2: Weight loss and corrosion rate of Mg sample in E_{ocp} & E_{pzc}

Sample Marker	Immersion Period (Day)	Weight Loss (± 0.5 mg)		Corrosion rate (mm/year)		Corrosion Protection Efficiency at E_{pzc} (%)
		E_{ocp}	E_{pzc}	E_{ocp}	E_{pzc}	
A_1	5	1896.1	118.6	19.28	1.21	93.7
A_t	20	3105.5	124.1	7.90	0.32	96.0
A_{t+1}	25	3458.3	127.0	7.03	0.25	96.3
B	5	333.4	20.4	3.39	1.26	93.9
A_2	5	352.8	2.9	3.59	0.00	99.2

The data obtained from Table 3 are found to be $B < A_1$ for both E_{ocp}, and E_{pzc}. This suggests that there is a clear reduction of liquid corrosiveness for both E_{ocp} & E_{pzc}. The lower liquid corrosiveness over the immersion time is due to the depletion of Mg ion (Mg^{2+}) concentration in NaCl solution. Additionally, the corrosion rate for immersion at E_{pzc} was 16 times lower than that of E_{ocp}. In contrast, the metal corrodibility under E_{pzc} shows a significantly reduced from -1562.7mm/yr to -98.2mm/yr. This is believed because of the absence of an electric double layer which resulting in no charge transfer on the metal-electrolyte interface.

Although Mg is a reactive metal that is susceptible to oxidation reactions, however, a long-term immersion causes Mg surfaces to be less stimulated with the formation of positively charged Mg ions. By comparing the data between B and A_2, the Mg metal corrodibility is increased in the E_{ocp} system (B <A_2), while the immersion in the E_{pzc}, showed the decreasing of metal corrodibility (B> A_2) (Table 3). The increasing in metal corrodobility in the E_{ocp} system is a natural process as the presences of Mg ions (Mg^{2+}) the presence of Mg ions (Mg^{2+}) in the electrode-electrolyte interface creates an electric double layer that constantly promotes a charge transfer between the Mg surface and NaCl solution, resulting in corrosion reaction to occur. Thus, at E_{ocp}, corrosion does occur due to charge transfer, meanwhile, the metal corrodibility under E_{pzc} was significantly reduced. It is believed caused by the absence of an electric double layer which resulting in no charge transfer on the metal-electrolyte interface [7].

Table 3: Corrosion analysis and findings through PIT

Parameter	E_{ocp}	E_{pzc}
B - A_1	-1562.7	-98.2
B - A_2	-19.4	17.5
Criteria	$A_1 > B < A_2$	$A_1 \geq B > A_2$
Description	The tendency of corrosion due to liquid corrosiveness is decreased, whereas metal corrodibility is increased.	The tendency for liquid corrosiveness is decreased whereas the metal corrodibility is decreased.

The morphology of the immersion effect on the Mg samples surface is shown by the stereomicroscope images in Fig. 2 and 3. From the top views, it is noted that Mg immersed in 3.5% NaCl solution under E_{ocp}, for 20 and 25 days suffered severe corrosion (Fig 2). The masking profile measured by volume shows A_t and A_{t+1} samples losing 1.15 and 1.28 g/cm^3, respectively. The increase of Mg volume loss due to corrosion reaction from day 20 to day 25 shows that Mg is a highly reactive metal in 3.5% NaCl solution. However, Mg soaked under E_{ocp} showed no corrosion effect (Fig 3 for the same period. It is also observed that there are still machining marks on the sample surface. These findings suggest that corrosion protection at E_{pzc}, has been proven to be a successful method for Mg oxidation mitigation.

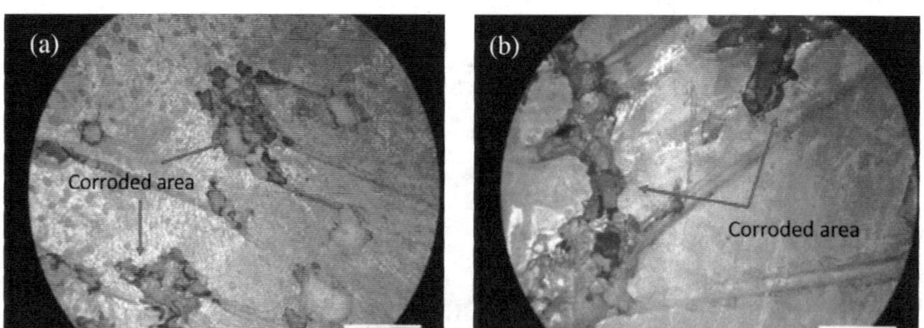

Fig. 2. Stereoscopic images of the Mg surface of the Mg under E_{ocp} exposed in 3.5% NaCl solution for (a) 20 and (b) 25 days.

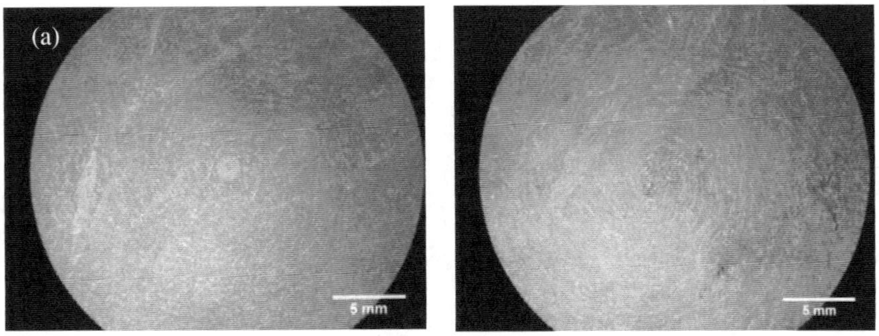

Fig. 3. Stereoscopic images of the Mg surface of the Mg under E_{pzc} exposed in 3.5% NaCl solution for (a) 20 and (b) 25 days

4 Conclusion

The efficiency of ZCCP at E_{pzc} for pure Mg metal in 3.5% NaCl solution is in the range of 93 to 96%. The protection efficiency of ZCCP is higher than 93% for iron-based metal or steel. The morphology of the immersion effect on the Mg immersed in 3.5% NaCl solution under E_{ocp}, suffered severe corrosion while Mg soaked under E_{ocp} showed no corrosion effect for the same period.

The author would like to acknowledge the support from the Fundamental Research Grant Scheme (FRGS) under a grant number of FRGS/1/2015/TK10/UNIMAP/02/5 from the Ministry of Education Malaysia. Thanks to the School of Materials Engineering & Universiti Malaysia Perlis for providing special financial support for enabling the publication of this manuscript.

5. References

1. S. Trasatti, E. Lust. *Modern Aspects of Electrochemistry*. 1-215 (2002)
2. M. Sajjadnejad, A. Mozafari, H. Omidvar, M. Javanbakht. Appl. Surf. Sci. **300**. 1-7 (2014)
3. F. Bimbi, P.E. Alvarez, H. Vaca, C.A. Gervasi. Corr. Sci. **92**. 192-199 (2015)
4. G. L. Song, Z, Shi. Corr. Sci. **85**. 126-140 (2014)
5. M. Curioni. Electroc. Acta, **120**. 284-292 (2014)
6. K. Gusieva, C.H.J. Davies, J. R. Scully, N. Birbilis. Inter. Mater. Rev. **60**. 169-194 (2015)
7. K. Ojha, N. Arulmozhi, D. Aranzales, M.T. Koper. *Angewandte Chemie Inter. Ed.* **59**. 711-715 (2020)

IOP Conf. Series: Materials Science and Engineering **957** (2020) 012024 doi:10.1088/1757-899X/957/1/012024

Effect of Sodium Bicarbonate Additions on the Physical, Mechanical and Bioactive Property of Sol-Gel Bioglass

Syed Adam Syed Nuzul Fadzli[1,*], *Alias* Suffi Irni[2], and *Zainuddin* Firuz[2]

[1]Faculty of Engineering Technology, Universiti Malaysia Perlis, Malaysia
[2]School of Materials Engineering, Universiti Malaysia Perlis, Malaysia

Abstract. Porous bioceramics are the materials of choice for non-load bearing bone implants. Calcium phosphates and bioglass are widely used due to excellent biocompability. The primary function of porous bioceramics are as filler material for bone defects. In this research, 10% amount of sodium bicarbonates (Na_2HCO_3) were mixed with sol-gel derived glass powder (SiO_2-CaO-P_2O_5) and sintered at 700 °C for 3 hours. It was found that, additions of sodium bicarbonate induced a foaming effect during sintering of bioglass thus increased the porosity content of the glass-ceramics obtained. However, the increased in porosity significantly reduced the compressive strength of the crystallized glass. The increased in porosity content and formation of sodium related phases within the crystallized glass matrix after sintering resulted in enhancement of its *in vitro* bioactivity property when tested in SBF solution.

1 Introduction

Macroporous structure in the range of 100 to 300 µm can promote tissue ingrowth and enable bone ingrowth as well as the formation of capillary to occur [1-2]. Bioglass with macroporous structure within this range of pore size is required to fulfil the requirement as an implant scaffold. This scaffold structure can be achieved by preparing porous bioglass by using several different methods such as polymeric templating and foaming agent [3-7].
The component of Na_2O in the sol-gel derived glass is important due to addition of network modifier cations; such as sodium (Na^{2+}), non-bridging oxygen (NBO) bonds can be introduced through the disruption of structure. This NBO bonds may increase dissolution from the glass surface promoting the formation silica layer thus improve the bioactivity of bioglass [3-4]. It is necessary for the deposition of Ca^{2+} and P^{5+} ionic species that lead to the bonding and crystallization of apatite layer. Other than that, presence of Na_2O composition in bioglass resulted in the formation of sodium-related crystalline phases after sintering. This crystalline phases which are hard yet biodegradable can increased the bioactivity property of sol-gel derived bioglass [8]. Na_2O also is typically used for lowering the melting temperature in conventional melt-derived glasses [9].

* Corresponding author: syed.nuzul@unimap.edu.my

Our previous studies has found that sintered sol-gel derived glass with macroporous structure and higher mechanical properties can be obtained by the presence of specific Na_2O content [10-11]. Typically, Na_2O component in bioglass composition was obtained by hydrolisis reactions of sodium nitrate precursor in acidic solution which is added during a sol-gel process. Sodium bicarbonate ($NaHCO_3$) is commonly known as baking soda, bread soda, bicarbonate of soda, and cooking soda. It has many uses, ranging from cooking to cleaning and personal hygiene. Its physical appearance is usually a white, odourless, fine powder and has alkaline or basic properties. It dissolves in water, so it is water-soluble and it is nonflammable [12]. Several previous studies were carried out on the preparation of scaffold material using sodium bicarbonate as foaming agent [13-15]. However, no studies has yet been done in the application of sodium bicarbonate as the source of sodium in sol-gel derived glass composition and at the same time acts as pore forming agent. Thus, the aim of this study is to investigate the effect of sodium bicarbonate additions on the physical, mechanical and bioactivity properties of ternary glass composition made by sol-gel method. In this study, the comparison between ternary sol-gel derived glass and sol-gel derived glass with added sodium bicarbonate were discussed and presented.

2 Methodology

Gel-glasses with novel compositions of 60% SiO_2 – 37.5% CaO – 2.5% P_2O_5 (in mol%) were initially prepared via acid catalyzed sol-gel method. The reagents and precursors used to prepare these gel-glasses includes nitric acid (HNO_3, 1 M, Sigma Aldrich), tetraethyl orthosilicate (TEOS, $Si(OC_2H_5)_4$, Sigma Aldrich (99.0%)), tetraethyl phosphate (TEP, $OP(OC_2H_5)_4$, Merck (99.0%)), calcium nitrate tetrahydrate ($Ca(NO_3)_2.4H_2O$, Merck (99.0% - 102.0%)) and sodium bicarbonate (Na_2HCO_3, Merck (96.5%)). At first, 1 ml of nitric acid was mixed with 40 ml of deionized water and stirred continuously using magnetic stirrer for 5 minutes. Then, followed by addition of precursor with the sequence of 10.0 ml TEOS, 2.0 ml of TEP and 8.0 gm of calcium nitrate tetrahydrate. Each of these reagents were dissolved into the solution one by one after being stirred for 40 minutes in between them. For Na-BG, 3.0 gm of sodium nitrate was added and dissolved into the solution. The sol then casted into teflon mold for gelation process. The gel obtained then been subjected to aging at 60 °C for 24 hours followed by drying at 120 °C for 48 hours. The dried gel then heated at 600 °C for 3 hours for stabilization before grinded into fine powder using agate and mortar.

The glass powders were then pressed into 12 mm diameter pellets via uniaxial hydraulic press and sinter at temperature of 700 °C with soaking time and heating rate of 3 hours and 5 °C/min respectively. For compression test, six pellets with 6.00 mm diameter and 10.00 mm height were prepared for each sample parameter. The compression test (Instron 8874) was done with maximum load capacity and speed of 5 kN and 20 seconds respectively. The sintered pellets were immersed into simulated body fluid (SBF) (prepared by following the ISO 23317:2007(E)) solution for 7 days for *in vitro* bioactivity test. Samples (before and after immersed in SBF solution) were characterized for surface morphology by using Scanning Electron Microscopy (SEM) model JOEL JSM6460LA)) and phases analysis using X-ray Diffraction (XRD) model BRUKER D2. The percentage of porosity and densification of glass-ceramic were obtained by dividing bulk density (obtained from Archimedes principle) with rule of mixture (obtained from qualitative analysis based on XRD). Meanwhile, percentage of weight loss and diameter shrinkage of samples were calculated based on the particular data measured before and after sintering process.

3 Results and discussions

The XRD pattern of BG (in figure 1) showed the presence of hydroxyapatite ($Ca_{10}(PO_4)_6OH)$) (JCPDS Card No. 9-432) as the major phase. While, β-tricalcium phosphate ($Ca_3(PO_4)_2$) phase and also known as (β-*TCP*) (JCPDS No: 70-2065) as minor phases. Song Wang *et al.,* (2017), also had found that when temperature reached to 700 °C the β-tricalcium phosphate (β-*TCP*) phase appeared [16]. However, the split of (211) and (112) reflections is not complete due to the fact that the HA is not fully crystallized. This phenomenon is due to the fact that the atoms in BG sample was not obtain sufficient energy for atom rearrangement at the HA lattice to induce complete crystallization [16]. The XRD pattern of Na-BG (in Figure 2) showed the formation of similar crystalline phases after sintering as BG, however, sodium calcium silicate ($Na_2Ca_2Si_3O_9$) (JCPDS No: 75-1685) and sodium calcium hydrogen phosphate ($NaCa(HPO_4)_2$) (JCPDS No: 50-1758) phases were formed. Sodium calcium silicate ($Na_2Ca_2Si_3O_9$) (known as combeite) can influence the bioactivity property [17]. Meanwhile, formation of sodium calcium hydrogen phosphate ($NaCa(HPO_4)_2$) which related to calcium phosphate in bioglass can enhanced the properties in bone tissue engineering [17]. Thus, this results showed that both BG and Na-BG samples has been crystallized and became glass-ceramic compound after sintering at 700 °C.

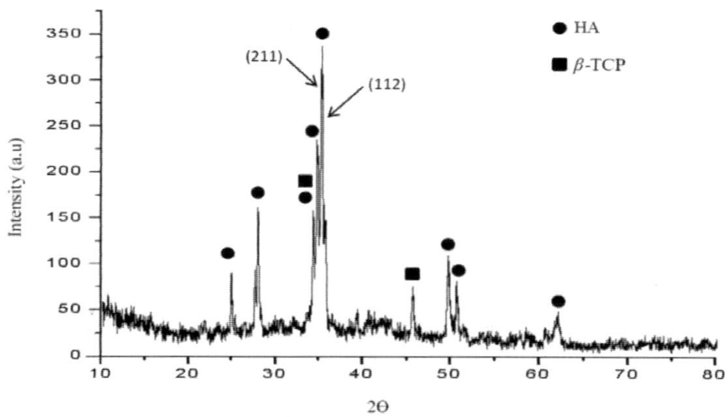

Fig. 1. XRD diffractogram for BG sample

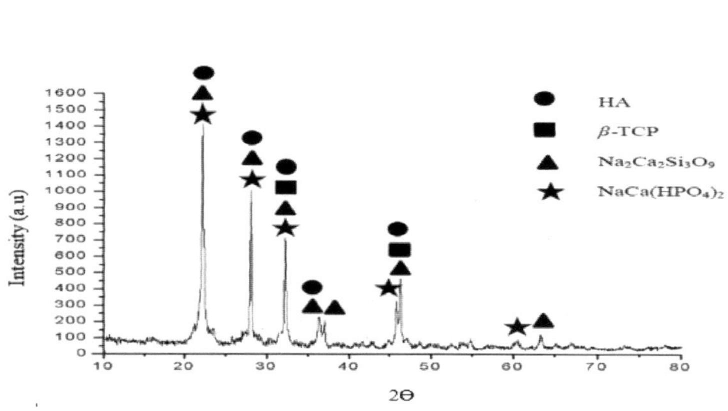

Fig. 2. XRD diffractogram for Na-BG sample.

Table 1 showed the physical and mechanical properties of BG and Na-BG samples. It was found that, the presence of sodium bicarbonate in bioglass affected in higher porosity, higher weight loss (40 wt.%) while reduced in body densification of the glass-ceramics. During sintering process, the decomposition of sodium bicarbonate should be occurred and foaming effect has been taken place. This is also explained by previous study where during sintering, the sodium bicarbonate was decomposed into sodium carbonate, carbon dioxide (CO_2) and vapour (H_2O) [18]. The release of carbon dioxide (CO_2) gaseous and vapour during sintering caused a foaming effect. The foaming effect occurred was in line with diameter expansion at 3.33% experienced by Na-BG compared with BG which exhibited diameter shrinkage at 6.38%. Furthermore, higher weight loss experienced by Na-BG also was due to the decomposition of sodium bicarbonate content in the glass matrice. As explained by Galusek *et al.,* (2013), during the intermediate phase of sintering, the microstructure of samples was undergoing rapid inter-particle neck growth and coalescence [18]. Atoms diffused into the empty space creating the grain boundaries interface. The diffusing action was driven by energy, usually provided by the high temperature during sintering in the form of heat energy reported by M. N. Rahman (2003). As a result, the movement of atoms get into the empty space will cause the sample to shrunk [19]. The compressive strength for BG and Na-BG were 3.3 MPa and 1.2 MPa respectively. The compression strength of Na-BG was apparently much lower compared to BG due to the porosity content of Na-BG was higher. This suggested that porosity of bioglass significantly affected its compressive strength property. The entrapped carbon dioxide gas produced during foaming effect created some holes and channel inside the Na-BG sample which then attributed to the relatively lower compression strength [20].

Table 1. Physical and mechanical properties of bioglass samples.

Properties	Sample	
	BG	Na-BG
Weight loss	24.0%	40.0%
Diameter shrinkage/ expansion	6.38%	-3.33%
Densification	70.0%	55.0%
Porosity	30.0%	46.0%
Compressive strength	3.3 MPa	1.2 MPa

Figure 3 (a) showed the morphology of BG which consisted of inhomogeneous grains, rough surface and apparent porous structure. This was due to the bioglass particles used for sample preparations were inhomogeneous in particle sizes, thus easily produced stress concentration on the surface. Meanwhile, the addition of sodium bicarbonate into bioglass matrice significantly formed glass-ceramic morphology with higher pores structure and larger pore sizes (Figure 3 (b)). This is in par with the sample physical properties which showed the addition of sodium bicarbonate into bioglass formed higher porosity and lower densification of the glass-ceramic obtained.

Fig. 3. Surface morphology of (a) BG and (b) Na-BG both at 500X magnification.

Higher porosity content formed on the Na-BG morphology was particularly due to the foaming effect resulted from release of gaseous during sintering. The high porous structure on the morphology should be the main factor which attributed to lower compression strength of Na-BG. The compressive strength of Na-BG however was under the minimum strength required as scaffold materials for bone implant materials. Because of that, this samples would serve as a better candidate as filler materials in bone defected area instead of scaffolding materials.

The morphology image of BG after immersed in SBF solution for 7 days is shown in Figure 4 (a). The morphology was accumulated by inhomogeneous grains, crack-like and severe pores structures. Under the effect of water erosion and ion exchange between bioactive glass sample and SBF solution, cracks were formed as result of the dual function of physical and chemical reaction [21]. It was clear that, no apatite-like layer was formed on its surface thus indicate the low bioactivity property of BG. Meanwhile, morphology of Na-BG was characterized by the formation of homogenous structure and clearly apatite-like layer was completely cover its surface after soaked in SBF solution for 7 days (Figure 4 (b)). This indicate that the ionic dissolution rate of Na-BG was significantly higher compared to BG. According to Liu *et al.* (2013), the addition of sodium in bioglass sample was important in favoring the apatite formation. The presence of sodium and calcium network modifier cations interrupt the bridging oxygen (BO) bond which causes the formation of non-bridging (NBO) bond that improved the dissolution and bioactivity of glass and fasten HCA nucleation [22]. Furthermore, ion exchange reaction between Ca^{2+} ions from bioglass network and H^{+} from SBF solution took place at the corner of crack easily [19]. Other than that, higher porosity is required to offer sufficient space to reactive site for tissue growth and nutrient supply within the bone graft [23]. This result showed that, increased in porosity structure will increase the bioactive property of bioglass-ceramics due to higher HCA formation.

Fig. 4. Surface morphology of (a) BG and (b) Na-BG after immersed in SBF solution for 7 days.

Conclusion

In conclusion, additions of sodium bicarbonate were succeeded in increasing the porosity content of sintered sol-gel derived glass besides allowing the presence of sodium related crystallized phases. Both of this factor has improved the sol-gel glass *in vitro* bioactivity property when tested in SBF solution. However, the increased in porosity content significantly reduced the compressive strength of the sintered glass. Because of that, sodium contained-BG samples have high potential as the material of choice for bioactive filler materials in bone defects problem.

References

1. V. Karageorgiou, D. Kaplan. Porosity of 3D biomaterial scaffolds and osteogenesis. *Biomaterials*, **26**(27), pp.5474-5491 (2005)
2. X. Liu, M. Rahaman, Y. Liu, B. Bal, L. Bonewald. Enhanced bone regeneration in rat calvarial defects implanted with surface-modified and BMP-loaded bioactive glass (13-93) scaffolds. *Acta Biomaterialia*, **9**(7), pp.7506-7517 (2013)
3. I. Sabree, J. Gough, B. Derby. Mechanical properties of porous ceramic scaffolds: Influence of internal dimensions. *Ceramics International*, **41**(7), pp.8425-8432 (2015)
4. X. Liu, M. Rahaman, Q. Fu. Bone regeneration in strong porous bioactive glass (13-93) scaffolds with an oriented microstructure implanted in rat calvarial defects. *Acta Biomaterialia*, **9**(1), pp.4889-4898 (2013)
5. Q. Chen, I. Thompson, A. Boccaccini. 45S5 Bioglass®-derived glass–ceramic scaffolds for bone tissue engineering. *Biomaterials*, **27**(11), pp.2414-2425 (2006)
6. J. Jones, L. Ehrenfried, L. Hench. Optimising bioactive glass scaffolds for bone tissue engineering. *Biomaterials*, **27**(7), pp.964-973 (2006)
7. Z. Wu, R. Hill, J. Jones. Optimizing the Processing of Porous Melt-Derived Bioactive Glass Scaffolds. *Bioceramics Development and Applications*, **1**, pp.1-4. (2010)

8. K. Xie, L. Zhang, X. Yang, X. Wang, G. Yang, L. Zhang, H. Shao, Y. He, Ji. Fu, Z. Gou. *Biomedical Glass* **1** (2015)
9. Q. Chen, Y. Li, L. Jin, J. M. W. Quinn, and P. A. Komesaroff: *Acta Biomater.* **6** (2010)
10. S.N.F. Adam., J. Banjuraizah, Z. Firuz. Sol Gel Synthesis and Preparation of Macroporous Glass: Effect of Sodium Nitrate Addition. *International Journal of Current Research in Science, Engineering & Technology*, **1** (2018)
11. Zainudin, S., S.N.F. Adam, D.S.Che Halin, M. Yusof, J. Banjuraizah Z. Firuz.. Preparation and Characterization of Macro Porous Glass-Ceramics as Bioactive Scaffold Material. *Solid State Phenomena*, **281**, pp.83-89 (2018)
12. Sodium Bicarbonate Supplements and Exercise Performance", Healthline. [Online]. Available: https://www.healthline.com/nutrition/baking-soda-and performance [Accessed: 09- Dec- 2018].
13. X. Shen, J. Ruan, Z. Zhou, H. Zhang. Fabrication of porous scaffolds using $NaHCO_3$ particulates as the porogen material. *Journal of Wuhan University of Technology-Mater. Sci. Ed.*, **22**(2), pp.279-283 (2007)
14. F. Dehghani, N. Annabi. Engineering porous scaffolds using gas-based techniques. *Current Opinion in Biotechnology*, **22**(5), pp.661-666 (2011)
15. M. Fauzi, N. Du, H. Osman, A.G. Supri. Effect of Sodium Bicarbonate as Blowing Agent on Production of Epoxy Shape Memory Foam using Aqueous Processing Method. *Sains Malaysiana*, **44**(6), pp.869-874 (2015)
16. I. Hung, W. Shih, M. Hon, M. Wang. The Properties of Sintered Calcium Phosphate with [Ca]/[P] = 1.50. *International Journal of Molecular Sciences*, **13**(12), pp.13569-13586 (2012)
17. J. Chen, L. Zeng, X. Chen, T. Liao, J. Zheng. Preparation and characterization of bioactive glass tablets and evaluation of bioactivity and cytotoxicity in vitro. *Bioactive Materials*, **3**(3), pp.315-321 (2018)
18. S. Wang, Y. Wang, K. Sun, X. Sun. Low temperature preparation of α-tricalcium phosphate and its mechanical properties. *Processing and Application of Ceramics*, **11**(2), pp.100-105 (2017)
19. M. N. Rahaman, *Ceramics Processing and Sintering*, 2003.
20. E. Aguilar-Reyes, C. León-Patiño, B. Jacinto-Diaz, L. Lefebvre. Structural Characterization and Mechanical Evaluation of Bioactive Glass 45S5 Foams Obtained by a Powder Technology Approach. *Journal of the American Ceramic Society*, **95**(12), pp.3776-3780 (2012)
21. R. Li, A.E. Clark, L.L Hench. An investigation of bioactive glass powders by sol-gel processing. Journal of Applied Biomaterials, **2**,231-9 (1991)
22. M. Trivedi. An Impact of Biofield Treatment on Spectroscopic Characterization of Pharmaceutical Compounds. *Modern Chemistry & Applications*, **3**(03) (2015)
23. N. Gupta, D. Santhiya, *Mesoporous bioactive glass and its applications*. Bioactive Glasses (Second Edition). pp. 63-85 (2018)

Effect of Photoetching Process onto Immobilized PANI/TiO_2 Films Towards Photocatalytic Degradation of RR2 dye

Sharin Razak[1,*], *Nazry* Salleh[1], *Wan Izhan* Nawawi[2], and *Nadiah* Natar[2]

[1]School of Materials Engineering, Universiti Malaysia Perlis, 02600, Arau, Perlis, Malaysia.
[2]Photocatalysis Laboratory, Universiti Teknologi MARA, Perlis, 02600 Arau, Perlis, Malaysia.

Abstract. Polyaniline (PANI) and P-25 Titanium dioxide (TiO_2) was fabricated as immobilized P-25TiO_2/PANI/ENR/PVC film and reported on the photocatalytic study. PANI and TiO_2 was immersed into the organic solvent and being coated onto glass plates under influenced of ENR/PVC polymeric binder by using a dip-coating method. It was found that, the removal of RR2 dye nearly good as suspension system after the pre-irradiation process. Based on COD and BET results, it showed that the optimum time of etching process was 7 hours irradiation time. This pre-irradiation process enhanced the removal of dye due to leached out of excessive binder and also exhibit visible light active after insertion of PANI into the immobilized system.

1 Introduction

The use of immobilized photocatalysts has been proposed as a method to overcome suspended system since 1993 [1]. Immobilization involves the use of inert supports, such as glass, activated carbon, silica-based materials and polymeric materials. However, the main drawbacks of using immobilized system for the photocatalytic degradation process are low stability during continuous applications. Moreover, the limitation of mass transfer due to the reduction of specific surface area also reduced its potential applicability to waste water treatment [2]. To solve those problems Nawi and Zain, (2012) and Nawi et al. (2012) had proven that immobilized TiO_2 system using ENR/PVC blend as the organic binder improved its photocatalytic activity once most of the organic binders were etched out using light irradiation. This pre-irradiation process allowed the regeneration of the surface area to within 86% of the pristine TiO_2 particles. The enhanced surface properties of the immobilized photocatalyst system generated a photocatalytic performance as good as the slurry method of the TiO_2 particles during the photocatalytic degradation of an aqueous MB dye solution [3].

* Corresponding author : fairulsharin@unimap.edu.my

In this study, a conjugated polymer polyaniline (PANI) was selected as amongst of versatile heterogeneous-conjugated polymers that exhibit good stability toward environment. Furthermore, it represents as a hole scavenger and improve photocatalytic activity when electrons are added into photocatalyst conduction band under visible light irradiation [4]. Regardless of having a good performance in photocatalytic activity, there are still no focus attention on sustainability and re-usability of the system on the longer operation time which can minimize the secondary cleaning process and also operational cost.

2 Methodology

The P-25TiO_2/PANI/ENR/PVC solution was prepared by addition of PANI soluble in chloroform prepared by slight modification of method described by Palaniappan et al.(2006) [5]. This soluble PANI was added onto an optimized ENR/PVC binder prepared exactly as reported of the work of Nawi et al. (2012). Then 6 g of P-25 TiO_2 was added into the mixture and followed by another sonication process for 8 h in order to homogenize the mixture. Complete homogenization was usually indicated by good homogenity of the PANI/TiO_2 nanoparticles inside the solution with additives of PVC/ENR as binders. Finally, the homogenized formulation was sealed up tightly and stored in an ambient atmosphere to avoid unnecessary evaporation. For immobilized purposes, P-25TiO2/PANI/ENR/PVC solution was coated onto glass plate by using a dip coating technique. The amount composite loading coated onto glass were fixed according to the parameter study.Meanwhile during irradiation, fluoresence lamp and sunlight was used as source of energy for the photocatalytic study. SEM micrographs and EDX elemental analyses were carried out by using scanning electron microscope (SEM) (LEO SUPRA 50 VP Carl-Ziess SMT) and energy dispersive X-ray spectrometer (EDX) (Oxford INCA 400 Energy). Meanwhile Chemical Oxygen Demand (COD) reagent contained about 0.1 g mercury sulphate will mix 2mL of each sample aliquot with 3 mL COD reagent. On the other hand, another test tube containing ultrapure water and the COD reagent was also prepared to serve as a blank solution. The mixture was then vigorously shaken before being refluxed in the HACH Cod reactor at 140°C for two hours. After the refluxing process, the mixture was cooled down to room temperature. The COD concentration was measured with a HACH DR2000 spectrophotometer at 620 nm [6].

3 Results and Discussion

ENR/PVC blend organic polymer was a good binder to fix the catalyst onto the glass plate. However, this binder could decompose by the photocatalytic reaction and leach out a dissolved organic matter into the treated water. In this regards, COD test was carried out to determine the amount of leachable organics from the plates. Table 1 shows the pre-irradiation process using P-25TiO_2/PANI/ENR/PVC system over the span of 10 h. As suspected, organic binders were etched out and was measured based on COD concentration. Based on Nawi and Zain, (2012), most of the contribution of COD values came from the ENR polymer. Furthermore, as reported by Nawi and Zain, (2012), it was observed that removal of ENR was high at the early stage of the irradiation period and achieved zero COD values after 4 h of irradiation respectively. This was due to the ENR binder which was unstable state during it's modification with a peracid. Nevertheless, P-25TiO_2/PANI/ENR/PVC system needed longer time to remove the leachable organic from the irradiated plate. Table 1 shows that the degradation of the polymer blend occurred significantly during the first 3 h of irradiation and it started to slow down until it achieved zero value at 7 h onwards. Hence, it can be concluded that the optimum period of the pre-irradiation process was 7 h and it was suspected at this period of time most of the organic binders were etched out from the system.

Table 1: The COD value of treated sample upon irradiation time by using P-25TiO$_2$/PANi/ENR/PVC plate.

Etching time (h)	0	2	3	4	5	6	7	8	9	10
COD (mg/L)	183	82	47	38	24	11	0	0	0	0

3.1 SEM-EDX and BET analysis

Based on the effect of the pre-irradiation process, the non-irradiated and irradiated plates were subjected to SEM analysis in order to observe the surface morphology of the coatings before and after 7 h of light irradiation. Figure 1 presents the SEM micrographs for freshly prepared P-25TiO$_2$/PANI/ENR/PVC plate and the 7 h pre-irradiated plate under fluorescent lamp source.It can be seen that, the surface morphology of freshly prepared P-25TiO$_2$/PANI/PVC was quite compact and dense due to TiO$_2$ particles that were enwrapped by the organic binders. The image also shows the particles were less porous due to most of the particles were bounded together. After the plates were subjected to 7 h irradiation, the particles were clearly visible with less bounded and become more porous since the organic binders were degraded and removed upon light irradiation. This observation is consistent with the COD concentration values as at 7 h and onwards gave zero COD which indicates that most of the binders were degraded and removed from the immobilized system [7-12].

Fig.1. SEM images at 1,000X; a) before irradiate , (b) after 7 hrs irradiation

Table 2: The surface area from initial to 7 hours irradiation and amount of % C content and within the P-25TiO$_2$/PANI/ENR/PVC wih time irradiation.

Etching time (h)	Weight % C	Surface area (s_{bet})
0	16.55	28
7	5.96	34

Table 2 illustrates the decrease of % C content within the P-25TiO$_2$/PANI/ENR/PVC with time of irradiation. Upon 7 hour of irradiation, % C decreased from 16.85% to 5.95 %. This decrease in C content can be due to the degradation of ENR, PVC or PANI. As indicated in Table 2, partial removal of the organic binders had increased the surface area from the initial 28 m^2 g^{-1} to 34 m^2 g^{-1}.

3.2 Photocatalytic Activity Study

Figure 2. shows the photocatalytic degradation of RR4 dye under UV and visible light conditions.As has been pointed out by many reports [14-18], doping of PANI into the TiO_2 catalyst enabled the resulted composite to be visible light active. Figure 2. shows the results obtained under both unfiltered and UV filtered (visible light irradiation) lamps for various photocatalyst systems in the degradation of RR4 dye. For comparison purposes, the results of photocatalytic degradation of RR4 by an equivalent amount of P-25TiO_2 powder in the slurry system are also shown in the Fig. The P-25 TiO_2 slurry system only required 15 minutes to decolorize RR4 dye under unfiltered 45-W fluorescent lamp. However, once the fluorescent lamp was attached with a UV filter, the P-25 slurry system was essentially inactive where only 20 % of RR4 was removed and this was mainly due to the adsorption process. Nevertheless, the immobilized P-25TiO_2/ENR/PVC-7h exhibits some visible light activity whereby the removal of RR4 dye after 60 minutes of irradiation was 35%. As suspected the P-25TiO_2/PANI/ENR/PVC-7h was clearly visible light sensitive where 85 % of RR4 dye was decolorized after 60 minutes of irradiation.

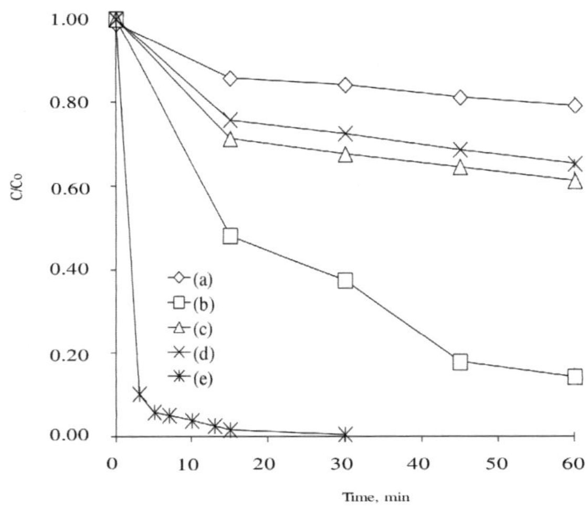

Fig. 2. Photocatalytic degradation of 30 mg L^{-1} RR4 dye under visible light irradiation for various photocatalyst systems: (a) TiO_2 slurry UV filtered; (b) P-25TiO_2/PANI/ENR/PVC; (c) P-25TiO_2/PANI/ENR/PVC-7h; d) P-25TiO_2/ENR/PVC-7h and (e) P-25TiO_2 slurry

4 Conclusion

Photocatalytic degradation of RR2 dye by using P-25TiO_2/PANI/ENR/PVC plate under 7 hours pre-irradiation time has been successfully improved the photocatalytic activity of the system whereby its photocatalytic activity was almost as good as the suspended system.The increment of the photocatalytic activity was in line with the increased BET surface area due to removal of excessive binders toward etching process prolong for 7 hours. Additionally, it had also clearly proven that the immobilized P-25TiO_2/PANI/ENR/PVC-7h photocatalyst systems were clearly visible light sensitive due to insertion of PANI conjugated polymer

We would like to thanks the Ministry of Higher education (MOHE) for providing generous internal financial support under FRGS grant: FRGS/1/2019/STG01/UNIMAP/02/2 in conducting this study under Universiti Malaysia Perlis and Universiti Teknologi MARA (UiTM), Arau for providing all the facilities.

References

1. A.Y. Shan, T.I.M. Ghazi, S.A. Rashid, Appl. Catal.A: Gen., **389** (1-2) (2010).
2. K. C Chen, and V. K. Pham, Forsight , **16(3)** (2014).
3. M.A. Nawi, M.S. Zain, Appl.Surf.Sci., **258**(2012)
4. S. Razak, M. A. Nawi, and K. Haitham, Appl. Surf. Sci., **319** (2014)
5. S.Palaniappan, C.A.Amarnath, React. Funct. Polym., **66**(12) (2006)
6. R. Sedghi, H.M. Moazzami, S.S.H. Davarani, M.R. Nabid, A.R. Keshtkar, A J. Alloys Compd, **695** (2017)
7. W.I. Nawawi, M.A. Nawi, Mol. Catal. A; Chemical, **383- 384** (2014)
8. W.I. Nawawi, M.A. Nawi, Mol. Catal. A; Chemical, **374- 375** (2013)
9. W.I. Nawawi, M.A. Nawi, Appl. Catal. A: Gen. **453** (2013)
10. N.N. Bahrudin, M.A. Nawi, W.I. Nawawi, Korean J. Chem. Eng **35** (2018)
11. N.A. Sabri, M.A. Nawi, W.I. Nawawi, Optical Materials, **48** (2015)
12. W.I. Nawawi, A.Y. Ani, M.A.M. Ishak, A. Ramli, M.S. Azami, F. Zaid, F. Desalin. Water Treat. **113** (2018)
13. M.A. Nawi, Y.S. Ngoh S.M. Zain,Hindawi (2012)
14. N.N. Bahrudin, M.A. Nawi, W.I. Nawawi, Mat. Res Bul. **106**
15. H. Bejbouj, L. Vignau, J.L. Miane, M. Dang, E. Oualim, M. Harmouchi, A. Mouhsen, Sol. Energy Mater. Sol. Cells, **94,** 2 (2010)
16. J.Y. Lin, W.Y. Wang, Y.T. Lin, Surf. Coat. Tech., **231** (2013)
17. K. Haitham, S. Razak, M.A. Nawi, Arab. J. Chem. (2014)
18. H. Zhang, R. Zong, J. Zhao, Y. Zhu, Environ. Sci. Technol., **42** (2008)

Preparation of a Visible light Active N-doped TiO_2 Photocatalyst using Microwave Irradiation Technique

Wan Izhan Nawawi[1], *Sharin* Razak[2,*] and *Nazeri*[1]

[1]Photocatalysis laboratory, Universiti Teknologi MARA, Arau Campus, 02600 Arau, Perlis, Malaysia.
[2]School of Materials Engineering, Universiti Malaysia Perlis, 02600, Arau, Perlis, Malaysia.

Abstract. The dye wastewater which directly discharges without treatment process will affect the environment. The semiconductor like titanium dioxide (TiO_2) has gained attention as most potential material for photocatalysis process to treat the dye wastewater with act as photocatalyst. However, this semiconductor has their limitation toward the photodegradation under visible light irradiation. In this study, TiO_2 has been modified by doping with nitrogen (N) to improve the effectiveness under visible light irradiation. This simple prepared N-doped TiO_2 was prepared by mixing of TiO_2 powder with urea as N precursor under microwave irradiation instead of using muffle furnace as heating media. The nitogen was chemically bonded with TiO_2 as proven by XPS analysis. N-doped TiO_2 sample show active under visible light irradiation due to lower band gaps energy of sample observed by UV/Vis-DRS. An active photo response under visible light was observed from N-doped TiO_2 with 80 minutes time irradiation to complete the reactive red 4 (RR4) color removal, while no photocatalytic degradation was observed from unmodified TiO_2 under the same irradiation.

1. Introduction

Advance oxidation process (AOPs) has gained attention by researchers and it give a great offers for degradation of dyes pollutant. AOPs technique employed from chemical oxidation process which requires hydroxyl radicals to remove the organic pollutant [1]. Heterogenous photocatalysis is one of famous applied processes in advanced oxidation and titanium dioxide (TiO_2) is one of promising materials due to its non-toxic, cheap material, and chemically stable [1-3]. Eventhough, this material has owns advantageous among others photocatalyst conversely it only can absorb (ultra-violet) UV light irradiation due to the wide band gap energy (3.1 eV). At this point, TiO_2 is not fully utilized the entire solar system in which consist only 5% of UV light region [4-5]. Efforts have been made to make this photocatalyst active under visible light solar spectrum by modifying TiO_2 photocatalyst [6-10].

* Corresponding author: fairulsharin@unimap.edu.my

Thus, the doping techniques have been introduced to enhance the photocatalysis toward the visible light. Non-metal element (N, C, S, F and P) have been discovered as great potential for enhance the visible light in TiO_2 materials [11-13]. Among all the non-metal doped TiO_2 materials, nitrogen doped TiO_2 have been discerned to exhibit superior photocatalytic activity under visible light irradiation. In previous works indicated that N doping supreme potential toward the visible-light absorption due to the metastable center formation, small ionization energy, and its comparable atomic size in oxygen [14]. Various procedures are already done with the intention of preparation N-doped TiO_2 such as solvothermal, ion implantation, sol-gel, calcination, sonochemical, and sputtering techniques. Conversely, these previous methods are very complicated due to the process either using specific equipment for modification or involved lots of chemical like a solvent to dissolve the precursor [15]. The drawback of previous studies can be overcome by preparing N-doped TiO_2 by using solid state preparation method under microwave irradiation. It was applied by using commercially available P25 and urea as precursor under microwave treatment method. From the best of our knowledge there is no study focused on the modification of N-doped TiO_2 using commercially available TiO_2 under microwave. In this work, N-doped TiO_2 sample was preparad by microwave irradiation. This prepared N-doped TiO_2 was applied to degrade reactive red 4 (RR4) dye for photocatalytic activity performance.

2. Methodology

TiO_2 Degussa P25 was synthesized by mechanical mixing using weight ratio TiO_2: Urea 7:3 for 5 min using ball mill. The mixed powder was placed into a conical flask in the semi-closed reactor and heated by using microwave at the power irradiation 800 Watt for 30 minutes under normal atmospheric condition. The sample was cooled down to the room temperature and the yellow light powder was produced and denoted as N-doped TiO_2. The band gap energy of photocatalyst was determined by using UV-Vis spectrophotometer, model Lambda 35 from Perkin Elmer equipped with diffuse reflectance attachment at the wavelength of 200-800 nm. X-ray photoelectron spectroscopy (XPS) measurement was used to detect the presence of nitrogen binding on the titania that is responsible for the visible absorption. The photocatalytic performances of N-doped TiO_2 and unmodified TiO_2 samples were investigated in the degradation of RR4 dye under LED light irradiation. The photodegradation activity was carried out by adding 0.030 g of photocatalyst sample into 25 ml of 30 mg L^{-1} RR4 dye to form suspensions. This suspension was then poured into a glass cell with dimension 50 mm × 10 mm × 80 mm (L × B × H) and irradiated with 20 Watt LED lamp with light intensity (Visible: 237 W m^{-2}) was used as the visible light source. An aquarium pump model NS 7200 was used as an aeration source for oxygen supply. During each photocatalytic experiment, the decolourization degree of RR4 was determined at specific time interval until complete decolorization was achieved. The absorbance was measured by using HACH DR 1900 spectrophotometer at 517 nm.

3. Result and Discussion

The XPS peaks of N-doped TiO_2 shows the presence of N 1s with binding energies at 404.8 and 400.2 respectively as shown in Fig. 1. Based on previous study, binding energy greater than 400 eV indicates there was interstitial N doping and formation of N-O-Ti species. According to Azami et al [16], the peak at 400.2 eV was assigned to the binding energy of N atom that chemically bonded to the TiO_2 as a Ti-O-N structural feature. Binding energy of 404.8 eV indicates existence of the N-O bond in the form of nitric oxide, nitrogen monoxide

(NO) or nitrite formations [17]. Based on Fig. 1, deconvolution of N 1s peak reveals the presence of two peaks with binding energy 404.8 and 400.2 eV in N-doped TiO_2 sample. This clearly indicated that a chemical interaction has occurred between N and titania forming N doped TiO_2.

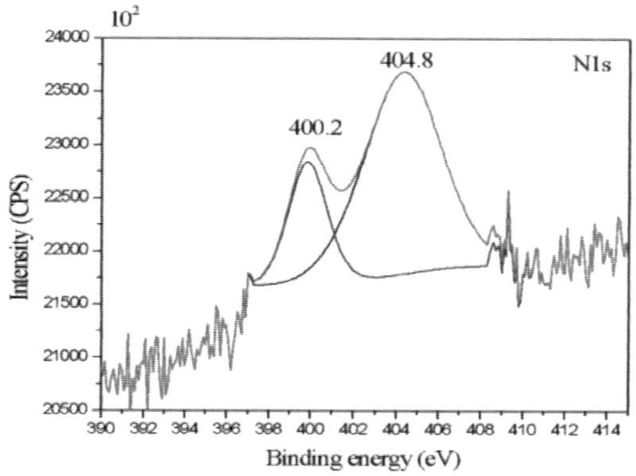

Fig. 1. XPS Spectra of EVONiC sample; N 1s.

The present of nitrogen in TiO_2 will narrow the band gap of TiO_2. Thus, the UV-Vis DRS analysis will prove there is narrowing band gap in the TiO_2. Fig. 2 shows the UV-Vis diffuse reflectance spectra for N-doped TiO_2 and unmodified TiO_2. Based on Fig. 2, the absorption edge of unmodified TiO_2 is around 400 nm (2.9 ev of bandgap energy calculated by using tauc equation) while for N-doped TiO_2 is around 430 nm (3.2 eV). A wavelength for UV light is from 200 to 400 nm while wavelength for visible light is from 401 to 700 nm [18]. N-doped TiO_2 sample shows the shifted absorption towards visible light region while unmodified TiO_2 in the UV region.

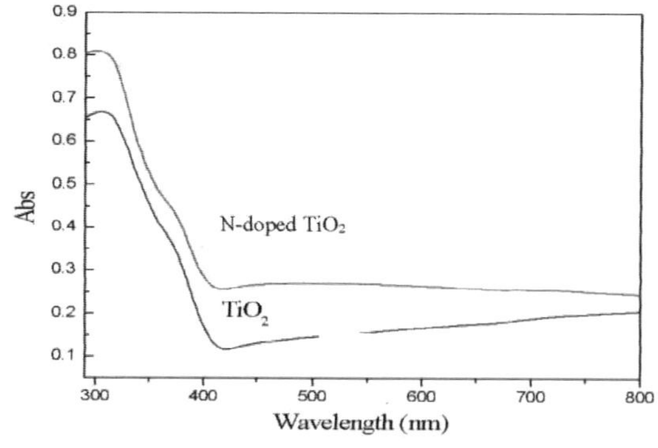

Fig. 2. Graph UV-VIS DRS of TiO_2 and N-doped TiO_2.

Fig. 3 shows the photocatalytic degradation of RR4 dye by N-doped TiO_2 and unmodified TiO_2 under LED light (237 W m^{-2}). It was found that N-doped TiO_2 took only 60 minutes to complete the degradation of RR4. No photocatalytic degradation under unmodified TiO_2 was observed. However, RR4 dye was removed about 40% for 30 minutes and no decreasing was detected after pro-long the irradiation and this trend is due to the adsorption effect of RR4 dye onto TiO_2 as been reported by others [5-7]

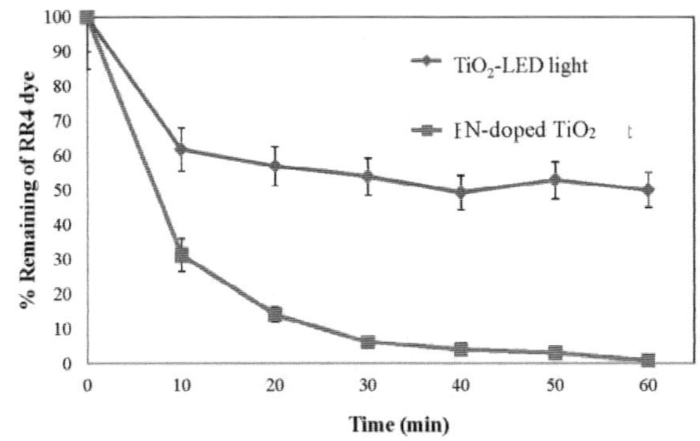

Fig. 3. Photocatalytic study of RR4 dye using TiO_2 and N-doped TiO_2 under LED light as a visible light source.

4. Conclusion

A simple technique was successfully applied for the preparation of visible light active N-doped TiO_2 by mixing urea and TiO_2 powder under microwave irradiation. The XPS analysis was chemically identified that the nitrogen incorporated with titania crystal to producing Ti-O-N linkage. The band gap energy of the N-doped TiO_2 was reduced to 2.9 eV recorded by UV-Vis DRS spectrum. In addition, the presence of N makes N-doped TiO_2 active under visible light irradiation. This N-doped TiO_2 allows the development of more environmental friendly to produce N-doped TiO_2 by utilizing urea as N source and more cost effective process by using commercially available photocatalyst.

We would like to thanks the Ministry of Higher education (MOHE) for providing generous internal financial support under FRGS grants: FRGS/1/2017/TK10/UITM/02/11 in conducting this study and Universiti Teknologi MARA (UiTM) and Universiti Malaysia Perlis for providing all the facilities.

References

1. W. I. Nawawi, M. S. Azami, L. S. Ang, M. A. M. Ishak, K. Ismail Water Quality Research Journal **52** (2017)
2. W. I. Nawawi, M. A. Nawi J. Mol Catal A Chem: Chemical **383** (2014)
3. M. A. Nawi, S. Sabar, and Sheilatina J. Colloid Interface Sci. **372** (2012)
4. Y. Cong,J. Zhang, F. Chen and M. Anpu J. Phys. Chem. C 2007, **111** (19) (2007)
5. W. I. Nawawi, M. A. Nawi J. Mol Catal A Chem: Chemical **374** (2013)
6. I. Ismail, W. I. Nawawi, M. A. Nawi, React. Kinet. Mech. Catal. (2014)
7. N. A. Sabri, M. A. Nawi W. I Nawawi Opt. Mater. **48** (2015)
8. W. I. Nawawi R. Zaharudin M. A. M Isha. K. Ismail A. Applied Sciences **7** (2016)
9. M. S. Azami S. K Ain. R. Zaharudin F,Bakar W. I. Nawawi MATEC Web of Conferences **47** (2016)

10. F. Bakar, M. S. Azami, S. K. Ain, R. Zaharudin, W. I. Nawawi Applied Mechanics and Materials **835** (2016)
11. J.P.Michael, W.D.Charles, P.Parkin Journal of Photochemistry and Photobiology A: Chemistry **281** (2014)
12. X. Wu,S. Yin, Q. Dong. Applied Catalysis B: Environmental **142-143** (2013)
13. M. S. Azami, W. I. Nawawi, M. A. M. Ishak, K. Ismail, Z. Ahmad A. H. Jawad MATEC Web of Conferences **47** (2016)
14. P. Miguel, T.N. Nicholas Appl. Catal. B **125** (2012)
15. T. H. Kim, Catal. Today **212** (2013)
16. M. S. Azami, S. K Ain, R. Zaharudin F. Bakar and W.I. Nawawi Applied Mechanics and Materials, **885** (2015)
17. S. Buzby, M. A Barakat. and H. Lin J. of Vacuum Sci. and Tech. **24** (2006)
18. J. Ananpattarachai, P. Kajitvichyanukul and S. J. of Hazard. Mater. **168** (2009)

IOP Conf. Series: Materials Science and Engineering **957** (2020) 012027 doi:10.1088/1757-899X/957/1/012027

Low temperature prepared N doped TiO_2 photodegradation of janus green B

Siti Aisyah Othman[1], *Siti Nurjannah* Zainon[1], *Mohd Fauzi.* Abdullah[1], *Wan Izhan* Nawawi[1], *Syahirul* Ikhwan[1], and *Sharin.* Razak[2]*

[1]Photocatalysis Laboratory, Universiti Teknologi MARA, Arau Brunch, 02600 Arau, Perlis, Malaysia.
[2]School of Materials Engineering, Universiti Malaysia Perlis, 02600, Arau, Perlis, Malaysia.

Abstract. N doped titanium dioxide (TiO_2) was prepared under low temperature calcination of TiO_2 Degussa P25 powder and urea at various mixing ratios. Muffle furnace was used as heating media with the temperature used in the range of 300 to 500 °C. The prepared N doped TiO_2 was characterized by using UV-Vis diffuse reflectance spectroscopy (DRS), Fourier transform infrared (FT-IR) and scanning electron microscopy (SEM). It was observed that N doped TiO_2 prepared at 0.5 g urea under the 400 °C calcination process (named as U2-400) is the optimum photocatalyst sample with the highest rate removal of Janus green B (JG) dye. The U2-400 had goes the photodegradation rate removal at 0.1071 min^{-1} where it is 3.5 times faster compared to unmodified TiO_2 sample. The high photodegradation rate removal for U2-400 was due to the N incorporated with TiO_2 in U2-400 photocatalyst which narrowed the bandgap energy from 3.2 to 3.0 eV as observed by UV-Vis DRS spectra. This shows the U2-400 sample has increased its photocatalytic activity since it has become more active under wider an optical response compared to the unmodified TiO_2

1 Introduction

Janus green B (JG) dye is one of the major contributors of the organic pollutants. This organic pollutant if discharged into the water will cause a dark blue to pink color and unpleasant taste of drinking water [1]. The JG dye is mostly used in industries as pigment agent in coloring the material compounds like textile and polymer materials. The contamination of water caused by JG dye has attracted the vast attention by many researchers due to its. toxicity to organisms and degrading difficulties [2]. Physical, biological or chemical treatment process is normally used in conventional wastewater technologies remediation [3]. [1]

Advanced oxidation process (OAPs) is a chemical treatment method to remove organic pollutants and break down all organic structures in water or wastewater by oxidation through reaction with hydroxyl radical (·OH). AOPs involves two steps which are hydroxyl

[1]*Corresponding author : fairulsharin@unimap.edu.my

radicals foundation and the reaction of these radicals with organic contaminants in wastewater. Among many AOPs, titanium dioxide (TiO_2) photocatalysis has gained more attention as one of the most environment friendly cleanup technologies, and the cost is lower than the conventional treatment process [4,5].

However, TiO_2 is less efficient under visible light region due to its wide bandgap energy. A pure TiO_2 has a large bandgap which means it can only be activated under ultraviolet (UV) region. Doped TiO_2 with non-metal elements such as boron (B), sulphur (S), carbon (C), flourine (F), and nitrogen (N) is an effective way to narrow the bandgap energy [6]. One of the promising non-metal dopant material is N due to its comparable atomic size with oxygen, small ionization energy, eco-friendly, higher stability and simple synthesis methods that can induce visible absorption [6]. N doped TiO_2 has becoming one of the most dynamic modifications of TiO_2 for photodegradation of organic pollutants. Several approaches have been employed for the preparation of N doped TiO_2 material which are the wet chemical and dry powder methods. However, most of the preparation of N doped TiO_2 are focus on wet chemical methods such as hydrothermal treatment, sol-gel processes, spray pyrolysis, and supercritical methods.

In this study, N doped TiO_2 photocatalyst sample was prepared by using dry powder method using urea as N source under simple calcination process at low temperature level. The prepared sample was studied for their photocatalytic activiy by measuring the effectiveness in removing the Janus Green B dye.

2 Experimental

2.1 Preparation of N doped TiO_2.

N doped TiO_2 was prepared by mixing TiO_2 (P25) with 0.2, 0.5, 1.0, 2.0 and 5.0 g of urea. The mixed powder was placed into a special glass tube reactor. The reactor was placed into a muffle furnace at temperatures ranging from 300 to 500 °C under normal atmospheric condition. The sample is cooled down to room temperature and the products was cleaned by sonicating the sample with 0.1N HCl then centrifuged to isolate the contamination. The modified sample was finally rinsed thoroughly using distilled water using 0.45 μm cellulose nitrate membrane filter. The product is named N doped TiO_2.

2.2 Photocatalytic degradation of janus green (JG)

In this study, for photocatalytic degradation of JG dye, a 0.024 g sample of N doped TiO_2 photocatalyst was placed inside a glass cell that contains 20 mL solution of 10 mg L^{-1} of JG solutions. The mixed photocatalyst sample powder and dye was sonicated for 30s to agitate the particle in the sample. The modified TiO_2 was transfered into a custom made glass cell with dimensions 50 mm×10 mm× 80 mm (L×B×H). An aquarium pump model NS 750 was used as the aeration source. A direct reading air flow meter was used to determine the aeration rate supply to the reactor. The TiO_2 suspendedwith 10 mg L^{-1} JG will then be irradiated with 55 Watt fluorescent lamps for every 5 minutes interval until 20 minutes. The degradation degree of JG is determined by spectrophotometer at 660 nm.

2.3 Characterization study

The immobilized N doped TiO_2 sample was characterized by using scanning electron microscope model JSM-6700F with an accelerating voltage of 30 kV. The sample was coated with platinum utilizing a GIKO IB-3 ion coater before observation. FTIR spectrometer brand

Perkin Elmer with model Frontier was used for determination of compound presented in the samples. UV-Vis DRS was used todetermine N present in the sample and estimate the bandgap energy of modified N doped TiO_2. About 0.2 g powder sample is placed in UV-Vis DRS with absorption mode ranging from 250-500 nm wavelenghts.

3 Results and discussion

3.1 Optimization Study

Fig. 1a shows the photodegradation rate of the prepared N doped TiO_2 under different amounts of urea at three calcination temperatures (300, 400 and 500 °C). It was observed that N doped TiO_2 prepared under 400 °C has the optimum temperature condition as compared with other temperatures (300 and 400 °C) with the pseudo rate constant was 0.051, 0.107, 0.055, 0.049 and 0.024 min-1 for 0.2, 0.5, 1.0, 2.0 and 5.0 g of urea respectively (Table 1). The photodegradation rate was increased by increasing amount of urea with the highest photodegradation rate was observed at 0.5 g. However, the photodegradation rate is become decreased beyond 0.5g of urea in prepared N doped TiO_2 sample.

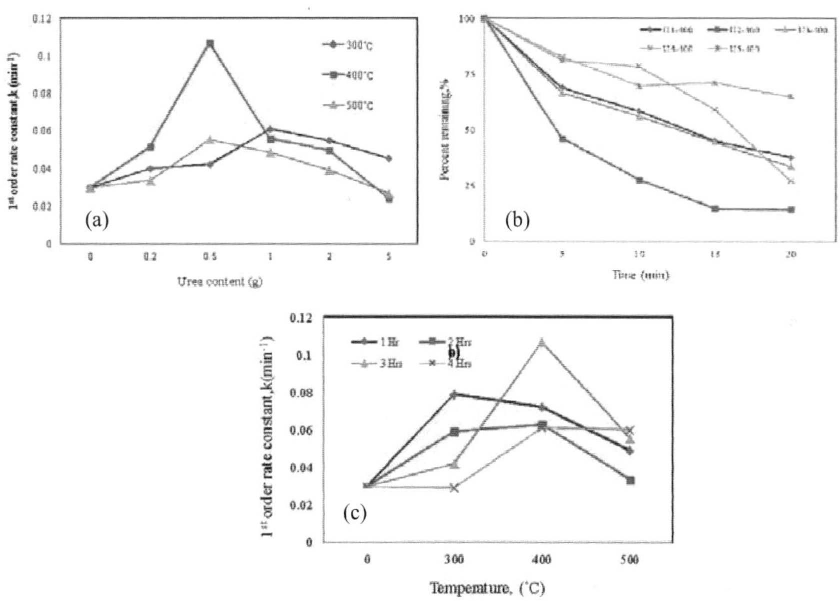

Fig. 1. Photodegradation rate of a) unmodified TiO_2 and various modified TiO_2 samples under different calcination temperatures, b) percent JB dye remaining in 400 °C at different urea loading and c) N doped TiO_2 samples at different calcination time under different temperatures.

The same observation was also seen for the samples calcined at 500°C where the photodegradation rate was increased up to 0.5 g urea and it becomes decreased beyond 0.5 g. However, photodegration rate for samples at 300°C calcination process have increased up to 1.0 g urea and it starts decreased beyond 1.0 g. The low photodegradation rate beyond the optimum amount of urea in modified samples is due to the increasing amount of N in prepared N doped TiO_2 where it becomes saturated and thus covered the TiO_2 surface and makes less penetration of light to reach on surface of TiO_2 photocatalyst.

Table 1. The photodegradation rate values of unmodified and modified TiO_2 samples under different calcination temperatures.

Sample name	Urea (g)	Temperature (°C)	Rate const., K (min^{-1})
Unmodified TiO_2	0	0	0.030
U1-300	0.2	300	0.040
U1-400	0.2	400	0.051
U1-500	0.2	500	0.033
U2-300	0.5	300	0.042
U2-400	0.5	400	0.107
U2-500	0.5	500	0.055
U3-300	1.0	300	0.061
U3-400	1.0	400	0.055
U3-500	1.0	500	0.048
U4-300	2.0	300	0.055
U4-400	2.0	400	0.049
U4-500	2.0	500	0.039
U5-300	5.0	300	0.045
U5-400	5.0	400	0.024
U5-500	5.0	500	0.027

Fig. 1b represents the percent remaining of JB for the batch of 400°C calcination temperature. As mentioned earlier, U2-400 has the optimum photodegradation behavior where it takes 20 minutes to remove 85.5 % of JG dye followed by U4-400, U3-400 and U5-400 where the photodegradation behavior was 72.9, 66.25 and 34.95 % respectively (Table 1). It clearly show that the U2-400 sample is the fastest modified TiO_2 to degrade JG dye compared with other samples and it is used as the optimum condition for N doped TiO_2. Fig. 1c shows that the prepared N doped TiO_2 using 0.5 g of urea under calcination temperature (U2-300, U2-400 and U2-500) at various calcination time. The N doped TiO_2 prepared samples have significant photocatalytic activity as compared with unmodified TiO_2 sample except for U2-300 at 1 hr calcination time where the photodegradation rate seems slightly lower. The same trend was observed for all calcination times where 400 °C is the optimum calcination temperature. The photodegration rates become decreased beyond 400 °C and 500 °C at all calcination times except for 4 hrs where photodegradation rate has become slightly increased.

3.2 Characterization Studies

The scanning electron microscopy images in Fig. 2 shows the morphology of unmodified TiO_2 and U2-400 photocatalyst samples. The images consist of irregular agglomerated nanoparticles of size ranging from 15 to 20 nm. It can be seen that both samples show the similar morphologies which is spherical shapes with hollow pits and spongy structured surface. The morphology of N doped TiO_2 was not changed by treatment of urea at 400°C due to comparable atomic size of nitrogen atom with oxygen [8]. The UV-Vis diffuse reflectance spectroscopy (UV-vis/DRS) was used to identify the absorption shift and bandgap energy of unmodified and optimum modified TiO_2 samples. The UV-Vis DRS in Fig. 3 shows absorption versus wavelength (nm). The bandgap sample was determined by the equation $Eg = 1239.8/\lambda$, where λ was determined by the slope extrapolation of the sample spectrum. From the calculation, the bandgap energy of N doped TiO_2 is 3.0 eV compared to unmodified TiO_2 3.2 eV that according to the literature report. It was proven that modified TiO_2 give better phocatalytic activity then unmodified TiO_2 due to the narrowed bandgap energy and it is confirmed that nitrogen was successfully doped. This narrowing of bandgap energy on N doped TiO_2 may be due to the mixing of N 2p orbitals with O 2p orbitals [9].

Fig. 2. Scanning Electron Microscopy of a) unmodified TiO_2 and b) U2-400.

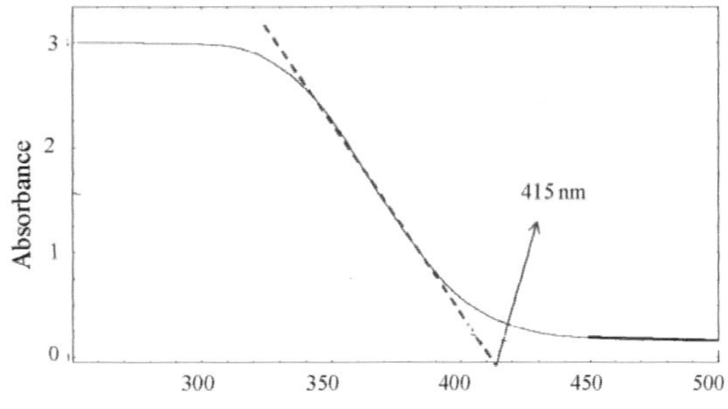

Fig. 3. UV-Vis DRS spectrum of N doped TiO_2 sample.

4 Conclusion

N doped TiO_2 was successfully prepared using calcination process at low temperature with various amounts of urea. The optimum condition for N doped TiO_2 was found at U2-400 where it is 3.5 faster as compared with unmodified TiO_2 for the photodegradation of JG dye. SEM, FTIR, and UV-Vis DRS studies confirmed that the nitrogen was incorporated in TiO_2. SEM image revealsthat no significance changes in terms of the morphology and size of the particle between U2-400 and unmodified TiO_2 where both samples have spherical shape with hollow pits and spongy structures. UV-Vis DRS shows that the bandgap energy of the N doped TiO_2 was 3.0 eV. This makes the U2-400 sample increased its photocatalytic activity since U2-400 has become more active under wider optical response compared to unmodified TiO_2.

We would like to thanks the Ministry of Higher education for providing generous internal financial support under FRGS grants: 600-IRMI/DANA 5/3/FRGS (1/2018) in conducting this study and Universiti Teknologi MARA (UiTM) and Universiti Malaysia Perlis for providing all the facilities.

References

1. D.Gómez, M. Gómez-García, S.M. López Zamora, E. Gilpavas, J. Bojarska, M. Kozanecki, J.M. Rynkowski, T *Comptes Rendus Chimie*, ***18*** (10), (2015) 1170–1182.
2. S. Jakar, R. Chaturvedi, M.K. Sharma, A Comparative Physico-Chemical Analysis Of Janus Green B Contaminated Water And Treated Water By Ammonium Phosphomolybdate, *12*(2) (2014) 573–582.
3. W.I. Nawawi, , M.A Nawi,., Journal of Molecular Catalysis A: Chemical, ***383–384*** (2014) 83–93.
4. H. Ling, K. Kim, Z. Liu, J. Shi, X. Zhu, J. Huang, Catalysis Today**, 258,** (2015) 96–102.
5. M.A. Nawi, I. Nawawi Applied Catalysis A: General, **453** (2013) 80–91.
6. M.A. Mohamed, W.N.W. Salleh, J. Jaafar, A.F. Ismail, N.A.M. Nor, Materials Chemistry and Physics, **162** (2015) 113–123.

A short review on polyaniline (PANI) based nanocomposites for various applications: enhancing the electrical conductivity

Nor Aisah Khalid[1,*], *Jeefferie* Abd Razak[1*], *Hazman* Hasib[1], *Mohd Muzafar* Ismail[2], and *Noraiham* Mohamad[1], *Ramli* Junid[3], *Poppy* Puspitasari[4]

[1]Center of Smart System and Innovative Design, Fakulti Kejuruteraan Pembuatan, Universiti Teknikal Malaysia Melaka, Hang Tuah Jaya, 76100, Durian Tunggal, Melaka, Malaysia.
[2]Fakulti Kejuruteraan Elektronik & Kejuruteraan Komputer, Universiti Teknikal Malaysia Melaka, Hang Tuah Jaya, 76100, Durian Tunggal, Melaka, Malaysia.
[3]Faculty of Mechanical Engineering, Universiti Malaysia Pahang, 26600, Pekan, Pahang, Malaysia.
[4]Mechanical Engineering Department, Fakultas Teknik, Universitas Negeri Malang, Jl. Semarang No. 5, Malang, 65145, Jawa Timur, Indonesia.

Abstract. This short review has summarized the significance of polyaniline (PANI) advanced polymer that focusing into their modification strategy, electrical conductivity and various potential applications. PANI is one type of conductive polymer that was synthesized by oxidative aniline polymerization with varied concentration of acid dopant. In recent year, many researches has been conducted specifically to enhance the electrical conductivity of PANI. There have been numbers of studies involving PANI that specially reported the electrical conductivity could be improved through proper dopant (acid) selection and robust composite strategy. The PANI based nanocomposite shows higher electrical conductivity by integrating it with nanofiller due to the filler-matrix interface contact. Therefore, by modifying the PANI properties, it could be benefited for various potential application in the future.

1 Brief on PANI

Over the past decade, most research has emphasized the use of polyaniline (PANI) as an advance polymeric material due to its rich chemistry, high conductivity and good processing ability [1]. Recent developments in PANI research, has increased the need of conducting polymer utilization for the purpose of electron activity, semi conducting properties and electrical activity [2]. Therefore, PANI was increasingly recognized in various application such as solar batteries, anticorrosive coating, flexible electrode, storage devices and electromagnetic screen [3]. Basically, PANI was resulted from three oxidation states, known as completely reduced leucoemeraldine base, fully oxidized pernigraniline base and the emeraldine base, which consists of alternating oxidized and reduced repeat unit structure, as depicted in Fig. 1 [4]. Emeraldine base is very stable at room temperature

*Corresponding author: jeefferie@utem.edu.my; 6noraisah@gmail.com

and emeraldine salt are highly electrical conductive. While leucoemeraldine and pernigraniline are poor conductors [5]. There has been quantitative analysis of electrical conductivity of PANI that basically was influenced by their ability to transfer electrons. This paper seeks to remedy this issue by analyzing related literatures on the properties, synthesis and application of PANI for specific purposes. Moreover, due to current trend of integrating PANI with various functional fillers or nanofillers through robust composite strategy, PANI based nanocomposites was given special attention for this short review at the following section.

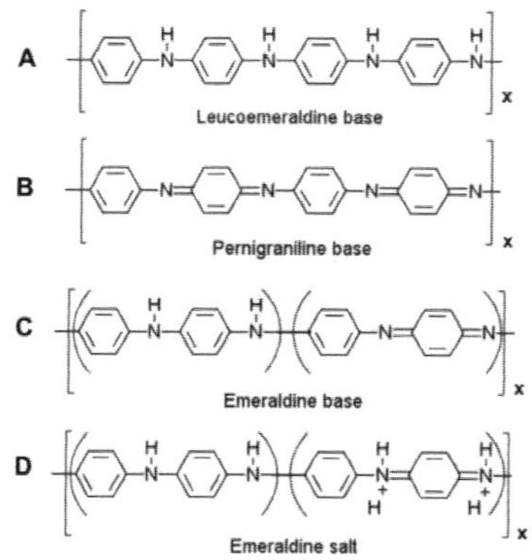

Fig. 1. Oxidation states of PANI, (A) the completely reduced Leucoemeraldine base, (B) the fully oxidized pernigraniline base, (C) the half oxidized-half reduced emeraldine base, and (D) the emeraldine salt [4].

2 The association between the synthesis and electrical conductivity of PANI

Previous researches has indicated that PANI can be synthesized by chemical or electrochemical processes. Electrochemical synthesis is the purest product which was free of admixture and does not require special procedure for purification [3]. The electrical conductivity properties of PANI are influenced by chemical and electrochemical redox reactions. Previous studies have reported that the conductivity values were increased by from 0.12 x10^{-3} Scm^{-1} to 0.13 x10^{-3} Scm^{-1} for the use of sulfonic aminophenol dopants [6]. Other studies have considered the relationship between the synthesis of PANI by using sulphuric acid which shows better electric conductivity compared than nicotinic acid (NA) and 2-methylnicotinic acid (MNA), due to large dopant size in polymeric matrix [7]. Similarly, Wang et al. (2017) found that the measured conductivity values of Molybdenum disulfide (MoS_2)/PANI nanocomposites shows an enhancement from 1.09x10^{-1} Scm^{-1} to 2.50x10^{-1} Scm^{-1} due to effects of dopant used. These values have increased with increasing amount of MoS_2 [8].

However, among the dopants used, fluorine doped tin oxide (FTO) shows the highest value of electrical conductivity readings reached into 4.76 Scm^{-1} for 3.00 mg of dopant weight. The electrical conductivity of PANI had decreased as the weights of dopant was decreased [9]. This result has supported further the hypothesis for the electrical properties of PANI which are a function of synthesis condition either oxidation or protonation. The following Table 1 provides the summary of synthesis methods using different types of dopants to give different results in term of their electrical conductivity. It was found that the selection of acid based dopant are crucially significance in influencing the big gap different of the resulted electrical conductivity of PANI.

Table 1. The acid dopant agent against the electrical conductivity of PANI.

Ref.	Synthesis method	Dopant	Sample	Conductivity (Scm^{-1})
[6]	PANI	aminophenol sulfonic	PANI co-polymer	$0.12x10^{-3}$ $0.13x10^{-3}$
[7]	PANI/H_2SO_4	sulphuric acid nicotinic acid 2-methylnicotinc acid	PANI/H_2 SO_4 PANI/NA PANI/MNA	1.09 0.63 0.089
[8]	MoS_2/PANI	molybdenum disulfide	PANI MoS_2/PANI-8 MoS_2/PANI-38	$1.09x10^{-1}$ $2.50X10^{-1}$ 2.38
[9]	FTO/PANI-GR	fluorine doped tin oxide (FTO)	FTO/ PANI FTO/PANIGR 1 mg FTO/PANIGR 3 mg	0.09 0.454 4.76

3 Nanofiller for PANI based nanocomposites

Over the past decade, most research in nanotechnology has emphasized the investigation on polymeric based nanocomposites. The manipulation of molecules in one dimension is about 1 to 100 nanometers (nm) that are available in nanocomposite. There are various types of nanofillers used such as clay nanoparticles [10], silver nanoparticle [11], oxide nanoparticles [12], graphene nanoplatelets [13] and carbon nanotubes [14]. Recent development in PANI research has confirmed the need to incorporate PANI with varies nano-fillers that may resulting in higher performance and good feature advantages for ideal generation of future energy storage devices [15]. Chatterjee et al. (2011) has demonstrated that the electrical conductivity has increased by adding the optimum nanofiller content. Electrical percolation could be achieved in the composites at between 1.50 – 2.00 wt% of nanofiller loadings, as clearly depicted in Fig. 2 [16]. It was presented that, different type of nanofillers would possessed different value of electrical conductivity when it was incorporated into PANI polymers at different amount of nanofiller loadings.

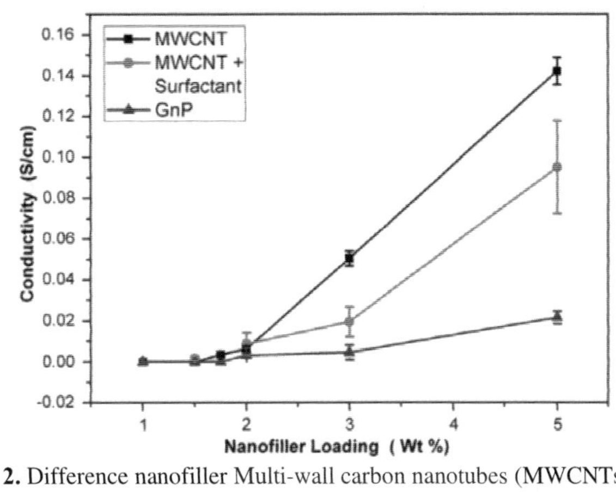

Fig. 2. Difference nanofiller Multi-wall carbon nanotubes (MWCNTs) and Graphene Nanoplatelets(GNPs) for electrical conductivity response.

Studies conducted by Jelmy et al. (2013) shows the produced polymer nanocomposites samples with nanofiller exhibits higher electrical conductivity. For instance, by adding Carbon Nanotube's CNTs into PANI has yielded about 2.28 Scm^{-1}, Carbon Black (CB) about 0.03 Scm^{-1} and graphite about 0.07Scm^{-1} as compared than PANI sample, without nanofiller addition of 0·007 Scm^{-1} [15]. Similarly, Imran et al. (2018) has found that the electrical conductivity of PANI-GNPs had increased steadily with increasing of GNPs loading, which are $2.00x10^{-4}$ Scm^{-1} (1.00wt%) and $3.16x10^{-2}$ Scm^{-1} (2.00 w%) respectively, as compared to pure PANI $4.38x10^{-17}Scm^{-1}$ [13]. Other studies also have considered the relationship between filler and electrical conductivity. It shows that Multi-wall carbon nanotubes(MWCNTs) nanofiller shows higher electrical conductivity as compared than pure PANI due to charge transfer effect from the quinoid rings of the PANI to the (MWCNTs) [17]. The most striking result to emerge from the data is that, the electrical conductivity of PANI sample filled by CNTs and Graphene(GN) shows the higher value of conductivities than PANI without filler [17]. As an output, the usage of carbon based nanofillers provides various potential applications depending on the appropriate electrical conductivity function. Overall, this result has indicated that the use of carbon based nanofiller was beneficial to influence the electrical conductivity of PANI. Debate has continues about the various application of PANI in term of electrical conductivity function.

Table 2. The Variable PANI System and nanofiller against reaction of Electrical Conductivity.

Ref.	PANI System	Filler	Sample	Conductivity (Scm^{-1})
[13]	PANI	unfilled	PANI	$4.38x10^{-17}$
	PANI/GNP	GNPs	PANI-GNP 1.00wt%	$2.00x10^{-4}$
			PANI-GNP 2.00wt%	$3.16x10^{-2}$
[15]	PANI	unfilled	P–H_2SO_4	0·01
	PANI/CB	carbon black (CB)	PCB	0·03
	PANI/GR	graphite (GR)	PGR	0·07
	PANI/MWCNT	MWCNT	PCNT–H_2SO_4	2·28
[17]	PANI/MWCNTs	unfilled	PANI	0.50
		MWCNT	PANI-CNT	1.95
[18]	PANI	unfilled	PANI	0.70
	PANI/CNT	CNT	PANI/ CNT	6.80
	PANI/GN	GN	PANI/GN	1.50

4 Application of PANI based nanocomposites

Conducting polymer in energy application has gaining special attention due to large specific capacitance properties. Questions have been raised about the limitation usage of PANI in engineering applications due to its unstable condition at neutral pH and at higher temperature. However PANI belongs to the organic conductive electro active polymers, where under ambient conditions may possessed electrical conductivity and the features are suitable to electrochemical storage application [19]. There was a large scale of published studies describing the utilization of PANI in electrical conductivity related applications such as transparent antistatic coating [20], semiconducting properties like solar batteries [21] and electrical activity for example, electrodes supercapacitor [22] and anticorrosive coating [23]. This has been summarised in the following Fig. 3.

PANI have been identified as major contributing application for increased of many electrical purposes and useful ways to suppress electrostatic charging. This conducting polymer have advantages in various fields such as for electrode material due to the nature of active material in energy storage application [22]. However, some disadvantages may include PANI electrode fragility and structural degradation under the harsh environments. A considerable amount of literature has been published regarding advantages and disadvantages of PANI in potential application of transparent antistatic coating [20], anticorrosive coating [23] and solar batteries [21], as summarized in the following Table 3.

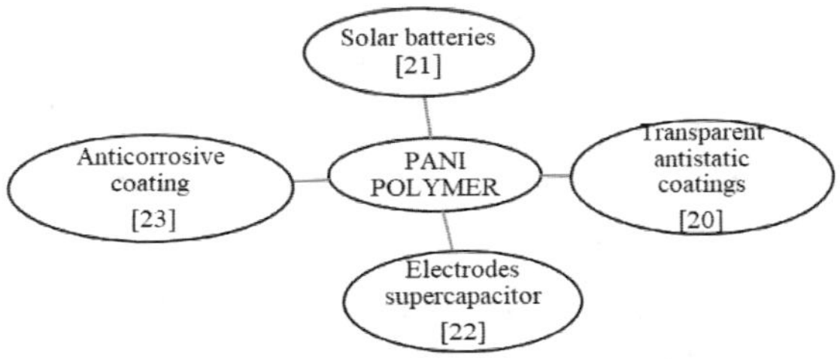

Fig. 3. The variable application of PANI in electron conductivity, semiconducting properties and electrical activity

Table 3. Various PANI application, advantages and disadvantages of PANI system against the electrical conductivity function.

Ref	PANI system	Potential Application	Advantages	Disadvantages
[20]	PANI/acrylic	transparent antistatic coatings	nanocomposite film is flexible	expensive
[21]	PANI-SnO_2 (Tin Dioxide)	solar batteries	higher efficiency under UV light irradiation	predominant factor for the photocatalytic activity
[22]	PANI/Graphene	electrodes supercapacitor	ability to deliver higher power density than rechargeable batteries	obtain both high power and energy density
[23]	PANI-An/G (Modified graphene)	anticorrosive coating	prevents the path surface of the substrate from being corroded	extent limits of anti-corrosion performance.

5 Durability and future development on PANI

Recently, simpler and more rapid test in the durability and consistency of PANI in terms of conductivity has been developed. Although substances can produce materials with high conductivity values, but their use at longer period of time should be noted. The Studies show the important of doped with phosphoric acid (PANI-PA) is capable of forming a protective oxide layer on the surface of the material that prevents the diffusion of corrosive ions [24]. Factor found to be influencing have been explored in several studies. In 20000 cycles, ionic and electronic resistances of PANI-based electrochemical was increased and show good charge capacity value of only 5% loss [25]. In addition, the nanostructure also proved to be effectively exhibited better properties. the next study should focus on the unique nanostructure of the electrical conductivity field [26]. Previous research has indicated that, in further work PANI needs to be designed in tailor-made derivatives for advanced performance in supercapacitor application [27]. PANI when combined with 3-D porous scaffolds can produce medium for transmission of myocardial infarct [28]. For the future development, the addition PANI has stability of conductivity value and electricity power at longer period of time.

6 Conclusions

In overall, the electrical conductivity of PANI has generally influenced by at least two important aspects which are the dopant agent and the nanofiller used. As the matter of fact, from this short review, PANI with dopant agent has shown increased of their electrical conductivity as compared than PANI without dopant addition. It has proven to be effective that the electrical conductivity was increased with the increase of dopant concentration. In another hand, the application of carbon based nanofiller not only shows the higher resulted electrical conductivity of PANI. In addition, due to the ability to modify the electrical conductivity of PANI through the utilization of acid dopant and carbon based nanofiller, has opened up various potential application for PANI based conductive polymer. However, it can be said that, this situation only can be achieved if the optimum amount of acid dopant and nanofiller added could be accurately determined.

Authors would like to acknowledge Universiti Teknikal Malaysia Melaka (UTeM) for sponsoring this research under the short term research grant: *PJP/2019/FKP (4B)/S01662*. Special appreciation to COSSID and FKP for facilities and technical support provided in completing this study.

References

1. Wankhede, Y.B., Kondawar, S.B., Thakare, S.R. and More, P.S. Adv. Mater. Lett. **4(1)**,89-93 (2013)
2. Al Sheheri, S.Z., Al-Amshany, Z.M., Al Sulami, Q.A., Tashkandi, N.Y., Hussein, M.A. and El-Shishtawy, R.M., Des Monomers Polym **22(1)**, 8-53 (2019)
3. Boeva, Z.A. and Sergeyev, V.G., Polyaniline, Polym Sci. Ser C **56(1)**, 144-153 (2014)
4. Qazi, T.H., Rai, R. and Boccaccini, A.R., Biomater **35(33)**,9068-9086 (2014)
5. Vyas, S., Shivhare, S. and Shukla, A., Int. J. Res. And Sci. Innov, ***4*(7)**,86 (2017)
6. Ibrahim, K.A., 2017. Arab. J. Chem. **10**,S2668-S2674
7. Alesary, H.F., Ismail, H.K., Khudhair, A.F. and Mohammed, M.Q., Orient. J. Chem. **34(5)**, 2525 (2018)
8. Wang, J., Wu, Z., Hu, K., Chen, X. and Yin, H., J. Alloys. Compd. **619**, 38-4 (2015)
9. Sayah, A., Habelhames, F., Bahloul, A., Nessark, B., Bonnassieux, Y., Tendelier, D. and El Jouad, M., J. Electroanal. Chem. **818**,26-34 (2018)
10. Jlassi, K., Chandran, S., Poothanari, M.A., Benna-Zayani, M., Thomas, S. and Chehimi, M.M., **32(14)**, 3514-3524 (2016)
11. Reda, S.M. and Al-Ghannam, S.M., Adv. Mater Phys Chem. **2(2)**, 75 (2012)
12. Kondawar, S., Mahore, R., Dahegaonkar, A. and Agrawal, S., Adv. Appl. Sci. Res. **2(4)**,401-406 (2011)
13. Imran, K.A. and Shivakumar, K.N., J. Reinf. Plast. Compos. **37(2)**,118-133 (2018)
14. Jelmy, E.J., Ramakrishnan, S., Rangarajan, M. and Kothurkar, N.K., Bull. Mater. Sci **36(1)**, 37-44 (2013)
15. Majumdar, D., In Polymer Nanocomposites for Advanced Engineering and Military Applications , 220-253 (2019)
16. Chatterjee, S., Nüesch, F.A. and Chu, B.T., Nanotechnol. **22(27)**, 275714 (2011)
17. Kondawar, S.B., Deshpande, M.D. and Agrawal, S.P., Int. J. Compos. Mater, **2(3)**,32-36 (2012)

18. Imran, S.M., Kim, Y., Shao, G.N., Hussain, M., Choa, Y.H. and Kim, H.T., J. Mater. Sci. **49(3)**,1328-1335 (2014)
19. Yesappa, L., Niranjana, M., Ashokkumar, S., Vijeth, H., Raghu, S. and Devendrappa, H., Polym Plast Technol Eng Mater **58(2)**,193-205 (2019)
20. Mirmohseni, A., Gharieh, A. and Khorasani, M., Iran. Polym. J. **25**(12),991-998 (2016)
21. Bui, Phat Dai, Huy Hong Tran, Fei Kang, Ya-Fen Wang, Thi Minh Cao, Sheng- Jie You, Nam Hoang Vu, and Viet Van Pham. ACS Appl. Nano Mater. **1**(10), 5786- 5794 (2018)
22. Moussa, M., El-Kady, M.F., Zhao, Z., Majewski, P. and Ma, J., Nanotechnol. **27(44)**,.442001 (2016)
23. Li, Y., Xu, Y., Wang, S., Wang, H., Li, M. and Dai, L., High Perform Polym. 0954008319839442 (2019)
24. Jafarzadeh, S., Claesson, P.M., Sundell, P.E., Tyrode, E. and Pan, J., 2016. Prog. Org. Coat. **90,**154-162 (2016)
25. Bélanger, D., Ren, X., Davey, J., Uribe, F. and Gottesfeld, S., J. Electrochem. Soc ***147*(8),** 2923-2929 (2000)
26. Minnich, A., Dresselhaus, M.S., Ren, Z.F. and Chen, G.,. Energy Environ. Sci. **(5)**, 466-479 (2009)
27. Banerjee, J., Dutta, K., Kader, M.A. and Nayak, S.K., Polym. Adv. Technol. **30(8),**1902-1921 (2019)
28. Qazi, T.H., Rai, R., Dippold, D., Roether, J.E., Schubert, D.W., Rosellini, E., Barbani, N. and Boccaccini, A.R., Acta Biomater. **10(6)**,2434-2445 (2014)

A short review on polyurethane-based nanocomposites for various applications

Marina Mohamad[1,*], *Jeefferie* Abd Razak[1,*], *Noraiham* Mohamad[1], *Sahrim* Haji Ahmad[2], *Ramli* Junid[3], and *Poppy* Puspitasari[4]

[1]Center of Smart System and Innovative Design, Fakulti Kejuruteraan Pembuatan, Universiti Teknikal Malaysia Melaka, Hang Tuah Jaya, 76100, Durian Tunggal, Melaka, Malaysia
[2]Faculty of Science & Technology, Universiti Kebangsaan Malaysia, 43600, Bangi, Selangor, Malaysia
[3]Faculty of Mechanical Engineering, Universiti Malaysia Pahang, 26600, Pekan, Pahang, Malaysia
[4]Mechanical Engineering Department, Fakultas Teknik, Universitas Negeri Malang, Jl. Semarang No. 5, Malang, 65145, Jawa Timur, Indonesia

Abstract. In recent years, polyurethane (PU) has attracted researchers' attention due to their easily tailored properties to meet specific application requirements. PU and PU-based composites are being developed to be applied in various demanding fields such as aerospace, automotive, and electronics. This is because PU-based nanocomposites have numerous types of functional nanofiller integration. This has led to the advancement of polymer nanocomposite of PU with the inclusion of nanosised fillers for various specific purposes. In this short review article, the authors haves reviewed the fundamental or basic understanding of PU and several types of nanofillers used such as clay, carbon nanotubes, and graphene for PU-based nanocomposites development, as well as its potential applications.

1 Polyurethane (PU) and its potential applications

Polyurethane (PU) is accessible in both thermoset and thermoplastic form. Historically, Otto Von Bayer and his co-workers discovered PU in 1937. Then, commercial PU was first developed at the beginning of World War II (WWII) as a replacement for rubber [1]. PU belongs to a block copolymer which contains low molecular weight of polyester or polyether segment that is covalently attached to urethane the group (-NH- C (=O) –O-). Polyester-based polyurethanes is more biodegradable than polyether-based PU [2]. These polymers are synthesised by step-growth polymerisation through the reaction with three basic components which are isocyanate, polyol, and low molecular weight of chain extenders [3]. Fig. 1. shows the reaction occurred between the polyols and isocyanate in a standard PU synthesis. The properties of the PU end product depend on the composition of isocyanate and polyol. The PU structure may consists of a hard and soft segment which can organised alternately as shown in Fig. 2. The hard segment is generally formed by the isocyanate which resulted in the rigid and fragile properties of PU, while the soft segment is formed by the polyols group which characterised the flexible polymer chain. Whether it is

* Corresponding author: jeefferie@utem.edu.my ; me.marina24@gmail.com

an aromatic, cyclic or aliphatic compound, the reactivity of the isocyanate differs [4]. Aromatic isocyanate is more reactive than aliphatic isocyanate [5] because aliphatic isocyanate does not have enough reactivity for foam application but could be used for other urethane application such as coating [6]. Diphenylmethane diisocyanates (MDI) and toluene diisocyanates (TDI) are the commonly used aromatic isocyanate in rigid PU production [5]. Rigid PU is obtained from low molecular weight polyols which have a few hundred units, while flexible PU are produced from high molecular weight polyols which possessed more than 10,000 units [6]. Table 1 summarised the physical form of PU foam using different types of isocyanate and polyol. The types of polyols used to manufacture polyurethane are polyester and polyether. Polyol is a monomer which contain two or more hydroxyl groups in their structure for an organic reaction [4].

It was found that the selection of isocyanate crucially influenced the end properties of PU. This situation may possibly open a very wide opportunity to design PU for various specific applications. Other than manipulating the use of isocyanate that are basically toxic, nowadays to design green PU, researchers have started to employ composite strategy. One of the method to design green PU with composite are by using Design of Experiment (DOE) [7] and Taguchi [8,9]. The following section summarised the effort to produce PU-based nanocomposite by utilising various types of nanofillers for various possible applications.

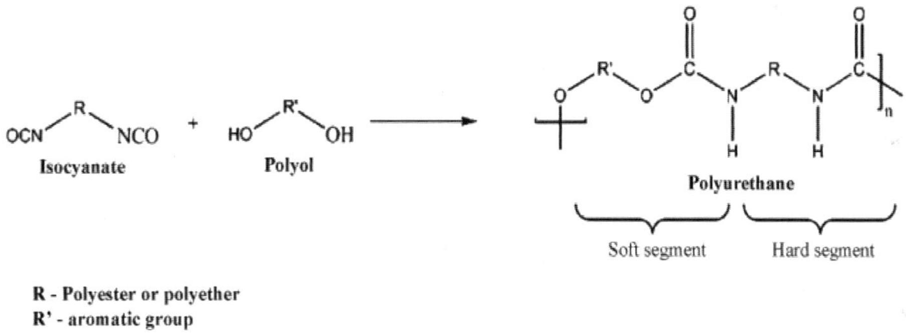

Fig. 1. Formation of PU linkage pathway [10].

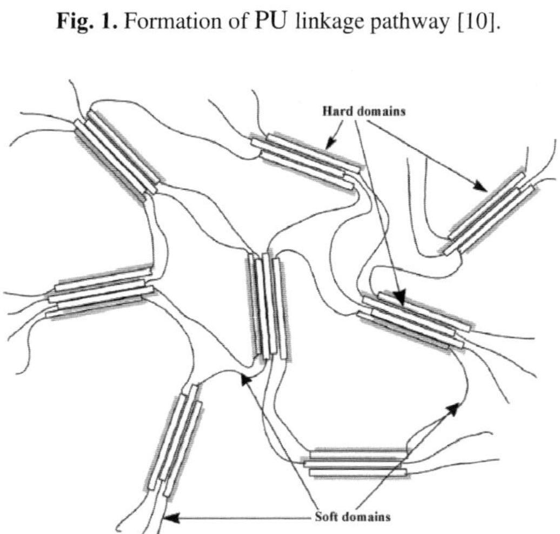

Fig. 2. Cross-linked PU elastomer [11].

Table 1. Comparison of different type of isocyanate and polyols for PU production.

Types of polyol	Isocyanate	Physical form	Reference
Recycled PU	Polymeric methylene-diphenyl diisocyanate (PMDI)	Flexible	[12]
Waste Cooking Oil (WCO) and post-consumer polyethylene telephthalate (PET)	Polymeric methylene-diphenyl diisocyanate (PMDI)	Rigid	[13]
Copolymerization of carbon dioxide	Diphenyl-methane diisocyanate (MDI)	Flexible	[14]
Rapeseed oil-based polyol	Polymeric methylene-diphenyl diisocyanate (PMDI)	Rigid	[15]
Soybean oil-based polyol	Toluene diisocyanate_(TDI)	Flexible	[16]

PU can be classified into several categories based on their desired properties and mechanical performances as summarised in Fig. 3. For instance, due to its remarkable mechanical properties and low density, rigid PU is widely used for domestic appliances and building insulation [17]. Other applications of PU involved building and construction, automotive, marine, coating, medical, apparel, wood composites, packaging and appliances [18]. The next section specifically summarised the PU-based nanocomposites and its potential application.

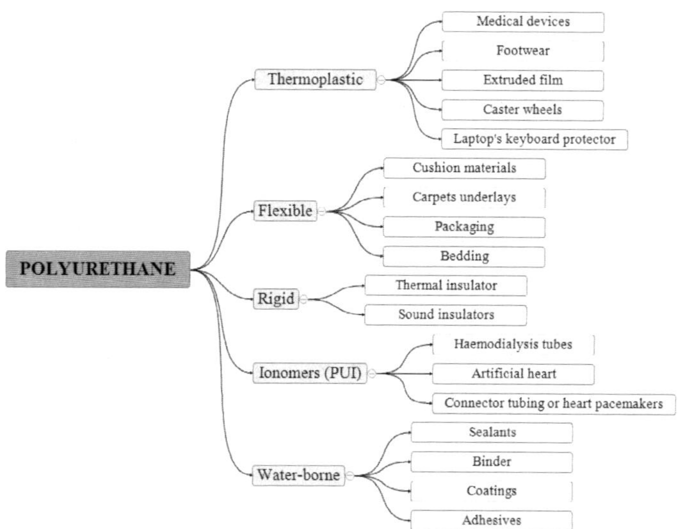

Fig. 3. Types of PU and their example application [3, 12, 14, 15, 19].

2 PU-based nanocomposites and its application

Nanocomposites possessed dimensions of at least one solid phase and are in nanometer size range (1-100 nm) [20]. According to M. Bhattacharya [21], nanoparticles plays an important role as reinforcing fillers in the polymer matrix. Thus, the nanofiller should have (i) outstanding mechanical properties in terms of strength and Young's modulus, (ii) high surface area and high aspect ratio to interact with the polymer, (iii) and should dispersed well to avoid agglomeration.

Nanoscale fillers that are utilised in PU-based nanocomposites have superior properties over conventional composites in terms of electrical, thermal, and chemical properties [22]. Table 2 shows some examples of nanofillers, previously used in the PU matrix. Common carbon-based nanomaterials used are carbon nanotubes (CNTs) [23] which consists of single-walled carbon nanotube (SWCNT) and multi-walled carbon nanotube (MWCNT) [24], nanoclay [25] such as montmorillonite, nanosilica [26] and bentonite [27], carbon nanofibers (CNFs) [28], and graphene [29]. They can be divided according to their dimension; one dimension such as nanotubes and nanowires [30, 31], two dimension such as graphene [32] and clays such as montmorillonite [33]. Graphene and nanotubes possessed excellent properties due to their high aspect ratio and high mechanical strength [23]. To further enhance the performance of PU-based nanocomposites, several effective chemical functionalisation strategies were performed. This resulted in excellent mechanical properties but destructive to electrical properties due to harsh treatment of carbon nanofiller surfaces [22].

The existence of nanoscale filler has intrinsically provide nanocomposites with unique properties and superior characteristics compared to conventional composites. Cell morphology improvement enhances the thermo mechanical properties of nanocomposites. Numerous nanoparticles were added to PU foams to improve thermal properties, heat distortion temperature, mechanical properties, foaming behaviour, barrier properties, biodegradation, electrical conductivity, and crystallisation kinetics such as nanoclay, nanosilica, graphene, and CNTs [34]. Physical and mechanical properties of carbon-based nanoparticles are relatively high and have the potential to increase material performances [35].

Table 2. Example of nanofiller used in PU matrix.

Matrix	Nanofiller	Particle size	Dispersion technique	Response	Reference
PU	rod-shaped CNFs	0.20 µm	sonication	high strength, stiffness, thermal degradation	[19]
PU	Bentonite Cloisite® Na$^+$ (MNa)	88 nm	waterborne dispersion	efficient as gas barrier	[27]
Shape-memory PU (s-PU)	graphene nanosheets (flake form)	10 µm	magnetic stirrer	electrical conductivity increase, mechanical and shape memory improved	[28]
Rigid PU (r-PU)	montmorillonite nanoclay (NCL)	50 nm	homogenizer	Flame retardancy increase	[36]
PU	MWCNTs	5–20 nm	solution casting	CNTs loadings dépendances	[37]

PU-based nanocomposites are used in various applications due to their versatile properties such as aerospace [38] and automotive industry, radar absorbing and EMI shielding, fire proof material, shape memory appliances, PU foam sensor [39], and biomedical [40]. Due to the versatility of PU-based nanocomposites and the ability for us to modify its properties, there are various possible demanding applications that could be actualised as shown in Fig. 4.

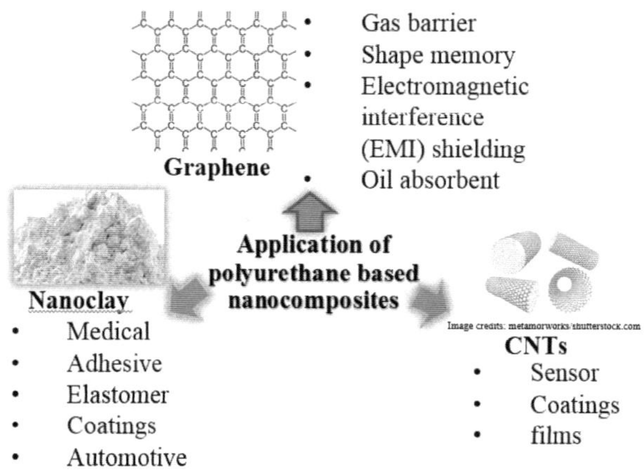

Fig. 4. Various application of PU based nanocomposites [21, 27, 30, 41].

Bahrami et al. (2019) [42] used graphene as the nanofiller for tissue engineering application including cardiac patch development and spinal cord injury treatment in which the main factor of the product's success is the scaffold conductivity. In this study, they discovered that the electroconductivity of the material increased significantly after using 2% of graphene concentration with electrospun mats. The results showed that the electrospinning technique distributed better compared to the solvent casting of graphene into the PU matrix.

On the other hand, Awad et al. (2018) [43] discovered that the most efficient solar thermal converter of water desalination is Plasmonic Graphene Polyurethane (PGPU) nanocomposite using 0.2 wt% Au and Ag nanoparticles. The solar thermal performance of Au/Ag-PGPU foams increased by 63%, 88%, and 96.5% under 1, 5 and 8 sun illumination where the water evaporation average rates were 1.00, 6.59, and 11.34 kg m-2 h-1 respectively. The results also showed that after the evaporation cycles were repeated 10 times, the PGPU foams displayed very stable evaporation rates without any failure. It can be concluded that PGPU nanocomposites is the best method for seawater desalination and solar-steam generation as they have a very high thermal evaporation efficiency, longer durability, and high solar thermal evaporation efficiency.

3 Conclusions

Overall, PU properties could be modified by manipulating the composite development strategy. Due to this factor, the utilisation of PU based composites could be diversified in accordance to specific applications based on the performance requirement. Various potential fillers can be used for this purpose. However, carbon-based nanofillers are highly preferable since they could offers outstanding improvements to PU-based nanocomposites likes enhancing mechanical, physical, and electrical properties. Hence, from this short review, it can be said that the future exploration for PU based nanocomposites are promising.

The authors would like to acknowledge Ministry of Education, Malaysia (KPM) and Universiti Teknikal Malaysia Melaka (UTeM) for sponsoring this research under the fundamental research grant scheme: *FRGS/2018/FKP-AMC/F00381*.

References

1. S. Suleman, S.M. Khan, N. Gull, W. Aleem, M. Shafiq, T. Jamil, Int. J. Innov. Sci. Res **12**, 165-169 (2014)
2. K. H. Badri, *Biobased Polyurethane from Palm Kernel Oil-Based Polyol* (Books on demand, 2012)
3. H. Janik, M. Sienkiewicz, J. Kucinska-Lipka, *Polyurethanes. In Handbook of thermoset plastics* (Elsevier Inc. Chapters, 2014)
4. H. AqilahHamuzan, K. H. Badri, AIP Conf. Proc. **1784**, 030019 (2016)
5. I. R. Clemitson, *Castable polyurethane elastomers* (CRC Press, 2015)
6. Z. S. Petrović, *Polym. Rev.* **48**, 109-155 (2008)
7. M. M. Kandar, H. M. Akil, Procedia Chem. **19**, 433-440 (2016)
8. E. Fallahiarezoudar, M. Ahmadipourroudposht, A. Idris, N. M. Yusof, Mater. Sci. Eng. C **76**, 616–627 (2017)
9. E. Oliaei, B. S. Heidari, S. M. Davachi, M. Bahrami, S. Davoodi, I. Hejazi, J. Seyfi, J. Mater. Sci. Technol. **32**, 710–720 (2016)
10. N. R. A. A. Norzali, K. H. Badri, Polym. Polym. Compos **24**, 711-718 (2016)
11. M. Ionescu, *Chemistry and technology of polyols for polyurethanes* (iSmithers Rapra, 2005)
12. N. Kraitape, C. Thongpin, Energy Procedia **89**, 186-197 (2016)
13. Y. Dang, X. Luo, F. Wang, Y. Li, Waste Manag. **52**, 360–366 (2016)
14. M. DeBolt, A. Kiziltas, D. Mielewski, S. Waddington, M. J. Nagridge, J. Appl. Polym. Sci. **133**, 1–9 (2016)
15. M. Kurańska, A. Prociak, Ind. Crops Prod. **89**, 182–187 (2016)
16. L. Zhang, H. K. Jeon, J. Malsam, R. Herrington, C. W. Macosko, Polymer **48**, 6656–6667 (2007)
17. K. H. Badri, A. M. Redwan, Sains Malaysiana **39**, 769–774 (2010)
18. B. Lin, G. A. Gelves, J. A. Haber, U. Sundararaj, Ind. Eng. Chem. Res. **46**, 2481–2487 (2007)
19. J. O. Akindoyo, M. D. H. Beg, S. Ghazali, M. R. Islam, N. Jeyaratnam, A. R. Yuvaraj, RSC Adv. **6**, 114453–114482 (2016)
20. M. C. Saha, E. Kabir, S. Jeelani, Mater. Sci. Eng. A **479**, 213–222 (2008)
21. M. Bhattacharya, Materials **9**, 2662 (2016)
22. Y. Li, X. Huang, L. Zeng, R. Li, H. Tian, X. Fu, Y. Wang,W. H. Zhong, J. Mater. Sci. **54**, 1036–1076 (2019)
23. M. Dong, C. Wang, H. Liu, C. Liu, C. Shen, J. Zhang, C. Jia, T. Ding, Z. Guo, Macromol. Mater. Eng. **304**,1900010 (2019)
24. J. N. Coleman, U. Khan, W. J. Blau, Y. K. Gun'ko, Carbon **44**, 1624–1652 (2006)
25. M. Špírková, L. Brožová, J. Hodan, J. Kredatusová, S. Krejčíková, A. Zhigunov, J. Pavličević, Polym. Compos. **40**, 4079–4092 (2019)
26. M. Santiago-Calvo, J. Tirado-Mediavilla, J. L. Ruiz-Herrero, M. Á. Rodríguez-Pérez, F. Villafañe, Polymer **150**, 138–149, (2018)
27. T. Rihayat, Suryani, Satriananda, Ridwan, Nurhanifa, A. Putra, N. Audina, M. Yunus, Sariadi, Safari, R. Jalal, *AIP Conf. Proc.* **2049**, 020020 (2018)

28. H. A. K. Toprakci, A. Turgut, O. Toprakci, Res. Eng. Struct. Mater, (2019)
29. R. L. M. Sofla, M. Rezaei, A. Babaie, M. Nasiri, Compos. Part B Eng **175**, 107090 (2019)
30. H. Duan, H. Zhu, J. Yang, J. Gao, Y. Yang, L. Xu, G. Zhao, Y. Liu, Compos. Part A Appl. Sci. Manuf. **118**, 41–48 (2019)
31. X. Yang, S. Fan, Y. Li, Y. Guo, Y. Li, K. Ruan, S. Zhang, J. Zhang, J. Kong, J. Gu, Compos. Part A Appl. Sci. Manuf. **128**, 105670 (2020)
32. Z. Gu, Y. Xu, Q. Lu, C. Han, R. Liu, Z. Zhou, T. Hao, Y. Nie, Phys. Chem. Chem. Phys. **21**, 6443–6452 (2019)
33. Y. Liu, Y. Wang, C. Zhang, T. Liu, APL Mater. **7**, 080904 (2019)
34. B. Aydoğan, N. Usta, J. Chem. Technol. Metall. **54**, 55–63 (2019)
35. M. H. Moghim, S. M. Zebarjad, Polym. Bull. **74**, 4267–4277 (2017)
36. C. C. Ibeh, M. Bubacz, J. Cell. Plast. **44**, 493–515 (2008)
37. A. Kausar, Polym. - Plast. Technol. Eng. **56**, 1468–1486 (2017)
38. V. T. Rathod, J. S. Kumar, A. Jain, Appl. Nanosci. **7**, 519–548 (2017)
39. S. Y. Hong, J. H. Oh, H. Park, J. Y. Yun, S. W. Jin, L. Sun, G. Zi, J. S. Ha, NPG Asia Mater. **9**, e448 (2017)
40. A. Kausar, Polym. - Plast. Technol. Eng. **57**, 346–369 (2018)
41. S. Das, S. Kumar, S. Mohanty, and S. K. Nayak, *Synthesis, characterization and application of bio-based polyurethane nanocomposites* (Springer, 2019)
42. S. Bahrami, A. Solouk, H. Mirzadeh, and A. M. Seifalian, Compos. Part B **168**, 421–431 (2019)
43. F. S. Awad, H. D. Kiriarachchi, K. M. AbouZeid, U. Özgür, M. S. El-Shall, ACS Appl. Energy Mater. **1**, 976–985 (2018)

IOP Conf. Series: Materials Science and Engineering **957** (2020) 012030 doi:10.1088/1757-899X/957/1/012030

The mixture material of dried straw and kapok for acoustic treatment for UTeM mosque.

Mohd Muzafar Ismail[1,*], *Jeefferie* Abd Razak[2] , *Ramli* Junid[3] , *Nor Aisah* Khalid[2] , *Faiz* Arith[4] , *Maizatul Alice* Meor Said[4]

[1]Fakulti Teknologi Kejuruteraan Elektrik dan Elektronik, Universiti Teknikal Malaysia Melaka
[2]Center of Smart System and Innovative Design, Fakulti Kejuruteraan Pembuatan, Universiti Teknikal Malaysia Melaka, Hang Tuah Jaya, 76100, Durian Tunggal, Melaka, Malaysia.
[3]Structural Materials and Degradation, Universiti Malaysia Pahang, 26600, Pekan Pahang
[4]Fakulti Kejuruteraan Elektronik & Kejuruteraan Komputer, Universiti Teknikal Malaysia Melaka, Hang Tuah Jaya, 76100, Durian Tunggal, Melaka, Malaysia.

Abstract. The acoustical requirements in terms of intelligibility and liveliness are dissimilar for both of sounds-recitation and speech on religious subjects at UTeM mosque. To attain an overall satisfactory acoustical performance, it is significant to find an optimum Reverberation Time (RT) and the best directivity, which is the major factors affecting intelligibility and liveliness. This experiment is conduct to gauge the effect of RT by making it a variable, while keeping other factors as non-variable and fulfilling ideal conditions for maximum intelligibility. Bearing in mind, the acoustical problems always occurred in any communication building. To improve the problem, this paper presents the new idea of improvement by using biodegradable sound absorber materials which is the mixture of dried rice straw and kapok.

1 Introduction

The concept idea of the acoustic treatment and speech intelligibility quality analysis in different dimension rooms in Masjid UTeM are an eminence of sound distribution. The important of loudspeaker reverberation time and absorption coefficient in respect of speech intelligibility will observed. In order to get an efficient learning and communication process, the room's clarity is one of the important factors [1-8]. It became more important when listeners need to understand every single word. In certain condition such as a well-built mosque, without electronic aid system, the ability to perceive information is limited. The purpose of this experiment is to study the effect of absorption coefficient and its reverberation time (RT) on speech intelligibility with respect to different room sizes and loudspeaker configurations.

There are several numbers of acoustical parameters that can gauge the quality of sound [9-12]. In a very limited budget, one of the parameter where practically used is loudspeaker. Different arrangements of loudspeaker affect the sound distribution. The other way to get a

* Corresponding author: muzafar@utem.edu.my

good sound production is sound absorber. In this experiments, the dried rice straw (*jerami padi*) & kapok (*Kekabu*) are used to observed either both materials are a good sound absorber or not.

1.1 Relationship between Reverberation Time (RT) and absorption coefficients (α)

Reverberation time (RT) can be classified as one of major factors that affect the transmission of the speech sound from speakers to listeners. It is functioning to amplify the sound in an enclosed spaced and may help the system maintain the sound pressure level from seat to seat. In open spaces, Reverberation Time can be ignoring since there is no sound reflection required. In this project the researcher chooses RT60 which represent the sound is decreased by 60 dB.

$$RT60 = (0.16V)/(S\alpha) \qquad (1)$$

Where:

V= Volume of the room, m^3

S = Total surface area of the room, m^2

α = absorption coefficient of the surface of absorber material

2 Methodology

This experiment was conducted in the Masjid Sayyidina Abu Bakar, UTeM. The plan drawing for the Masjid given from the Development of UTeM. **Fig. 1.** Shows the sample of plan layout of the mosque and the placement of loudspeaker. The sound source is placed at three different areas to observe the sound distribution. First loudspeaker was placed at the centre of the room. The loudspeaker spreads a sound wave in a spherical surface form. The spherical distribution will give the directivity of sound at various places unidirectional. The second loudspeaker was positioned at the flat surface area as shown at the back area of the mosque will produce the hemispherical surface of sound wave. While, the third loudspeaker was set at the edge of the mosque, the loudspeaker will produce the quarter of spherical forms of sound propagation.

Fig. 2. shows the steps of implementation the sound absorber. In this experiment, the anechoic chamber is used to carry out because the material of the chamber is acrylic which helps sound reflection. Thus, the sound produced by the loudspeaker is only distributed and travel in the box which slightly same with an enclosed space. Before dried rice straw used as sound absorber, it need to be rendered under the sundried about two weeks. The implementation of the mixture of dried rice straw and kapok in the modified anechoic chamber will yield the value of the RT60. All the data were collected by using NTI AL1 acoustilyzer meter.

IOP Conf. Series: Materials Science and Engineering **957** (2020) 012030 doi:10.1088/1757-899X/957/1/012030

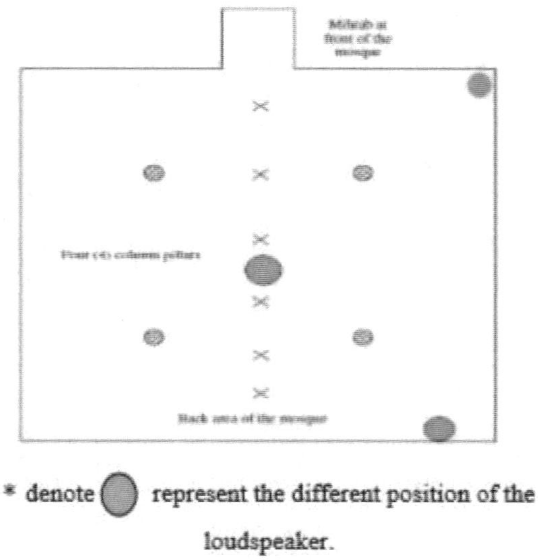

Fig. 1. Plan layout and loudspeaker placement of UTeM mosque.

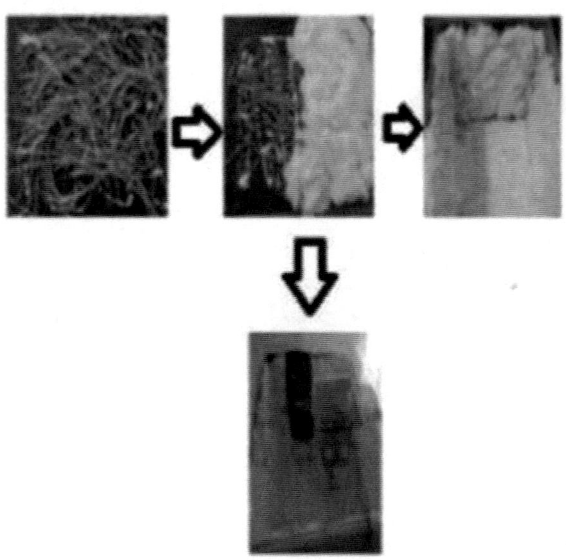

Fig. 2. The process flow for implantation sound absorber using mixture material (kekabu and kapok).

IOP Conf. Series: Materials Science and Engineering **957** (2020) 012030 doi:10.1088/1757-899X/957/1/012030

3 Result and discussion

In this section, the parameters and components that used in this experiment were described in detail. This project separated in two major parts, first, the quality analysis of speech intelligibility and second, the implementation of new sound absorber.

The two rooms in the mosque are used as an experiment area. The first floor, which is the main praying area and the second floor, refers to the muslimah praying room area. Different room's design will give a different reverberation time calculation. It is due to the size of the room space and amount of the absorptive surfaces within the space that give effect to the reverberation time. The value of absorption coefficient for each material was taken at 2 kHz frequency. During this experiment, the speech predictor used is %Alcon's and the maximum frequency required for this speech predictor is only 2 kHz. Material density gives effect on sound dispersion; hence, it also affects the total of the absorbed sound liveliness. By referring **equation (1)** the reverberation time can be calculated by using the above parameters values shows in **Table 1**. The effectiveness on rice straw mixed kapok as acoustic materials can be shown in the higher value of the absorption coefficient. Sound absorption coefficients of the 0.25 and 0.6 specific gravity rice-straw mixed kapok were higher and significant than the others bio gradable-based materials.

Table 1: The parameters for calculation reverberation time (RT).

Surface	Material	Area, S, (m^2)	Absorption coefficient, (α)	A= Sα
Ceiling	Lime, ciement, plaster	409	0.07	29
Floor	Tufted pile, Carpet on felt	409	0.25	102.25
Walls	Painted concrete	210	0.09	18.9
Kiblah Niche	Plaster	57	0.04	2.28
Door	Plywood	30 in 9 units	0.10	2.0
Ceiling design	Underplayed perforated	43	0.9	38.7
			Total	194
			Volume (m^3)	1178m^3

4 Conclusion

From the results, it can be concluded that the loud speaker arrangement for each room effects the perception of the signal at each transmits point. The closer listeners to the loudspeaker, the better perception and reception of the signal. This is due to excessive reverberation time, the correlation between SPL and RT60 was explored. The higher quality of sound, Q, covers the larger region of perception in each classroom. In this experiment, it is also proved that the sound absorber which is the mixture of dried rice straw and kapok are one of the best acoustical treatment solution. This is due to higher value of the absorption coefficient in between 0.25 to 0.6.

Acknowledgement

Author would like to thanks Universiti Teknikal Malaysia Melaka for their support in doing research.

References

1. M. Mehta, J. Johnsson, J. Rocafort (1999), *Architectural Acoustics: Principles and Design,* (1st ed.) Upper Saddle River, New Jersey. : Prentice-Hall (1999).
2. M. P. Paulraj, Sazali Yaacob, A. Nazri Abdullah, M. Thagirarani and M. Rodhwan Tamjis, *Classroom Speech Intelligibility Prediction Using Elman Neural Network*, University Utara Malaysia, Malaysia (2010).
3. T. Houtgast, H. J. M. Steeneken, *The Roots of the STI Approach*, International Symposium on Past, Present and Future of Speech Transmission Index (STI), Soesterberg, Holland (2002).
4. Z. Maekawa, and P. Lord, *Environmental and Architectural Acoustics.* E & FNSPON. London (1994).
5. D. (Ed.) Templeton, P. Sacre, P. Mapp, D. Saunders, *Acoustics in the Built Environment*. Oxford : Butterworth-Heinemann Ltd. 122 (1993).
6. Christopher A. Egan, Richard D. Cookson, Applied Acoustic, Sustainable acoustic absorbers from the biomass, David J. Oldham, Acoustics Research.
7. Y. Wang, C. Zhang, L. Ren, M. Ichchou, M.-A Galland. And O. Bareille, *Influences of rice hull in polyurethane foam on its sound absorption characteristics,* Polym. Compos. 34 1847-1855 Nov (2013).
8. J. G. Gwon, S. K. Kim and J. H. Kim, *Sound absorption behavior of flexible polyurethane foams with distinct cellular structures,* Mater. Des. 89 448-454 Jan (2016).
9. A. Gupta, Industrial safety and environment (New Delhi: Laxmi Publications (P) LTD) (2006).
10. C. Mediastika, *Potensi Jerami Padi Sebagai Bahan,* Baku Panel Akustik, (Yogyakarta) 35 183-189 Desember (2017).
11. A. Putra , Y. Abdullah , H. Efendy, W.M.F.W. Mohamad. And N. L. Salleh, *Biomass from Paddy Waste Fibers as Sustainable Acoustic*, Material Adv. Acoust. Vib. 2013 1-7 (2013).
12. L. Peng, B. Song, J. Wang and D. Wang, *Mechanic and Acoustic Properties of the Sound-Absorbing Material Made from Natural Fiber and Polyester*, Adv. Mater. Sci. Eng. 2015 1-5 (2015).

IOP Conf. Series: Materials Science and Engineering **957** (2020) 012031 doi:10.1088/1757-899X/957/1/012031

Effect of using Palm Char and Coke as a Reductant in Production of Ferrosilicon

Nur Farhana Mohd Yunos[1,3,*], *Muhammad Asri* Idris[2,3], *Sri Raj Rajeswari* Munusamy [2,3] and *K.* Perumal[2]

[1]Institute of Engineering Mathematics, Universiti Malaysia Perlis, Pauh Putra, 02600 Arau, Perlis, Malaysia
[2]School of Materials Engineering, Universiti Malaysia Perlis, 02600 Arau, Perlis, Malaysia
[3]CoE Frontier Materials Research, Universiti Malaysia Perlis, 02600 Arau, Perlis, Malaysia

Abstract. This research focused on the synthesis of ferrosilicon alloy by coke and palm char as a reductant. Raw materials which were iron ore, silica sand, coke and palm char mixed and compacted into pellets. The pellets were reduced at temperature of 1300°C using horizontal tube furnace with nitrogen gas flow. The pellets after reduction process were characterized by X-ray fluorescence (XRF), X- ray diffraction (XRD), Scanning electron microscope equipped with energy dispersive spectra (SEM/EDS). The results from XRD analysis indicated that the formation of FeSi, Fe_3Si and SiC phases were appeared at 1300 °C for both reductants. The brighter phases were observed in SEM/EDS mapping analysis represents as iron and silicon. The EDX analysis showed the high amount silica developed for coke compared to palm char due to high content of silica and carbon. The results found that palm char as carbon reductant has the potential to produce ferrosilicon and silica carbide in sustainable way.

1 Introduction

Ferrosilicon (FeSi) and silicon carbide (SiC) production processes are all carbothermic reduction processes, where carbon materials are used as reductant to produce the ferroalloys. For ferroalloy and silicon producers, the carbon materials used in the process have a large effect on the total cost, energy consumption, and stability of the operation and, ultimately, on the quality of the final product [1]. There is hence a continual quest for the optimal carbon material available. The determination of the industry to reduce carbon dioxide emissions has directed people's attention even more toward carbon materials. For many producers, the main carbon material is using metallurgical coke that produced from coal. However, the possibility of using alternative reductants such as charcoal from palm shell waste is becoming increasingly popular. Hence, the aim of this study is to investigate the effect of using different carbon reductants in ferrosilicon synthesis where coke as carbon-based material and compare it to renewable carbon material like palm char.

*Corresponding author: farhanadiyana@unimap.edu.my

It is well known that, biochar, a coal-like material obtained by slow pyrolysis of biomass, can be used as a renewable reductant and may partly or fully replace fossil agents [2-4]. Biochar pellets were investigated as renewable reducing agents in substitution of coal and coke in the silicon/ ferrosilicon and iron making production, where a high reactivity, good mechanical properties and low feedstock costs are appreciated [5-8]. In the silicon and ferrosilicon production, the silicon monoxide (SiO) reactivity is considered. Compared to fossil fuel, biochar has higher reactivity [9]. Previous researched by Liu, et al. reported approximately 20% waste carbon material (woodchips) of the total carbonaceous reductants (petroleum coke and coal) was applied to a submerged arc furnace to identify better carbonaceous reductants for the silicon production process. The results indicated that the silicon yield reached a maximum value of 0.945 higher than other reductants [10]. The low ash and high carbon content of macadamia shell waste showed as an attractive carbon source for synthesising high purity SiC and Si_3N_4 nanopowders in the size range 20-80 nm and 40-100 nm respectively [11].

The most important role of carbonaceous material in ferrosilicon/silicon synthesis is to react with the SiO gas to produce SiC [12]. The reactivity of conventional reduction material like coke/coal is directly correlated to the material properties and leads to high silicon (Si) recovery and economical with good furnace operation. A number of studies investigated the effect of coke and coal properties in the production of high Si alloys focusing on the SiO reactivity [13, 14]. Farzana et. al. found that the initial volatiles generation from bakelite as renewable reductant accelerated the rate of silica reduction compared to coke, particularly the formation of silicon carbide (SiC) through gas phase reduction reactions of silica (SiO_2) [14]. However, higher crystallinity of graphite compares to coke facilitated the faster carbon dissolution into iron. This solute carbon increased the rate of ferrosilicon alloy formations were detected by XRD [15]. A complete reduction of silica and alumina (Al_2O_3) was observed in the Fe_2O_3-SiO_2-C system, along with the formation of phases of FeSi and SiC at temperature of 1550 °C were found in XRD pattern and SEM respectively [16].

Even the number of studies on using graphite, plastic waste from bakelite and charcoal/macadamia shell as renewable reductant is increasing by year, it is crucial to investigate the effect of palm char as renewable reductant in ferrosilicon synthesis considering the reduction capability, reactivity and environment impact. Thus, in this paper, the influence of using different carbon reductants is established and investigated. This study has led to a sustainable alternative solution by transforming agricultural waste into a reducing material for ferrosilicon synthesis.

2 Experimental Procedures

2.1 Preparing palm char via pyrolysis technique

Palm shell used as a raw material which act as reductant in ferrosilicon process was collected from Palm Oil Mill Pahang. The palm shell is crushed and ground by jaw crusher to attain particle size of less than 1 mm. The palm char is prepared via pyrolysis technique by using a horizontal tube furnace with injection of nitrogen gas (flow rate of 0.1 L/min) at temperature 450°C for 2 hours to produce palm char.

2.2 Sample preparations and characterizations

Iron ore as an iron oxides (Fe_2O_3) sources was collected from Kulim Kedah while coke as carbon control from Southern Steel Sdn Bhd, Penang. Silica sand was used as the main sources of SiO_2 for ferrosilicon synthesis. All raw materials were crushed and sieved to get fine particles less than 63μm in powder form. Next, the palm char, iron ore, coke and silica sand are added into a ball milling machine using the mass ratio of SiO_2:C (coke):Fe_2O_3 and SiO_2:C (palm char):Fe_2O_3 which were 0.5:0.7:0.3 and 0.3:1.0:0.2, respectively based on reducible carbon and oxygen ratio calculation [19]. The speed of ball milling is set to 400 rpm for 2 hours to complete the mixing process homogeneously. The mixture is transformed into pellets by adding distilled water. Then, the pellets are dried in an oven for 12 hours at 90°C.

The proximate and ultimate analysis of coke and palm char was conducted using CHONS elemental analyzer and the results are presented in Tables 1 and 2. The chemical composition for all materials used in ferrosilicon synthesis were detected by using XRF and presented in Table 3.

Table 1. Ultimate analysis for carbonaceous materials.

Ultimate analysis [wt. %]					
	Carbon	Hydrogen	Nitrogen	Sulfur	Oxygen
Coke	83.36	1.14	0.39	0.30	14.82
Palm char	59.35	2.98	0.30	-	37.37

Table 2. Proximate analysis for carbonaceous materials.

Proximate analysis [wt. %]				
	Moisture	Ash	Volatile matter	Fixed carbon
Coke	2.4	8.0	6.5	83.1
Palm char	6.2	5.9	6.6	81.3

Table 3. XRF analysis for all materials used in ferrosilicon synthesis.

Composition [wt. %]	Coke	Palm char	Silica	Iron ore
SiO_2	43.36	26.80	53.9	16.4
Fe_2O_3	9.60	52.87	5.3	53.9
Al_2O_3	30.51	nd	16.4	5.3
TiO_2	7.07	0.44	nd	nd
P_2O_5	0.60	0.66	nd	nd
MnO	0.30	0.64	9.69	0.25
CaO	4.1	3.43	3.7	4.80
K_2O	0.70	9.75	nd	1.32
SO_3	0.73	nd	nd	9.69
Other oxides	3.03	6.05	11.01	8.34
*nd (not detected)				

2.3 Ferrosilicon synthesis

Horizontal tube furnace was used in the ferrosilicon synthesis process. The experimental setup was explained in previous works [4]. Once the temperature was set at 1300°C, then the pellets were placed in a ceramic crucible and pushed in the hot zone of the tube for 1 hour of reaction. The nitrogen gas was supplied at a rate of 0.1 L/min with the heating rate of 10 °C/min.

IOP Conf. Series: Materials Science and Engineering **957** (2020) 012031 doi:10.1088/1757-899X/957/1/012031

2.4 Phases and Microstructures Transformation Analysis

The phases transformation of the reduced ferrosilicon was identified using XRD (Bruker D2 Phaser M32002) with CuK_α beam radiation. The scanning range, 2θ was set from 20° to 90° with a step size 0.02° at rate of 0.1s/step. The XRD pattern analysis was then analysed using X'pert High Score Plus software to identify the phases present in the samples and referred to Inorganic Crystal Structure Database (ICSD)[16]. While the changes of microstructure were observed using SEM/EDS (JSM6460LA) operating at an accelerating voltage of 10 kV.

3 Results and discussion

Fig. 1 shows the XRD patterns of the reduced pellets (coke and palm char) at 1300 °C. SiO_2 was the main reaction product for both reductants showed that the reaction is not fully reduced due to low temperature. It was found that SiO_2 will fully reduce to SiC at above temperature of 1300 °C [20]. Fe_3Si was detected at angle of 45° for both carbon reductants, which indicated that reduction of iron oxide was quicker than silica [21]. Iron oxide reduction followed reaction in Eq. 1 and the Boudouard reaction (Eq. 2):

$$Fe_2O_{3(s)} + 3C_{(s)} \rightarrow 2Fe_{(l)} + 3CO_{(g)} \quad (1)$$

$$C_{(s)} + CO_{2(g)} \rightarrow 2CO_{(g)} \quad (2)$$

The generated CO from iron oxide reduction facilitated gas phase reduction of silica through intermediate phase of SiO (Eqs. 3 and 4).

$$SiO_{2(s)} + CO_{(g)} \rightarrow SiO_{(g)} + CO_{2(g)} \quad (3)$$

$$SiO_{2(s)} + C_{(s)} \rightarrow Si_{(l)} + CO_{(g)} \quad (4)$$

$$SiO_{2(g)} + 3C_{(s)} \rightarrow SiC_{(s)} + 2CO_{(g)} \quad (5)$$

At higher temperatures (1300 °C), the reduction of silica takes place by carbon. The silica is completely volatile and it gets vaporized significantly during high temperature reaction [22]. The reaction between silica in sand and carbonaceous material initiates at the contact point between SiO_2 and C. Once both SiO and CO are formed, the reactions proceed at the surfaces of both carbon and silica particles via gas phases. Some amount of SiO will be used in the system, and the remainder of the SiO will escape in the gas phase. Thus, some silicon loss could be attributed to the SiO gas escaping from the system prior to its conversion to SiC [14]. Some reduced silica in the form of silicon, dissolved into the molten iron, while the rest of the silica reduced to SiC.

SiC formation also started to occur below 1400°C through a solid-state reaction of silica and carbon (Eq. 5) [23]. SiC peaks were detected at angles of 35.7° and 60° for both reductants. When iron and silicon are formed due to inter-diffusion during solidification, a number of irons silicides like Fe_3Si and FeSi can be formed [24]. Therefore, more silica is reduced and formation of SiC phase and transformation of Fe_3Si to FeSi was evident where FeSi phase was found in both reductants (Fig. 1 (a) and (b)).

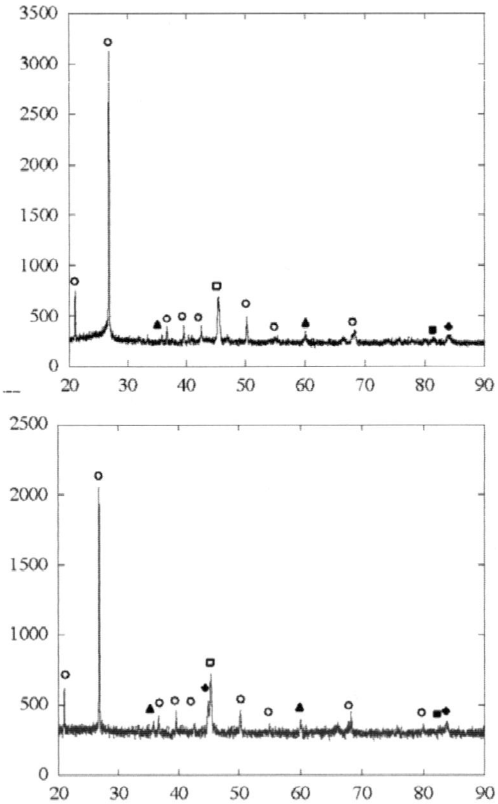

Fig. 1. XRD pattern for ferrosilicon synthesis using different reductants (a) coke and (b) palm char at 1300 °C.

Fig. 2 shows the SEM images with EDS mapping scan of the synthesized ferrosilicon alloy by using both coke and palm char as carbon reductant. The darker phase represents the higher concentration of carbon, the brighter phase remarks as silicon and iron were found in both reductants. Coke showed higher concentration of silicon (41.14 wt. %) compared to palm char, while palm char contained higher iron concentration with amount of 30.99 wt.% (Table 4). Higher carbon content (Table 1 and 4) of coke facilitated the faster carbon dissolution into iron from coke compared to palm char derived carbon. This solute carbon, because of higher carbon pickup by iron from coke, increased the alloy formation [15]. Therefore, predominantly Fe_3Si, FeSi along with SiC peaks were detected in XRD analysis (Fig. 1 (a)) was in a good agreement with SEM/EDS mapping analysis (Fig. 2 (a)) showed better silica reduction by coke compared to palm char (Fig. 2 (b) and Table 4). However, by considering the product formation (FeSi and SiC) and environmental aspects, palm char and generally agricultural waste could be considered as an alternative reductant in the ferrosilicon/silicon industry.

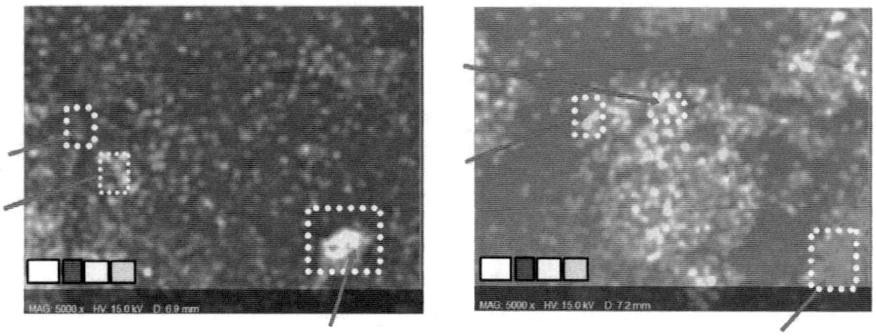

Fig. 2. SEM/EDS mapping for ferrosilicon synthesis using different reductants (a) coke and (b) palm char at 1300 °C.

Table 4. SEM/EDS mapping with elemental analysis for ferrosilicon synthesis using different reductants.

Element (wt. %)	Coke	Palm char
Carbon	56.79	51.78
Iron	2.06	30.99
Silicon	41.14	17.23

Conclusions

XRD analysis confirmed that Fe_3Si, FeSi and SiC phases were developed during ferrosilicon synthesis at 1300 °C using coke and palm char as reductants. The brighter phase was observed as iron and silicon in SEM analysis. Coke showed better silica production due to high silica and carbon content in the sample compared to palm char. In conclusion, waste carbon material from agricultural industry can be successfully used for silicon production and will help build a foundation for future steps with new carbonaceous reductants in the silicon production process.

We acknowledge the financial support from the Fundamental Research Grant Scheme (FRGS/1/2013/SG06/UNIMAP/02/4) from the Ministry of Higher Education Malaysia (MOHE), Research Management and Innovation Center (RMIC), Institute of Engineering Mathematics and School of Materials Engineering, Universiti Malaysia Perlis, (UniMAP) for this research.

References

1. M. Tangstad et al., 14 - Coal-based Reducing Agents in Ferroalloys and Silicon Production, New Trends in Coal Conversion, Woodhead Publishing (2019).
2. A. Adrados et al. Materials, **9**, 1 (2015).
3. N. H. Najmi et al., J. of Mat. Research and Tech. **8**, 2 (2019).
4. M. Yunos, N. F. et al., J. of Ther. Ana. and Calorimetry. **138**, 1 (2019).
5. L. Riva et al. Energy. **181** (2019).
6. N. H. Najmi, et al. Solid State Phenomena. **280** (2018).
7. N. F. M. Yunos, et al., Materials Science Forum. **819** (2015).
8. M. Zaharia, N. F. M. Yunos, and V. Sahajwalla, From Mat. to Structures: Advancement through Innovation. (2013).

9. L. Wang et al., Energy and Fuels. **30**, 10 (2016).
10. J. Liu. et al. Silicon. **10**, 6 (2018).
11. R. Rajarao and V. Sahajwalla, J. of Cleaner Production. **133** (2016).
12. L. Holappa, Chapter 2 - Basics of Ferroalloys, in Handbook of Ferroalloys, M. Gasik, Butterworth-Heinemann, Oxford. (2013).
13. T. Videm Buø, R. J. Gray, and R. M. Patalsky, Int. J. of Coal Geology. **43**, 1 (2000).
14. R. Farzana, R. Rajarao, and V. Sahajwalla, Materials Letters. **116** (2014).
15. R. Farzana, R. Rajarao, and V. Sahajwalla. Ind. and Eng. Chem. Research. **53**, 51 (2014).
16. M. Ikram-Ul-Haq, P.S. Mukherjee, and R. Khanna. Metals. **7**, 10 (2017).
17. J. Deng et al., Carbon. **96** (2016).
18. X. Liao, et al. J. of Alloys and Compounds. **708** (2017).
19. R. Farzana and V. Sahajwalla, J. of Sustainable Metall. **1**, 1 (2015).
20. R. Farzana et al., J. of Cleaner Production. **188** (2018).
21. A. E. Anikin, G. V. Galevskii, and V. V. Rudneva. Steel in Translation. **47**, 2 (2017).
22. R. Farzana, R. Rajarao, and V. Sahajwalla, Waste Manage. Res. **34**, 2 (2016).

IOP Conf. Series: Materials Science and Engineering **957** (2020) 012032 doi:10.1088/1757-899X/957/1/012032

Sugarcane Bagasse Reinforced Polyester Composites: Effects of Fiber Surface Treatment And Fiber Loading on The Tensile And Flexural Properties

Muhammad Salihin Zakaria[1,2,*], *Luqman* Musa[1,2], *Razif* Muhammed Nordin[3,4] and *Khairul Anwar* Abdul Halim[1,2]

[1]Center of Excellence Geopolymer and Green Technology (CEGeoGTech), School of Materials Engineering, Universiti Malaysia Perlis (UniMAP), 01007, P.O Box 77, D/A Pejabat Pos Besar, Kangar, Perlis, Malaysia
[2]School of Materials Engineering, Universiti Malaysia Perlis (UniMAP), 01007, P.O Box 77, D/A Pejabat Pos Besar, Kangar, Perlis, Malaysia
[3]Department of Chemistry, Faculty of Applied Sciences, Universiti Teknologi MARA, Perlis Branch, Arau Campus, 02600 Arau, Perlis, Malaysia
[4]Green and Functional Polymer Research Group, Universiti Teknologi MARA, 40450 Shah Alam, Selangor, Malaysia

Abstract. The effect of fiber surface treatment and fiber loading on the tensile and flexural properties of untreated and treated fiber-based sugarcane bagasse reinforced polyester (SBRP) composites have been investigated. Sugarcane bagasse fibers were alkalized with 1, 3 and 5% concentration of sodium hydroxide (NaOH) solution combined with different fiber loading formulation 0, 20, 30 and 40 php, and hot-pressed to form natural fibers composite were prepared. The results indicated that the incorporation of bagasse fiber composites alkalized with 3% NaOH solution improved the tensile strength and modulus of reinforced composites meanwhile the untreated bagasse gave the lowest value. However, the elongation at break (E_b) indicated that treated bagasse fiber composites show a lower value compared to untreated composites. A similar trend also obtained for flexural strength and modulus.

1 Introduction

Fiber-reinforced plastic composites have long performed a leading role in a wide range of applications with their high relative strength and modulus. Fiber integrated plastics were very common because of their versatility, their lightness and the ease of manufacturing complicated shapes with economical savings compared to fiber-reinforced metals/alloys. Therefore, these composites can easily replace traditional materials in various sectors such as the construction industry, transport and consumer goods. Some of the recent attempts to

* Corresponding author: salihin@unimap.edu.my

use natural fibers by composite material manufacturing have shown their promise in many applications as replacements for traditional materials such as wood and glass fiber reinforced plastics (GFRP) [1].

Natural fillers, including cellulose-based fibers, have largely taken ground on the industry landscape in previous years in replacing synthetic materials [2]. Cellulose-based fibers (natural fibers) arise through relatively affordable and sustainable resources; and the addition of natural fibers such as wood fibers, flax or sisal to polymer matrices can lead to composites that are feasible from a mechanical and economic perspective [3]. Natural fiber-based composites, such as fiberglass, are biologically superior to those based on synthetic fibers. Natural fiber products' incineration requires 45% less energy and leads to lower air pollution as reported in recent studies [4].

Among the diverse agro-industrial residues, sugarcane bagasse is detached to be a highly produced residue containing silica (2.0%), ash (2.4%), fat and wax (3.5%), lignin (19.95%), hemicellulose (24.5%), and cellulose (46.0%) [5]. Bagasse is a plant fiber consisting primarily of cellulose, a glucose polymer with a significantly high modulus, frequently found in association with lignin as a fibrillar portion of many naturally occurred composites (wood, sugarcane straw and bagasse) [6].

Alkali treatment is considered to be one of the commonly used chemical treatment techniques for the surface modification of cellulosic fibers to enhance their adhesion properties with specific thermosetting binders [7, 8, 9]. This treatment involves partial removal from the fiber surfaces of the cementing materials (hemicellulose and lignin) and other impurities [10]. A similar change due to alkali treatment at limited NaOH concentrations has been reported to improve some of the mechanical properties of composites bonded with phenol-formaldehyde [7]. Alkali treatment of cellulosic fibers, also known as mercerization, enhances the fiber's surface adhesive properties by eliminating organic and inorganic impurities, creating a rough topography.

The mechanical properties of sugarcane bagasse reinforced polyester (SBRP) composites were assessed in this work. The effect of NaOH treatments and fiber content has been studied on the tensile and flexural properties of untreated and treated fiber-based composites.

2 Materials and Methods

2.1 Materials and preparation of composites

The bagasse was obtained from the Kilang Gula Felda Perlis Sdn Bhd (KGFP) and sieved into a diameter size average of 1.18 mm and 600 μm. Unsaturated Polyester (UP) resin was used with the grade Reversol P9509 supplied by Echemo Trading Sdn. Bhd., Penang. Tert-butyl perbenzoate (Fluka 77200) was used as an elevated temperature curing catalyst supplied by Sigma Aldrich. Sodium hydroxide (NaOH) pellets extra pure obtained from Hamburg Chemicals. Zinc strearate supplied by Sigma-Aldrich and was used as a mould release agent (lubricant).

In each 1, 3 and 5% concentration of NaOH solution, bagasse fibers were soaked at room temperature for 1 hour, maintaining a 20:1 liquor ratio. The fiber was then washed several times with distilled water to remove any NaOH solution that stuck to the fiber and the fibers were oven-dried for 72 hours at 70 °C [11]. The bagasse fibers were then kept in desiccators before compounding.

All materials were weighted by following the formulation to produce composite with percentage fiber that mentioned in Table 1. The composites were then prepared by compounding bagasse fiber, unsaturated polyester, tert-butyl perbenzoate and zinc stearate with different fiber loading using high dispersion impeller (Electronic Laboratory Stirrer, Qualtech Products Industry). The catalyst was then put in the beaker and the resin followed. The stirrer was switched on to start the compounding process. The lubricant was then placed in the beaker and the fiber proceeded to obtain a homogeneous compound. The compounding process is conducted for 30 minutes. The compounds were then transferred into the mould for moulding process.

Table 1 Composites Formulation

Materials	Untreated	NaOH Treated
Unsaturated Polyester (php)	100	100
Sugarcane Bagasse (php)	0, 20, 30, 40	0, 20, 30, 40
Tert-Butyl Perbenzoate [a]	1	1
Zinc Stearate [b]	6	6
NaOH [c]	-	1, 3, 5 %

[a] 1 php from the weight of unsaturated polyester
[b] 6 php from the weight of unsaturated polyester
[c] 1, 3, 5 php from weight filler

The samples were hot-pressed by using a hydraulic hot press machine Technopress 50HC – β model. The compression was done at 110 °C for 5 minutes, and the samples were allowed to cool for 5 minutes by the cooling process. The pressure was set at 1 MPa. A hollow rectangular mold was used in the compression molding, which has a size of 170 mm×135 mm× 3mm.

2.2 Tensile Test

An Instron 5569 tensile tester was employed to determine the tensile properties of the sample according to ASTM D638. The rectangular test specimens were tested at a crosshead speed of 5 mm/min under a cell load of 2500N. A total of 8 samples have been tested for each formulation to ensure consistency.

2.3 Flexural Test

In compliance with ASTM D790 three-point flexural tests were performed. The support range had been set to 50 mm. Tests were carried out at a crosshead speed of 2 mm/ min on the same Instron 5569 tensile testing machine used for tensile tests. A total of 8 samples have been tested for each formulation to ensure consistency.

3 Results and Discussion

3.1 Tensile Properties

Figure 1 shows the effect of NaOH solution concentration at 20, 30 and 40 wt % on tensile strength of sugarcane bagasse reinforced unsaturated polyester, respectively. The use of treated bagasse fiber has been found to increase the tensile strength of the polyester composites as opposed to the untreated bagasse fiber. This means that after alkali treatment, most of the pectin, lignin and fat covered by the fiber surface are eliminated, the fiber adhesive character enhanced in combination with the matrix [12].

The results show that all treated fibers have a higher tensile strength from those of untreated fibers, with considerably higher properties shown in 3% concentration of NaOH solution treated fibers. The fibrillation of fibers has developed superior interfacial adhesion in the composites [11]. Fiber fibrillation breaks the composite fiber bundle into smaller fibers, thus increasing the optimal surface area available for interacting with the wet matrix [13]. Nonetheless, the mechanical properties decreased with higher fibrillation due to the relatively larger fiber ends necessary for crack initiation for the 5% NaOH treated fiber composites. This could lessen the effectiveness of stress transfer at the interface [11]. The maximum tensile strength improvement was observed in NaOH-treated fiber composites of 3%. Better interfacial adhesion along with better fibrillation of these fibers was considered to have effectively attributed to improving tensile strength.

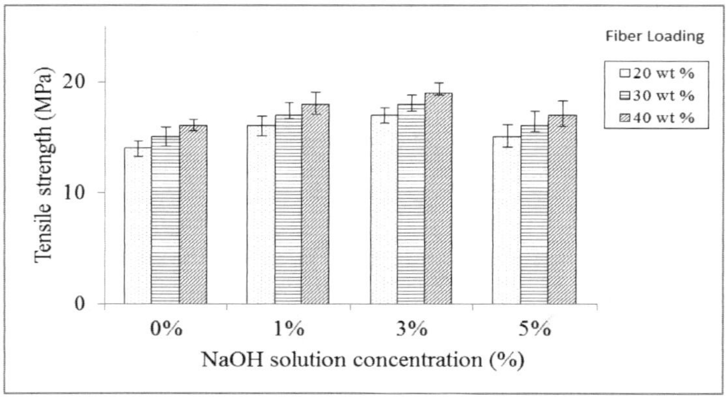

Figure 1 The effect of different NaOH solution concentration on tensile strength of sugarcane fiber reinforced unsaturated polyester composites with different fiber loading

The effect of NaOH solution concentration at 20, 30 and 40 wt % on elongation at break of sugarcane bagasse reinforced unsaturated polyester composites are illustrated in Figure 2. The graph indicates that elongation at the break of treated bagasse fiber composites has a lower value than untreated composites. There can be two hypotheses; (A) Fibers are failed points which could trigger a crack, and (B) The fibers are sliding in respect of the matrix. The elongated materials have ductile fractures and the composite failure points have the brittle fracture [14]. This was probably because modified fibers were made more hydrophobic and consequently increased compatibility between fiber and matrix resin. Enhancing fiber-matrix bonding will improve composite strength, rigidity and interfacial adhesion. This was also due to the improved adhesion of the fiber-matrix (to give the composites more stiffness) and since modified fiber was susceptible to splitting and falling apart, fiber after modification was more brittle. We can, therefore, infer that improved wetting of the treated fibers through the matrix was responsible in this case [15].

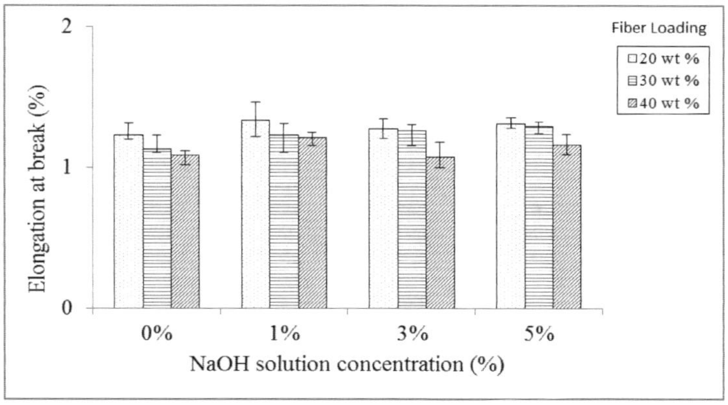

Figure 2 The effect of different NaOH solution concentration on elongation at break of sugarcane fiber reinforced unsaturated polyester composites with different fiber loading.

Figure 3 shows the effect of NaOH solution concentration at 20, 30 and 40 wt % on the modulus of elasticity of sugarcane bagasse reinforced unsaturated polyester composites. The Young's modulus of the composites increased with increasing the filler loading and NaOH solution concentration. The increase of Young's modulus with an increase in filler loading is expected as the addition of filler increases the composites' stiffness. It can be seen that 40 wt % of bagasse fiber of 3% NaOH solution concentration has the highest value of modulus of elasticity. Here 40 wt % treated bagasse fiber with 3% NaOH solution concentration – reinforced composites showed 28% of the increase in Young's modulus. Nevertheless, the mechanical properties of the 5% NaOH-treated fiber composites decreased with greater fibrillation due to the relatively larger fiber ends required for crack initiation. This could lower the effective stress transfer at the interface [11].

Such difference can be described as follows: The NaOH reacts with hydroxyl groups of the cementing material hemicelluloses during mercerisation, and it induces the deterioration of the cellular structure, and thus the fibers fragmented into filaments. This phenomenon is called fibrillation, which breaks down the untreated bundle of fibers into smaller ones by dissolving the hemicelluloses. This fibrillation increases the optimal surface area available for matrix interaction [11]. The deletion of lignin and hemicelluloses affects the fibers' tensile characteristics. When the fractions of hemicelluloses are withdrawn, the interfibrillar area is likely to be less dense and rigid, which may make the fibrils more capable of rearranging themselves along the tensile deformation direction. As reported, the alkaline treatment of sisal fibers causes fibrillation and the cellular structure collapsed due to the removal of the cementing materials [16]. Mercerisation also improves the percentage of fibers' crystallinity index due to the removal of cementing compounds, resulting in a better packaging of cellulose chains.

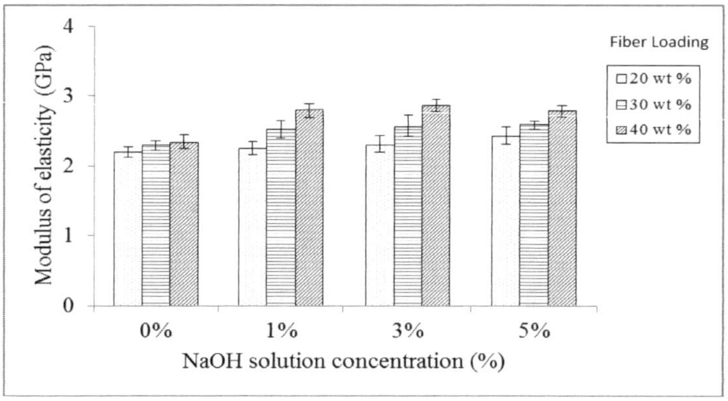

Figure 3 The effect of different NaOH solution concentration on the modulus of elasticity of sugarcane fiber reinforced unsaturated polyester composites with different fiber loading

3.2 Flexural Properties

The flexural strength and modulus of bagasse fiber reinforced polyester composite are shown in Fig. 4 and 5. It can be seen from the figure that the flexural strength of the composites increases with the increasing filler loading and the concentration of NaOH solution. The results in Fig. 4 showed that composites with 3% NaOH solution concentration have higher flexural strength than other concentrations. The fibrillation offered a superior interface adhesion in the composites thus enhances the flexural strength compared to the lower fibrillation of 5 percent NaOH solution concentration, due to the relatively larger fiber ends needed for crack initiation.

Figure 5 shows that the composites' flexural modulus increased due to the concentration of the NaOH solution. The effect on the flexural modulus was more substantial than on the flexural strength, as the integration of modified fiber was still able to give the stiffness of the composites. The composites with 3% NaOH solution concentration demonstrated higher stiffness as opposed to those with NaOH modification. This could be attributed to the enhanced compatibility between the resin and the fiber. The increased compatibility leads to a continuous interfacial region, which enabled a better and more stress transfer efficiency in the samples [15].

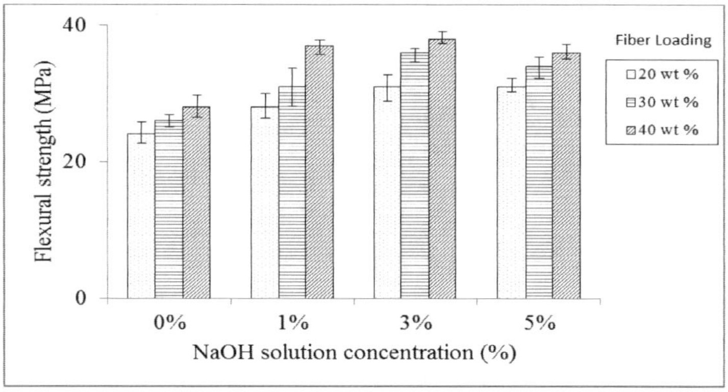

Figure 4 The effect of different NaOH solution concentration on flexural strength of sugarcane fiber reinforced unsaturated polyester composites with different fiber loading

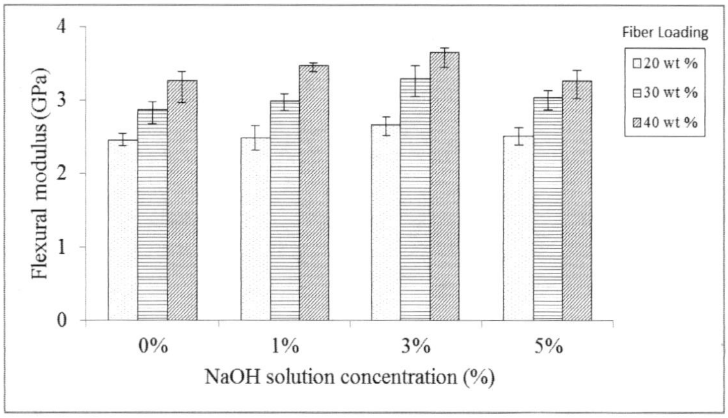

Figure 5 The effect of different NaOH solution concentration on flexural modulus of sugarcane fiber reinforced unsaturated polyester composites with different fiber loading

4 Conclusions

The following conclusion can be drawn from this study;

a. Tensile strength, Young's modulus, and flexural strength value were increased with an increasing solution concentration of NaOH. However, 5% NaOH solution concentration modification shows the decreasing value of tensile strength, Young's modulus and flexural properties of the composites.
b. Elongation at break for the sugarcane bagasse reinforced unsaturated polyester composites decreased with increasing bagasse fiber content in both composites.

The authors acknowledge the School of Materials Engineering, Universiti Malaysia Perlis for the lab facilities. Special thanks to those who contributed to this project directly or indirectly.

References

1. K.G. Satyanarayana, K. Sukumaran, P.S. Mukherjee, C. Pavithran, S.G.K. Pillai, Cem. Concr. Compos., **12**, 2 (1990)
2. A. Pandey, C.R. Soccol, O. Nigam, V.T. Soccol, Bioresour. Technol., **74**, 69–80 (2000)
3. A.S. Luyt, , M.E. Malunka, Thermochim Acta, **426**, 101–107 (2005)
4. S.V. Joshi, L.T. Drzal, A.K. Mohanty, S. Arora, Composite Part A, ***35***, 371–376 (2004)
5. L. Sene, A. Converti, M.G.A. Felipi, M. Zilli, Bioresour. Technol., **83**, 153–157 (2002)
6. P.W. Alonso, P. Garzone, G. Cornacchia, Waste Manage., **27**, 869–885 (2007)
7. B.S. Ndazi, S. Karlsson, J.V. Tesha, C.W. Nyahumwa, Composites Part A, **38**, 925-935 (2007)
8. D. Ray, B.K. J. Sarkar, Appl. Polym. Sci., **80**, 1013-1020 (2001)
9. E.T.N. Bisanda, M.P. Ansell, Compos. Sci. Technol., **41**, 2 (1991)
10. C. Albano, M. Ishazo, J. Gonzalez, M. Delgado, R. Poles, Mater. Res. Innovations, **4**, 284-293 (2001)
11. Y. Cao, S.S. Shibata, I. Fukumoto, Composites Part A, **37**(3): 423-429 (2006)
12. H. Gu, Mater. Design, **30**, 3931-3934 (2009)
13. Y.Z. Lai, *Chemical modification of lignocellulosic materials* (Marcel Dekker Inc., New York, 1996)
14. L.C. Sawyer, D.T. Grubb, *Polymer microscopy* (Chapman & Hall, London, 1996)
15. A. Vazquez, V.A. Dominguez, J.M.J. Kenny, Thermoplast. Compos. Mater., **12**, 477-497 (1999)

IOP Conf. Series: Materials Science and Engineering **957** (2020) 012033 doi:10.1088/1757-899X/957/1/012033

The effect of graphite flakes (GFs) and hybrid graphene nanoplatelets (GNPs) particles to the mechanical properties of epoxy composites

Nurhazwani Abu Bakar[1*], *Mohd Arif* Zulkarnain[1], *Wan Anuar* Wan Hassan[1], *Ramli* Junid[1*], *Jeefferie* Abdul Razak[2] and *Mohd Muzafar* Ismail[3]

[1]Faculty of Mechanical Engineering, University Malaysia Pahang, 26600 Pekan Pahang, Malaysia.
[2]Center of Smart System and Innovative Design, Fakulti Kejuruteraan Pembuatan, Universiti Teknikal Malaysia Melaka, Hang Tuah Jaya, 76100, Durian Tunggal, Melaka, Malaysia.
[3]Fakulti Kejuruteraan Elektrik & Elektronik, Universiti Teknikal Malaysia Melaka, Hang Tuah Jaya, 76100, Durian Tunggal, Melaka, Malaysia.

Abstract. This paper reports the effects of epoxy when reinforced with graphite flakes (GFs) and was compared to epoxy reinforced by hybrid reinforcements consisted of graphite flakes (GFs) and graphene nanoplatelets (GNPs). DGEBA, the type of epoxy which has been used in this work was added with 2, 4 and 6 wt.% of reinforcement respectively, relative to the total weight of the mixture. Nanocomposites was prepared using mechanical stirrer, stirred at 2000 rpm for 30 minutes followed by curing in the oven. The flexural testing shows that the epoxy/GFs composites has higher modulus compared to epoxy/GFs+GNPs. At 4wt.% filler loading, epoxy containing GFs exhibited 132% enhancement of modulus relative to neat epoxy. At the same filler amount, epoxy containing GFs+GNPs demonstrated slightly lower magnitude than system containing GFs with only 27% increase in modulus. However, addition of higher filler loading to the epoxy resin caused the modulus to decrease in magnitude speculated due to agglomeration of particles within the host matrix. The Charpy impact testing indicated similar pattern with epoxy/GFs system exhibited higher capability in absorbing energy than epoxy/GFs+GNPs where the peak was obtained at 4wt.% filler loading. At this amount of filler, 18.36 J/m increase in energy absorbed was recorded for epoxy/GFs compared to 2.13 J/m increased for epoxy/GFs+GNPs composite. Higher amount of filler loading added into epoxy only deteriorate the impact energy absorb by the sample. The hardness test show similar pattern of result with epoxy/GFs shows higher resistance in scratching compared to epoxy/GFs+GNPs.

1 Introduction

Nowadays, there are many areas in which polymer composite materials has been used due to advantages it can offers for example high specific strength and stiffness, high wear

* Corresponding author: nurhazwanibakar@gmail.com, ramli@ump.edu.my

resistance, excellent corrosion, chemical resistance, high dimensional stability, reduces noise and flexibility shapes [1]. The best example of polymer composite materials is epoxy resin that widely adapted in many application such as adhesive, aerospace composite, automotive components, used as the binder in countertops or coatings for floors, offshore equipment, boats, pipes, and pressure vessels [2]. Epoxy resin has good in mechanical properties, high adhesion strength, good heat resistance, and high electrical resistance [3].

Nevertheless, despite giving benefit in strength, epoxy resin is known to be disadvantage in ductility. It offers poor brittleness characteristics. Over the years, there are many research works which has been carried out to modify the properties of epoxy by addition of fillers to ameliorate the properties matrix dominated composite [1]. The typical filler content needed for significant enhancement of these properties can be as high as 10-20% by volume [4].

Nanomaterials are basic of nanoscience and nanotechnology [6]. They are materials or chemical substances which function at a very small scale. Nanoscale materials are substances where at least one dimension is less than approximately 100 nanometers. Nanometer is approximately 100,000 times smaller than diameter of a human hair which means of one millionth of a millimetre[5]. It can be found in sporting goods, stain resistant clothing, tires, electronics, sunscreens, cosmetics, as well as many other items, and are also used in medicine, imaging and drug delivery. Nanomaterials have several advantages compared to the same materials without nanoscale features such as high strength, high chemical conductivity and reactivity [6].

Graphene is a two-dimensional platelet consisting of carbon atoms, arranged in a hexagonal shapes or honeycomb structure that have attracted the great deal of attention due to excellent physical and electronic properties [7]. GNPs consists of small stack of graphene that can be replace carbon fiber, nano clays, carbon nanotubes or any other compounds in many composite applications. GNPs is also known as graphite nanosheets, graphene nanosheets (GNs), graphite nanoplatelets and graphite nanoflakes (GNFs) [8]. The improved mechanical properties of nanocomposites is due to the uniform dispersion of functionalised graphene and strong interfacial bonding between modified graphene and epoxy resin where this can be confirmed by microscopy observations [9]. Graphene have the incredibly high specific surface area, unique graphitised plane structure and extremely high charge mobility[9]. Because of their unique nanoscale size, shape, and material composition, graphene nanoplatelets can be used to improve the properties of a wide range of polymeric materials, including thermoplastic and thermoset composites, natural or synthetic rubber, thermoplastic elastomers, adhesives, paints and coatings [10]. The extraordinary properties of graphene could only happen if graphene nanoplatelets are well spread in matrix and that there is a strong interfacial adhesion between graphene nanoplatelets and polymer matrix [9].

GFs is a three-dimensional form of carbon atoms [11]. Basically, it is a layered material and also can be observe as a two-dimensional graphene crystal that weakly joined [12]. Graphite colour is dark gray to black and became black gray colour when it was crushed to powder. The graphite crystal lattice consisted of sp2-hybridized carbon atoms which bonded covalently in hexagonal rings [13]. The layers between atom which eventually forms graphite are bonded to each other by weak Van der Waals forces. Graphite has been widely used in variety of the areas of industry, transport, energy, defense and medical. It is due to its prominent structural, electrical and mechanical properties.

From the previous works, various particles and effect of nanoreinforcements to the properties of polymer matrix had been studied. For example, Chatterjee [7] studied the mechanical reinforcement and thermal conductivity in expanded graphene nanoplatelets reinforced epoxy composites. Their group processed the particles from natural graphite flakes (NG) and converted into expanded graphene nanoplatelets (EGNPs) using acidification procedures. Embedded in the bisphenol A type of epoxy as matrix, they later characterised the performance of the composite by mechanical testing. The mechanical testing that they have used for characterisations are three-point bending test and fracture toughness test. They demonstrated that significant improvement in mechanical properties for all samples with EGNP addition compared to neat epoxy. By increasing the concentration of EGNP in the composite, fracture toughness (K_{IC}) increases steeply, reaches a highest at some amount of loading and decreases subsequently at higher EGNP loadings due, speculated due to agglomeration. In addition, the load transfer from the matrix to the EGNP fillers shows increase in flexural modulus when stress is being applied [7] [14].

A composite can be termed as hybrid, if two or more types of fillers are combined in a common matrix to produce a composite that drives benefits from each of the individual fillers and exhibits a synergetic response [15]. Basically, it will increase the elastic modulus, decreases brittleness controls cracks initiation and its subsequent growth and propagation. The advantage of hybrid reinforced composite is to provide an arrangement where one type of filler, which is stronger and stiffer, ameliorate the first cracks stress and ultimate strength, and another type of filler, which is more flexible, and ductile, leads to reinforced toughness and strain in the existed cracking zone [16]. Previously, the effects of hybrid nanoreinforcements to the properties of polymer matrix were reported. For example, Subhani et al [17] developed composites containing multiwalled carbon nanotubes (MWCNTs) and nanodiamonds (NDs) to explore the hybridized effect of nanoreinforcements on the mechanical performance of nanocomposites [17]. Hardness test was carried out to investigate the mechanical properties of nanocomposites. It was discovered that neat epoxy showed a hardness value of 11.5 ± 0.4Hv, which increased to 14.1 ± 0.3 Hv (22% rise) by adding 0.1 wt.% MWCNTs and 12.8 ± 0.6 Hv (11% rise) by the addition of 0.1wt.% NDs. The effect of the incorporation of low content of MWCNTs in epoxy matrix is more pronounced in comparison to NDs. The combined effect of each NDs and MWCNTs at a loading of 0.05 wt.% showed a rise in the hardness value, 13.6 ± 0.6 Hv (18% rise), which is greater than the effect of NDs and less than the individual influence of MWCNTs on hardness but it shows the hardness values is continuously increase with contents of MWCNTs and NDs, 0.1 wt.% and 0.2 wt.% each which are 16.4 ± 0.7 Hv (42% rise) and 17.3 ± 0.8 Hv (50% rise). However, they stated that, generally, increasing the concentration of nanoreinforcements will cause the formation of particles agglomeration [17].

2 Experiment

2.1 Materials

The first stage of this study consists of selection of materials and preparation of the materials. GNPs was purchased from Sigma Aldrich. The density of graphene is from 0.03-0.1 g/cm^3. Graphite bulk was ground by using mortar to turn into GFs. Epoxy resin (Araldite LY 556) was used as matrix combined with hardener, Triethylenetetramine (TETA) to give great binding properties between the filler and the matrix [18]. Curing of epoxy resin was at room temperature in this work. Types of epoxy resin were Araldite LY 556 and Diglycidal Ether of Bisphenol A (DGEBA), and the hardener or curing agent was HY 951, Triethylenetetramine (TETA). This type of hardener was employed to improve the interfacial adhesion and impart good strength to the composite [19]. The mixing ratio of resin to hardener is 2:1 as suggested by manufacturer to obtain the optimum degree of crosslinking.

2.2 Composites Fabrication

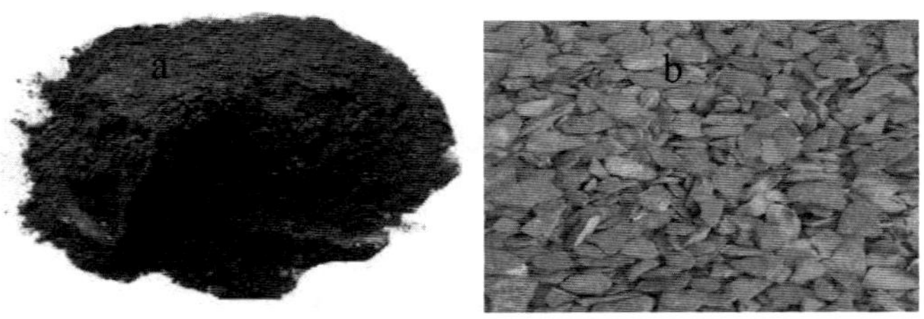

Fig. 1. The filler (a) Graphene Nanoplatelets (GNPs) and (b) Graphite bulk (GFs).

The composite was prepared in different compositions of filler (GNPs and GFs) which were 0wt.%, 2.0wt.% (1.0wt.% GNPs/1.0wt.% GFs), 4.0wt.% (2.0wt.% GNPs/2.0wt.% GFs) and 6.0wt.% (3.0wt.% GNPs/3.0wt.% GFs). For single filler, the composition will be the 2.0wt.% GFs, 4.0wt.% GFs and 6.0wt.% GFs. Firstly, the desired amount of epoxy and hardener were added into a cup. After that, the certain amount of compositions of filler was placed on the mixture of epoxy and hardener. Then, the mixture was stirred by using mechanical stirrer at 2000 rpm for 30 minutes. After that, the mixture was degassed using vacuum oven for 15 minutes at 50°C to eliminate the air bubbles [20]. It is important to remove the air bubbles because their presence will significantly affect the mechanical properties of the composite. Afterwards, the mixture of epoxy and filler was poured into the mould. Before fabrication of composite, the mould was sprayed with the mould release as a release agent to ensure that the composite and the mould will not stick and get hard for withdrawal. The mould would also be coated by the plastic wrap tape. The moulding will be left in the mould with load placed on top for about 24 hours to allow the settlement within mould container and for appropriate curing at room temperature. After the composite was cured, the casted product was withdrawn from the mould. The samples were cut to the required dimensions as per individual test requirements and example of the specimens are indicated as shown in Fig. 2a and 2b.

Fig. 2. Sample for (a) Flexural Test (b), Impact Test.

2.3 Mechanical testing

There are three different types of mechanical testing which was carried out in this work which includes hardness test, impact test and flexural test. The specimen is conducted in the standard laboratory atmosphere at temperature 24 ± 5 °C.

2.3.1 Hardness Test

The Vickers hardness test was carried following standard ASTM E384. It uses a square base diamond pyramid as the indenter. The included angle between the opposite faces of the pyramid is 136°. As a result of the indenter's shape, the impression on the surface of the specimen will be a square. The loads are between 1 and 1,000 g which referring to micro hardness test. In this hardness testing, 100g load was used with 5 seconds of dwell time.

2.3.2 Impact Test

This test provides a measure of energy required to break a material under impact loading. The test consists essentially of a hammer with a given amount of energy striking a notched test piece of fixed dimensions and recording the energy required to fracture the specimen at a specific temperature and recording whether the fracture mode was ductile or brittle. This test is used for determining behavior of material subjected to a shock load. Here the quantity measured is the energy absorbed by the specimen after being struck by a pendulum in a single blow. Normally, the impact strength of the samples is largely determined by the filler composition and strength regardless of the resin used. An impact testing machine with the Charpy arrangement is employed to perform the test having a striking angle of hammer is 124.4° and use 1J hammer strike. The specimens testing are prepared according to ASTM D256 [21] with dimension of 80 mm length, 10 mm width and 4 mm thickness with v-shaped notch depth 2mm.

2.3.3 Flexural Test

Flexural testing or also known as bending test is used to determine the bending properties of a material as well as known as three-point bending that consists in applying a point load at the centre of the material specimens. According to the ASTM D790 standard the dimensions of specimen used are 125 mm length 12.7 mm width and 3.2 mm thickness [23]. The span distance (distance between two support points) is 52 mm. The test speed was 2 mm/min [23].

3 Results

3.1 Mechanical Testing Results

3.1.1 Hardness Test

Fig. 3 shows the graph for micro vickers hardness test result of epoxy resin with different content and different composition of particles. From Fig. 3, the hybrid nanoreinforcements shows the smaller vickers hardness than the epoxy/GFs composite reinforcements. It shows that the 2wt.% of epoxy/GFs+GNPs (15.69HV) increase than the value of neat epoxy (13.86HV). The highest value for the epoxy/GFs+GNPs for 4wt.% is 18.525HV. The hybrid nanoreinforcement for 6wt.% of epoxy/GFs+GNPs shows the decreases in vickers hardness test which is 16.15HV. This may be due to, at higher wt.%, the reinforcements act as flaws and it were not perfectly aligned with matrix. For the epoxy/GFs, the vickers hardness test also increase at 2wt.% (16.385HV) GFs compared to neat epoxy (13.86HV). The vickers hardness value remain increase when 4wt.% of GFS was added yield 23.61HV but decreased when 6wt.% of GFs was added. It shows that adding a small amount of reinforcements into epoxy resin could potentially enhance the scratch resistance which optimum at 4wt.%. Adding more particles beyond 4wt.% indicated that the hardness was dropped potentially due to agglomeration of particles which caused uneven dispersion within epoxy hence indentation becomes unstable during testing. Subsequently, lower reading was recorded when particles more than 4wt.% was added into the epoxy.

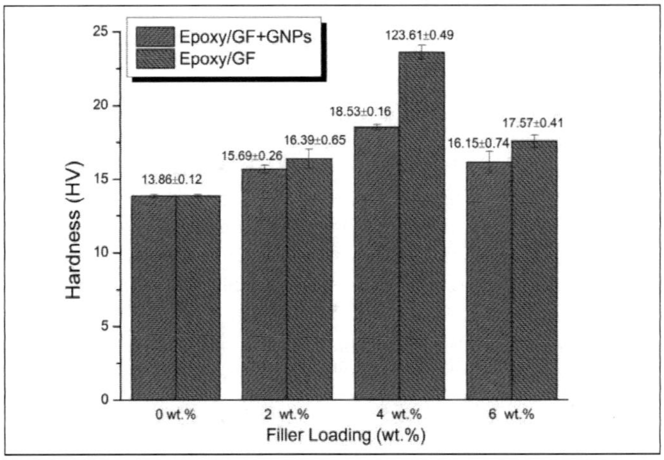

Fig. 3. Graph of micro vickers hardness test.

3.1.2 Flexural Test

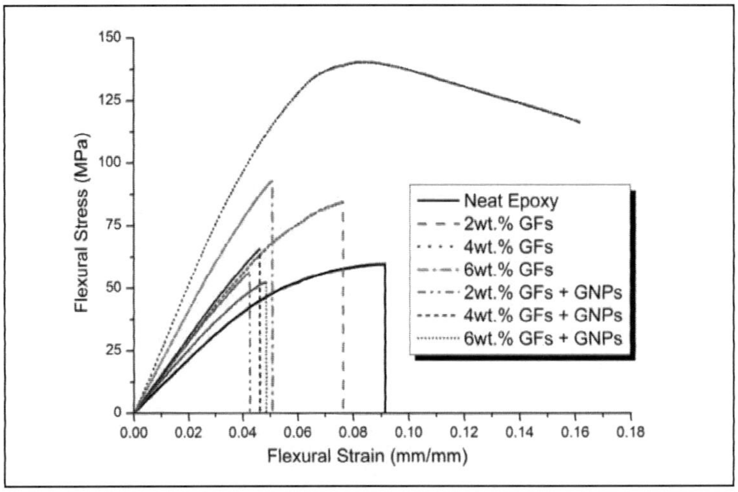

Fig. 4. Stress versus strain for each specimen as a function of reinforcement content.

Table 1. Flexural Properties of Composite (n = 3 to 6).

Flexural Test	**Modulus (MPa)**			
Sample	**0wt.%**	**2wt.%**	**4wt.%**	**6wt.%**
Epoxy/GFs+GNPs	1110.07 ± 65.96	1410.41 ± 80.88	1572.54 ± 157.57	1367.39 ± 82.49
Epoxy/GFs	1110.07 ± 65.96	1523.44 ± 103.09	2577.56 ± 26.62	2174.81 ± 147.49

Fig. 4 shows the flexural stress – strain curve of each different composition while Table 1 displays the summary of flexural modulus. From the graph and table, the 4wt.% epoxy/GFs shows the highest value of the modulus, 2577.56MPa. similarly, for epoxy/GFs+GNPs, 4wt.% is the highest amount of the weight percentage which is 1572.53MPa. if more reinforcement was added into epoxy beyond 4wt.%, the modulus of the composite decreased perhaps due to agglomeration. Amount of the GFs and GNPs is excessive as well as the epoxy cannot combined well and start to re-aggregate. Same with epoxy-GFs, the modulus will decrease if the parameter is more than 4wt.% which is 6wt.% at 2174.81MPa. Experimental results show that, micro sized particles have effect on modulus and strength of epoxy such as increasing modulus and strength but at certain limit it will decrease the mechanical properties of material. In addition, the flexural strength will increase if the number of filler increases. This is due to the increasing number of filler will increase the barrier in the composite to resist the crack [21, 22]. At certain percentage of fillers, the flexural strength of the specimen will decrease due to the agglomeration of fillers.

3.1.3 Impact Test

Table 2 shows the summary of impact strength of epoxy resin with different content and different composition of particles. From the Table 2, the composite containing hybrid particles shows the smaller impact strength than the composite containing GFs alone. For the epoxy/GFs+GNPs, the impact strength increases when 2wt.% (11.25J/m) of filler was mixed to the epoxy compared to neat epoxy (10.29J/m). The impact strength value was also increased when the hybrid reinforcement composites increased for 4wt.% (12.42J/m) but decreased for 6wt.% (11.13J/m) of hybrid composite. For epoxy/GFs, the value of impact strength also increases when 2wt.% of GFs (19.53J/m) was added into the composite of compared to the neat epoxy (10.29J/m). The energy absorbed was found increased continuously for reinforcements of epoxy/GFs at 4wt.% (28.65J/m) but decreased at 6wt.% (17.13J/m). The highest value was achieved for epoxy/GFs at 4wt.% which is 28.65J/m. For the epoxy/GFs+GNPs, the highest value is from 4wt.% specimen which is 12.42J/m. This mechanical testing shows that the epoxy/GFs has higher impact strength than the epoxy/GFs+GNPs. The hybrid particles of GFs+GNPs can improve the mechanical properties of neat epoxy on the impact strength but not as much as in the case for epoxy/GFs. As the conclusion for impact test, the epoxy/GFs absorbs more impact strength than epoxy/GFs+GNPs.

IOP Conf. Series: Materials Science and Engineering **957** (2020) 012033 doi:10.1088/1757-899X/957/1/012033

Table 2. Result of impact strength.

Impact Test		Impact Strength (J/m)		
Sample	0wt.%	2wt.%	4wt.%	6wt.%
Epoxy/GFs+GNPs	10.29 ±0.31	11.25±1.13	12.42±0.31	11.13±0.66
Epoxy/GFs	10.29±0.31	19.53±0.78	28.65±1.95	17.13±2.18

4 Conclusion

This study shows that GFs and hybrid particles of GFs and GNPs can affect the properties of epoxy. The present of hybrid particles and nanoreinforcements composite exhibited enhancement in mechanical properties of the epoxy. The finding indicate that these particles are effective to become obstacles to resist crack/failures. The higher the weight percentage of particles within the system, the better the mechanical properties of the composite but there was an optimal limit in which beyond the optimum limit, the particles will deteriorate the mechanical performance of composites probably due to agglomeration.

We acknowledge Universiti Malaysia Pahang for funding provided under the grant code RDU190324 and PGRS1903163.

References

1. El Wafa. M. A.,123-135 (2011)
2. Sharma, B., R. Chibber, and R.J.M.T.P. Mehta, **5**(1),276-286 (2018)
3. Jin, F. L.L., Xiang, Park, Soo Jin, J Ind Eng Chem **29**, 1-11 (2015)
4. Zhou, Y.W., Everett, S. Jeelani, and A. Tuskegee, Nanotech (2007)
5. Alagarasi, A., *Chapter-Introduction to Nanomaterials.* (2011)
6. Iqbal, P.P. and P.M.J. Jon A Mendes, *Nanotechnology: The "Top-Down" and "Bottom-Up" Approaches* (2012)
7. Chatterjee, S., et al., **531**,6-10 (2012)
8. Bin Junid, R., *Multiscale carbon fibre composites with epoxy-graphite nanoplatelet matrices*. University of Manchester (2017)
9. Naebe, M., et al., **4**, 4375 (2014)
10. Mittal, G., et al., Compos. B. Eng. **138,**122-139 (2018)
11. Wu, Y., et al., **184**, 92-100 (2018)
12. Geckeler, K.E. and H. Nishide, Adv. Nanomater. **2 (**2010)
13. Krupa, I.C.. **37**(11), 2159-2168 (2001)
14. Jumahat, A., et al., **41**, 1634-1640 (2012)
15. Baptista, R., et al., Procedia Structural Integrity, **1**, 74-81 (2016)
16. Rostamiyan, Y., et al., **56**,1068-1077 (2014)
17. Subhani, T., et al., **87**,436-444 (2015)
18. May, C., *Epoxy resins: chemistry and technology*. 2018: Routledge.
19. Storer, J.W., et al., *Hardener composition for epoxy resins*. 2012, Google Patents.
20. Salom, C., et al., *M.* Int J Adhes Adhes ,**84**, 119-125(2018)
21. Malhotra, N., et al., **3**(1),75-80 (2012)
22. Baley, C., **33**(7), 939-948.(2002)

IOP Conf. Series: Materials Science and Engineering **957** (2020) 012034 doi:10.1088/1757-899X/957/1/012034

Morphology and Conductive Properties of Carbon Black- and Graphite- Filled Conductive Epoxy Micro-porous

Anusha Leemsuthep[1,2] , *Zunaida* Zakaria[1,2] , and *Du Ngoc* Uy Lan[3*]

[1]School of Materials Engineering, Universiti Malaysia Perlis (UniMAP), Komplek Pusat Pengajian Jejawi 2, 02600 Arau, Perlis, Malaysia.
[2]Centre of Excellence Geopolymer and Green Technology (CEGeoGTech), Universiti Malaysia Perlis (UniMAP), Komplek Pusat Pengajian Jejawi 2, 02600 Arau, Perlis, Malaysia.
[3]Department of Polymer Engineering, University of Bayreuth, Universitatsstrafte 30, 95447 Bayreuth, Germany.

Abstract. Conductive composite requires a well dispersion of conductive filler, which is this difficult to be achieved low content of filler for high density conductive filler. This work investigates the effect of carbon black and graphite on the morphology and conductive properties of conductive epoxy micro-porous. The conductive epoxy micro-porous was prepared by using single emulsion technique. It involved the drop of epoxy-hardener-blowing agent-conductive filler mixture into corn oil at 160 °C and followed by the leaching process to remove excess corn oil. Results show the addition of conductive filler in epoxy matrix lead to increases porous structure in epoxy micro-porous. Carbon black filled epoxy micro-porous possessed smaller particles compared to graphite filled epoxy micro-porous. For skeleton density and total pore, both of the value increase with filler loading due to packing sphere effect. However, the skeleton density for carbon black filled epoxy micro-porous is higher compared to graphite filled epoxy micro-porous. For electrical conductivity, graphite filled micro-porous was higher than that of carbon black filled epoxy micro-porous.

1 Introduction

Nowadays, conductive polymer composites have gained a lot of interests in many applications such as electromagnetic interference (EMI) shielding, conductor, anti-static materials and sensor [1-3]. The application of bipolar plates in polymer electrolyte membrane fuel cells (PEMFCs) requires compound with a high conductive filler loading to achieve high electrical conductivity [4]. However, high filler loading can reduce both mechanical strength and process-ability of conventional composite. In order to overcome this problem, metal coated cenospheres is considered as one of the most advanced

* Corresponding author: epoxylan@gmail.com

conductive materials to reduce the conductive filler used in conductive composites [5]. Nevertheless, there are some limitions in applications of these conductive cenospheres as well as the difficulties encountered during the process. Metal coated cenospheres have two layers, namelycenosphere wall and metal coated layer. The adhesion of metal coated layer and cenospheres is expected to be strong enough to avoid the debonding mechanism of metal particles at the coated layer during mixing process and preparation of composites. Moreover, the use of metal coated cenospheres have to be cared on the wetability of metal surface with the matrix binder. In addition, the metal coating process onto the cenospheres such as electroless plating, magnetron sputtering deposition and fabrication of cenospheres (ceramic microballoons) are complex [6-8]. This research proposes one step process to produce conductive epoxy micro-porous through single emulsion technique. The process is simple and economic. The conductive epoxy micro-porous have one cell wall containing dispersed conductive filler, hence it could provide good wettability and well adhesion to epoxy matrix. Moreover, the effect of different type of conductive fillers used on morphology and conductive behavior is also investigated.

2 Experimental

2.1 Materials

Epoxy resin DER 331 and polyamide A062 from DOW chemical were used. Epoxy resin DER 331 has epoxide equivalent weight of 182-192, density and viscosity of 1.16 g/cm3 and 11-14 Pas at 25 °C. Polyamide A062 has equivalent weight per H active of 110, density and viscosity of 0.96 g/cm3 and 35–45 Pas at 25 °C, respectively. White crystalline powder of sodium bicarbonate from HmbG Chemicals was used as the blowing agent with density of 2.22 g/cm3. Corn oil was purchased from Yee Lee Corporation Sdn. Bhd. and contained free polyunsaturated fatty acids. Carbon black N300 (HAF) was purchased from Chemicals. Inc. Carbon black with particles size 100 µm has iodine absorption number of 82 ± 5 g/kg and CTAB surface area of 79~87 10^3m2/kg. Graphite powder from Sigma Aldrich was purchased from AR Alatan Sdn Bhd. Graphite powder was a synthetic graphite with < 20 µm particles size. For conductive testing, epoxy resin DER 331 and polyamine were used as binder. Polyamine has equivalent weight per H active of 95, density and viscosity of 1.03 g/cm3 and 3–6 Pas at 25 °C, respectively.

2.2 Preparation of Conductive Epoxy Micro-porous (CEMP)

In this study, two different conductive filler content (10 phr and 20 phr) and two types of conductive filler (carbon black and graphite) were used. The formulation and nomination of conductive epoxy micro-porous are depicted in Table 1. The ratio of epoxy and polyamide was 1E:2P (as hundred part resin - phr). Epoxy was mixed with 4 phr of sodium bicarbonate by using over-head stirrer with the speed of 300 rpm for 3 minutes. Conductive filler was varied from 10 phr and 20 phr and added and stirred for 2 minutes at the same speed. Then, polyamide was added and mixed at the speed of 300 rpm for another 3 minutes. Epoxy mixture was dropped into corn oil (weight ratio of 1:9) at 160 °C and stirred at 1000 rpm for 1 hour. The mixture was then allowed to cool down to 80 °C. The micro-porous was filtered and washed five times with detergent solution at 60 °C. The ratio of detergent to water of 1:20 was used. The detergent solution was changed every 15 minutes. Drying and post-cure process were carried out at 80 °C for 4 hours in air-circulating oven.

IOP Conf. Series: Materials Science and Engineering **957** (2020) 012034 doi:10.1088/1757-899X/957/1/012034

Table 1. Nomination of conductive epoxy micro-porous.

Nomination	Control	CBEMP 10	CBEMP 20	GPEMP 10	GPEMP 20
Epoxy: Polyamide	1 : 2	1 : 2	1 : 2	1 : 2	1 : 2
Sodium Bicarbonate (phr)	4	4	4	4	4
Carbon Black (phr)	-	10	20	-	-
Graphite (phr)	-	-	-	10	20

*CB= carbon black, GP= graphite, EMPn= epoxy micro-porous with n phr of conductive filler content, phr= part per hundred resins.

2.3 Preparation of Epoxy Composite Filled CEMP for Conductive Testing

The amount of CEMP in 20 phr was added into 100 phr of epoxy-polyamine binder (weight ratio of 10:5). The mixture was casted into PP mold with dimension of 24 mm x 20 mm x 4 mm and allowed to harden at room temperature for 24 hours.

2.4 Characterization and Testing

Particle density of CEMP was measured by using gas pycnometer density analyser, AccuPyc II 1340 V1.0.5. Testing was carried out at room temperature using the chamber of 10 cm^3 and Helium gas.

The morphology analysis of CEMP was carried out using a scanning electron microscope, model JSM 6260 LE JOEL. The CEMP was coated with gold/platinum layer to avoid electrostatic charging during examination.

Electrical conductivity test of CEMP was obtained by using Keithley 4200 Semiconductor Characterization System. The samples were measured using two probes I-V measurement system. The distance between two probes was 5 mm. The voltage was varied from 0 to 10 V.

3 Results and Discussion

3.1 Morphology

Figure 1 shows the variation morphology of carbon black filled epoxy micro-porous (CBEMP) and graphite filled epoxy micro-porous (GPEMP) at different concentration. As seen in Figure 1(a), epoxy micro-porous without adding conductive filler possessed large micro-porous particle which adhered to one another [9]. In contrast, the addition of conductive filler led to the reduction of the particle size of CEMP as shown in Figure 1. Results also indicated that the particle size of CEMP was smaller than that of control sample. The addition of conductive filler increased the viscosity of epoxy mixture droplets and consequently the matrix less expanded. An increased viscosity of epoxy mixture droplets caused it difficult to expand when sodium bicarbonate decomposed [10]. However, as the content of conductive fillers increased from 10 phr to 20 phr, the particle size of the CEMP reduced. This was due to strong interaction between fillers which complicated the separation of epoxy droplet [11]. Hence, CEMP with 20 phr conductive fillers has smaller particle size compared to CEMP with 10 phr conductive fillers. Furthermore, higher conductive filler loading increased the viscosity of the epoxy mixture. The increased viscosity of epoxy mixture in emulsion system will also affect the distribution of droplet size and increase the tendency of the droplet to aggregate [12]. On the other hand, GPEMP20 produced large particle compared to CBEMP20. It can be implied that the reduction of particle size of conductive filler will result in increased viscosity of epoxy mixture droplets and thus difficult to blow [13].

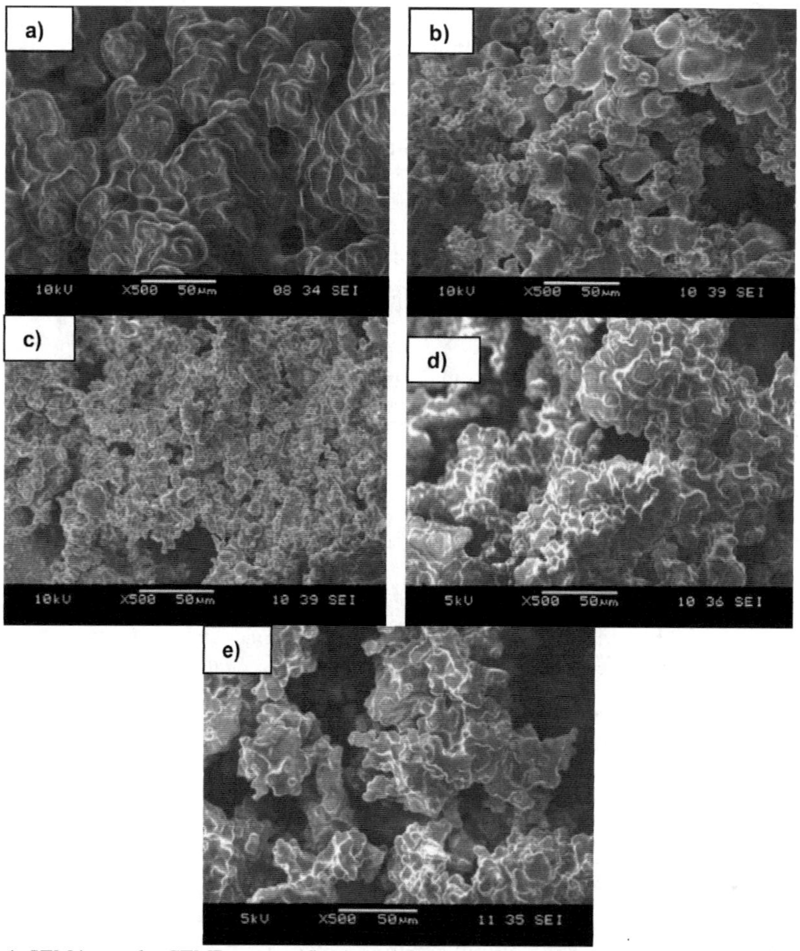

Fig. 1. SEM image for CEMP at magnification of 500x: (a) Control, (b) CBEMP10, (c) CBEMP20, (d) GPEMP10, (e) GPEMP20.

3.2 Density

Figure 2 shows the skeleton density and total pore volume of CEMP with variation of conductive filler and content. It showed that an increase of conductive fillers content from 10 phr to 20 phr increased the micro-porous skeleton density. The increased micro-porous density was due to high loading of conductive fillers. Besides, CEMP with carbon black of 10 phr showed the highest skeleton density than that of CEMP with graphite of 10 phr. This result could be influenced by larger particle size of carbon black than the graphite. The total pore volume of CEMPs increased compared to that of control sample. In both types of conductive filler, an increase in conductive filler content increased total pore volume. This was greatly related to the packing sphere effect [14]. Furthermore, total pore volume GPEMP was slightly increased compared to CBEMP due to high density of graphite which led to increment of total pore volume compared to carbon black.

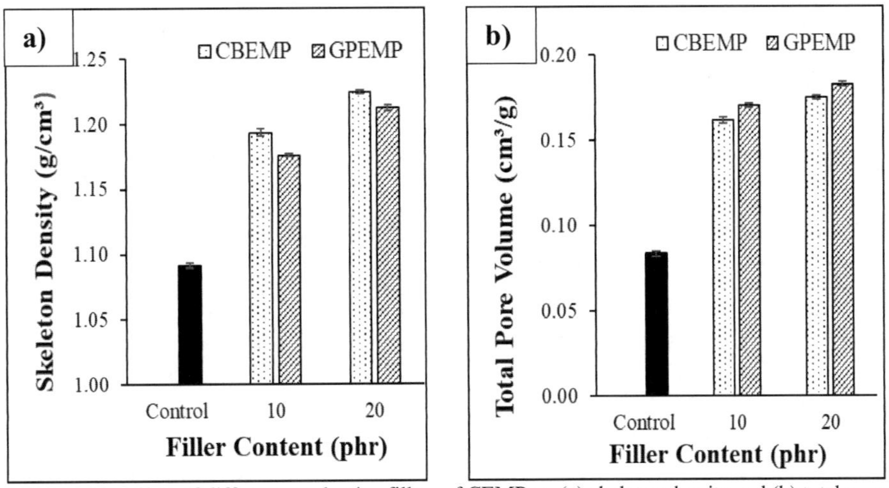

Fig. 2. The effect of different conductive fillers of CEMP on (a) skeleton density and (b) total pore volume.

3.3 Conductivity Behavior

The electrical conductivities of epoxy composite filled CEMP (epoxy/CEMP) are presented in Figure 3. The figure depicted that electrical conductivity increased with the increasing of conductive fillers content which reduced the particles size of CEMP as reported in Figure 1.

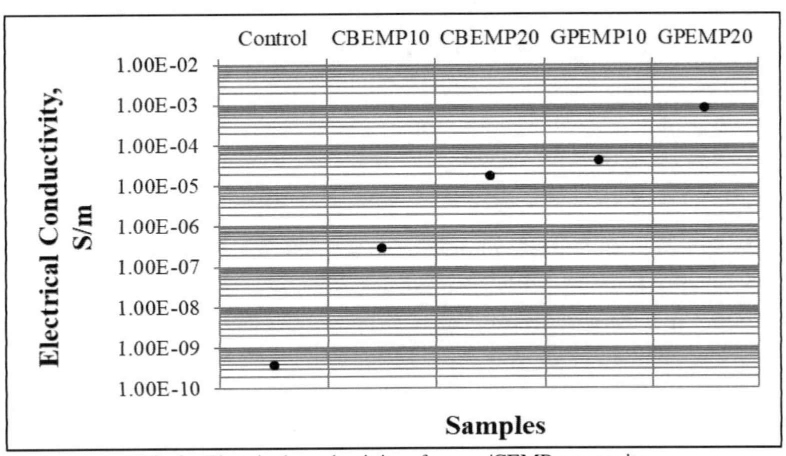

Fig.3 . Electrical conductivity of epoxy/CEMP composite.

The reduction of particles size of CEMP promotes higher surface area and better particles dispersing in polymer matrix [15]. Hence, more electrical pathway provides better conductivity behaviour. Moreover, the use of different types of conductive filler also affected the electrical conductivity properties. GPEMP20 showed higher electrical conductivity compared to CBEMP20because of different chemical structure bonding arrangement between carbon black and graphite [16]. Graphite has each atom connected evenly to three carbons formed sp2 hybridization. The sp2 set forms the hexagonal lattice typical of a layer of graphite [17]. On the contrary, carbon black has disorder crystal structure [18]. Hence, the bonds between carbon in graphite were stronger compared to carbon black. Therefore, GPEMP20 has better electrical conductivity properties.

4 Conclusions

CEMPs were prepared through single emulsion technique. Higher conductive fillers loading increased the skeleton density, and contributed to high total pore volume and excellent electrical conductivity of CEMP. Besides, high density of graphite powder resulted in higher skeleton density and total pore volume of GPEMP compared to CBEMP. The dispersion of conductive fillers has been proven by electrical conduction of epoxy composite filled CEMP. It was found that GPEMP produced excellent electrical conductivity compared to CBEMP.

The authors would like to acknowledge Ministry of Higher Education Malaysia for providing the ERGS grant (grant no: 9010-00030) that has made this work possible.

References

1. A. Murat, T. Karazehir, SA. Sezai SA. Curr Phys Chem ,**2**,224-240 (2012)
2. Y.S. Jun, S. Sy, W. Ahn, H. Zarrin, L. Rasen, R. Tjandra, B.M. Amoli, B. Zhao, G. Chiu, A. Yu, Carbon, **95**, 653-658 (2015)
3. V. Kumar, T. Yokozeki, T. Goto, T. Takahashi, J Reinf Plast Compos, **34**,16: 1298–1305 (2015)
4. E. Planes, L. Flandin, N. Alberola, Energy Procedia, **20**, 311-323 (2012)
5. M. Supriya, M. Sashikala, M.S., Khowdwe, S. Krishnan, S. Vynatheya, Int J Recent Sci Res, **4**, 1736-1742 (2013)
6. Z. Aixiang, X. Weihao, X. Jian, Surf Coat. Technol, 197(**2-3**), 142-147 (2005)
7. X. Yu, Z. Shen, Z. Xu, Nucl Instrum Methods Phys Res Sect B; 265(**2**), 637-640 (2007)
8. X. Meng, D. Li, X. Shen X, et al. Preparation and magnetic properties of nano-Ni coated cenosphere composites. Appl Surf Sci, 256 (**12**),3753-3756 (2010)
9. A.A. Bakar, D.N.U. Lan, B. Z.M. Ariff, Mater Lett , 63(11), 827-829 (2009)
10. H. Oxfall, G. Ariu, T. Gkourmpis T , R.W. Rychwalski, M. Rigdahl, Express Polym Lett , 9(**1**), 66–76 (2015)
11. H. Zhang, G. Zhao, Z. Zhang, Y. Liu, Micro Nano Lett,11(**7**), 402–406 (2016)
12. I.N.A. Anisa, H.A. Nour, Engineering and Technology, 38(**2**), 213–216 (2001)
13. N. Mangesana, R. Chikuku, A. Mainza A. J South Afr Inst Min Metall, 108(**4**), 237-243(2008)

14. L. Burtseva, B.V. Salas, F. Werner, V. Petranovskii, VRevista Mexicana de Fisica E, 61 (**1**), 20–27 (2015)
15. S. Chuayjuljit, A. Imvittaya, N. Naranong N, P. Potiyaraj, J Metals Materials and Minerals, 12 (**1**), 51-57 (2002)
16. V. Gunasekaran, K. Sang-Jae K. J the Korean Phys Soc, 55(**3**), 1102-1105 (2009)
17. V. Linda, W.H. Daniel, M. Gad, Adv Colloid Interface Sci, 37(**46**), 128-130 (2006)
18. A. Venkateshwar, University of Massachusetts Lowell (2008)

Effect of Porosity and Water Absorption on Compressive Strength of Fly Ash based Geopolymer and OPC Paste

Siti Aisyah Razak[1,*], *Farah Farhana* Zainal[1], and *Shaiful Rizam* Shamsudin[1]

[1]Center of Excellence Geopolymer Geopolymer and Green Technology (CEGeoGTech), School of Materials Engineering, University Malaysia Perlis (UniMAP) P.O. Box 77, D/A Pejabat Pos Besar, 01000 Kangar, Perlis, Malaysia.

Abstract. The fly ash based geopolymer is a promising binder by activation of fly ash with an alkaline activating solution. The fly ash based geopolymer prepared was characterized by several methods. The experimental result, studies effect of the porosity and water absorption on compressive strength of fly ash based geopolymer and Ordinary Portland Cement paste for comparison. The porosity studies were determined using the Brunauer, Emmett and Teller method included nitrogen adsorption/ desorption plots. Then followed by water absorption and compressive strength tested at 7 and 28 days curing time. The result shows that the porosity of fly ash based geopolymer paste was in the lower surface area, pore volume and pore size compared to Ordinary Portland Cement paste. The small pore size of the fly ash based geopolymer had a significant proportion of a micropores whilst Ordinary Portland Cement paste pores were mostly mesopores .The highest compressive strength of fly ash-based geopolymer can be achieved up to 76.723 MPa at 28 days when less of pore size and water absorption. Therefore, the paste based on geopolymeric materials is a better durability and high resistance to aggressive environment compared Ordinary Portland Cement paste.

1 Introduction

The fly ash based geopolymer binder provides an alkaline environment that protects the steel from corrosion by forming a protective ferric oxide film. The reinforcement bar embedded in geopolymer concrete had a low corrosion rate because silicate membrane is coated strong compared to Ordinary Portland Cement (OPC) concrete [1]. But the corrosive like carbonation and chloride ions can absorb through the concrete pores to the rebar oxide layer.

The passive layer will deteriorate and leaving the rebar vulnerable to chloride attack. It is due to exposed in highly corrosive environments, thereby requiring additional help to

* Corresponding Author: ctaisyahrazak@gmail.com

prevent the corrosion occurred [2,3]. Mustafa et al. stated the primarily durability of concrete due to the deterioration corrosion of reinforced concrete, it depends on the permeability of the structure which are the damage can reduce when sealed concretes can resist the ingress of aggressive ions into the concrete [4]. Therefore, corrosion performance of geopolymer and the compressive strength of concrete were related to the porosity and water absorption percentage.

The porosity is the quantity of the total volume occupied by the pores of the sample. If the pores are disconnected and less of pore size, the percentage of water absorption is respectively low. The Brunauer, Emmett and Teller (BET) method allows measuring the pore structure of cement based materials completely and generally used this method includes nitrogen adsorption/ desorption [5]. The internal surface area of pores can be evaluated from the BET multilayer adsorption theory. Meanwhile, the pore size distribution can be obtained using Barrett–Joyner– Halenda (BJH) interpretation based on capillary condensation [6]. The pore structure of samples is characterized by the pore internal surface area, pore volume and pore size distribution. For more details into the pore size distribution, according to Tabe 1, the measured pore distribution is divided into four size ranges[5].

Table 1. Range of pore size distribution.

Pore distribution	Size range
Micropores	< 4.5 nm
Mesopores	4.5–50 nm
Middle capillary pores	50–100nm
Large capillary pores	> 100 nm

According to Alehyen et al. [7] the fly ash based geopolymers using The BET method obtained porosity of cement paste. The result shows average pore sizes, total pore volumes and surface area of pores was lower than OPC. The average pore size was 1.1738 nm obtained the fly ash based geopolymer is lesser than OPC was 5.252 nm means that the geopolymeric matrix is a microporous material size range distribution. Therefore, the porosity of geopolymer is smaller than OPC.

The small pores volume and surface area for the geopolymer indicates that the geopolymer matrix is dense, which means that its permeability will be low and consequently its durability will be high [8]. Fly ash based geopolymer paste showed considerably lower pore volume than OPC paste, meaning that geopolymer paste is more compacted and dense than the OPC.

Same previous study by Mustafa et al. was observed that the porosity and water absorption of fly ash-based geopolymer concrete was in the lower range compared OPC concrete [9]. Furthermore, Thokchom et al. stated decreases porosity and water absorption can increase the compressive strength [10]. According to Suresh et. al., it is because the calcium content performs in fly ash played an important role in increasing the compressive strength and the geopolymer provided faster hardening in shorter curing time due to calcium ion produced a quick reactivity [11].

The compressive strength of concrete is the most important property. It can also be affected by these pores [12]. It can be expected that the mechanical strength of the geopolymer will be greater than that of OPC. In fact, the porosity has been shown an inverse relationship to compressive strength, with a pore size reduced, due to increases in a material strength [13]. Farah Farhana et al. reported that the less of water absorption and porosity towards the highest compressive strength of the geopolymer paste at day 90 [8, 14].

2 Methodology

2.1 Sample Preparation

2.1.1 Geopolymer paste

The 12 molar sodium hydroxide (NaOH) solutions were prepared by dissolving the NaOH pellets in distilled water. Then, an alkaline activator was prepared 24 h prior to use with the ratio of the mixture of sodium silicate (Na_2SiO_3)/NaOH is 2.5. The optimum ratio is important in order to obtain a homogeneous solution. The prepared mixtures fly ash was placed in mould and compacted. The samples were kept at ambient temperature in the mould and taken out from the mould after 24h and then cured at 60°C in the oven for 24h.

2.1.2 OPC paste

The prepared mixture OPC with the optimum ratio of water content is 0.5. Then was placed in mould and compacted. The sample was kept at ambient temperature in mould and taken out from the mould after 24h and then cured in water for 24h.

2.2 Brunauer, Emmett and Teller

The BET method used to determine the powders of surface area and porous materials [15]. Nitrogen gas is generally employed as the probe molecule and exposed to a solid under investigation at liquid nitrogen conditions [13]. Nitrogen adsorption/desorption plots 1 g of powder were carried out under continuous adsorption conditions. Then the powders of geopolymer and OPC were heated and gas adsorption analysis in the relative pressure was used to determine the total specific area. The total pore volume and micropore volume of the powders were calculated using by t-plot analysis. The BJH method was used to obtain pore size distribution curves [15].

2.3 Water absorption

The water absorption tests were conducted according to ASTM C642-13 [16]. The percent of water absorption for these samples were determined by Eq. (1).

$$\%\ \text{Water Absorption} = \frac{\text{Wet weight} - \text{Dry weight}}{\text{Dry weight}} \times 100\% \quad (1)$$

2.4 Compressive Strength

The compressive strength test was on fly ash based geopolymer and OPC samples using mechanical testing in order to obtain the ultimate strength were measured according to BS 1881-116:1983. The samples were compressed with 50 kN load and 5 mm/min loading rate.

3 Result and discussion

In this study, Brunauer-Emmett-Teller surface (BET) method was used to determine the pore structure of the sample. Table 2 shows parameter obtained by nitrogen absorption on fly ash based geopolymer and OPC pastes, while Fig. 1(a)(b) shows the representative pore size distribution for geopolymer and OPC pastes.

Table 2. Parameter obtained by BET method -nitrogen absorption/ desorption on fly ash based geopolymer and OPC paste.

Parameter	Geopolymer		OPC	
Time (Days)	7	28	7	28
Surface Area (m^2/g)	2.3822	2.2719	3.5712	3.1264
Pore Volume (cm^3/g)	0.0007	0.0006	0.0015	0.0014
Median Pore Size (nm)	1.9080	0.8235	5.3284	4.5720

According to Table 2, the fly ash based geopolymers show the data of surface area, pore volumes and median pore sizes are lower than OPC paste. The pore size 1.9080 nm and 0.8235 nm at 7 and 28 days obtained for the fly ash based geopolymer is lesser than 4.5 nm means that the geopolymer matrix is mainly a microporous material. Meanwhile, the pore size of OPC paste is 5.3284 nm and 4.5720 nm at 7 and 28 days were more than geopolymer paste. So the pore size is over than 4.5 nm, indicating that the OPC.paste is mostly composed of mesopores. Furthermore, fly ash based geopolymer paste showed lower pore volume than OPC paste, it is due to geopolymer paste is more dense and compacted than the OPC paste. The small pores volume and surface area for the geopolymer paste indicates that the geopolymer matrix is dense, which means that its permeability was low and consequently its increase the durability of paste.

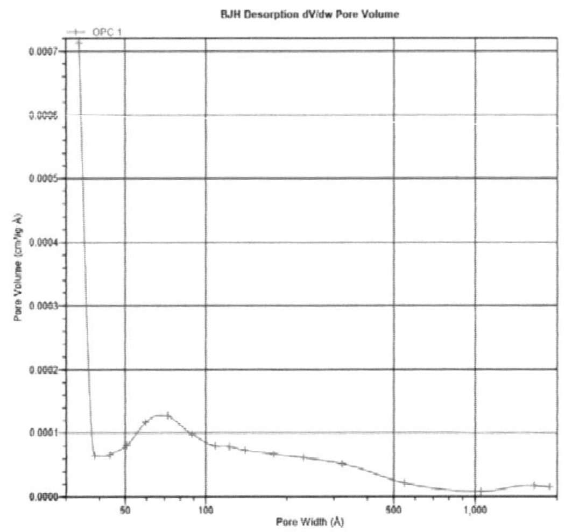

Fig. 1(a). Pore size distribution for geopolymer pastes.

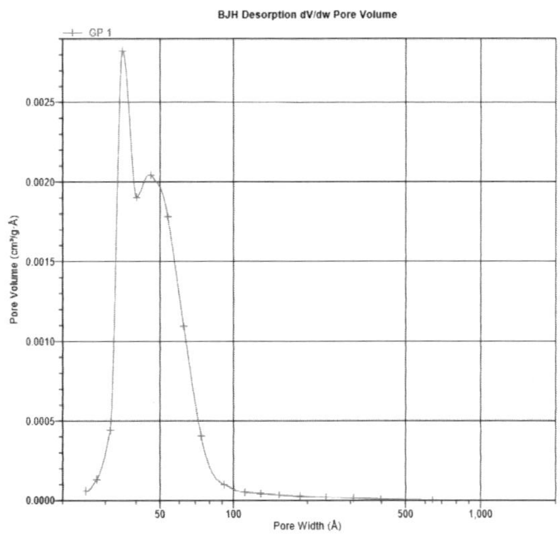

Fig. 1(b). Pore size distribution for OPC.

Fig. 1(a)(b) shows that the pore size distribution of the fly ash based geopolymer and OPC are unimodally distributed. Both of samples present an optimum pore size at 3.5710 nm and 7.0000 nm respectively meaning that the fly ash based geopolymer largely on microporosity material while mesoporosity is mostly in the OPC. More the creation of mesoporosity in OPC will be increasing the surface area and reducing the strength of cement. Besides that, the pore size also can affect the capacity of the water absorption for concrete structure. Water absorption can be used to present an open pore of concrete paste. The measurement by the difference percentage of dry weight of specimen and wet weight for fully saturated conditions.

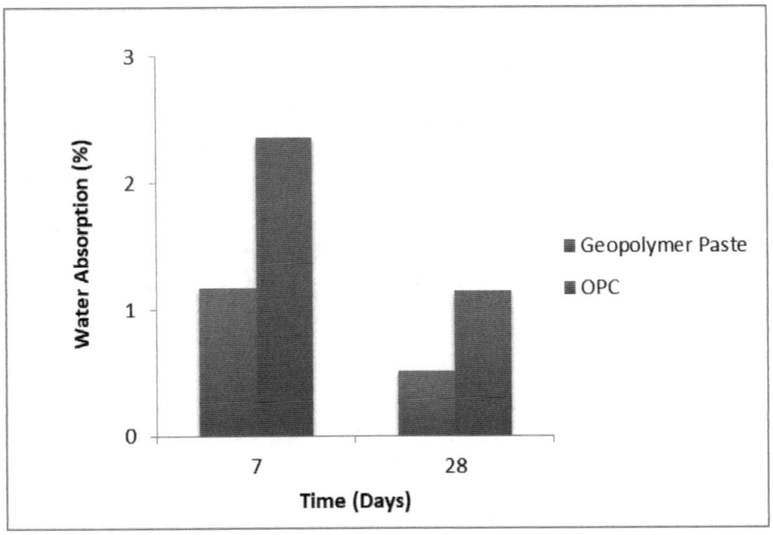

Fig. 2. The water absorption percentage of geopolymer and OPC pastes.

Fig. 2 shown that geopolymer paste has smaller water absorption percentage than OPC paste.The water absorption percentage of fly ash-based geopolymer was 1.1720 % compared to OPC was 2.3570% at day 7 and reduced respectively 0.5130% and 1.1460 % at 28 days. It has concluded, that the percent geopolymer paste of water absorption was reduced more with duration time. This show, the inclusion of less of pore size will be decreasing of water absorption capacity. The different percentage of water absorption for both samples at 7 days was 1.1850% and reduced was 0.6330% at 28 days, means that the OPC paste is more porous than geopolymer and durability can affect at early ages.

There are relationship between porosity and water absorption towards mechanical strength. The compressive strength tests of both samples were performed at 7 and 28 days and the results produced are shown in Fig. 3.

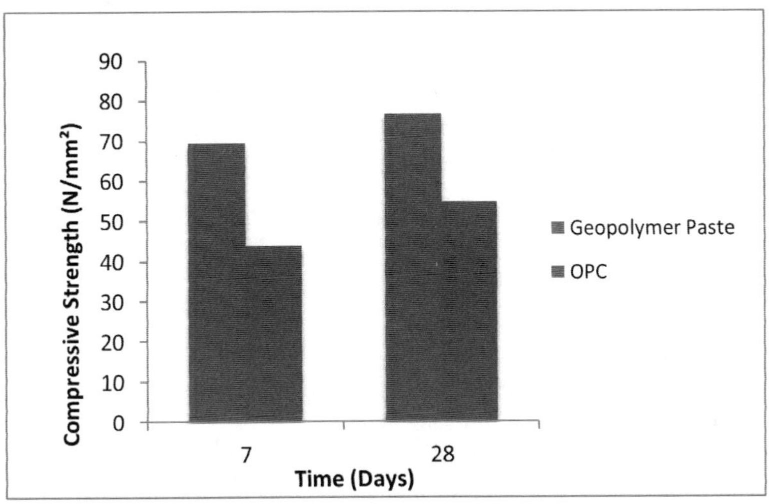

Fig. 3. Compressive strength of geopolymer and OPC pastes.

The result shown the compressive strength of fly ash based geopolymer is higher than OPC. The compressive strength simply comparable between both of the sample which is increasing steadily compressive strength with curing time. The smaller pore size and water absorption of geopolymer can increase the compressive strength greater than OPC, which is fly ash based geopolymer was 25.4240 N/mm^2 higher than OPC at early ages. Meanwhile, the result obtained at 28 days of testing, the highest compressive strength 76.7230 N/mm^2 was produced compared to OPC was 54.7490 N/mm^2. It was concluded that the fly ash-based geopolymer produced higher compressive strength than OPC. This has been proved through finding in this study, which is the samples with less of porosity and water absorption of geopolymer paste produced high compressive strength than OPC.

4 Conclusions

The fly ash based geopolymer with lowest porosity and water absorption had highest compressive strength while the OPC with highest porosity and water absorption had lower compressive strength. Thus, the compressive strength is related to porosity and water absorption. It can be concluded that pastes based on geopolymer materials possess is a high resistance to aggressive environment and higher compressive strength compared to OPC.

The authors would like to thank the staffs of Centre of Excellence Geopolymer & Green Technology (CEGeoGTech), School of Materials Engineering, Universiti Malaysia Perlis (UniMAP) for their involvement in the research. This work was supported and funded by the Fundamental Research Grant Scheme (FRGS) under a grant number of **FGRS/1/2018/TK06/UNIMAP/02/2** from the Ministry of Education Malaysia.

References

1. Farhana, Z., Kamarudin, H., Rahmat, A., Al Bakri Abdullah, M., & Norainiza, S. Adv. Mater. Res. **795,** 509-512 (2013)
2. Reiner, l., MCI 2022 (2003).
3. S. Razak, F. Zainal and S. Shamsudin, IOP Conf. Ser. Mater. Sci. Eng. **701**, 012027 (2019)
4. Al Bakri Abdullah, M., Hussin, K., Binhussain, M., Khairul Nizar, I., Abd Razak, R., & Zarina, Y. Adv. Mater. Res. **476-478**, 2173-2180 (2012)
5. Metha P.K., Monterio P.J.M., "*Concrete, Microstructure, Properties and Materials, McGraw-Hill, London*", (2006).
6. E. P. Barrett, L. G. Joyner, and P. P. Halenda, J. Am. Chem. Soc. **73**(1), 373–380 (1951)
7. S. Alehyen, M. Zerzouri, M. ELalouani, M. EL Achouri, M. Taibi JMES, **8 (10)**, 3676-3689 (2017).
8. Farhana, H. Kamarudin, A. Rahmat and A. Al Bakri, Mater. Sci. Forum, **803,** 166-172 (2014)
9. Mustafa Al Bakri, A.M., Kamarudin, H., Bnhussain, M., Khairul Nizar, I., Rafiza, A.R. and Zarina, Y. J. Rev Adv Mater Sci, **30**, 90-97 (2012).
10. Tokchom, Partha Ghosh and Somnath Ghosh. J. Eng. Appl. Sci. **4,** 28-32 (2009)
11. J. G. S. Van Jaarsveld, J.S. J. Van Deventer and G. C. Lukey. Mater. Lett **57(7),** 12721280 (2003)

IOP Conf. Series: Materials Science and Engineering **957** (2020) 012035 doi:10.1088/1757-899X/957/1/012035

12. Asmara, Y., Siregar, J., Tezara, C., Nurlisa, W., & Jamiluddin, Int. J. Corros. 1-5 (2015)
13. Asthana R., Kumar A., Dahotre N. Materials Processing and Manufacturing Science , First Edition. United Kingdom: Butterworth- Heinemann. 656 (2006).
14. Farhana, Z., Kamarudin, H., Rahmat, A., & Al Bakri, A. Key Eng. Mater. **594-595**, 1112-1116 (2013)
15. Adamson R.W., A.P. Gast A.P. *Physical chemistry of surfaces* Ch. 17, 6th ed., Wiley, New York, (1997).
16. ASTM C642-13, Standard Test Method for Density, Absorption, and Voids in Hardened Concrete, ASTM International, United States (2013).

Preparation of Bimetallic Catalyst: Gold-Copper (Au-Cu) Nanoparticles for Catalytic Reduction of *p*-Nitrophenol

Norizwan Nordin[1], *Nurul Adilah* Md Noor[1], *Najwa Afifi* Abdul Wahab[1], *Hanani* Yazid[1] and *Abdul Mutalib* Md Jani[2,*]

[1]Faculty of Applied Sciences, Universiti Teknologi MARA Perlis Branch, Arau Campus, 02600 Arau, Perlis, Malaysia.
[2]Faculty of Applied Sciences, Universiti Teknologi MARA Perak Branch, Tapah Campus, 35400 Tapah, Perak, Malaysia.

Abstract. The combination of gold (Au) and copper (Cu) nanoparticles can enhance their chemical and physical characteristics for the application in catalysis. This combination of Au and Cu reduce the cost of Au and improve the stability of Cu towards oxidation. This study involves the preparation of Au-Cu NPs for the application in the reduction of *p*-nitrophenol (*p*-NP). The Au-Cu NPs catalyst was synthesized with different Au:Cu ratio and addition hexadecylamine (HDA). The highest rate constant (*k*) of 8.2 x 10^{-3} s^{-1} was obtained over 1.5 Au-Cu HDA2 catalyst. UV-Vis spectrophotometer and Transmission Electron Microscope were used to characterize the prepared sample.

1 Introduction

In recent years, there has been an increasing interest in the application of gold nanoparticles (Au NPs) as a catalyst for reduction of *p*-nitrophenol (*p*-NP) [1; 2] and *p*-nitroaniline [3]. The Au NPs was widely studied either as monometallic or bimetallic NPs. Wang et al. (2013) mentioned that the performance of monometallic Au NPs is limited by its particles size and the preparation procedure, while an additional of second metal reduced these limitations [4]. The combination of Au NPs with other transition metals such as Au-Pt, Au-Cu, Au-Pd, and Au-Ag has better catalytic activities than monometallic Au NPs. Specifically for Au-Cu NPs, Rout et al. (2017) emphasized that it is a good bimetallic alloy since it has better resistance toward oxidation, thus extended the catalyst's lifetime [5].

Therefore, this work proposes the preparation of Au-Cu NPs for catalytic reduction of *p*-NP. Several parameters such as the Au/Cu ratio and capping agent, HDA were varied to optimally produce Au-Cu catalyst of high activity towards reduction of *p*-NP.

*Corresponding author: izwan722@gmail.com

IOP Conf. Series: Materials Science and Engineering **957** (2020) 012036 doi:10.1088/1757-899X/957/1/012036

2 Methodology

2.1 Synthesis of Au-Cu

Au-Cu NPs were synthesized based on the chemical reduction technique. An aqueous solution of $CuCl_2 \cdot 2H_2O$ (50 mM) was mixed with 0.3 mL $HAuCl_4 \cdot 3H_2O$ (50 mM) and 0.28 mL of glucose (0.5 M) in a glass vial containing the HDA solution of 22.5 mg in 2 mL distilled water [6]. The vial was magnetically stirred at room temperature overnight followed by heating treatment at 100 °C for 10 mins. This sample is denoted as 0.3 Au-Cu HDA. Similar procedures were repeated for 1.5 Au-Cu HDA and 1.5 Au-Cu HDA2 but using 1.5 mL of $HAuCl_4 \cdot 3H_2O$ precursor and addition of 15 mg of HDA after heating, respectively. The detail preparation parameters were summarized in Table 1.

Table 1. Summarized parameters for synthesis of Au-Cu.

Sample notation	Parameters						
	0.5 M glucose	50 mM $CuCl_2$	50 mM $HAuCl_4$ (0.3 or 1.5 mL)	22.5 mg HDA	Stirred (rt, overnight)	Heating time at 100 °C	15.0 mg HDA
0.3 Au-Cu HDA	√	√	0.3 mL	√	√	√	x
1.5 Au-Cu HDA	√	√	1.5 mL	√	√	√	x
1.5 Au-Cu HDA2	√	√	1.5 mL	√	√	√	√

2.2 Catalytic reduction of *p*-nitrophenol

Accurately, 20 μL of Au-Cu HDA catalyst was put into a cuvette containing freshly prepared $NaBH_4$ (1.5 mL, 1.5 x 10^{-2} M) and *p*-nitrophenol (1.5 mL, 0.05 mM) solution. The reduction of *p*-NP was monitored by using UV-Vis spectrophotometer (Lambda 25, Perkin Elmer) at 400 nm based on a pseudo first-order model. The rate constant (k) value was then calculated.

3 Findings

3.1 Au-Cu HDA

The aqueous phase synthesis of bimetallic Au-Cu involves $CuCl_2$ and $HAuCl_4$ as the precursors while glucose and hexadecylamine (HDA) as the reducing and capping agent, respectively. The colour of the aqueous mixture (0.3 Au-Cu HDA) turns from kelly green to black after heating at 100 °C (Fig. 1.). Meanwhile, for 1.5 Au-Cu HDA formulation (Refer to Table 1), the colour change from kelly green to dark brown (Fig. 2.). Figure 3 shows the UV-Vis spectra of 0.3 and 1.5 Au-Cu HDA with the surface plasmon resonance (SPR) band at 520 nm and 565 nm, respectively. According to Pal et al., monometallic Au NPs has the SPR band around 521 nm while monometallic Cu NPs is at 573 nm. The bimetallic Au/Cu of 1:1 ratio shows a peak at 548 nm [7]. Therefore, the 1.5 Au-Cu HDA with Au/Cu ratio of 2.5:1 shows positive confirmation of producing Au/Cu as the appearance at 565 nm. Besides, the colour change from kelly green to dark brown was also an indication of Au/Cu [6].

(a) (b)

Fig. 1. Photograph of 0.3 Au-Cu HDA (a) after stirring overnight and (b) after heating at 100 °C.

(a) (b)

Fig. 2. Photograph of 1.5 Au-Cu HDA (a) after stirring overnight and (b) after heating at 100 °C.

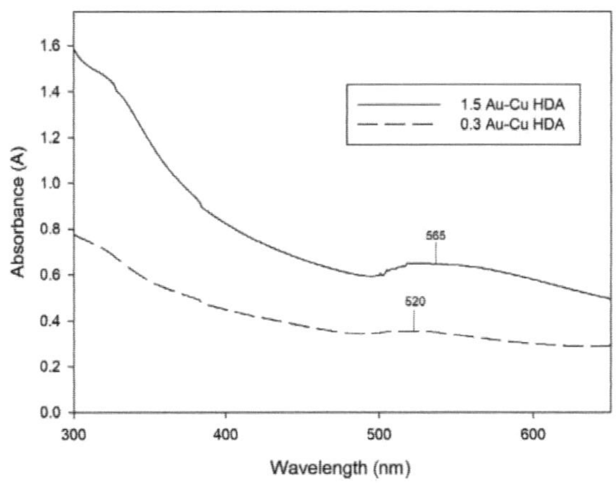

Fig. 3. UV-Vis spectra of (a) 0.3 and (b) 1.5 Au-Cu HDA formulations.

3.2 Effect of addition HDA

The addition of HDA into the sample solution gives an important impact on the size of Au-Cu NPs [6]. Fig. 4 shows the photograph of 1.5 Au-Cu HDA2 before and after addition of HDA. Positive indication of Au/Cu was observed as the solution turn to deep brown (Fig. 4b). Fig. 5 shows UV-Vis spectra of 1.5 Au-Cu HDA2 at 552 nm, shifted towards shorted wavelength compared to 1.5 Au-Cu HDA formulation discussed above. The finding indicates the formation of a smaller diameter of bimetallic Au/Cu. According to the previous report of spherical gold-copper nanocubes (NCs), the nanocubes highly depends on its size in which increasing size of NCs will cause red shift of the SPR band wavelength [8]. The TEM image in Fig. 6 depicts Au/Cu nanocubes of 116 ± 29 nm.

(a) (b)

Fig. 4 Images of 1.5 Au-Cu HDA2 (a) after stirred overnight and (b) after addition of 15 mg HDA.

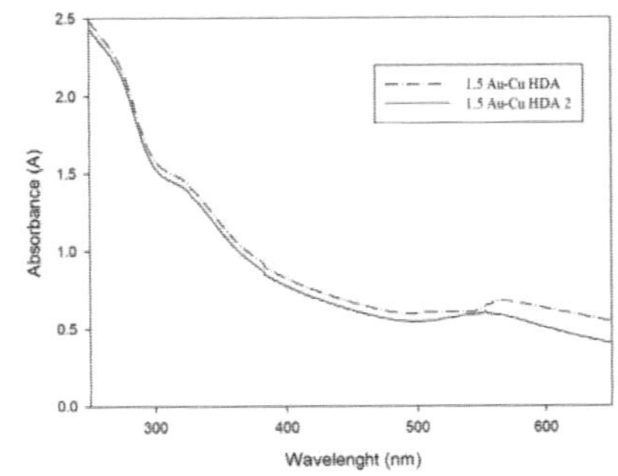

Fig. 5. UV-Vis spectra of 1.5 Au-Cu HDA and 1.5 Au-Cu HDA2.

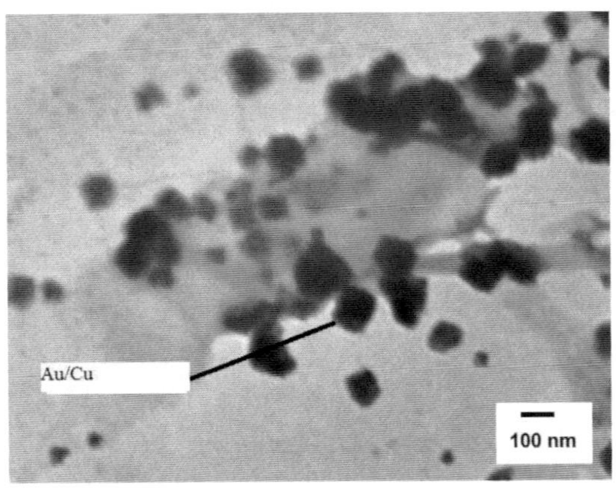

Fig. 6. TEM images of 1.5 Au-Cu HDA2.

3.3 Catalytic reduction of *p*-nitrophenol

The reduction of *p*-nitrophenol (*p*-NP) to *p*-aminophenol (*p*-AP) in the presence of $NaBH_4$ was used as a model reaction to evaluate the catalytic activity of the prepared catalysts [3]. This method is easy to be monitored by a simple spectroscopic technique that is UV-Vis spectroscopy [9]. In all catalytic runs, the experimental conditions at a molar ratio of $NaBH_4$: *p*-NP at 300:1 were kept constant. From the observations, after the addition of a fresh aqueous solution of $NaBH_4$ into an aqueous solution of *p*-NP, the absorption peak of *p*-NP shifted from 317 to 400 nm due to the formation of *p*-nitrophenolate ions [10]. The significant change of solution colour has occurred from light yellow to dark yellow. Fig. 7 shows the UV-Vis spectra of catalytic reduction of *p*-NP over 0.3, 1.5 Au-Cu HDA and 1.5 Au-Cu HDA2. The incomplete reduction was obtained for 0.3 Au-Cu HDA while 1.5 Au-Cu HDA and 1.5 Au-Cu HDA2 has the k value of 3.0×10^{-3} and 8.2×10^{-3} s^{-1}, respectively. Plot ln $(A-A_\infty)$ vs time shows linear correlation inconsistent with pseudo first-order kinetic reaction (Fig. 8). The 1.5 Au-Cu HDA2 requires only 6 mins to completely reduce *p*-NP which, illustrates a highly active catalyst for this reaction. Overall k values are summarized in Table 2.

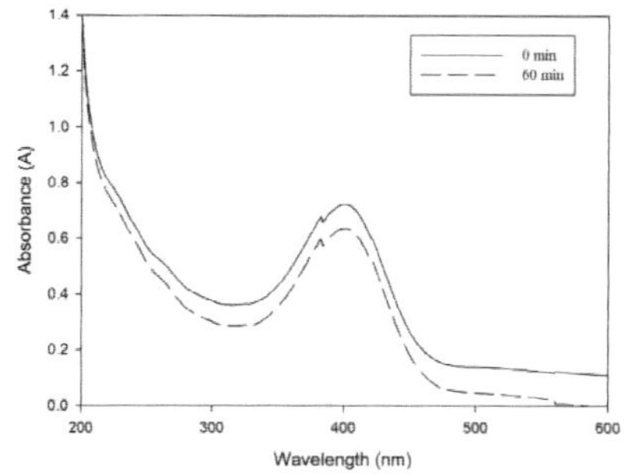

(a)

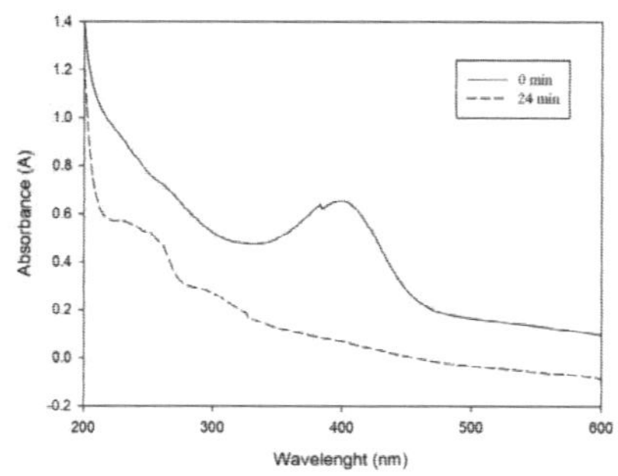

(b)

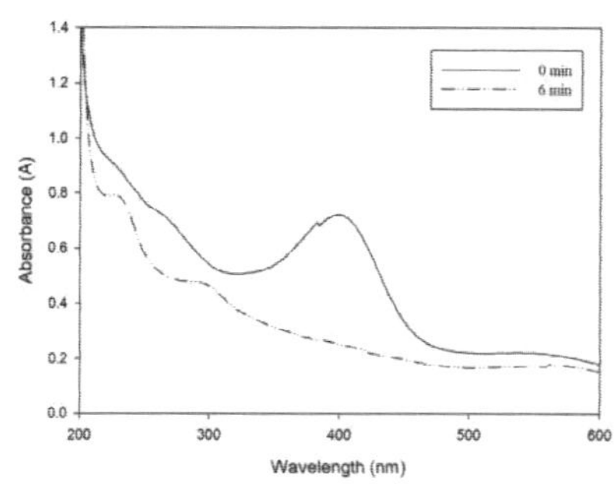

(c)

Fig. 7 : UV-Vis spectra of catalytic reduction of *p*-NP over different formulation of Au-Cu (a) 0.3 Au-Cu HDA, (b) 1.5 Au-Cu HDA and (c) 1.5 Au-Cu HDA2.

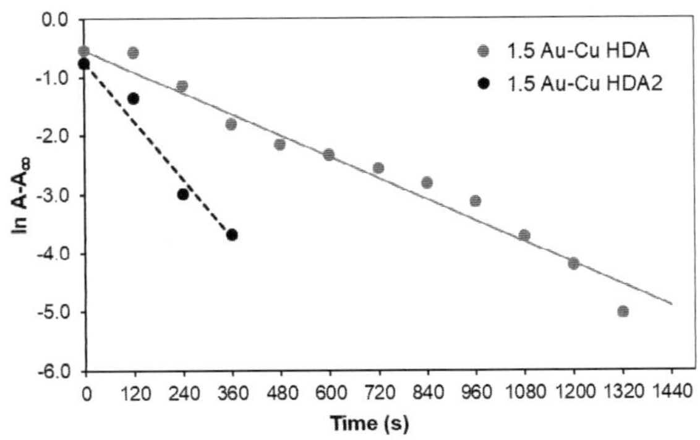

Fig. 8. Plot Plot ln A/A∞ vs time of 1.5 Au-Cu HDA and 1.5 Au-Cu HDA2.

Table 2: Comparison of the catalytic reduction of *p*-NP by using Au-Cu HDA catalysts.

Samples	**Time for reduction of *p*-NP to complete (min)**	**Rate constant (*k*) value (s^{-1})**	**R^2**
1.5 Au-Cu HDA	24	3.0 x 10^{-3}	0.8378
1.5 Au-Cu HDA 2	6	8.2 x 10^{-3}	0.9599

4 Conclusions

The Au-Cu nanocubes shape of 116 ± 29 nm size was successfully prepared by using chemical reduction technique. The highest *k* value of rate constant (*k*) is 8.2 x 10^{-3} s^{-1}. The efficiency of this prepared catalyst might be due to the addition of HDA as the capping agent and Au:Cu of 2.5:1 ratio.

We would like to thanks the Ministry of Higher education (MOHE) for providing generous internal financial support under FRGS grants: 600-RMI/FRGS 5/3 (119/2019) in conducting this study and Universiti Teknologi MARA (UiTM) for providing all the facilities.

IOP Conf. Series: Materials Science and Engineering **957** (2020) 012036 doi:10.1088/1757-899X/957/1/012036

References

1. M. Zhang, X. Lu, H-Y. Wang, X. Liu, Y. Qin, P. Zhang, Z-X. Guo, RSC Adv. (2016)
2. M. Ramakrishna, D.R. Babu, RM. Gengan, S. Chandra, GN. Rao, J. Nanostructure Chem. (2016)
3. M.A. Bhosale, SSR Gupta, BM Bhanage, Polyhedron (2016)
4. A. Wang, X.Y. Liu, C-Y. Mou, T. Zhang, J. Catal. (2013)
5. L. Rout, A. Kumar, R.S. Dhaka, G.N. Reddy, S. Giri, P Dash. Appl. Catal. A: Gen. (2017)
6. R. He, Y-C. Wang, X. Wang, Z Wang, G. Liu, W. Zhou, L. Wen, Q. Li, X. Wang, X. Chen, Nat. Commun. (2014)
7. U. Pal, J.S. Ramirez, H. Liu, A. Medina, J. Ascencio, Appl. Phys. A (2004)

IOP Conf. Series: Materials Science and Engineering **957** (2020) 012037 doi:10.1088/1757-899X/957/1/012037

The effect of choline chloride: acetamide deep eutectic solvent to physicochemical and mechanical properties of pectin-based bioplastic

Mohamad Azhar Azman[1], *Rizana* Yusof[1,*], *Dalina* Samsudin[1] and *Roziana* Mohamed Hanaphi[1]

[1]Faculty of Applied Sciences, Universiti Teknologi MARA, 02600 UiTM Arau, Perlis, Malaysia.

Abstract. The rapid increment of synthetic plastic wastes in our environment has triggered in production and development of new biomaterial for resolving problems such as pollution as well as weak mechanical properties. Pectin polysaccharides are recognized as a new potential biomaterial in developing bioplastic. The main purpose of this study is to develop and characterize citrus pectin bioplastic using deep eutectic solvents (DES) of choline chloride: acetamide (1:2), CC:Ac as a plasticizer. The potential of produced bioplastic (BP-CC:Ac) was studied through investigation of their physicochemical and mechanical properties which include pH, thickness, opacity, solubility, tensile strain, tensile stress and Young Modulus. A citrus pectin-based bioplastic without the plasticizer (C1) and with the conventional plasticizer of pure acetamide (C2) were prepared as controls. The produced BP-CC:Ac showed a higher value of thickness (0.2658±0.0015mm), opacity (0.492±0.02A), and solubility (23.26±0.03%) than the control, C1 (0.1963±0.0032mm, 0.407±0.01A, and 16.42±0.01% respectively) and C2 (0.1833±0.001mm, 0.447±0.01A, and 17.88±0.05% respectively). The high acidity of BP-CC:Ac (pH 3.21) provide a strong inhibition toward bacteria which is highly demanding for medical purposes. The tensile strain, tensile stress and Young Modulus of BP-CC:Ac were measured at 5.080 ± 0.04 %, 2.221 ± 0.08 MPa and 113.946 ± 0.08 MPa, respectively. It was found the BP-CC:Ac was more flexible and have smooth surface texture than both controls, C1 and C2 which suitable for plastic wrapping. Other than being natural and biodegradable, the addition of CC:Ac DES had improved the physicochemical and mechanical properties of pectin based-bioplastic. In conclusion, the addition of CC:Ac DES has enhanced the properties of BP-CC:Ac and increased the potential of plastic as an ecofriendly and sustainable future material.

1 Introduction

The everyday uses of petrochemical-based plastic have led to large-scale demand for mass-produce product, which then rapidly accumulated and ends up as non-biodegradable trash in landfills and the ocean for centuries [1]. The generic growth of these unsustainable products

*Corresponding author: azharazman93@gmail.com

was continuously damaging the environment and human while researchers were struggled in finding the solution to the problems.

There is still a lot of sustainable material that can be used to encounter the problem, such as polysaccharide from sugar cane or potato[2]. The development of the bioplastic for future applications is normally focused on the ability of bioplastic to sustain. One of the material bases for producing a sustainable bioplastic is a polysaccharide which can be disposed of in the environment and can easily degrade through the enzymatic actions of microorganisms[2]. The presence of enzymes in bacteria known as *Ideonella sakaiensis* have converted the plastics into an intermediate substance and also plastic building blocks[3].

A common additive such as phthalate used in plastic production has become one of the causes of health problem, which leaking of this endocrine disruptors can cause cancerous tumour and other development disorders[4]. Therefore, a green solvent has been introduced to replace the conventional and toxic additive, which is known as deep eutectic solvent (DES). DES is a better extension of the ionic liquid model that made up from the formation of two or more compounds; a quaternary ammonium salt and either a metal salt or a hydrogen bond donor[5]. Previous studies have shown the abilities of DES as a solvent in various fields such as organic reactions[6], extraction of metals, polyaromatic hydrocarbons and DNA[7]–[9], electrochemistry [10] and bioplastic development[11]–[13].

Hence, this study was conducted to investigate the role of DES as a plasticizer in the production of pectin-based bioplastic by characterizing their physicochemical and mechanical properties. The effect of addition DES to the produced plastic were evaluated according to the changes observed in both physicochemical and mechanical properties. The unique properties of DES itself such as low cost of components, easy to prepare, tuneable physicochemical properties, negligible vapour pressure, non-toxicity, non-reactive with water, non-flammability, high conductive, bio renewability and biodegradability have promised great potential for enhancing the bioplastic properties[5], [14]–[16].

2 Method

2.1 Synthesis of DES

The DES (CC:Ac) was synthesized by heating choline chloride (CC) (20.1g) and acetamide (Ac) (17.72g) in 1:2 moles ratio at 60 to 80 °C until the colourless liquid was formed. Both chemicals were bought from Acros, Belgium with 99% purity.

2.2 Bioplastic preparation

30 ml of 2% of w/w citrus pectin solution were poured in a cast. The sample of bioplastic BP-CC:Ac was added with 3% of DES CC:Ac. Two controls were prepared; which were the pectin bioplastic without plasticizer as control one (C1) and pectin bioplastic with acetamide as control two (C2). The bioplastics were prepared in triplicates for each test [17].

2.3 Physicochemical characterization of DES

2.3.1 pH measurement of bioplastic

The pH of pectin-DES mixtures was measured using a pH meter (EL20-Kit Benchtop, Mettler Toledo, USA). The pH probe was immersed into the non-solidified bioplastic prior to casting in the beaker. The pH measurements were conducted in triplicates.

2.3.2 Thickness determination of bioplastic

The thickness of each produced bioplastics was measured using digimatic thickness gauge (Model 547-301, Mitutoyo, Japan). The thickness of the bioplastic was measured at the centre and the reading was repeated for three times.

2.3.3 Opacity of bioplastic

The opacity of each bioplastic was measure using UV visible spectrometer (SPECTRONIC™ 200 Spectrophotometer, ThermoFisher Scientific). The light transmittance value for each bioplastic was recorded three times. The highest and lowest value of absorbance were taken for the indication of high and low opacity of bioplastic[18].

2.3.4 Solubility (S) of bioplastic

Each of the bioplastics was cut into a standard size of 2 cm x 2 cm. Thus, the bioplastic was incubated in a drying and sterilization oven (Fisherbrand™ Isotemp™). Then the sample was weighed (W_i), immersed and stirred in a beaker filled with 50 ml of distilled water at 175 rpm for 6 hours. The solubility of the bioplastic was measured based on weight loss after 6 hours. The samples were then removed from the water and the dry mass (W_f) was determined under the same described conditions to obtain the initial dry mass. The solubility in water was expressed as a percentage of solubilized material, calculated from the results in triplicate, as shown in Eq. (1).

$$S = \frac{Wi - Wf}{Wi} \times 100 \qquad (1)$$

2.4 Mechanical characterization of DESs

The tensile properties of bioplastic which include tensile stress, tensile strain and Young modulus analysis were measured at room temperature using Instron Universal Testing Instrument (Instron 3365, Instron, USA) equipped with a 1kN load cell according to method ASTM D882-02 (ASTM, 1992).

3 Results and Discussion

3.1 Physicochemical properties

Fig. 1 shows the results obtained from the testing conducted on three produced bioplastics, BP-CC:Ac, C1 and C2. It can be observed that all of the produced bioplastics were relatively acidic (pH 2.98 to 3.28) and the most acidic bioplastic is the bioplastic made of pure pectin without acetamide, C1 at pH 2.98±0.05. The acidity of pectin bioplastic was influenced by the depolymerization through a treatment with sodium hydroxide and hydrochloric acid in the extraction of pectin process. Besides that, pectin is a low ester compound that has a pH range from 2.6 to 7.0[17]. Both of BP-CC:Ac and C2 show a small increase of pH than C1 due to the presence of substances other than pectin which changed the concentration of proton and reduced the acidity of solution. The acidic properties of all produced bioplastics showing that they can provide a strong inhibition toward bacteria which suitable as a material for medical application, such as anti-microbial appliance.

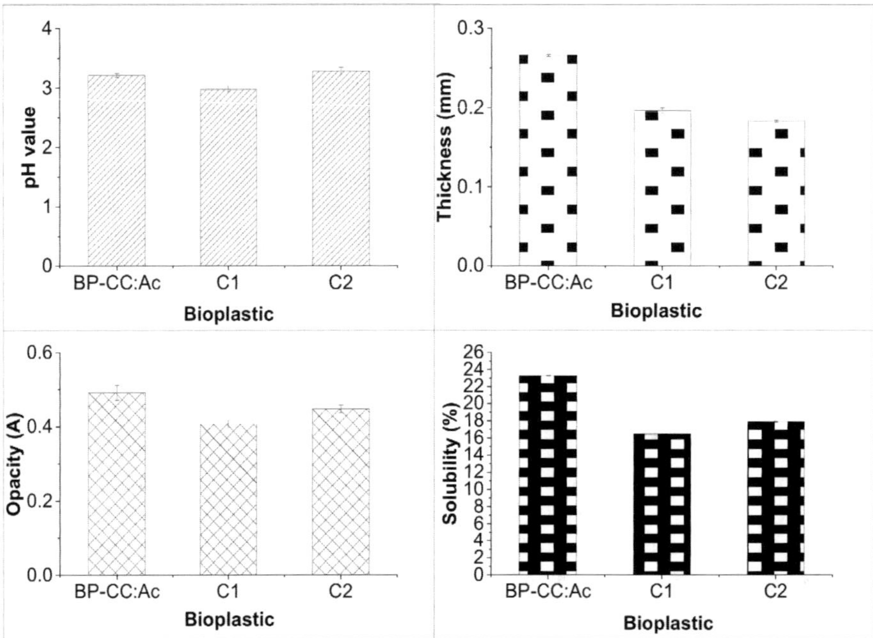

Fig. 1. Physicochemical properties of BP-CC:Ac, C1 and C2 bioplastic

Thus, in Fig. 1 a linear effect of the DES on the thickness was observed where the width of BP-CC:Ac (0.2658±0.0015 mm) was measured thicker than controls (C1:0.1963±0.0032 mm; C2: 0.1833±0.001 mm). The increase in the thickness of BP-CC: Ac was caused by the interaction of plasticizer added to the pectin solution. This results can be related to the study done by Cabello et al., [19] who reported that the interaction between the ion pairs of the carboxylic group on pectin with amine from chitosan produced a diminution, which resulted for reducing numbers of hydroxyl groups and caused the hydrophilicity followed by swelling of the membranes. Hence, since the result of the study have shown a similar trend with the past research, it can be concluding the size of the plastic with DES become thicker than controls due to the interaction of plasticizer with pectin structures.

BP-CC:Ac was found to have the lowest opacity with the absorbance of 0.492±0.02A, which is lower than the controls. The additional of DES add up the number of components in the bioplastic structure; thus it decreases the space for light to be transmitted [20]. The high value of absorbance reading proves the number of particles that can absorb light were higher in BP-CC:Ac.

The solubility of the bioplastics was ranged from 16.42 to 23.26%. The solubility rates for BP-CC:Ac was the highest when compared to the controls. The hygroscopic choline chloride from DES has increased the rate of solubility as it easily absorbs water [21]. The addition of DES plasticizer increases the solubility rates of the bioplastic as the two hydrogen atoms that were bonded to the nitrogen in acetamide carbonyl group able to form hydrogen bonding with water. However, all the produced bioplastics are considered to have high solubility rates. From the other point, this result has shown with the addition of DES; the produced bioplastic was more easily to degrade and the sustainability of bioplastic was not disturbed.

3.2 Mechanical properties

Fig. 2 shows the tensile strain of BP-CC:Ac (6.771±0.23%) was higher than control (C1: 2.153±0.15%; C2: 5.067±0.16%). The tensile strain of the bioplastic was increased when adding the plasticizer and having a greater thickness. In general, tensile strain is the relative length of deformation exhibited by bioplastic when a force applied. According to Shafie et al., [11], there was an interruption of the polymer chains interactions when the plasticizer added. Besides that, the complex structures between pectin and plasticizer created when the plasticizer was added and its eventually increased the strength of the bioplastic thus increase the tensile strain percentage.

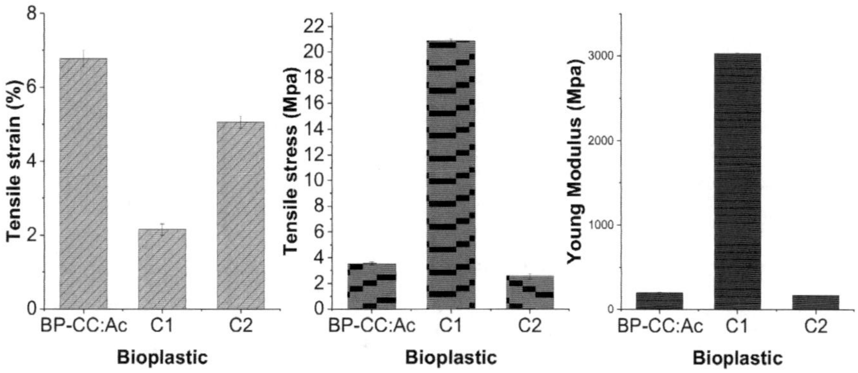

Fig. 2. Mechanical properties of BP-CC:Ac, C1 and C2 bioplastics.

The tensile stress was analysed to measure the maximum stress of bioplastics can stand before it breaks. As depicted in Fig. 2, the tensile stress of BP-CC:Ac and C2 were lower than C1 as the added plasticizers in both BP-CC:Ac and C2 caused the interaction of hydrogen bonding in the bioplastics. The added of DES (CC:Ac) and acetamide have decreased the contact between the polymer matrix and the adjacent segments of polymer chains, therefore increasing the segmental mobility and free volume[11]. Meanwhile, a very high value of tensile stress (20.874±0.132 MPa) in C1 caused the bioplastic with pure pectin was too stiff and easily broken.

Hence, the addition of plasticizer could cause the larger inter-molecular area between the polymer chains, thus preventing them from gathering close to each other. Consequently, it restricted the compactness and cohesive strength of the molecular chains in bioplastic with plasticizer. The trend of Young modulus was found to be similar with tensile stress. The interaction of plasticizer was beneficial in reducing the stiffness of polymeric materials which allow it to be more flexible.

Conclusion

The bioplastics were successfully characterized and developed by measuring physicochemical and mechanical properties. The addition of DES as plasticizer into bioplastic had overcome the brittleness and enhance the structures of the bioplastic. Apart from that, the addition of DES also retains the sustainability of BP-CC:Ac as it was easily dissolved in water. Therefore, based on the results of the physicochemical and mechanical properties in this study BP-CC:Ac can be suggested as new sustainable material to be used in multiple types of application such as food packaging materials

This study was carried out with the support of a Fundamental Research Grant Scheme provided by the Ministry of Education, Malaysia, FRGS 5/3 (113/2019) and facilities in Universiti Teknologi MARA, Perlis branch.

References

1. Carvalho, J. *et al.*. **2**, 712–718 (2019)
2. Muhammad Shamsuddin, I. Adv. Biosci. Bioeng. **5**, 63 (2017)
3. Wei, R. & Zimmermann, W. Microb. Biotechnol. **10**, 1302–1307 (2017)
4. Yang, J., Song, W., Wang, X., Li, Y. & Sun, J. Food Addit. Contam. Part B **0**, 1–8 (2019)
5. Abbott, A. P. The plasticisation of plastics using Deep Eutectic Solvents Supervisor. (2015)
6. Qu, W. The plasticisation of plastics using deep eutectic solvents. *Master of Philosophy* University of Leicester (2015)
7. Smith, E. L., Abbott, A. P. & Ryder, K. S. Chem. Rev. **114**, 11060–11082 (2014)
8. Das, A. K., Sharma, M., Mondal, D. & Prasad, K Carbohydr. Polym. **136**, 930–935 (2016)
9. Rogošić, M. & Kučan, K. Z. J. Ind. Eng. Chem. **72**, 87–99 (2019)
10. Abbott, A. P., Alabdullah, S. S. M., Al-Murshedi, A. Y. M. & Ryder, K. S. Faraday Discuss. **206**, 365–377 (2018)
11. Shafie, M. H., Samsudin, D., Yusof, R. & Gan, C. Y. Int. J. Biol. Macromol. **118**, 1183–1192 (2018)
12. Galvis-Sánchez, A. C., Castro, M. C. R., Biernacki, K., Gonçalves, M. P. & Souza, H. K. S. Food Hydrocoll. **82**, 478–489 (2018)
13. Santana, R. F. *et al.* J. Food Sci. Technol. **55**, 278–286 (2018).
14. Tomé, L. I. N., Baião, V., da Silva, W. & Brett, C. M. A. Appl. Mater. Today **10**, 30–50 (2018)
15. Zdanowicz, M. & Johansson, C. Carbohydr. Polym. **151**, 103–112 (2016)
16. Gómez, A. *et al.* J. Braz. Chem. Soc. **30**, 717–726 (2019)
17. Bátori, V. *et al.* Int. J. Polym. Sci. **2017**, (2017)
18. Nisar, T. *et al.* CYTA - J. Food **17**, 695–705 (2019)
19. Cabello, P., Takara, N. A. O. E. A. & Cabello, P. Carbohydr. Polym. (2016)
20. Shamsuri, A. A. & Daik, R. BioResources **7**, 4760–4775 (2012).
21. Shekaari, H., Zafarani-Moattar, M. T., Shayanfar, A. & Mokhtarpour, M. J. Mol. Liq. **249**, 1222–1235 (2018)

Effect of glutaraldehyde as crosslinker on the properties of cellulose nanocrystal/chitosan films

Pei Gie Gan[1], *Sung Ting* Sam[1,*] , *Mohd Firdaus* Omar[2*] and *Mohammad Faiq* Abdullah[1]

[1]School of Bioprocess Engineering, Universiti Malaysia Perlis (UniMAP), Kompleks Pusat Pengajian Jejawi 3, 02600 Arau, Perlis
[2]School of Materials Engineering, Universiti Malaysia Perlis (UniMAP), Kompleks Pusat Pengajian Jejawi 2, 02600 Arau, Perlis

Abstract. Cellulose nanocrystals (CNC) were isolated from the empty fruit bunch as the reinforcing filler to enhance the tensile and thermal properties of chitosan films. The addition of CNC has greatly increased the tensile strength of the chitosan films from 32.9 MPa for neat chitosan film to 50.9 MPa for 5 wt% CNC/chitosan films. The addition of glutaraldehyde as crosslinker has further improved the tensile property of chitosan films. However, the incorporation of CNC and glutaraldehyde has a negative effect on the elongation at break of the films due to restricted polymer chain mobility. Besides, the addition of CNC also enhanced the melting temperature of the chitosan films. The crosslinking process has further increased the melting temperature to 112.8°C with the addition of 5 wt% CNC. Therefore, the crosslinked CNC/chitosan films display greater property reinforcements than non-crosslinked films.

1 Introduction

Increasing of public concerns regarding the environmental pollution throughout the world has led to the utilization of biodegradable and environment friendly polymer such as chitosan. Chitosan can be used as an alternative for the fabrication of biodegradable polymer composites to replace the synthetic polymers. However, chitosan is brittle and has poor mechanical properties. Therefore, cellulose nanocrystal (CNC) is introduced as the reinforcement filler to enhance its properties.

There are various methods for preparing CNC, with sulphuric acid hydrolysis being the most commonly used technique [1]. Isolation of CNC using sulphuric acid forms a negatively charged crystallite surface resulting from esterification by sulphate ions which lead to a more stable CNC dispersion. Isolation of CNC from renewable sources has earned the greatest interest lately due to its excellent mechanical properties (high specific strength and modulus), exceptional crystallinity, light weight and high aspect ratio compared to micro-sized cellulose [2]. The concept of using nanocellulose as reinforcement material in polymer matrix has been proven as the efficient approach for enhancing the tensile

* Corresponding author: stsam@unimap.edu.my, firdausomar@unimap.edu.my

IOP Conf. Series: Materials Science and Engineering **957** (2020) 012038 doi:10.1088/1757-899X/957/1/012038

properties of biopolymers by forming the hydrogen bonding network between the filler and the matrix [3].

Aside from adding reinforcing filler, the performance of chitosan can also be enhanced through crosslinking reaction. Crosslinking is an important procedure in the production of composite films as the properties such as mechanical strength are dependent on the crosslinking degree of composites [4]. The crosslinking reaction normally takes place between the functional group of crosslinker and the amine group of chitosan [5]. Among the crosslinkers, glutaraldehyde (GA) is the most common crosslinker used in chitosan matrix due to its high stability and proven ability in improving the tensile and thermal properties of chitosan films [6, 7]. Most of the crosslinkers require heating for the crosslinking reaction to take place. The thermal crosslinking of film is generally performed by using conventional heating in which the thermal energy is directly transmitted to the film via conduction. The properties of crosslinked films can be further enhanced through the conventional heating by forming a more stable network between the crosslinker and polymeric matrix [8].

The aim of this study is to fabricate the glutaraldehyde-crosslinked CNC/chitosan films with different CNC content using conventional heating. The tensile and thermal properties of both non-crosslinked and glutaraldehyde-crosslinked CNC/chitosan films were investigated and compared.

2 Methodology

2.1 Materials

Empty fruit bunch (EFB) was provided by Taclico Company Sdn. Bhd. Chitosan powder was supplied by Cleo Shanghai Pharmaceutical, China. Acetic acid and glutaraldehyde were purchased from Sigma-Aldrich whereas sulphuric acid and glycerol were purchased from Merck. Sodium chlorite was supplied by Acros Organics. Potassium carbonate was provided by HmbG Chemicals.

2.2 Isolation of CNC from EFB

EFB fibres were treated with deep eutectic solvent (DES) at 80 °C for 2 hours. DES was made up of potassium carbonate and glycerol at a molar ratio of 1:7. Bleaching was then carried out at 80°C for 2 hours. Both treatments were repeated thrice. Acid hydrolysis was performed by using 60 wt% sulphuric acid at 45°C for 1 hour. The suspension obtained was washed with distilled water and centrifuged repeatedly till the supernatant become turbid. The suspension was then sonicated and stored in refrigerator at 4 °C.

2.3 Preparation of CNC/Chitosan Films

Chitosan powder was dissolved in 2 v/v% acetic acid solution and heated at 60 °C for 4 hours. Glycerol (10 wt%) and CNC were added into the chitosan solution and heated for another 1 hour. The solution was poured into a plastic mold and dried in a non-air-circulating oven at 50°C. For crosslinked composites, 5wt% of glutaraldehyde was added. Conventional heating was carried out at 110 °C for 20 minutes using a Memmert UF100 universal oven.

3 Characterization

3.1 Tensile properties

The tensile test was performed using Instron universal tensile testing machine (UTM). A crosshead speed of 10mm/min was used in the test. Samples of size 10 x 1 cm were cut from the films. The thickness of specimens was measured randomly using a micrometer and the average thickness was recorded. The data were averaged over at least 5 samples.

3.2 Differential Scanning Calorimetry (DSC)

DSC was performed using a Shimadzu DSC-60 to investigate the melting behaviors of composites. The scanning was performed in a temperature range between 30 to 200°C at a heating rate of 10°C/min under nitrogen gas flow of 50 mL/min.

4 Results and Discussion

4.1 Tensile properties

Table 1 tabulates the tensile strength and elongation at break of non-crosslinked and crosslinked chitosan films with different CNC loading. The non-crosslinked neat chitosan film exhibited the lowest tensile strength. The tensile strength increased significantly with the addition of CNC from 37.4 MPa (1 wt% CNC) to 50.9 MPa (4 wt% CNC). The enhanced tensile strength of chitosan films was effective stress transfer between CNC and chitosan matrix.[9]. However, the incorporation of 5 wt% CNC has reduced the tensile strength to 37.9 MPa. The reduction in tensile strength could be resulted by the agglomeration of CNC which acted as the stress concentrator and resulted in the uneven stress transfer between the filler and polymeric matrix [10].

Table 1: Tensile strength and elongation at break of non-crosslinked and glutaraldehyde-crosslinked CNC/chitosan films.

Samples	CNC Content (wt%)	Tensile Strength (MPa)	Elongation at Break (%)
Non-crosslinked chitosan films	0	32.9	38.6
	1	37.4	34.7
	2	41.7	30.3
	3	45.8	26.8
	4	50.9	24.4
	5	37.9	21.5
Glutaraldehyde-crosslinked chitosan films	0	44.1	30.3
	1	48.5	37.1
	2	53.4	24.3
	3	55.1	21.4
	4	60.8	28.9
	5	50.6	16.5

The crosslinking of the films with glutaraldehyde was found to increase the tensile strength of the chitosan films significantly. The enhancement in strength could be attributed to the formation of stable crosslinked network between aldehyde group of glutaraldehyde and amine group of chitosan which enhanced the intermolecular reaction within chitosan matrix [11]. This results was in agreement with Jose and Al-Hathi [12] in which the tensile strength increased remarkably by the addition of crosslinker.

The addition of CNC and glutaraldehyde had a negative impact on the elongation at break of the chitosan films. The addition of rigid CNC reduced the polymer chain mobility and made the film become stiffer [13]. Hence, the CNC/chitosan film become stiffer and less stretchable compared to the neat chitosan film. Furthermore, the addition of glutaraldehyde has further reduced the elongation at break of the films. This was due to the formation of the more compact and rigid crosslinked network which reduced the mobility of polymer chains and resulted in the declination in elongation at break [14]. The result was in concurrent with Yang et al. [15], who reported the reduction of elongation at break in bacterial cellulose/poly(vinyl alcohol) composite after crosslinking.

4.2 Differential Scanning Calorimetry (DSC)

The DSC thermogram of CNC/chitosan films is depicted in Figure 1. The non-crosslinked chitosan films with 0 wt%, 1 wt% and 5 wt% CNC were labelled as XC0, XC1 and XC5, whereas the glutaraldehyde-crosslinked chitosan films with addition of 0 wt%, 1 wt% and 5 wt% CNC were labelled as CH0, CH1 and CH5.

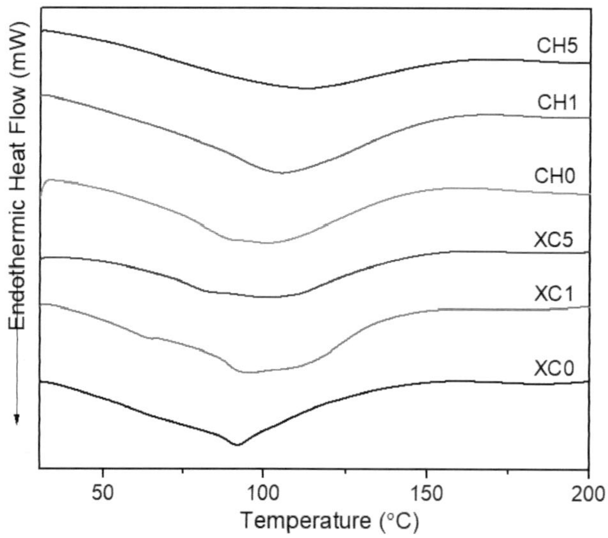

Figure 1: DSC Thermogram of non-crosslinked and glutaraldehyde-crosslinked CNC/chitosan films.

The DSC thermogram of neat chitosan composite film presented an endothermic peak at 92.4°C which was attributed to the dissociation process of intermolecular hydrogen bonding of chitosan [16]. The addition of CNC enhanced the melting point (T_m) from 95.7°C to 100.4°C as the content of CNC increased from 1 to 5 wt%. The enhancement in T_m could be attributed to the addition of highly crystalline CNC which acted as the nucleating agent and promoted the crystallization of polymeric composite films [17]. This result was concurred with Atef, Rezaei, & Behrooz [18] in which the incorporation of CNC greatly increased the T_m of agar composite film.

The crosslinking reaction has further enhanced the melting temperature. The glutaraldehyde-crosslinked chitosan composite films have a denser crosslinked network and stronger intermolecular bonding compared to non-crosslinked composite films, which required a higher temperature and heat energy to disrupt the crosslinked network [19].

5 Conclusion

CNC/chitosan film were successfully fabricated and chemical crosslinked using glutaraldehyde by conventional heating. The tensile and thermal properties of chitosan films were greatly improved by the addition of CNC and crosslinker. However, the incorporation of CNC and glutaraldehyde has a negative impact on the elongation at break of the chitosan films due to the reduction in polymer chain mobility.

The authors would like to thank the Ministry of Higher Education for the Fundamental Research Grant Scheme (FRGS/1/2017/TK05/UNIMAP/02/10) for financial support.

References

1. N.F. Vasconcelos, J.P.A. Feitosa, F.M.P. da Gama et al. Carbohydr Polym **155**,425 (2017)
2. W.P. Flauzino Neto, M. Mariano, I.S.V. da Silva et al. Carbohydr Polym **153**,143 (2016)
3. S.S. Nair, H. Chen, Y. Peng et al. ACS Sustain Chem Eng **6**,10058 (2018)
4. P. Falamarzpour, T. Behzad, A. Zamani. Int J Mol Sci **18**,1 (2017)
5. R.A. Lusiana, D. Siswanta, Mudasir. Indones J Chem **16**,144 (2016)
6. C.M. Yeng, S. Husseinsyah, S.S. Ting. Polym Bull **72**,791 (2015)
7. J. Menegotto, F. Alan, A. Liliane et al. J Polym Environ **26**,2748 (2018)
8. X. Jin, L. Li, R. Xu et al. Polymers (Basel) **10**,1 (2018)
9. Y. Han, M. Yu, L. Wang. Ind Crops Prod **117**,252 (2018)
10. M.K. Mohamad Haafiz, A. Hassan, H.P.S. Abdul Khalil et al. Polym Test **48**,133 (2015)
11. C.M. Yeng, S. Husseinsyah, S.S. Ting. BioResources **8**,2910 (2013)
12. J. Jose, M.A. Al-Harthi. Iran Polym J **26**,579 (2017)
13. S. Yang, H. Li, H. Sun. Iran Polym J (English Ed **27**,645 (2018)
14. A. Mittal, S. Garg, D. Kohli et al. Carbohydr Polym **151**,926 (2016)
15. L. Yang, Q. Yang, D. Lu. Monatshefte Fur Chemie **145**,91 (2014)
16. M.M. Nasef, E.A. El-Hefian, S. Saalah, A.H. Yahaya. E-Journal Chem **8**,409 (2011)
17. M.K.M. Haafiz, A. Hassan, H.P.S.A. Khalil et al. Int J Biol Macromol **85**,370 (2016)
18. M. Atef, M. Rezaei, R. Behrooz. Int J Biol Macromol **70**,537 (2014)
19. Y. Liu, Z. Cai, L. Sheng et al. Carbohydr Polym **215**,348 (2019)

Optimization of Injection Moulding Processing Parameters for LDPE/RH Composites Tensile Strength Through Full Factorial Experiment

Haliza Jaya[1,2], *Nik Noriman* Zulkepli[1,2], *Mohd Firdaus* Omar[1,3*], *Shayfull Zamree* Abd Rahim[1,4], *Khairul Anwar* Abdul Halim[1,3]

[1]Centre of Excellence Geopolymer and Green Technology (CEGeoGTech), School of Materials Engineering, Universiti Malaysia Perlis, Kompleks Pengajian Jejawi 2, 02600 Arau, Perlis, Malaysia
[2]Faculty of Engineering Technology (FETech), Universiti Malaysia Perlis (UniMAP), 01000, P.O. Box, D/A Pejabat Pos Besar, Kangar, Perlis, Malaysia
[3]School of Materials Engineering, Universiti Malaysia Perlis, Kompleks Pengajian Jejawi 2, 02600 Arau, Perlis, Malaysia
[4]School of Manufacturing Engineering, Universiti Malaysia Perlis (UniMAP), Kampus Alam Pauh Putra, 02600 Arau, Perlis, Malaysia

Abstract. Optimal process setting implementation critically influences productivity, quality, and cost of production in the injection moulding industry. Precedently, trial and error method were the most common method for the production engineers to identify the optimal process parameter setting. Inappropriate injection moulding parameter settings can lead to poor production and quality. This paper presents a statistical technique on the optimization of injection moulding process parameters through Design of Experiments method. In this study, an understanding of the injection moulding process and consequently its optimization is carried out by Design of Experiment method based on three parameters (melt temperature, packing pressure, and cooling time) which influence the tensile strength of rice husk reinforced low density polyethylene composites. Statistical results and analysis are used to provide better interpretation of the experiment. The models are form from analysis of variance and the model passed the tests for normality and independence assumptions.

Keywords: Injection Moulding (IM); Design of Experiments (DoE); Factorial; Optimal Design; Tensile Strength

1 Introduction

Optimization is the process of making something better. It can be defined as the process of adjusting the inputs or characteristics of a device, mathematical process, or experiment to find the minimum or maximum output [1]. Optimization methods such as Factorial, Taguchi, and RSM are few methods which designed to provide the best values of system design ad operating policy variables – values that will lead to the highest levels of system performance [2]. These methods, combined with more detailed and accurate simulation methods, results in high quality and high performance of the desired output with cost effective within a short period of time, and also less in waste materials [3]. At this date, nearly everyone interested in injection moulding has either read the power of flow analysis for injection moulding, or used the technology to aid in the design of injection molded products. It is commonplace for new designs to be numerically simulated [4] or even optimized via computer simulation before the first piece of steel is cut for the mold. Rather than expand time and money building a series of prototypical tools to fine-tune a new product, computer simulation has dramatically affected the way new parts are designed. Thus, in this research, full factorial experiment has been utilized to provide the best combination values of injection moulding processing parameters towards the optimum tensile strength specifically for LDPE/RH composites moulded parts [1].

*Corresponding author: firdausomar@unimap.edu.my

2 Equipment

The results obtained from the experiment for tensile strength were analyzed using Design Expert software. All the results from the experiments were arranged in a list of experiments and used to generate a relationship between input parameters setting and output responses which is tensile strength of the specimen. In this study, 2-levels of full factorial design were selected to analyze using three factors which are melt temperature (°C), packing pressure (MPa), and cooling time (s). Tensile strength is selected as the response. List of experiments generated in Design Expert Software based on three factors and four center points is shown in Table 1. The insignificant parameters or interaction between parameters were removed and excluded from the generated model for tensile strength.

Table 1. List of experiments for full factorial design

Run	Melt temperature (°C)	Packing pressure (MPa)	Cooling time (s)
1	170	15.93	23.03
2	160	12.96	26.74
3	180	18.89	19.33
4	180	12.96	26.74
5	160	18.89	19.33
6	170	15.93	23.03
7	160	12.96	19.33
8	170	15.93	23.03
9	180	12.96	19.33
10	170	15.93	23.03
11	160	18.89	26.74
12	180	18.89	26.74

3 Results and discussion

3.1 Analysis results of tensile strength

Table 2 shows the ANOVA result for the tensile strength of LDPE/RH composites. From Table 2, it shows that the model was significant where the p-value is below than 0.0001. Melt temperature (A), packing pressure (B), and cooling time (C) are the main effects to the response of tensile strength. AB and BC interactions were also considered as a significant effect towards the tensile strength as the p-value is less than 0.05. The curvature in this model was also significant where the p-value is <0.0001 [5]. Table 3 shows the summary for ANOVA result of tensile strength for LDPE/RH composites, where the value of R^2 is 0.9998 and the difference between adjusted R^2 and predicted R^2 is below 0.2 which is 0.0031. This can be proved that the model had good relations with the experimental results. Furthermore, for the adequate prediction, the value is above 4 which is 169.908 which indicates that the model was adequate [6].

Table 2. ANOVA for tensile strength

Source	Sum of Squares (SS)	df	Mean Square	F value	p-value Prob > F	Summary
Model	0.64	5	0.13	4039.92	< 0.0001	Significant
A-Melt temperature	0.58	1	0.58	18287.06	< 0.0001	
B-Packing pressure	0.057	1	0.057	1800.19	< 0.0001	
C-Cooling time	0.001081	1	0.001081	34.38	0.0020	
AB	0.001711		0.001711	54.41	0.0007	
BC	0.0007411		0.0007411	23.57	0.0047	
Curvature	0.057	1	0.057	1804.30	< 0.0001	Significant
Residual	0.0001573	5	0.00003145			
Lack of fit	0.0001573	2	0.00007863			
Pure error	0.000	3	0.000			
Cor total	0.69	11				

Table 3. Summary of ANOVA for tensile strength

Description	Value
Standard Deviation	0.005608
Mean	10.81
R^2	0.9998
Adjusted R^2	0.9995
Predicted R^2	0.9964
Adequate Precision	169.908

Model adequacy checking can be observed in Figure 1 and Figure 2. From Figure 1, it was observed that the residual was normally distributed for the response model of tensile strength [7]. For the plots of residual versus run order in Figure 2, it is shown that the plots are scattered and no point lies outside the limits which not exceed below or above the red line. Besides, the data are randomly distributed with no obvious pattern plotted. Thus, the model can be accepted as an adequate model.

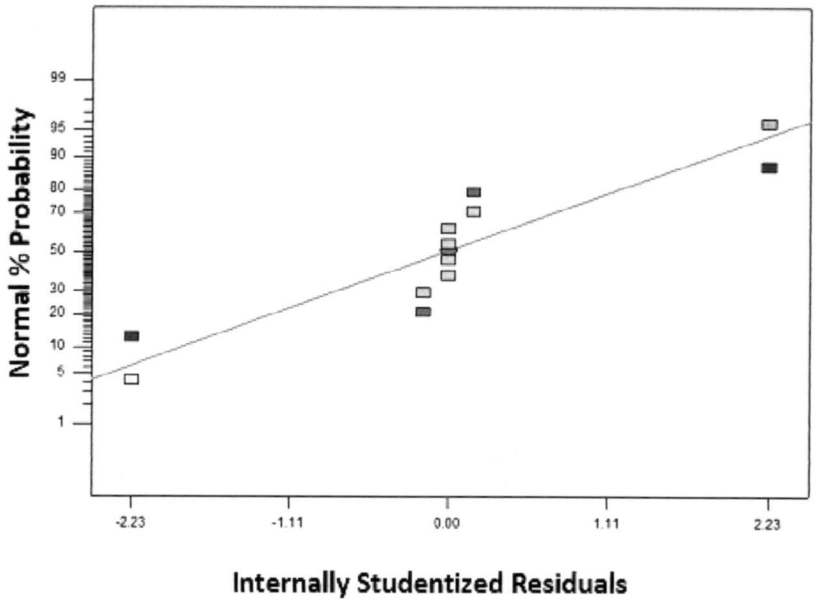

Fig. 1. Normal probability plots of residual for tensile strength

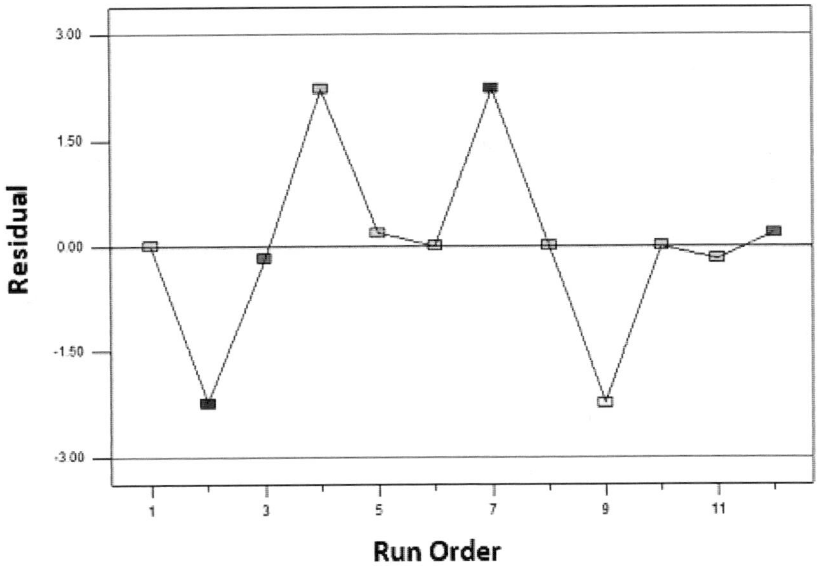

Fig. 2. Plots of residual versus run order for tensile strength

The statistical model in terms of actual factor of the tensile strength of the moulded part based on full factorial and center point experiments is shown in Equation (1) as follow:

$$Tensile\ strength = 4.11887 + 0.034668A + 0.13241B + 0.017091C - 0.000493255AB - 0.000876169BC \quad (1)$$

Where, A is the melt temperature (°C), B the is packing pressure (MPa), and C is cooling time (s). The experimental and predicted results versus run order for the tensile strength is shown in Figure 3. The graph shows a variation of tensile strength from predicted model and experimental results. From the model, the maximum predicted strength is 11.198 MPa by using the combination of 180°C of melt temperature, 18.89 MPa of packing pressure and 26.74 s of cooling time.

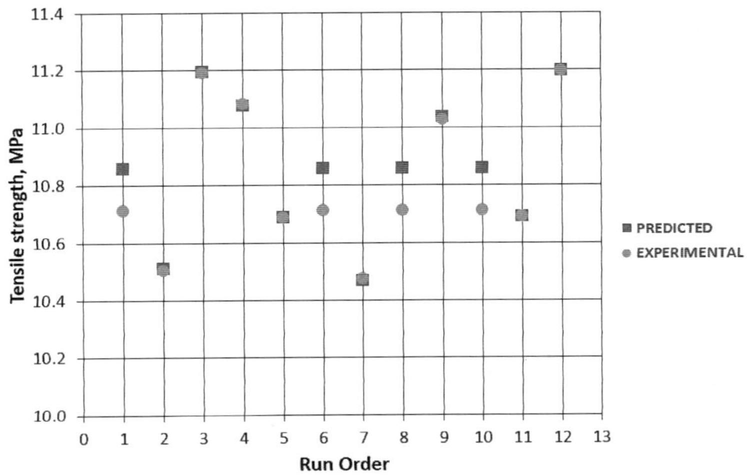

Fig. 3. Experimental and predicted result of tensile strength

IOP Conf. Series: Materials Science and Engineering **957** (2020) 012039 doi:10.1088/1757-899X/957/1/012039

Conclusion

Evaluation of tensile strength response for the LDPE/RH composite moulded parts was performed based on the full factorial design of experiment. Melt temperature was found to be the most significant factor that affected the tensile strength, and then followed by packing pressure and cooling time. Hence, the objective which is to investigate the significant factors that affects the tensile strength of the moulded parts for LDPE/RH composites through experimental works has been achieved.

References

[1] S. Elsheikhi and K. Benyounis, 45-50 (2016)
[2] D. P. Loucks and E. Van Beek, 81 (2017)
[3] M. F. Omar, H. M. AkilZ. A. Ahmad and S. Mahmud, Polym. Eng. Sci., **54**, pp. 949–960 (2014)
[4] M. F. Omar, H. M. AkilZ. A. Ahmad and S. Mahmud, Polym. Compos., **33**, pp. 99–108 (2012).
[5] D. C. Montgomery, 179-275 (2017)
[6] M. J. Anderson and P. J. Whitcomb, 53-57 (2016)
[7] M. F. Omar, H. M. AkilM. F. A. Rasyidand J. M. Sharif, J. Compos. Mater., **49**, 1195–1209 (2015)

Synthesis and Characterizations of Fe_3O_4 Added with Al_2O_3 Nanoparticles via Sol-Gel Technique

Mohd Shahadan Mohd Suan[1,*], *Chew Ke* Chin[1], *Jeefferie* Abd Razak[1], *Hazman* Hasib[1], *Mohd Asyadi 'Azam* Mohd Abid[1], *Irwan* Nurdin[2]

[1]Faculty of Manufacturing Engineering, Universiti Teknikal Malaysia Melaka, 76100 Durian Tunggal, Melaka, Malaysia
[2]Chemical Engineering Department, Lhokseumawe State Polytechnic, 24301 Aceh, Indonesia

Abstract. A single step sol-gel has been established as a new route for synthesizing Al_2O_3-added Fe_3O_4 nanoparticles. In this work, the effects of Al_2O_3 composition towards the microstructure and magnetic properties of Al_2O_3-added Fe_3O_4 samples were investigated. A series of Fe and Al nitrate mixture solutions were prepared by varying the Al nitrate composition from 2.0 wt. % to 10.0 wt. %. A dropwise of propylene oxide has dimed out the clear red mixture solution with a significant release of heat before turned into a gel. The gel was dried at 90 °C for 48 h in a ventilated oven which then pulverized into a powder by using mortar grinder. The powder was calcinated at 900 °C for 1 h in a normal furnace atmosphere. The XRD results revealed that the existence of polycrystalline Fe_3O_4 were completed at calcination temperature 900 °C. The SEM images of the Fe_3O_4 with 10.0 wt.% of Al_2O_3 samples showed a distinct nanoparticles microstructure due to the increased of Fe_3O_4 nucleation initiated by Al_2O_3 nanoparticles. The Vickers microhardness of the Al_2O_3 added Fe_3O_4 samples were significantly increased as the Al_2O_3 content was increased up to 10.0 wt.% attributed from good distribution of Al_2O_3 nanoparticles in the sample. Subsequently, the magnetic properties of Fe_3O_4 investigated by alternating gradient magnetometer, AGM was insignificantly reduced by the addition of Al_2O_3 nanoparticles. This work indicated that the sol-gel reaction is an effective method to achieve uniform distribution and high purity of Al_2O_3-added Fe_3O_4 nanoparticles with optimum microhardness and ferromagnetic performances.

1 Introduction

Magnetite, Fe_3O_4 nanoparticles have been extensively studied because of the novel applications of structural and functional elements. Recently, research regarding Fe_3O_4 has been received attention. Due to their potential uses such as not only in medical applications, but also in radiofrequency hyperthermia, pigment, cancer treatment, medical diagnostics, and magnetic resonance imaging [1-3]. However, the potential utilizations of the nanoscale particles are rigorously determined by their fundamental magnetic characteristics, for

* Corresponding author: mohdshahadan@utem.edu.my

instances, hard magnets used for data storage as well as soft magnetic materials applied for magnetic switches. In the other perspective, Fe_3O_4 was reported to be as the magnetic reinforce nanoparticles in aluminium structure for automobile and aerospace applications [4, 5]. Hence the mechanical improvement of this nanoparticles must be made without major effects towards its magnetic properties. Addition of the reinforced phase is the simple yet effective method to increase the mechanical hardness of Fe_3O_4 where the Al_2O_3 nanoparticles has attracted much attention due to its interesting properties such as high insulation, high hardness, high stability and transparency [6-8]. However, the conventional solid-state method for synthesizing Fe_3O_4 and adding in the Al_2O_3 have been reported to introduce impurities, high temperature requirements, time for washing and aging or gelation, and the use of expensive materials which lower the properties of nanoparticles. In continuing research on nanoparticles and new organic compounds, herein, sol gel, a simple and efficient method for synthesizing Al_2O_3 and Fe_3O_4 nanoparticles is established. The sol gel method produces the large number of metal oxide nanoparticles with high purity, higher surface area, good uniformity and low processing temperatures [9-11]. Therefore, Fe_3O_4 added with Al_2O_3 nanoparticles will be simultaneously synthesized by using a single step sol-gel reaction. By applying this method, uniform or well distribution and high purity of Al_2O_3 nanoparticles throughout Fe_3O_4 can be achieved. The effect of the difference compositions of Al_2O_3 nanoparticles on the properties of Fe_3O_4 nanoparticles will be investigated. The combination of Al_2O_3 added Fe_3O_4 nanoparticles through this method could contribute to optimum mechanical and magnetic properties of Fe_3O_4 nanoparticles.

2 Methodology

Appropriately, the calculated amount of Fe $(NO_3)_3.9H_2O$ and Al $(NO_3)_3$ reagent powders were weighed and mixed to obtain difference compositions of Al $(NO_3)_3$ from 2.0 wt. % to 10.0 wt.%. The mixture of powders then was dissolved in ethanol (99.0 % pure) and stirred by using magnetic stirrer until apparently clear red solution was formed. Dropwise as a proton scavenger, 9.6ml of propylene oxide was added into the solution. The solution dimed quickly with significant heat release and therefore the gel was formed. In all experiments, the gel was placed at room temperature for approximately 90 seconds to age. After that, dried gel samples were obtained by drying the gel at 90 °C for 48 hours in a ventilated oven. Then, the dried gel is pulverized in a mortar grinder. Then the powder was calcined in the furnace at 900 °C for 1 hour under normal atmosphere to obtain a stable powder. The powder sample was self-cooled to room temperature. The calcination process has transformed the brownish flaky powder into dark black fine powder. The X-ray diffraction (XRD) for the calcined powder phase identification and crystallite size estimation was carried out using Ni-filtered Co Kα radiation by Bruker D8-advanced XRD machine. The peak position and intensities were obtained between 20° and 80° with a velocity of 0.02° s^{-1}. The hardness of the sample was tested using Vickers microhardness Mitutoyo MV-2 machine. For this purpose, the powder was compressed into 10 mm diameter x 2 mm of thickness pellet by applying 12.4 MPa load and sintered at 900 °C for 1 h under normal atmosphere. Scanning electron microscope (SEM) Zeiss Ultra 40XB was used to observe the sample's microstructure. The behaviour of the synthesized Al_2O_3 added Fe_3O_4 nanoparticles under a magnetic field was investigated using an alternating gradient magnetometer (AGM). The device was highly sensitive and could measure magnetic properties such as saturation magnetization, coercivity and remanence for various sample types and strengths. AGM exerted a force on the synthesized Fe_3O_4 nanoparticles using alternating gradient field, resulting the synthesized samples to deflect within the field. The electrical signal was processed by a computer connected to the magnetometer. Analysis of the magnetic properties were performed by a software package.

3 Results and Discussions

Structural behaviours of the Fe_3O_4 added with Al_2O_3 samples represented by the XRD pattern is shown in Fig. 1. The significant peaks in each XRD patterns appeared at the same Bragg diffraction angle for the cubic structure Fe_3O_4 represented by red bars at the bottom of the figure. This indicates the complete transformation of the amorphous precursor gels into the polycrystalline Fe_3O_4 after the calcination process. The most significant peaks of Fe_3O_4 samples are attributed to the diffraction from (111), (311), (400), (422), (511), and (440) planes which in agreement to those have been reported by Chaki et al. [12]. It also can be seen that there is no significant Al_2O_3 peak presence in the pattern which has been expected due to its apparent low compositions. Detailed observation of the patterns showed the intensity difference of the peaks at the same 2Θ where the analysis of it revealed the variation of crystallite sizes.

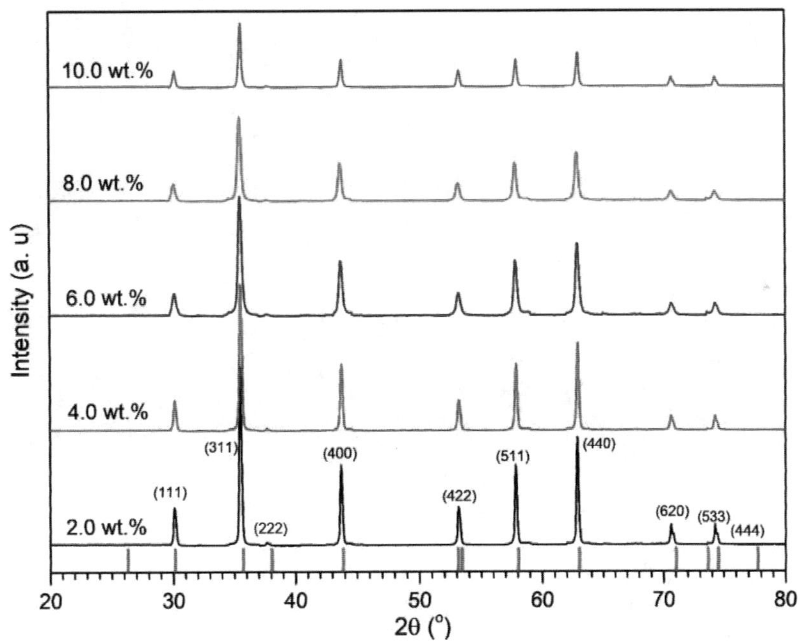

Fig. 1. XRD pattern of Fe_3O_4 added with Al_2O_3 nanoparticles samples. Bars at bottom indicated Bragg diffraction angle for Fe_3O_4.

Fig. 2 shows the SEM image of Fe_3O_4 and Fe_3O_4 added with Al_2O_3 nanoparticles. The pure Fe_3O_4 nanoparticles shown in Figure 2 (a) was observed to have larger and non-uniform grains ranged from 100 – 500 nm compare to the other samples. Additions of Al nitrate in the gel which contributed to the existence of Al_2O_3 nanoparticles have resulted with the decreased of Fe_3O_4 grain size as shown in Fig. 2 (b) – 2 (f). The grains in these samples are nearly uniform. It also can be seen the existence of spherical shape Al_2O_3 nanoparticles at the inter-grains of Fe_3O_4. The samples with 10.0 wt. % of Al_2O_3 yielded the smallest and uniform particles size around ~50 nm as shown in Fig. 2 (f). The reduction of the Fe_3O_4 particle size can be attributed to the increase of nucleation initiated at Al_2O_3 nanoparticles [11]. The increased of Al_2O_3 additions was also developed the pores located at the inter-grains boundary which then developed into a grain-boundary gaps in the Fe_3O_4 microstructure by further Al_2O_3 additions. Therefore, the Fe_3O_4 of sample with 10.0 wt. % of Al_2O_3 were existed with distinctive nanoparticles.

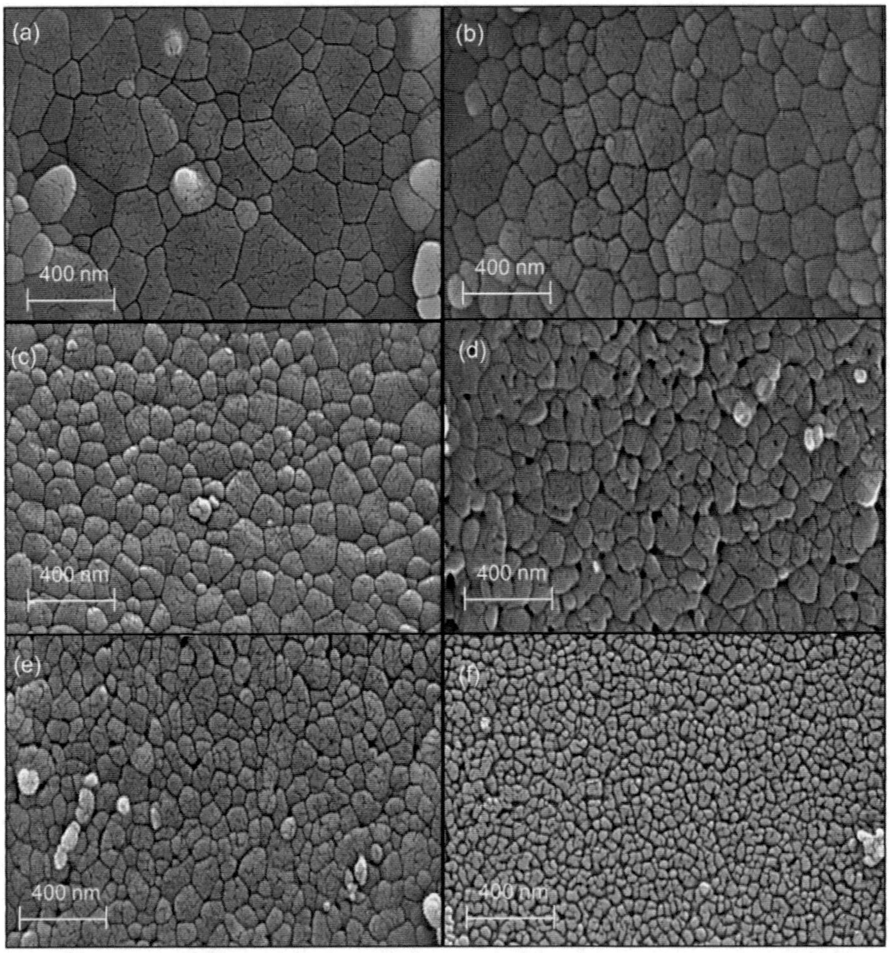

Fig. 2. SEM images of Fe_3O_4 and Fe_3O_4 added with Al_2O_3 nanoparticles samples; (a) Pure Fe_3O_4 (b) Al_2O_3 = 2.0 wt.%, (c) Al_2O_3 = 4.0 wt.%, (d) Al_2O_3 = 6.0 wt.%, (e) Al_2O_3 = 8.0 wt.%, and (f) Al_2O_3 = 10.0 wt.%,

Fig. 3 shows the Vickers microhardness profile for Fe_3O_4 and Fe_3O_4 added with Al_2O_3 nanoparticles samples as a function of the Al_2O_3 compositions. It can be observed that the microhardness of the Fe_3O_4 sample was enhanced from 5.3 GPa to 5.7 GPa by the addition of 2.0 wt.% Al_2O_3. The microhardness values keep increasing as the Al_2O_3 nanoparticles were increased and achieved the highest value of 11.2 GPa in 10.0 wt.% sample. It is notable that these values rose with low standard deviation. The variation of microhardness can be attributed to strengthening mechanism provided by nanoparticles reinforced phase which has been intensely discussed by Ohji et al., [13], the mechanism is possible since the Al_2O_3 nanoparticles were successfully yielded as the reinforce nanoparticles phase in Fe_3O_4 samples. The Al_2O_3 nanoparticles have been known to have excellent hardness property up to 200 GPa will elastically interact with dislocations within the Fe_3O_4 grains and consequently, prevent the dislocations movements during indented. In addition, well distribution of Al_2O_3 nanoparticles in the samples achieved by the synthesis method has resulted in uniform microhardness value. The hardening mechanism is maximized at 10.0 wt.% sample as more dislocation movements are prevented by more Al_2O_3 nanoparticles.

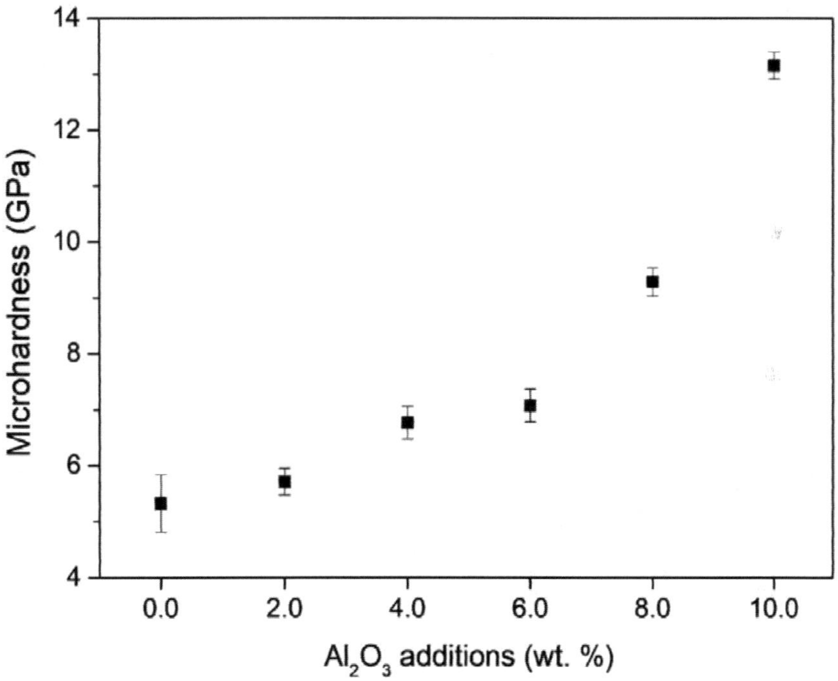

Fig. 3. Vickers microhardness of Al_2O_3-added Fe_3O_4 nanoparticles samples.

The field dependence of magnetization curve of Fe_3O_4 and Fe_3O_4 added with Al_2O_3 nanoparticles measured at 27 °C in a maximum applied field of ±10 kOe are shown in Fig. 4. It can be observed that each sample showed a hysteresis loop of a typical ferromagnetic behaviour which is in consistent with the fact that Fe_3O_4 nanoparticles would exhibit ferromagnetism. The AGM results show a decrease pattern of saturation magnetization, M_s with Al_2O_3 addition. The M_s decreased from 0.30 emu/g to 0.20 emu/g between the pure Fe_3O_4 and 10.0 wt.% of Al_2O_3 added samples respectively. The drop was, however, not directly proportional with the additions of Al_2O_3 shown by gradually decreasing value from 0.02 emu/g to 0.01emu/g to 0.005 emu/g for samples with 2.0 wt.%, 6.0 wt.% and 10.0 wt.% of Al_2O_3 addition respectively. The charge transfer model of ferromagnetism associated with defects in the bulk or at the surface of nanoparticles is a good approach to understand the phenomena [14]. As for the Al_2O_3 added Fe_3O_4 samples, exchange interactions may occur between the localized electron spin moments resulting from the non-magnetic performance of Al_2O_3 nanoparticles hence reduced the magnetization saturation. The decrease of M_s shown in each Al_2O_3 added samples is insignificant and maintained in a range for ferromagnetic applications.

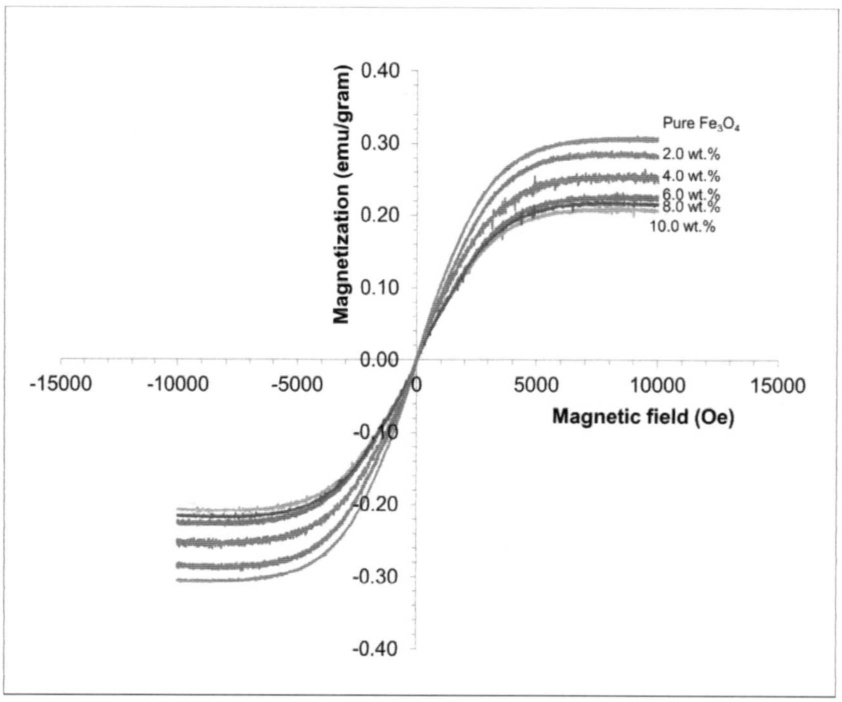

Fig. 4. Hysteresis loop for Fe_3O_4 and Fe_3O_4 added with Al_2O_3 nanoparticles samples.

4 Conclusions

High purity of Al_2O_3 added Fe_3O_4 nanoparticles were successfully synthesized via a single step sol gel method. The increased of Al_2O_3 additions has significantly improved the microstructure and tripled the microhardness of Fe_3O_4 by adding 10.0 wt.% of Al_2O_3

nanoparticles. The magnetic saturation was sustained in a range from 0.2 emu/g to 0.3 emu/g, indicated ferromagnetic behaviour of each samples.

Authors would like to acknowledge Universiti Teknikal Malaysia Melaka (UTeM) for sponsoring this research under the short term research grant: *PJP/2019/FKP(4B)/S01662*. Thanks to COSSID and FKP for facilities and technical support in completing this study. Special thanks also for UMP for some assistance under the grant code RDU190324.

References

[1] Pylypchuk I. V., Kołodyńska, D., Kozioł, M., & Gorbyk, P. P. Nanoscale Research Letters, **11**(1) (2016).
[2] Choi K, Nam K, Kim H, Min J, Uhm H J. Korean Phys. Soc. **65** (10), 1658-166 (2014).
[3] Bao, Y., Wen, T., Samia, A. C. S., Khandhar, A., & Krishnan, K. M. J. Mater. Sci., **51**(1), 513–553. (2015).
[4] Bayraktar E., & Katundi D., J. Achiev. Mater. Manuf. Eng 38 (1) 7-14 (2010).
[5] Ferreira L-M-P., Bayraktar E. & Robert M-H. Advances in Materials and Processing Technologies **2** (1) 165-173 (2016).
[6] Popat, K. C., & Desai, T. A.. Biomaterials Science, 162–166 (2013).
[7] Jung, K.-W., Choi, B. H., Ahn, K.-H., & Lee, S.-H. Applied Surface Science, 423, 383–393 (2017).
[8] Krivoshapkin, P. V., Mikhaylov, V. I., Krivoshapkina, E. F., Zaikovskii, V. I., Melgunov, M. S., & Stalugin, V. V. Micropor Mesopor Mat, **204**, 276–281 (2015).
[9] Kumar A, Nishtha Yadav, Monica Bhatt, Neeraj K Mishra, Pratibha Chaudhary and Rajeev Singh. J. Chem. Sci. **5**(12), 98-105 (2015).
[10] Ali A, Zafar H, Zia M, Ul Haq I, Phull AR, Ali JS, Hussain A. Nanotechnol. Sci. Appl., **9**, 49–67 (2016).
[11] Bhosale, R. R., Kumar, A., AlMomani, F., & Alxneit, I. Ceram. Int, **42**(6), 6728–6737 (2016).
[12] Chaki, S. H., Malek, T. J., Chaudhary, M. D., Tailor, J. P., & Deshpande, M. P. Adv. Nat. Sci-Nanosci, **6**(3), 035009 (2015)
[13] Ohji, T., Jeong Y. K., Choa Y. H. and Niihara K. J. Am. Ceram. Soc **81**, 1453-1458 (1998).
[14] Eichholz C., Knoll J., Lesche D., Magn. Magn. Mater. 368 139–148 (2014).

IOP Conf. Series: Materials Science and Engineering **957** (2020) 012041 doi:10.1088/1757-899X/957/1/012041

Preliminary Study of Hydrothermal Synthesis of TiO_2–GO Composites as a High Performance Photocatalyst

N.M.N Azira[1,2,*], *D.* Murizam[1], *D.S.C.* Halin[1], *M.A.* Idris[1] and *N. F. M.* Yunos[3]

[1]School of Materials Engineering, Kompleks Pusat Pengajian Jejawi 2, Universiti Malaysia Perlis, 02600 Jejawi, Arau, Perlis, Malaysia.
[2]Centre of Diploma Studies, Universiti Malaysia Perlis (UniMAP), Kampus UniCITI Alam Sungai Chuchuh, 02100 Padang Besar (U), Perlis
[3]Institute of Engineering Mathematics, Universiti Malaysia Perlis, Pauh Putra, 02600 Arau, Perlis

Abstract. In this study, the addition of graphene oxide (GO) into TiO_2 was investigated. GO was prepared by modified Hummer method before it was added into TiO_2 via hydrothermal method. The graphite and GO was characterized by Raman spectra, Fourier transform infrared spectroscopy (FTIR), X-ray diffraction (XRD) and scanning electron microscope (SEM) along with the TiO_2-GO composites also have been characterized. The morphology of TiO_2 deposited on the surface of the GO sheet was observed by SEM. the phase formation of anisotropic anatase (TiO_2) of the TiO_2-GO was detected in XRD. The photocatalytic activity was determined by calculating the photodegradation efficiency of methylene blue (MB) under UV light irradiation. The photodegradation of MB were increased with time for TiO_2-GO compared to pure TiO_2. The results indicated that TiO_2-GO composites was successfully produced using hydrothermal method for the photocatalytic application.

1 Introduction

Titanium dioxide (TiO_2) is viewed as one of the most promising catalyst materials due to its small band gap, superior photocatalytic performance, easy availability, long-term stability, and nontoxicity. Photocatalysis has become a method that can effectively degraded and mineralize a variety of pesticide contaminants in the presence of TiO_2 and TiO_2 based nanocomposite [1]. However, it is well known that the size and morphology of TiO_2 particles play crucial roles in its properties and applications [2]. There a number of ways to improve the rate of photocatalytic activity and producing a TiO_2 photocatalyst that utilises both UV and visible light. One solution to these problems is the use of a TiO_2 heterojunction photocatalyst. While anatase is commonly considered the most photocatalytically active phase of TiO_2. Previous studies used two phases (anatase and brookite or anatase and rutile) as a heterojunction had improved the photocatalytic activity when compared with the use of anatase alone [2].

*Corresponding author: noorazira@unimap.edu.my

Graphene is ideally suit the implementation in electrochemical application due to its reported large electrical conductivity, vast surface area, unique heterogeneous electron transfer and charge carrier rates, widely applicable electro-catalyc activity and low production cost [3]. The TiO_2-C composites can be generally categorized into three kinds: TiO_2-mounted activated carbon, carbon-doped TiO_2, and carbon coated TiO_2, and each of them exhibits good photocatalytic activity [4]. However, several problems still hinder further promotion of efficiency of the present TiO_2-C composites, such as the marked decrease of the absorptivity during photodegradation, the weakening of the light intensity arriving at catalysts' surface, and the lack of reproducibility due to the preparation and treatment variation. Sohail, M et. al reported the photocatalytic activities of $TiO_2^{@}$rGO nanocomposites are higher than pure TiO_2. The $TiO_2^{@}$rGO nanocomposite with 84wt% of TiO_2 is well responsible for enhancement in photocatalytic [5]. Therefore, it is a great significance to obtain a TiO_2-C composite possessing photocatalytic activity well beyond pure TiO_2 with reproducibility and controllability, which could be potent in environment remediation.

Different methods including chemical vapor deposition, sol-gel, spray pyrolysis and sputtering have been reported for the synthesis of TiO_2 in literature to date. Despite all the astonishing properties of TiO_2, it suffers from the band gap of 3.2 eV, which occupies only about 4% of the sunlight and the low separation probability of photo-induced electron-hole pairs in photocatalysts [2]. Various modification methods have been proposed via either doping or compositing to narrow its band gap and to enhance the photocatalytic activity in the range of under the visible light radiation. In the present paper, a simple hydrothermal synthesis is reported which TiO_2 structures were attached to the GO sheet under hydrothermal treatment.

2 Experimental

The TiO_2-GO composites were prepared using hydrothermal method. Graphene oxide (GO) was prepared using modified Hummers method by using graphite powder as a precursor.

2.1 Synthesis of GO

The first stage of synthesis involving oxidation of graphite where H_2SO_4:H_3PO_4 (180:20 ml), graphite powder (1.5 g), and $KMnO_4$ (9.0 g) were mixed using magnetic stirrer. The mixture was stirred for 6 h, at a constant speed before it was poured into ice water (~200 ml) and then added with H_2O_2 to stop the oxidation process. The GO formed after the resultant mixture solution was centrifuged (5000 rpm) for 7 minutes for one time with HCl and the supernatant was decanted away. Then, washing process was repeatedly done with distilled water until a pH of 4–5 was achieved. Next, it was dried at 80 °C in an oven overnight.

2.2 Synthesis of TiO_2

The TiO_2-GO composites were prepared using titanium (IV) sulfate ($Ti(SO_4)_2$) and GO aqueous suspension as the starting materials. In a typical preparation process, 0.12 ml $Ti(SO_4)_2$ was added into 5 ml of ethanol and then a certain amount of GO suspension was gradually dropped into the solution. Subsequently, deionized water was added to get a total volume of 10 ml. After stirring for 1 h, the obtained solution was transferred into a 5 ml Teflon-sealed autoclave and maintained at 180°C for 9 h in a furnace. Finally, the resulting products were isolated by centrifugation at 6000 rpm for 10 min, washed with deionized water and ethanol several times and dropped on the silica subtracts. The composites dried in an oven at 70°C for 12 h.

IOP Conf. Series: Materials Science and Engineering **957** (2020) 012041 doi:10.1088/1757-899X/957/1/012041

2.3 Characterizations

Raman spectra analysis that was conducted to ensure that graphite was transform to graphene oxide using this method. Raman spectra were carried out by Renishaw invia Raman Microscope using a 100× objective lens with a 532 nm laser excitation (WITEC, Germany). The phase structure of the samples as characterized by X-ray diffraction (XRD) on a Bruker D8 diffractometer using Cu Ka (k = 1.5406 Å) radiation at 40 kV and 40 mA. Scanning electron microscopy (SEM) images were taken on a Hitachi S-4800 field emission scanning electron microscope at a voltage of 10 kV.

2.4 Photocatalytic Activity Measurement

The photocatalytic activity was determined by calculating the photodegradation efficiencies of methylene blue (MB) under UV light irradiation. In this experiment, the TiO_2-GO composites were placed in aqueous of MB.

3 Results and discussion

GO samples were synthesis using modified Hummer method. The Raman spectra of graphite and graphene oxide (GO) shows in Fig. 1. Highly ordered graphite has three of Raman-active bands visible in the spectra, the in-phase vibration of the graphite lattice (G band) at 1551 cm^{-1} as well as the (weak) disorder band caused by the graphite edges (D band) at approximately 1330 cm^{-1} and 2655 cm^{-1}. The GO showed a G-band at 1599 cm^{-1} and a D-band at 1350 cm^{-1}, the findings value was similar to result obtained by Chen et. Al [6]. The G-band is associated with graphitic carbons and the D-band is related to the structural defects or partially disordered graphitic domains [7]. The D-bands in both spectra were strong, confirming the lattice distortions of graphene basal planes. While the lattice vibration with the frequency of the D band is, in fact, present in the infinite graphene sheet, it remains Raman-inactive until the edges "light" it up due to the symmetry breaking and the corresponding change in the selection rules. Both the G and the D bands undergo significant changes upon amorphization of graphite as amorphous carbon contains a certain fraction of sp^3 carbons. The higher disordered in graphite led to a broader G band, as well as to a broad D band of higher relative intensity compared to that of the G band was observed.

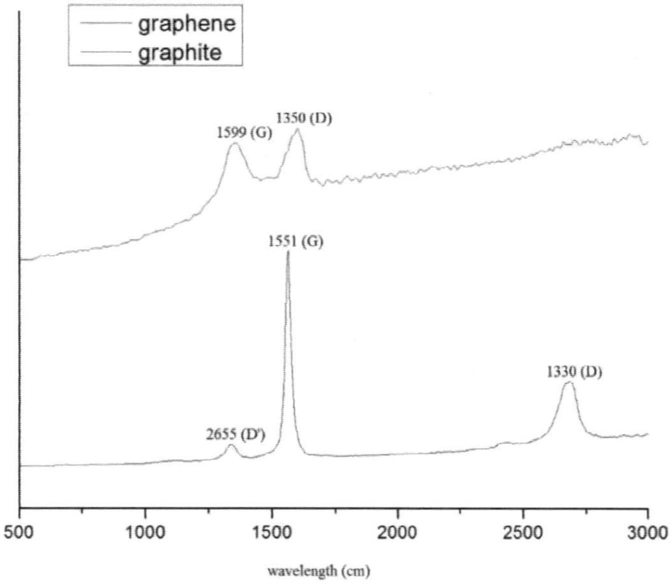

Fig. 1. Raman Spectra for graphite and GO.

Furthermore, XRD patterns of the raw graphite, synthesis GO and TiO_2-GO composite were shown in Fig. 2. It was obviously that the graphite has a peak at $2\Theta = 27°$ meanwhile the GO has peak at $2\Theta = 11.9°$. GO peak corresponding to the [001] interlayer spacing of 7.43 Å. The XRD pattern of TiO_2–GO show clearly diffraction peak of anisotropic anatase TiO_2. However, the [001] reflections of GO are not observed in the XRD pattern of TiO_2–GO composite because the regular stack of GO was destroyed by the intercalation of TiO_2 [8].

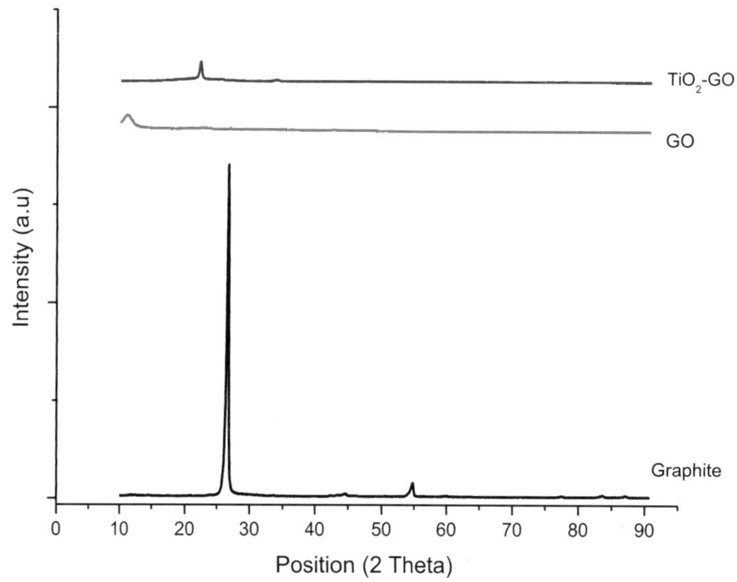

Fig. 2. XRD patterns of the raw graphite, synthesis GO and TiO_2-GO composite.

Figure 3 shows the morphology of graphite powder and graphene oxide which observed under SEM. There were differences between graphite and graphene oxide where for the graphite the sheets were stalked together. But for graphene oxide it was clearly seen the sheets were exfoliated. These was proved by Paulchamy et al. where the graphite was clearly seen exfoliated after oxidation process [9].

(a) (b)

Fig. 3. SEM for (a) graphite as starting material and (b) GO after synthesis using modified Hummer method

The morphology of TiO_2-GO composite using hydrothermal method was further observed by SEM. It was easy to find many platelets of TiO_2-GO composites, along with large aggregated TiO_2 particles. (Fig 4). The as-prepared GO was composed of several sheets and maintained a layered structure with micrometres-long wrinkles on the surface [10]. The TiO_2 were well distributed on the both sides of the GO, due to self-assembling of the TiO_2 on the GO during hydrothermal proses.

Fig. 4. SEM for TiO_2-GO composite

The Raman spectra, XRD and SEM analysis confirmed the effective self-assembly of TiO_2 at GO sheet grow as a composite on the silica substrate using hydrothermal method. The prerequisite of successful self-assembly was good dispersion of both TiO_2 and GO. As reported by Liu, J et. al, GO and Degussa P25 are used to fabricate Go-P25 composites because of P25 was not dispersed in DI water well, only some of the P25 nanoparticles are anchored to the GO sheet [8]. Moreover, the hydrothermal method was easy to control the TiO_2/GO ratio by decreasing the TiO_2 quantity.

Fig. 5 shows the photodegradation efficiencies vs irradiation time of MB under UV light for TiO_2-GO composite, where T the degradation% decreased when the times increased. During the photodegradation process, MB molecules were absorbed on the surface of GO and then reacted with photocarrier transferred from TiO_2. Therefore, GO was taken as a carrier of photocatalysis which effectively accelerated the photocatalytic reaction rate as reported by Zhang & Pan [11]. It was obvious that GO could enhanced the photocatalytic activity of TiO_2 under UV light irradiation due to a synergy effects including the formation of π-π conjugations between dye molecules and aromatic rings and retarding the charge recombination. From the result, no degradation of methylene blue was observed over a long period. Such behavior was probably due to the low crystallinity present in these materials, and to surfactant residue impurities in their pore walls [12].

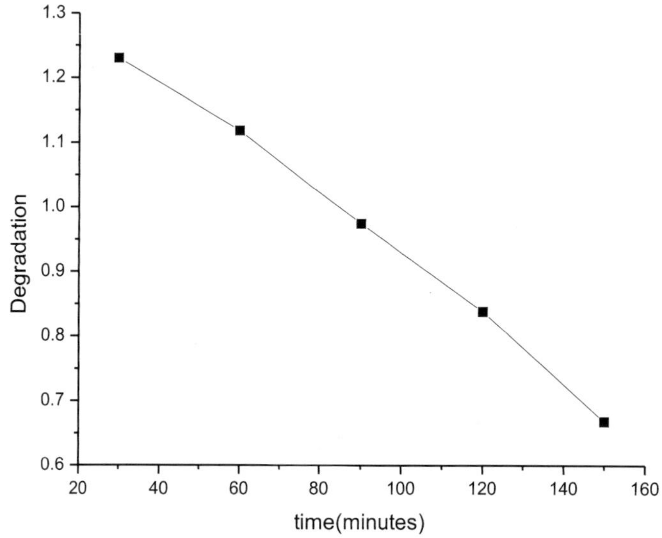

Fig. 5. Photodegradation efficiencies vs irradiation time of MB under UV-vis irradiation for TiO_2-GO composite

In this study, the graphene oxide worked as adsorbent, electron acceptor, photosensitizer and enhanced the photodecomposition efficiently. Since GO has the similar structure to TNT, we speculated that the electron can transport along the GO sheet and then reacted with absorbed O_2 to form ·OH for the further photocatalytic degradation of MB. It was well known that the carbonaceous materials had extraordinary adsorption properties which were used in various environmental applications. Most industrial dyes and related pollutants were usually aromatic in nature and had the ability to create stacking interactions with graphene aromatic domains reported by Wang et al. [2012].

4 Conclusion

In conclusion, we have successfully grown TiO_2 on GO using one step method hydrothermal method. This method produced a TiO_2-GO composites with strong interactions between the two components. The resulting composite material shows superior photocatalytic activity to other forms of TiO_2. Our method could be further extended to grow other functional materials on GO for photocatalytic [13].

References

1. Byrne, C., Subramanian, G., Pillai, S. C.. J. Environ. Chem. Eng., **November 2016** 0–1. (2017)
2. Najafi, M., Kermanpur, A., Rahimipour, M. R., Najafizadeh, A. J. Alloys Compd., **722**, 272–277 (2017)
3. Brownson, D. A. C., Kampouris, D. K., Banks, C. E. J. Power Sources, 196 **11**, 4873–4885 (2011)
4. Zhang, H., Lv, X., Li, Y., Wang, Y., & Li, J. ACS Nano, **4(1)**, 380–386 (2010)

5. Sohail, M., Xue, H., Jiao, Q., Li, H., Khan, K., Wang, S., Zhao, Y. In Materials Research Bulletin, **90**, 125–130 (2017)
6. Chen, J.. An improved Hummers method for eco- friendly synthesis of graphene oxide synthesis of graphene oxide. Carbon, **64**, 225–229 (2013)
7. Kudin, K. N., Ozbas, B., Schniepp, H. C., Prud'homme, R. K., Aksay, I. A., & Car, R. Nano Letters, **8(1)**, 36–41 (2008)
8. Liu, J., Bai, H., Wang, Y., Liu, Z., Zhang, X., Sun, D. D. Adv. Funct. Mater., **20(23)**, 4175–4181. (2010)
9. Paulchamy, B., Arthi, G., Bd, L.. A Simple Approach to Stepwise Synthesis of Graphene Oxide. **6(1)**, 2–5 (2015)
10. He, Y., Zhang, Y., Huang, H., Zhang, R. Micro and Nano Letters, **8(9)**, 483–486. (2013)
11. Zhang, Q., He, Y. Q., Chen, X. G., Hu, D. H., Li, L. J., Yin, T., Ji, L. L. Chinese Science Bulletin, **56(3)**, 331–339 (2011)
12. Kim, D. S., & Kwak, S. Y. Applied Catalysis A: General, **323**, 110–118 (2007)
13. Wang, G., Xu,. Lin, Zhang. J., Yin, T, Han, D Int. J. Photoenergy., **2012** (2012)

IOP Conf. Series: Materials Science and Engineering **957** (2020) 012042 doi:10.1088/1757-899X/957/1/012042

Building Information Modelling (BIM) as an UAV Information Processor for Generating an 'As Built Drawing'

Norsyakilah Romeli[1,*], *Hazry* Desa[1], and *Muhammad Azizi* Azizan[1,2]

[1]Centre of Excellence for Unmanned Aerial System, UniMAP
[2]Faculty of Engineering Technology, UniMAP

Abstract. The heritage building is a valuable asset to humans and countries due to the grandeur symbolization and the dignity representation of a country. However, the dramatic changes on the historical building physical feature has been documented poorly. Subsequently, the depletion of the building information has making the maintenance works intricate. On site surveying might difficult due to the safety issue that putting manpower into risky circumstances. Advance technology may allow many to obtain information using aid of remote sensing, photogrammetry, geographic information system (GIS) and 3D laser scanning. The projection of the As Built Drawing might possible without utilisation of BIM as an analyser for the raw information congregated. Thus, the objective of this research is to develop an 'As Built Drawing' for heritage building using Building Information Modelling (BIM) as an information processor. The methodology involved the Unmanned Aerial Vehicle (UAV) application and Terrestrial Laser Scanner, photogrammetry image then been analysed using BIM. The identification of design properties of heritage building was obtained which consisted of architectural properties and structural properties. Therefore, the development of 'As Built Drawing' will be produced completely using BIM and in the other hand solving the problem on facilities management on the historical building.

1 Building Modelling Information (BIM) as an Information Processor for Generating an 'As Built Drawing'

The importance of 'As Built Drawing' (ABD) in present day creates adequate task control and fundamental action plan for heritage building. Handling a construction mission involves the usage physical information inclusive of architecture, engineering, production, and operation (AECO) raw information [5]. Due to the changes of time, depletion of information on existing building are one of the hurdle the maintenance

*Corresponding author: norsyakilah@unimap.edu.my

work delayed for majority vast of the conservation projects. Nowadays, BIM Software provides an extensive data extraction that covers from the actual constructing strategies all through until the optimum life cycle span of the building. BIM software layout information shields the architectural and structural, that crucially required to assist the photogrammetry data rose from the unmanned on site building surveying [2], [3]. Despite the accuracy of BIM projecting the measurement of the building, the precision properties of BIM Software were highly acclaimed by architects or engineers, that revving continuously to be implement when the freshness of the data contained can be synchronized with the continued building system. A BIM is thus prone to being 'blind and deaf' to ongoing construction processes, whilst an 'as-built' version can extra reliably and usefully assist information exchange and decision-making in the course of the project life-cycle [1], [2] & [4]. Conceptual framework highlighting the theoretical perspective of bridging BIM and building, which connecting the implying records contained in BIM with bodily constructing techniques to make BIM reflecting actual-lifestyles situations. Such an instrument ought to help summarize information work in a building, which has evolved in a piecemeal fashion to date. It is able to additionally make it viable to theorize a constructing as a cluster of data [3], [5]; currently accomplished through distinctive information and communication technologies (ICTs), has virtually become a general research question at the fore of information management for decision-making in present-day project management and major conservation project. Similarly, an integrated conceptual framework could assist structure strategic development directions in this area, allowing future research on bridging BIM and building to proceed on a greater stable footing [4].

2 Research Methodology

The process of gathering the data ignited with Stage 1: Planning, which comprised of planning the location of the study. Masjid Alwi in Kangar was built in 1931, considering it is a heritage building that required a preservation act in Malaysia. Stage 2: Data Collection portrayed the Unmanned Aerial Vehicle (UAV) and Terrestrial Laser Scanner (TLS) application flight take off to provide a photogrammetry data. Then, Stage 3: Data Extraction and Processing which is the crucial of this research where the software used to process the photogrammetry image for this study is Pix4D Mapper, Civil Design Software, ArcMap and AutoCAD 2014 software and BIM Software in order to achieve the objective. Later the output presented in Stage 4: Results. Fig.1 explained the flow of the research.

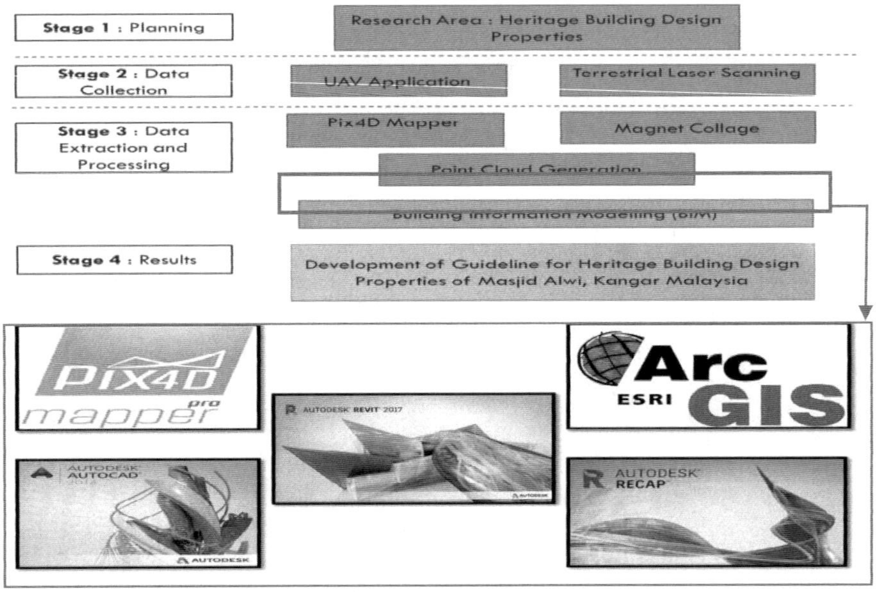

Fig. 1. Research Methodology.

3 Results and Discussion

In BIM Processing Code of Conduct; the initial processing, Pix4Dmapper first computes key points on the images. It uses these key points to find matches between the images. From these initial matches, the software runs an Automatic Aerial Triangulation (AAT) and Bundle Block Adjustment (BBA). If initial processing has already been done and is restarted, the software uses the existing key points and matches and starts the AAT and BBA. If from one processing to another the project type has changed, additional matches might be computed as the matching procedure is different for each project type. Before proceed using TLS scanner to scan the internal and external of Alwi Mosque, traversing is a must in order to determine the coordinate and station point of the scanner. There are at least 28 points that had been marked for the external site of the mosque. A total of 32 points also had been marked for the internal site of the mosque. Most of the marking point indicate the scanner place throughout the scanning process.

An Automatic by default one ground sampling distance is selected. One can also easily change the resolution to multiples of the ground sampling. Use Noise Filtering: The generation of the Point Cloud can lead to noisy and erroneous points. The noise filtering corrects the altitude of these points with the median altitude of the neighbouring points. Geo - reference file obtained. For most projects, the Topographic Elevation is split into several tiles and one geo – reference file is generated per tile. The method used for Raster Topographic Elevation generation. The method will affect the processing time and the quality of the results.

Later, the Topcon GLS 2000 use its latest point cloud processing software named Magnet Collage. All the related point cloud data capture during the scanning process has been transferred into the software. The data capture has been categorized into three differences site: - internal scanning, external scanning and combine both internal and external. Combination of internal and external scanning data using Terrestrial Laser Scanner has produces astounding 3D mesh model of Masjid Alwi. However, there are some modifications needed to restructure the pillars, windows, doors also the inside structure of the mosque. So, the point cloud has been export into the Autodesk Revit for BIM modifications. Using Revit can alter and develop much better 3D model for Masjid Alwi (See Fig.2). After finishing the modifications, the as-built drawing layout plan can be produced with detail dimensions of Masjid Alwi. The final output of this project is the integration of both unmanned and scanner data of Alwi Mosque in the form of 3D point cloud format. The as-built drawing of Alwi Mosque also has been produced in order to achieve the objective of this research.

Fig.2 3D Model Using BIM Software.

Existing 3D model input provides information on the Alwi Mosque built in 1931, with wall thickness of 37.50 centimeters and enlarged the building to the right of the mosque with a wall thickness of 25.00 and 30.00 centimeters. The thickness wall of mosque, Brickwall Set A is 37.50 centimeter, Brickwall Set B are 30.00 centimeter and a Brickwall Set B are 25. 00 centimeter. The architectural design properties and physical changes of Masjid Alwi recorded and tabulated in Table. 1.

Table 1. Physical Changes derived by BIM Software.

*(/) available
*(X) absent

BIM Software Processing (Physical Changes)	Architecture	Remarks	Structural	Remarks
Brick wall	/	The types of brick wall has a significant changes which portray an evolution of Middle Eastern brick wall masonry design influence to Malaysia	/	Difference types of dimension and sizing of the brick wall, that differs from the Prayers Area, Ceremony Area, and Utilities Room
Column	/	The pillar used Roman Style, required preservation of the column style, to maintain and match the original column types for conservation works	/	Deterioration in Roman pillar in Open Area, and Toilets (Wet Area) may cause structurally sound for the function ability of the Masjid
Roof	/	Minor deterioration of dome structure, making the architectural prestige value decreased	/	Open Area Roof exposed to windy weather may causing fatality to the occupants.
Window	/	Obsolete design of the Window may cause difficulties to find the finishing accessories in current market	X	Physical Changes subjected to not affecting structural properties
Door	/	Obsolete design of Door may causing door parts required custom design	X	Physical Changes subjected to not affecting structural properties

The mosque is constructed of 33 concrete columns such as 16 round columns is 32 cm in diameter and 17 rectangular columns is 8 columns in size (30 X 30 cm), 3 columns in size (50 X 50 cm), 3 columns (50 X 37 cm), 2 numbers of columns (15 X 25 cm) and 1 column (15 X 15 cm). In addition, there are 117 roman column used at the entrance and for the corridor.

The 'As Built Drawing' was successfully generated using BIM Software that has optimum usage in maintenance work of the heritage building. The drawing help local agencies such as the Fire Department, Construction Team, and Building Surveyors to plan a framework to tackle issues in heritage building to ensure the novelty still be preserved, the function ability of the building, and business generation from the heritage building will be safeguarded.

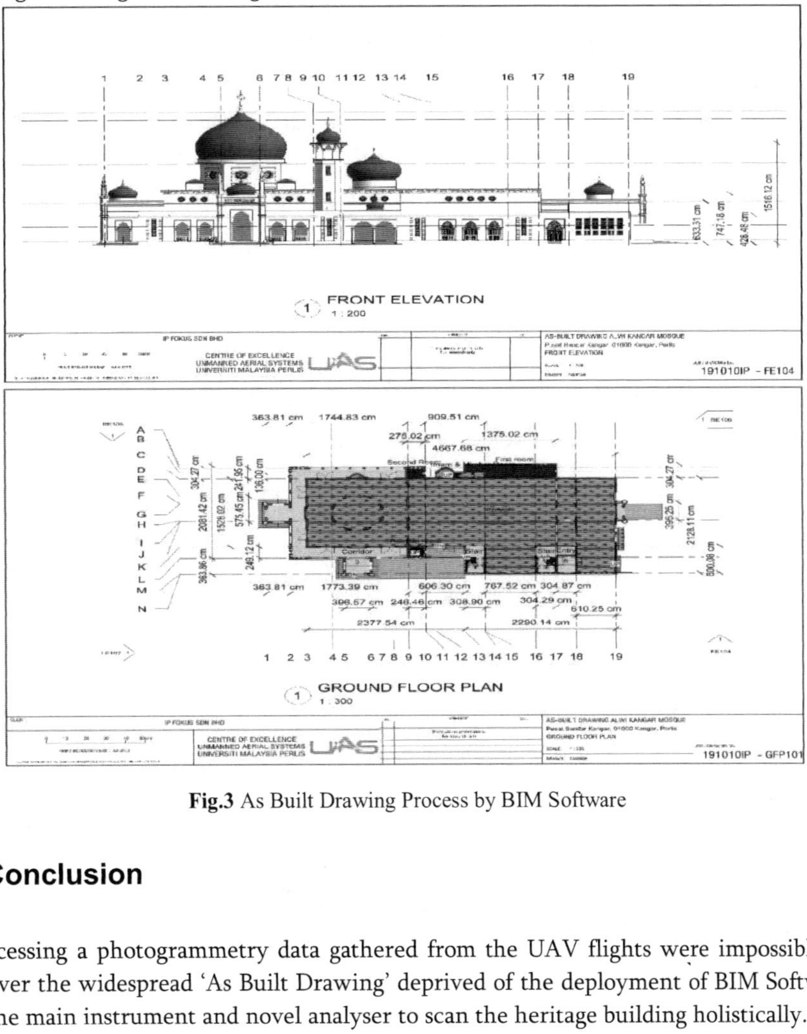

Fig.3 As Built Drawing Process by BIM Software

4 Conclusion

Processing a photogrammetry data gathered from the UAV flights were impossible to deliver the widespread 'As Built Drawing' deprived of the deployment of BIM Software as the main instrument and novel analyser to scan the heritage building holistically.

References

1. B. Bortoluzzi, I. Efremov, C. Medina., D. Sobeiraj, J.J Mc Arthur, Auto. In Constrctn. **105**, 7 (2019)
2. S. Daftry, C. Hoppe, H. Bischof, Intl. Conf. Robotic & Automation (2015)
3. H. Eisenbeiss, L. Zhang, ISPRS Commission V Symposium, **5** (2010)
4. Q. He, G. Wang, L. Luo, Q. Shi, J. Xie, X. Meng, Intl Jour of Prjct Mgmt, **35** (4) (2017)
5. N. Snavely, S. Seitz, M & R. Szeliski, Intl Jour of Comp Vis, **80** (2) (2008)

IOP Conf. Series: Materials Science and Engineering **957** (2020) 012043 doi:10.1088/1757-899X/957/1/012043

Development of iron thin films by electron beam physical vapour deposition (EBPVD): A Review

Nur Izzati Muhammad Nadzri[1,2,3*], *Dewi Suriyani* Che Halin[1], *Athirah* Khemar[1,2], *Mohd Asyraaf* M. Hasbi[1,2] and *Yufridin* Wahab[2]

[1]Center of Excellence Geopolymer & Green TEchnology (CEGeoGTech), School of Materials Engineering, Universiti Malaysia Perlis (UniMAP), 02600 Jalan Kangar-Arau, Perlis, Malaysia
[2] Ambiance UniMAP, School of Microelectronic, Univerisiti Malaysia Perlis (UniMAP), SME Bank Industrial Estate, 02000, Kuala Perlis, Malaysia
[3] Department of Materials Engineering, Imperial College, London

Abstract. This review paper study about the possibilities of results obtained from conducted experiment of iron (Fe) thin films by electron beam physical vapor deposition (EBPVD). Previous studies showed that by exposing the substrate of the thin film to pre-heat environment, the changes of morphology and adatom mobility is expected. Furthermore, annealed influence in the thin film also will give better-quality thin films as the surface are expected to be smoother and flat. Three different annealed temperature is conducted on the samples, which are 400°C, 800°C and 1200°C. Structural changes such as transition from alpha phase to beta phase, is possible due to the presence of high temperature.

1.0 INTRODUCTION

Iron (Fe) is a readily well known Earth's material that has been widely used in many sectors and industries. The crystal structure of iron itself which is capable to extinguish in many forms by depending their phase present, makes iron of the interesting earth elements to study. Bulk Fe exists in alpha, beta, gamma and delta phase. In alpha phase for example, the crystal structure is in body-centered cubic (bcc), however, when subjected to temperature or other growth parameter, the phase can change to beta with possess a face centered cubic (fcc) structure). These different crystal structures gives different mechanical properties of the iron performance in materials and/or samples which is very crucial in determining specification of the performance when it comes to heavy industries.

For the last decade, the focus on developing iron-based materials is to study their behavior under magnetization. The presence of phase difference in iron makes it interesting, as it is capable to change from ferromagnetism to anti-magnetism. To date there are more than 20 known iron based oxides, iron based nitrides, iron based hydroxides and iron based oxides-hydroxides. However, as the demand for technology increases, researches have diverted their attention on bulk Fe to thin films Fe.

Thin films are of interest as the growth of thin films give similar crystal structure as their parent material, but it has better performance in term of mechanical and physical properties. As the demand from industries such as micromechanical system (MEMS), semiconductor, hard coatings and many, thin films mechanical properties have become indispensable where these thin films are used for not only structure materials as well as electrical materials. Therefore, it is crucial to understand the characteristics of thin films in order to produce high performance thin films [1].

Metal based thin films and alloys are commonly produced by techniques like sputtering, molecular beam epitaxy and many other methods. As for iron thin films, our main aim is to develop an ultra-fine film as the changes in physical properties exhibit smaller grain size which effect the ferromagnetic properties of the material via electron beam physical vapor deposition (EBPVD). Fe based materials are one of the most promising candidates for high-density recording media to its high saturation magnetization and coercivity. Our main aim for

*Corresponding author: n.muhammad-nadzri13@imperial.ac.uk

this paper is to get the idea, understanding of metal-based thin films generally, and Fe specifically in term of their properties such as roughness, morphologies and structural changes [2,3,4].

2.0 EXPERIMENTAL PROCEDURE

2.1 Thin film deposition

All Fe thin films were deposited by electron beam physical deposition (EBPVD) KES-200. The thin films were grown on silicon substrate with (100) plane. This machine carried out the growth at a base pressure of $<10^{-5}$ mbar and the target to substrate distance is around 50 cm. All depositions were carried out at about 80°C to 100°C with the current of 7.17 A. The deposition rate is about 2.16 nm/s with the average thickness for thin films of 220±10 nm. For the first batch of thin films, the substrate was pre-heated up to 300°C before the deposition takes place while the second batch is following a normal procedure. The deposition rate is not affected by the condition of the substrate (preheat or not). Instead, the deposition rates depend solely on the ion bombardment of the atoms on the substrates.

2.2 Different temperature on the thin films

Thin films with two different condition (pre-heated substrate and normal substrate) are subjected to another annealing condition, which is from 400°C to 1200°C. Higher temperature also used in order to study the difference in term of the surface roughness of the thin films. Not only the thin films are subjected to different deposition exposure (due to substrate temperature effect), the thin films are also believed to have changes its phases and morphology when it is subjected to temperature.

3.0 RESULTS AND DISCUSSION

3.1 Effects of pre-heating substrate temperature on thin films

For the morphology, we believed the effect of temperature after the thin films deposition would be affected for all the thin film samples. The surface morphology of thin films at different substrate heat treatment temperature changes depending on the temperature given. This is in agreement with Jugyoon et al. [2] which the results are shown in Fig. 1. Four thin films that were subjected to different substrate temperature which were room temperature, 150°C, 200°C and 300°C. Based on this research, the influence of pre heating the substrate before deposition can be seen to have an impact on thin films surface. For substrate without any heating influence, fine grain morphology is observed. However, line-fibers arrangement can be seen on the substrate temperature of 200°C and higher.

A similar trend can also be seen in (atomic force microscopy) AFM images of zinc oxide (ZnO) thin films (Fig. 2) conducted by Li et al. [4], where the substrate was preheated to three different temperature, which were 100°C, 200°C and 300°C respectively. The substrate in the room temperature shows a coarse surface with larger grain size. However, as the preheating temperature increased, the grains becomes denser and resulting in smaller grain size, which leads to smoother surface.

Therefore, for our project, similar microstructure on Fe thin films are predicted. As regards to thin film thicknesses, the decrease in thin film is expected when preheating the substrate. This is because the evaporation rate of initial product increased with higher substrate temperature. When this occur, mass transport towards the hot substrates will be reduced hence leading the thinner thin film thickness [1]. However, this is contradict with our preliminary experiment that shows thicker samples for the unheated substrates.

The increased in surface roughness is a results of volatilization of the particles. Enough energy is obtained from the sputtered particles that leads to the formation of better compact in thin films. This is only possible to achieve with the influence of high temperature. Moreover, higher temperature of substrate gives perfect crystals with bigger crystal size due to the enhancement of adatom mobility during the growth process. Thence, giving higher surface roughness with temperature increment. However, issues like inhomogeneity [4] should also be taken into account when preheating the substrate. If the surface of the substrate is not fully exposed to the temperature, it is not possible to obtain a homogeneous thin film. This is because; only area that was exposed to temperature will have a better atoms interaction between the metal target and substrates[5, 6].

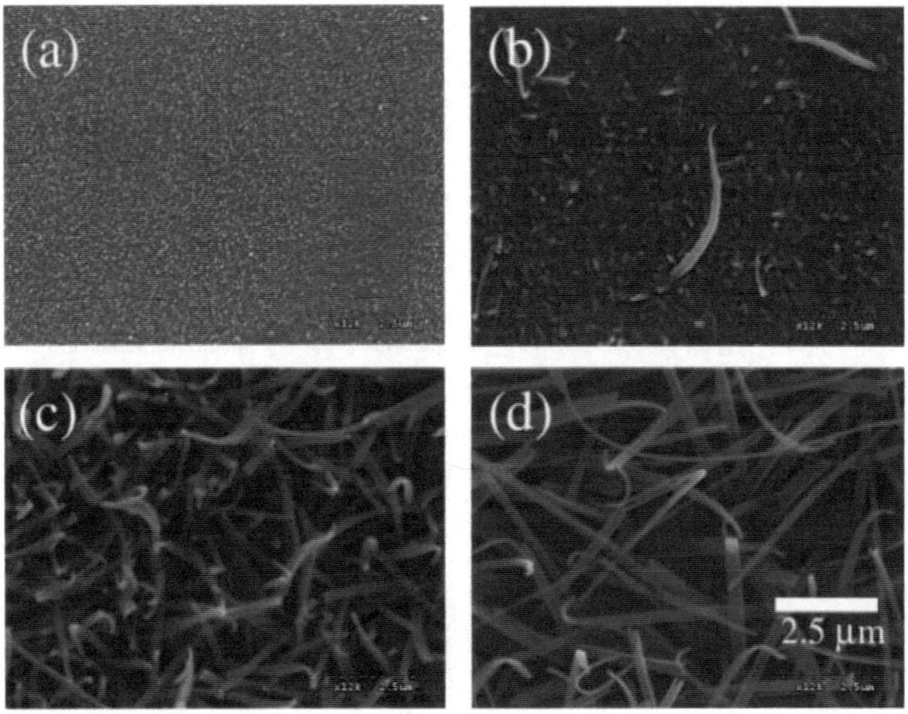

Fig. 1. Adapted from J.E.S Kim et al.[2] where substrate of the Cu-Pc thin films deposited at (a) room temperature; (b) 150°C, (c) 200°C and (d) 300°C.

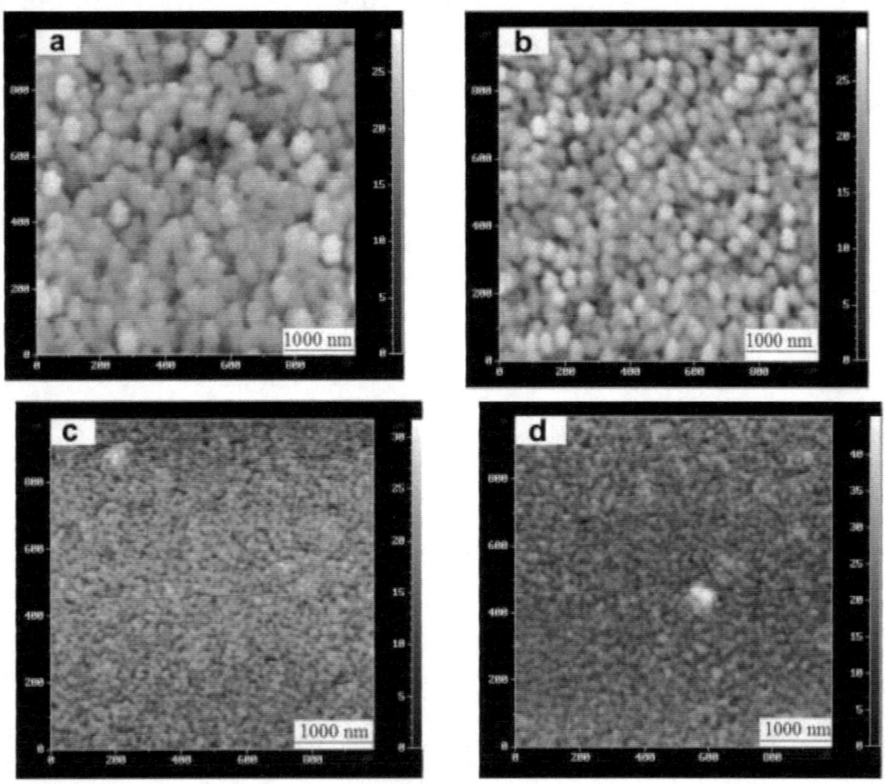

Fig. 2. Adapted from Li et al. AFM images of ZnO thin films deposited at various substrate temperature of (a) room temperature, (b) 100°C, (c) 200°C and (d) 300°C [4].

3.2 Effect of annealing temperature on thin films

In addition, after the thin film deposition, these films are then subjected to different annealing temperature for 1 hour. This method is similar to this project, which used similar temperature given to the pre-heating substrate range of 150°C, 200° and 300°, which is presented in Fig. 3. The heat treatment given to thin film is expected to result in the formation ow metastable and stable phases.

In our case of Fe thin films, the grains and grain boundary is expected to change as higher temperature is given to the films. This is because the diffusion of atoms along the grain boundaries may be happening thus leading to denser grains and flatter surface. This is also supported by Gmter and Myer [7] which stated that with the increase of substrate temperature, mobility in increased that leads to better crystallinity of the film and in agreement that the grain size is also affected where with the higher temperature the shape of particles gave more surface areas to the microstructure compared to lower temperature.

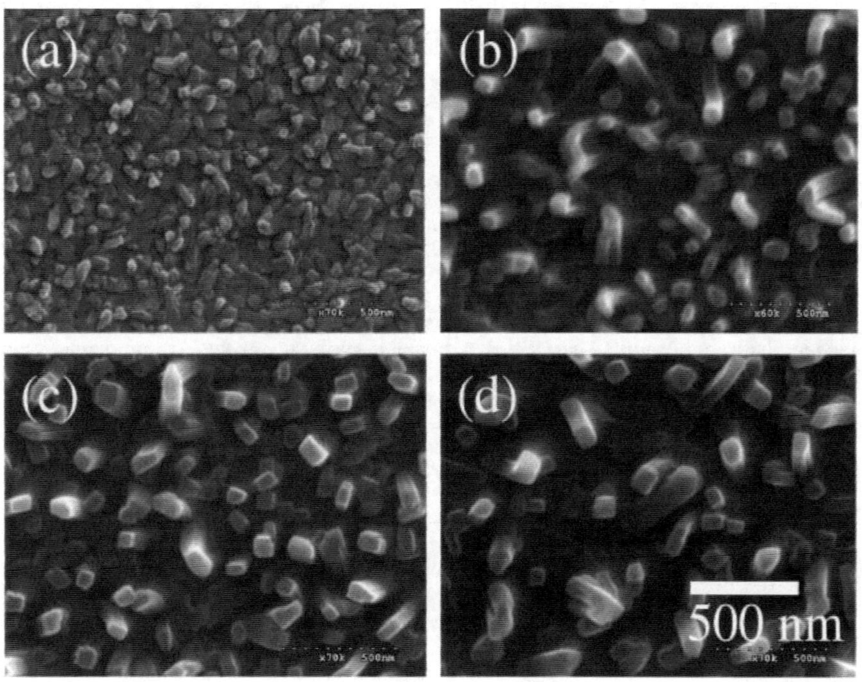

Fig. 3. Adapted from J.E.S Kim et al.[2], where the Cu-Pc thin films are annealed at (a) room temperature; (b) 150°C, (c) 200°C and (d) 300°C.

One of the main issue in Fe thin films is poor surface/interface of the materials which leads to depleted performance in their magnetic and electronic transport properties [8,9]. The higher temperature given to the thin films also contributed to better quality of thin film crystallinity. This is because, increasing temperature will leads to not only larger particle size but also increase in grain size, thus reducing grain boundary. Lower grain boundary will give an insight of better carrier mobility which will greatly influenced the sample electrical properties such as resistivity [10-12].

The phase of the thin films is expected to change once the sample is subjected to different temperature. For Fe thin films, phases like alpha or beta might be present after the heat treatment. Similar results were showed by Chen et al. [13] in their study of effect of thermal annealing of zinc oxide thin films. Based on Fig. 4, the x-ray diffraction (XRD) diffractogram, peak shifts are present as the temperature increased up to 950°C. As the temperature increased, there were two new peaks exist especially at the highest temperature (950°C). These additional peaks and peak shifts in the diffractogram suggested that the sample has undergone structural changes, which will attribute to their mechanical properties.

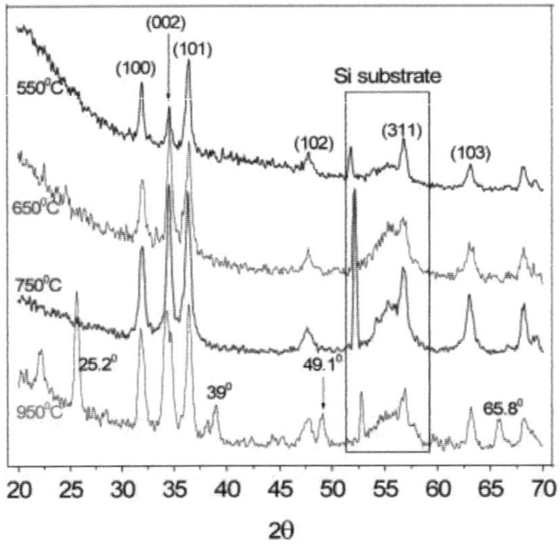

Fig.4. Adapted from Chen et al. [13] X-ray diffraction patterns of the ZnO thin films at different annealing temperature

4.0 CONCLUSION

In summary, the formation of Fe thin film on silicon is expected to experience some mechanical changes as their structural and morphologies changed. There are two reasons contributing to this; (i) the pre-heating of silicon substrate prior to deposition by EBPVD and (ii) the annealed temperature given to both samples with pre-heated substrate and unheated substrate. Results such as better quality of thin film and structural changes can be obtained by exposing these films to higher temperature. The outcomes can be observed by the crystallize size and roughness of the sample which is obtained by XRD and AFM.

REFERENCES

1. P. S. Patil, R. K. Kawar, T. Seth, D. P. Amalnerkar, P.S. Chigare, Ceram Int, **29**, 7, (2003)
2. J.E.S. Kim, E. Lim, K. Lee, D. Cha and B.Friedman, Appl. Surf. Sci, **205**, 4, (2002)
3. X.Y.Li et al., Opt. Commun, **282**, 2, (2009)
4. J. L. Meng, M. Andritschky, M. P Santos, Thin. Sol. Films. **223**, 2, (1993)
5. S.R. Kappaganthu and Y. Sun, Surf. Coatings Technol, **167**, 2, (2003)
6. P. Arunkumar et al., RSV Adv, **4**, 63, (2014)
7. J.M Gmter and J. Mayer, Sci., 167, (1988)
8. H. Masushima, Y. Fukunaka, Y. Ito, A. Bund, W. Plieth, J. Electoanal. Chem 587, **1**, (2006)
9. G. Palasamtzas, J. T. M. De Hosson, J. Barnasa, Sirf. Sci, 507, **5**, (2002)
10. M. Nisha, S. ANusha, A. Antony, R. Manoj, M. K. Jayaraj, Appl. Surf. Sci. 252, **5**, (2005)
11. M. Ohyama, H. Kozuka, T. Yoko, Thin Solid Films 306, 1, (1997)
12.Y.S. Kim, W.P,Tai, S.J. Shu, Thin Solid Films 491, 1, (2005)
13. Y. Chen, N. Jyoti, H. U. Ko, and J. Kim, J. Phys, Chem, Solids, **73**, 11, (2012)

IOP Conf. Series: Materials Science and Engineering **957** (2020) 012044 doi:10.1088/1757-899X/957/1/012044

The Correlation between Palm Shell Char Properties and the Production of Metallic Iron in EAF Steelmaking Slag Reduction Reaction

Anis Nadhirah Ismail[1,*], *Nur Farhana* M. Yunos[2], *M. Sri Raj* Rajeswari[1] and *M. A.* Idris[1]

[1]School of Materials Engineering, Universiti Malaysia Perlis, 02600 Arau, Perlis, Malaysia
[2]Instititе of Engineering Mathematics, Universiti Malaysia Perlis, 02600 Arau, Perlis, Malaysia

Abstract. Palm shells wastes generated from oil palm processing are in abundance in landfills every year thereby posing environmental problems. Enormous amount of wastes generated by agro-industry has previously studied as carbon source in steelmaking hence providing solution to environmental problems. This paper studied on the conversion of palm shell waste into carbon material via physical and chemical activation method for metallic iron extraction. Physical char was prepared by pyrolyzed in nitrogen atmosphere at 450°C while chemical char was impregnated in phosphoric acid before pyrolyzed. Composite pellets of EAF slag (43.18 %Fe_2O_3) with physical and chemical char were rapidly heated at temperature 1550°C within 20 minutes under argon flow. All reduced samples were analyzed on the weight loss, degree of reduction, iron recovery and phase analysis using X-ray diffraction (XRD). The results indicated that chemical/slag showed higher weight loss (38.8%) and excellent degree of reduction (29.94%) compared to physical/slag due to higher volatile matter content (9.8%) and larger surface area (562.14m^2/g). It was found that the production of metallic iron particles after the reduction process and indicated that chemical char achieved higher iron recovery (15.48%) compared to physical char due to higher total carbon content (60.28%). XRD and Rietveld refinement analysis confirmed that the iron phase was a major component in metallic iron particles for physical/slag and chemical/slag samples. This elucidated that the iron oxides in EAF slag was completely reduced into iron by using palm shell chars as carbon materials. This finding indicates that palm shell chars potentially act as carbon materials in steelmaking applications according to their good characteristics.

1 Introduction

Agricultural wastes could be an alternative carbon sources in steelmaking thereby providing the solution to environmental problems. According to Babich et al., agricultural waste needs to be converted into char before being used in steel production such as pyrolysis [1].

* Corresponding author: anis_nadhirah90@yahoo.com

Hidayu et al. reported that the derivation of waste into char intended to enrich the carbon content in the carbon material and develop porosity to enhance iron oxide process [2]. Char that contained high carbon content, great porosity, lower sulphur and good reducing capacity was a potential carbon reductant in the steelmaking industry [3]. High carbon content in carbonaceous material could lead to faster rate of reduction [4]. When the carbon/slag samples are heated at a certain temperature, the gas-solid reduction process occurred where higher surface area increased the gas molecules diffusivity thus promoting the reduction reaction [5].

The preparation of char involved two crucial processes such as carbonization and activation. Physical activation is the simplest method because it involved only a single step compared to chemical activation with two steps and more expensive however containing good properties that may contribute to the carbon/slag reactivity [6].

Palm shells are among the primary agricultural waste in Malaysia which is increasing steadily over the years [7]. Previously, palm shell char produced by physical activation method used as carbon reductant in reducing the iron oxide has been reviewed by Yunos et al., Zaharia et al. and Najmi et al. indicating that palm shell char had shown better performance in term of iron oxide reduction and off gas generation [8-10]. The reaction between iron oxide in steelmaking slag and carbonaceous materials are as follows: the formation of CO and CO_2; the reduction of Fe_2O_3 to FeO and finally to Fe (metallic iron) as expressed in Eqs (1-4) [3].

$$C_{(s)} + Fe_2O_{3(l)} \rightarrow 2FeO_{(l)} + CO_{(g)} \quad (1)$$
$$C_{(s)} + Fe_2O_{3(l)} \rightarrow 4FeO_{(l)} + CO_{2(g)} \quad (2)$$
$$C_{(s)} + Fe_2O_{3(l)} \rightarrow Fe_{(l)} + CO_{(g)} \quad (3)$$
$$C_{(s)} + 2FeO_{(l)} \rightarrow 2Fe_{(l)} + CO_{2(g)} \quad (4)$$

At high temperature reaction, the carbon was expected to be converted into CO and CO_2 as expressed in Eqs (5 and 6) [11]. Since the CO and CO_2 were the primary products of the reactions in Eqs (1-6), therefore the following reactions of Eqs (7) and (8) may also occur where CO has reacted with iron oxide in slags [3, 12].

$$C_{(s)} + \tfrac{1}{2}O_{2(g)} \rightarrow CO_{(g)} \quad (5)$$
$$C_{(s)} + O_{2(g)} \rightarrow CO_{2(g)} \quad (6)$$
$$CO_{(g)} + Fe_2O_3 \rightarrow 2FeO_{(l)} + CO_{2(g)} \quad (7)$$
$$CO(_{g)} + FeO_{(l)} \rightarrow Fe_{(l)} + CO_{2(g)} \quad (8)$$

The palm shell char has beneficial characteristics such as containing high total carbon, lower sulphur content, high volatile matter content and larger specific surface area. Therefore, in the present study, palm shell chars will be prepared by physical and chemical activation method through pyrolysis due to its nature of enriching the carbon content and highly porous with a larger specific surface area that enhances the reduction reaction process [13, 14]. The reduction study was focusing on the potential of the palm shell chars in producing the metallic iron particles from EAF slag.

2 Methodology

2.1 Sample preparations

The palm shell wastes generated from palm oil mill supplied by Felda Palm Industries Sdn Bhd, Pahang were used in this study as carbonaceous material. Prior to the experiment, the palm shells were firstly dried at room temperature and then oven-dried for 24 hours to reduce the moisture content. The palm shell wastes were then crushed into particle size of 2 mm using grinder machine. The carbonaceous material from palm shell wastes was prepared by using physical and chemical activation methods. For physical activation, palm shells was pyrolyzed at 450ºC in electric tube furnace within 2 hours with a heating rate of 10ºC/min in nitrogen atmosphere whereas for chemical activation, the palm shells were impregnated in phosphoric acid (H_3PO_4) first before pyrolyzed at a similar condition as physical activation. The chars from physical and chemical activation then were grinded into fine powder below 63μm using a ring mill to minimize the particle size for optimum binding during mixing process.

2.2 Preparation of composite pellet

The electric arc furnace (EAF) slag provided by Perwaja Steel Sdn. Bhd., Terengganu was used in this study as iron oxide source. The EAF slag was crushed and ground into fine powder for approximately 63 μm. The palm shell chars (physical and chemical char) and EAF slag were weighed and mixed at ratio a 1:3 palm shell chars to EAF slag. The mixture of constituent powder was milled using planetary mill machine at 200 rpm for 1 hour. After mixed, the samples were compacted in a die by applying a load of 78kPa by using uniaxial compact machine to produce cylindrical composite pellets with a dimension of 20 mm in diameter and 5 mm height.

2.3 Reduction reaction of pellet

The reduction of EAF slag was conducted in a horizontal tube furnace at 1550ºC under argon atmosphere with a flow rate of 1L/min (Fig. 1). After the furnace being heated to 1550ºC, the sample was rapidly inserted and heated for the reduction process within 20 minutes. The reacted sample was quenched by rapidly being withdrawn from hot zone into cold zone of the furnace. The metallic iron particles and slag layer was possibly found on the crucible after the reduction process. The metallic iron particles were characterized using X-ray Diffraction (XRD) and Rietveld refinement method to identify and quantify the phase existed in the samples after reduction.

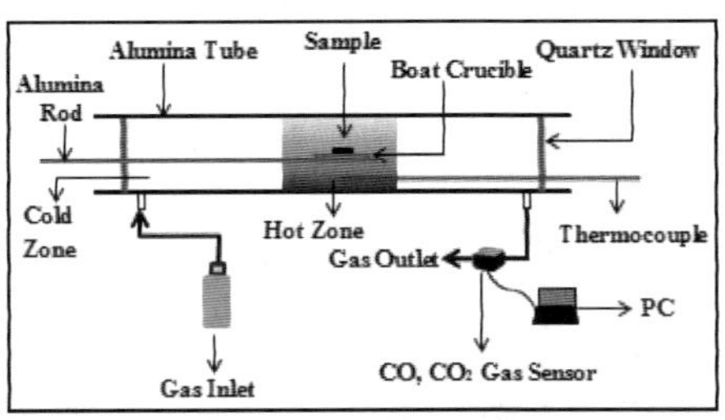

Fig. 1. Experimental arrangement for reduction reaction study.

3 Results and discussion

3.1 Materials characterizations

The ultimate, proximate and specific surface area analyses for raw palm shell and palm shell chars are presented in Table 1. From Table 1, it indicates that the carbon content had risen after the pyrolysis. The specific surface area had also drastically increased after the conversion process from 1.03 to 359.03m^3/g for physical char and to 562.14m^3/g for chemical char. The higher carbon and specific surface area of chemical char compared to physical char was due to the activating agent (H_3PO_4) which enhanced the development of porous structure and enlarging the existing pores [15]. Fig. 2 demonstrates the pore distribution before and after the activation. Micrograph of raw palm shell showed some pores available with hexagonal structure (Fig. 2 (a)). After 2 hours of activation via pyrolysis, the hexagonal structure disappeared had followed by the development of mesopores and micropores (Fig. 2 (b) physical char and (c) chemical char). Compared to physical activation, chemical activation method showed the development of new pores and widening of existing pores leading to a production of higher specific surface area. H_3PO_4 appeared to act as acid catalyst to promote the pyrolytic decomposition and reactant in the formation of cross-linked structure [16].

Table 1. Proximate, ultimate and specific surface area analysis of carbon reductants.

Materials	Proximate analysis (wt%)			Ultimate analysis (wt%)			Specific surface area (m^2/g)
	Fixed carbon	Volatile matter	Ash	C	H	N	
Raw palm shell	80.3	8.3	3.7	45.67	5.86	0.37	1.03
Physical char	81.3	6.6	5.9	59.35	2.98	0.30	359.03
Chemical char	68.5	9.8	12.8	60.28	1.85	0.44	562.14

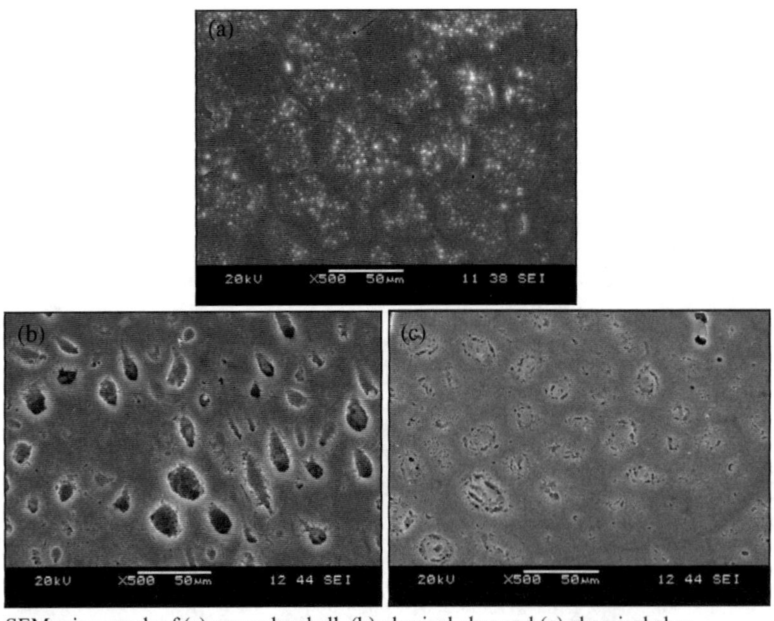

Fig. 2. SEM micrograph of (a) raw palm shell, (b) physical char and (c) chemical char.

The elemental analysis of the EAF slag was carried out by X-Ray fluorescence (XRF) as presented in Table 2. The result indicates that the primary component of EAF slag was Fe_2O_3 accounting to 43.18wt% of the slag.

Table 2. Composition of EAF slag.

Compound	Fe_2O_3	CaO	SiO_2	Al_2O_3	MnO	MgO	TiO_2
(wt%)	43.18	27.67	16.87	5.77	2.55	2.27	0.54

3.2 Weight Loss and Degree of Reduction

Table 3 shows the weight loss and the degree of reduction of the physical/slag and chemical/slag samples after the reduction reaction at temperature 1550°C. The difference in carbonaceous material has influence the weight loss and degree of reduction. The weight loss was calculated by the weight change before and after the reduction while the degree of reduction was obtained from the removal of oxygen content combined with iron in the samples [17, 18]. From the table, the weight loss for physical/slag was 35.2% and for chemical/char was 38.8%. Similar trend was presented by the degree of reduction where the reduction degree of chemical/slag (29.94%) was higher than physical/slag (25.06%). Chemical/slag showed better performance compared to physical/slag due to its good characteristics such as higher in volatile matter (9.8 wt%) and having larger specific surface area (562.14 m^2/g). Volatile matter presence in chemical char encouraged the gasification thus contributed to reducing gases production therefore improving the degree of reduction. According to Zuo et al., high amount of volatiles were released from the carbon reductant especially agricultural wastes participated in enhancing the iron oxides reduction reaction. This is also supported by El-Tawil et al. which reported that the decomposition of volatile matter, CH_4 into H_2, CO and CO_2 at high temperature indicating assisted that reduction reaction of iron oxide in the slag [19]. The larger surface area of chemical char has provided availability for carbon dissolution for more reducing gases production which attributed to higher degree of reduction. Safarian et al. stated that better surface area characteristic of carbonaceous material has contributed to a better reactivity due to higher gas bubble formation (CO, CO_2) during the reduction process [20].

Table 3. Weight loss and degree of reduction of physical/slag and chemical/slag after reduction reaction.

Samples	Weight loss (%)	Degree of reduction (%)
Physical/slag	35.2	25.06
Chemical/slag	38.8	29.94

3.3 Production of metallic iron

Fig. 3 shows the reduced iron after the reduction reaction at 1550°C for 20 minus and the metallic iron particles after being removed from the crucible. The reduced metals of metallic iron have spherical shape with various sizes ranging from 1 – 4 mm elucidated formed by diffusion of small particles and gradually formed bigger particles [11, 21]. The metallic iron particles produced were also weighed and the percentage of iron recovery was calculated according to the ratio of iron transferred to metallic iron particles and the initial amount of iron in the slag before reduction [22]. From Table 4, the utilization of chemical char (15.48%) as carbon materials has produced higher iron recovery from EAF slag compared to physical char (15.38%). As expected, the iron recovery of chemical/slag was slightly higher than physical/slag according to an excellent degree of reaction (Table 3).

According to Najmi et al., high amount of carbon content is desired in order to complete the reduction process. It can be clarified that higher total carbon of chemical char (60.28 wt%) has the tendency to form a bigger lump of metallic iron. It was elucidated that besides volatile matter and specific surface area properties, total carbon content of carbonaceous material can influence the reaction process.

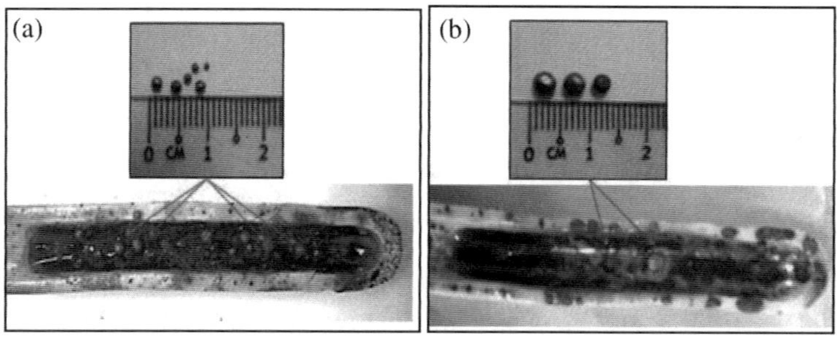

Figure 3. Images of metallic iron particles produced after the reduction reaction at 1550°C for (a) physical/slag and (b) chemical/slag

Table 4. Iron recovery (%) obtained after the reduction of EAF slag by physical and chemical char.

Samples	Metallic iron produced (g)	Iron recovery (%)
Physical/slag	0.303	15.38
Chemical slag	0.305	15.48

3.4 Qualitative and Quantitative analysis of metallic iron produced

The metallic iron particles produced were characterized by XRD to justify the phase existed and the Rietveld refinement method was performed to quantify the phase as shown in Fig. 4 and Table 5. The refined parameter for the reduced samples included background, phase and profile parameter. In terms of Rietveld refinement, the both physical/slag and chemical/slag has R_{wp} value lower than the recommended valued (12), R_{wp} closed to R_{exp} and the goodness of fit (GOF) value lower than 4. Thus the phase identification and quantification performed are acceptable [23, 24]. From the XRD pattern, two sharp and distinctive peaks corresponding to iron (Fe) appear at position 2θ (45.13° and 65.41°) for physical/slag and (44.91° and 65.66°) for chemical/slag (iron; PDF file: 01-087-0722). The weight percent of iron for physical and chemical/slag was found 100wt%. It was indicated that physical and chemical char has successfully reduced the iron oxide from EAF slag into iron suggesting that palm chars can be potentially used as carbon material.

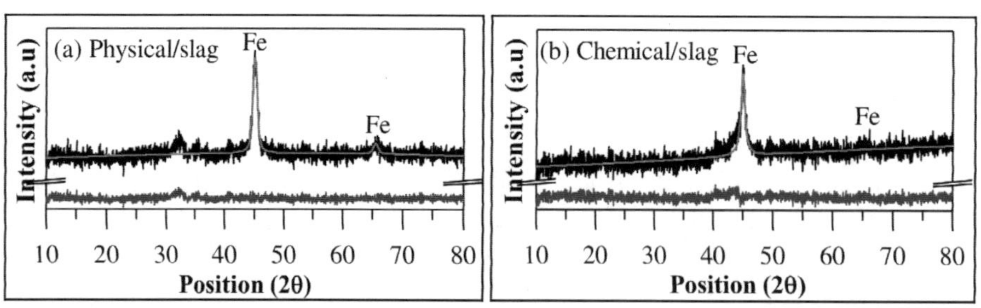

Fig. 4. Observed XRD pattern (black), calculated XRD pattern (red) by Rietveld method and the difference pattern between the observed and calculated pattern (blue) of (a) physical/slag and (b) chemical/slag at temperature 1550°C.

Table 5. Rietveld quantitative phase of physical/slag and chemical/slag reduced samples.

Materials	Iron content (wt%)	R_{wp} (%)	R_{exp} (%)	GOF
Physical/slag	100	2.71	2.98	0.83
Chemical/slag	100	2.30	2.57	0.90

4 Conclusion

The conversion of palm shell waste using physical and chemical activation method has produced different properties of palm char which was used as carbonaceous material for iron oxide reduction from EAF slag at temperature 1550°C. Results revealed that the higher volatile matter and larger specific surface area of chemical char has contributed to the higher weight loss and degree of reduction. The formation of metallic iron particles on the crucible after the reduction process can also be observed. It was indicated that higher total carbon content of chemical char has resulted in a higher percentage of iron recovery from EAF slag. XRD pattern and Rietveld refinement analysis has confirmed that the iron phase is dominant in the metallic iron particles for both physical/slag and chemical/slag samples since there was only iron peak existed on the pattern. It was elucidated that the iron oxide contained in EAF slag was completely reduced by physical and chemical char. Thus, it was suggested that palm shell chars can potentially be used as carbon materials for metallic iron extraction where the palm shell prepared via chemical activation method is more effective.

This research has been funded by Fundamental Research Grant Skim (FRGS) (9003-00366). The financial support for this research was also provided by the University Malaysia Perlis and Ministry of Higher Education, Malaysia.

References

1. A. Babich, D. Senk, M. Fernandez, ISIJ Int. **50**, 8 (2010)
2. A.R. Hidayu, N.F. Muhamad, S. Matali, A.S.A.K. Sharifah, Procedia Eng. **68**, 6 (2013)
3. N.F.M. Yunos, N.H. Najmi, S.R.R. Munusama, M.A. Idris, J. Therm. Anal. Calorim. 138, 9 (2019)
4. J.R. Dankwah, P. Koshy, N.M. Saha-Chaudhry, P. O'Kane, C. Skidmore, D. Knights, V. Sahajwalla, ISIJ Int. **51**, 10 (2011)
5. D. Guo, M. Hu, C. Pu, B. Xiao, Z. Hu, S. Liu, X. Wang, X. Zhu, J. Hydrogen Enenrgy **40**, 8 (2015)
6. V.H. Montoya, J.G. Servin, J.I.B. Lopez, Ins. Tech. Aguascalientes, 18 (2012)
7. S. Zafar, BioEnergy Consult, 11 (2014)
8. N. F.M. Yunos, M. Zaharia, K.R. Ahmad, D. Nath, M. Iwase, V. Sahajwalla, ISIJ Int. **51**, 9 (2011)
9. M. Zaharia, N.F.M. Yunos, V. Sahajwalla, Proc. of ACMSM 2012, 8 (2013)
10. N.H. Najmi, N.F.M. Yunos, N.K. Othman, M.A. Idris, J. Mater. Res. Technol. **8**, 9 (2019)
11. N.H. Najmi, N.F.M. Yunos, N.K. Othman, A.N. Ismail, M.A. Idris, J. Eng. Appl. Sci. **11**, 5 (2016)
12. P.L.W. Jr., R.L. Taylor, J.M. Ranish JM, Carbon **29**, 11 (1991)
13. N.K.E.M. Yahaya, M.F.P.M. Latiff, I. Abustan, M.A. Ahmad, Int. J. Eng. Tech. **10**, 5 (2010)
14. P.B. Elham, The Production of Palm Kernel Shell Charcoal By The Continuous Kiln Method (Universiti Putra Malaysia, 2001)

15. C. Anyika, N.A.M. Asri, Z.A. Majid, A. Yahya, J. Jaafar, Nanotechnol. Environ. Eng. **2**, 25(2017)
16. Y. Li, X. Zhang, R. Yang, G.L.C. Hu, RSC Adv. 11 (2015)
17. P. Mbele, J. South Afr. Inst. Min. Metall. **112**, 8 (2012)
18. A.J.B. Muwanguzi, A.V. Karasev, J.K. Byaruhanga JK, P.G. J'onsson, ISRN Mater. Sci. 9 (2012)
19. A.A. El-Tawil, H.M. Ahmed, A.A. El-Geassy, B. Bjorkman, (2015)
20. J. Safarian, M. Tangstad, 12th Int. Ferroalloys Congr. 12 (2010)
21. M. Ozawa, S. Kitagawa, S. Nakayama, Y. Takesono, Trans. Iron. Steel Inst. Japan. **26**, 8 (1986)
22. J.H. Heo, B.S. Kim, J.H. Park, Met. Mater. Trans. B. **44**, 12 (2013)
23. S.M. Al-Jubouri, B.I. Waisi, S.M. Holmes, J. Eng. Sci. Tech. **13,** 12 (2018)
24. P.T. Teo, A.A. Seman, P. Basu P, N.M. Sharif, Procedia Chem. **19**, 5 (2016)

A Recent blend of thermoplastic polyurethane (TPU)

Nor Azwin Ahad[1,*]

[1]School of Materials Engineering, Kompleks Pusat Pengajian Jejawi 2, Taman Muhibbah, Universiti Malaysia Perlis,02600 Jejawi Perlis

Abstract. Polymer blending is a well-developed manufacturing process with features of efficient production, ease of processing and good controllability. It can effectively modify and adjust the properties of polymer in order to have greater mechanical, thermal and physical properties, and thus has been commonly used to make functional polymer materials. However, the differences in polarity as well as poor interfacial between each polymer in the blend will promote the difference in polarity between the immiscible and incompatible polymers, and then the blends did not succeed as a good blend. The blending of thermoplastic polyurethane (TPU) with other polymer has been widely study with focus on the improvement of mechanical properties, wear resistance, thermal stability, toughness and other related properties. This review only focuses on the latest studies on blending TPU with other polymer, just in the range of 3 years from 2017 to 2019.

1 Introduction

The polymer blend method using commercial existing polymers is generally believed to be the most effective way to develop new with high flexibility and versatility, compared with the synthesis of a new functional polymer, based on the practical and economical viewpoints. The blending of TPU with other polymers has been widely studied with different focus. It can effectively modify and adjust properties of the polymer in order to have greater mechanical, thermal and physical properties, as well as used to make functional materials.

Thermoplastic polyurethane is a thermoplastic elastomer (TPE) that consists of micro-phase separated blocks of amorphous and crystalline segments in the molecules. Pondong [1] reported that, the soft segments of TPU consist of either polyether or polyester long-chain macro molecules. The hard segments are combinations of isocyanate-based aromatic or aliphatic structures with short-chained diols. Typically, the crystallization in the hard segments of TPU acts as physical crosslinking and reinforcement that promotes excellent performance properties, while the soft segments contribute to high flexibility of the TPU. The first commercial thermoplastic polyurethane was manufactured by Bayer-Farbenfabriken and B.F. Goodrich. The blending of TPU with other polymer has been widely study with focus on the improvement of mechanical properties, wear resistance, thermal stability, toughness and other related properties.

This review only focuses on the latest studies on blending TPU with other polymer, just in the range of 3 years which only about TPU with nine polymers.

2 TPU and its blend

2.1 Natural rubber

Previously, the author has started the research on thermoplastic polyurethane and natural rubber (TPU/NR) blends [2], which are investigated with two different mixing temperature; 180°C and 190°C. With the increasing of NR, tensile strength also decreased because this will leading to the formation of a softer blend which exhibits lower tensile strength. Though the blends of TPU/NR are believed to be compatible, further improvement with respect to miscibility can be achieved by incorporating LNR to the blend. The tensile properties of the blend are determined by the homogeneity of mixing; the properties are better with higher homogeneity. The tensile strength decreased with temperature 190°C. In blending two or more polymer, too high a processing temperature also causes the materials (especially the NR) to degrade from oxidation. The higher temperatures are undesirable due to the possibility of oxidation of NR. For the swelling test

*Corresponding author: norazwin@unimap.edu.my

in toluene, the equilibrium swelling degree of TPU and its blend increase with increasing of NR (Fig. 1). Swelling degree reduces with the increase of network because of the difficulty for solvent to diffuse into the gap between rubber molecules.

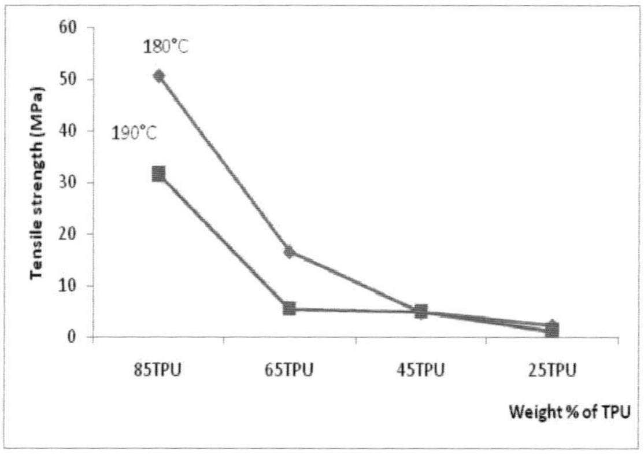

Fig. 1. Variation of tensile strength with weight % of TPU at 180°C and 190°C [2].

2.2 Acrylonitrile-butadienestyrene (ABS)

De Leon [3] was prepared a novel material suitable for additive manufacturing using ABS and. They decided to set printing temperature to 230°C which led to printable object for all material used ABS and TPU. From FTIR result, it can be observed that TPU characteristic peaks increase proportionally with the TPU content in the blends. This is the first indication of good compatibility between ABS and TPU, since cannot identify separated regions of ABS and TPU. The blends also present a homogeneous distribution of ABS and TPU. For the mechanical properties; it can be observed that Young's modulus and yield strength, decreasing gradually when TPU content increased. Elongation at break was decreasing slightly at 10wt% of TPU, but then increasing gradually for higher TPU content. It was shown that blends containing 10-20wt% TPU led to enhanced adhesion between layers without loss in yield strength, while 30wt% TPU led to good adhesion between layers. SEM of the fracture surface has showed the roughness of the surface rises when TPU content is increased (Fig. 2). The roughness seems confirm a better interaction, as well as compatibility of ABS and TPU is high.

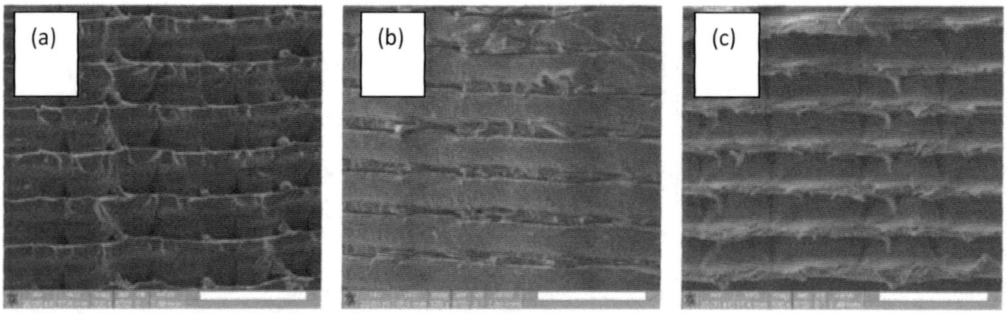

Fig. 2. TPU/ABS blends in various composition (a) 10TPU (b) 20TPU (c) 30TPU [3]

2.3 Poly(ethylene oxide) (PEO)

Solid polymer electrolyte (SPE) membranes composed of TPU, PEO and lithium bis (trifluorimethane sulfonamide) (LiTSI) are prepared and investigated in Tao's work [4]. The preparation step is represented in Fig. 3. With the increase of TPU amount in the blend polymer, the SPE membranes become more homogeneous.

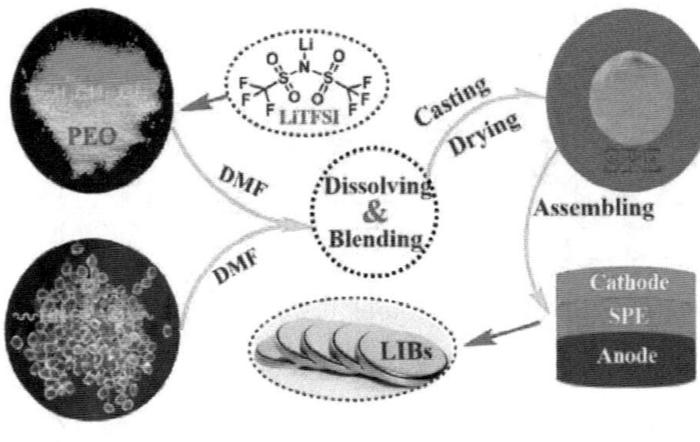

Fig. 3. The preparation of TPU/PEO blends [4]

From FTIR spectra in Fig. 4, it can be easily found the characteristic absorption peak of TPU and PEO. The –CO stretching vibration of the blend SPEs were more pronounced with the increasing of TPU amount, however these stretching vibrations were slightly shifting to higher frequency compared to TPU. This may be attributed to the hydrogen bonding among the hard segments are weakened by the blending of PEO.

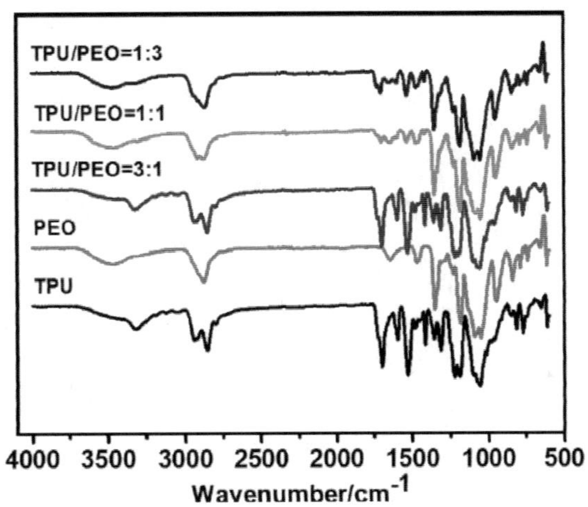

Fig. 4. FTIR spectra of TPU, PEO and TPU/PEO blend film [4]

From the DSC curves, a single T_g showed in all the blend polymer, indicating that TPU and PEO were thermodynamically miscible. The tensile strength and elongation at break were enhanced with the increase of TPU content. The physical crosslinked network could be formed with the addition of TPU. When blending TPU to PEO, the hydrogen bond interaction can be formed by the –NH group in TPU chains and the –C-O-C- group in PEO chain. This interaction weakened the crystallinity and enhanced the mechanical strength of PEO effectively.

2.4 Poly(butylenes terephtalete) PBT

The blends were prepared by melt blending technique. Hao [5] presented, PBT was immiscible with TPU and the crystalinity of PBT decreased with increasing of TPU content, as well as blending with TPU did not modify the crystal of PBT. But, an obvious improvement in toughness of PBT was achieved with addition of TPU.

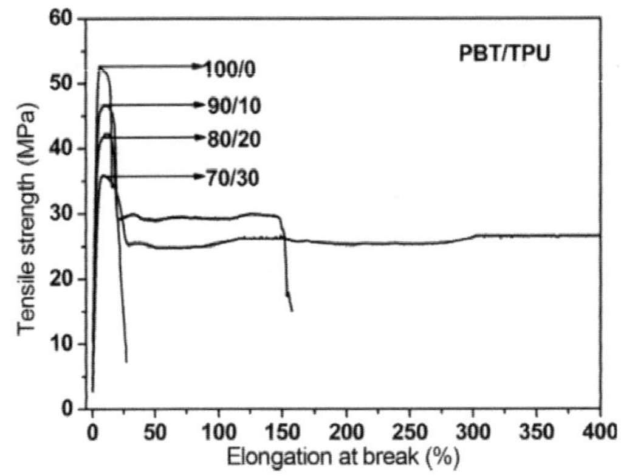

Fig. 5. Tensile strength vs. Elongation at break for TPU/PBT [5]

Neat PBT was very rigid and brittle, with tensile strength around 52.8 MPa (Fig. 5). The blend sample has very drastically increased elongation and the elongation continuously increased with increasing of TPU. On the other hand,the tensile strength and modulus of the PBT/TPU blends decreased with increasing TPU content, as shown in Table 1. This attributed to the lower yield stress and tensile modulus of the TPU elastomer as compared to those of PBT. The presence of TPU elastomer domains in PBT, acting as stress concentrator, resulted in yielding at an overall stress lower than that of neat PBT.

Table 1. Mechanical properties of TPU and TPU/PBT blends [5]

PBT/TPU (w/w)	Young's modulus (MPa)	Tensile strength (MPa)	Elongation at break (%)	Impact strength (kJ/m^2)
100/0	1340±37	52.8±5.6	15±0.8	2.9±0.1
90/10	1227±30	46.7±3.0	27±1.5	4.4±0.4
80/20	1066±19	42.3±2.6	159±5.1	10.9±1.2
70/30	811±11	36.9±1.9	410±10.6	62.9±6.1

2.5 Polymethylmethacrylate (PMMA)

An increasing content of TPU in combination with PMMA leads to decreasing tensile strength and tensile modulus. These phenomena have been observed by Seeger [6]. The blends with 10 wt% TPU shows a four time higher elongation at break and the blends with 20 wt% TPU shows approximately eight times elongation at break compared to the neat PMMA as well as 5 wt% TPU blends. We can conclude that, bending PMMA and TPU resulted in an enhanced impact strength and elongation at break. This blend also changed the behavior, from brittle towards ductile break up.

Blending PMMA with TPU resulted in an enhanced impact strength and elasticity, although decreased the tensile properties. The enhancement on elasticity prevented it from cracking and from material removal. Unfortunately, adding more TPU led to a changed in opacity making it impossible to compare the scratch visibility.

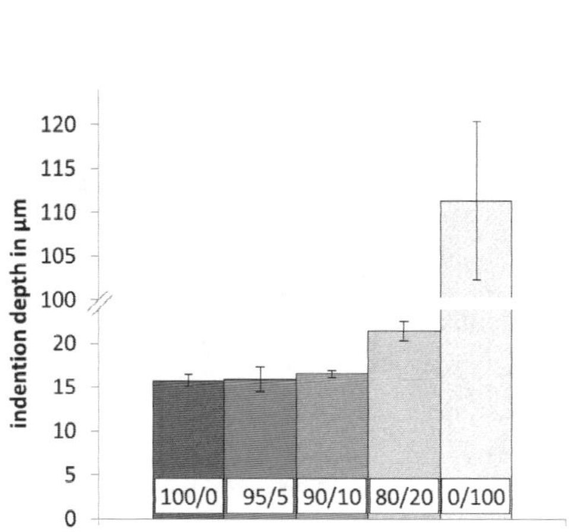

Fig. 6. The scratch behavior of the tested blend indentation depth [6]

The scratch behavior of the tested blend indentation depth of the scratch tip was recorded as in Fig. 6.Since the PMMA/TPU blend is a two phase system, the content of TPU plays an important role for the elasticity of the whole blend. With an increasing of the surface, an increasing of indention depth can be measured.

2.6 Polylactic acid (PLA)

The TPU/PLA blends which have reported by Liu [7] could possess unique properties, including excellent shape memory performance. The viscosity properties of PLA/TPU blends were enhanced as the TPU content increased (range are between 5% to 20%), due to the interfacial interaction between PLA phase and TPU phase were strengthen with the hydrogen bond effect. The binary TPU/PLA show two transition peaks (refer to Fig. 7),with little shifts towards each other demonstrate the immiscibility between TPU and PLA.

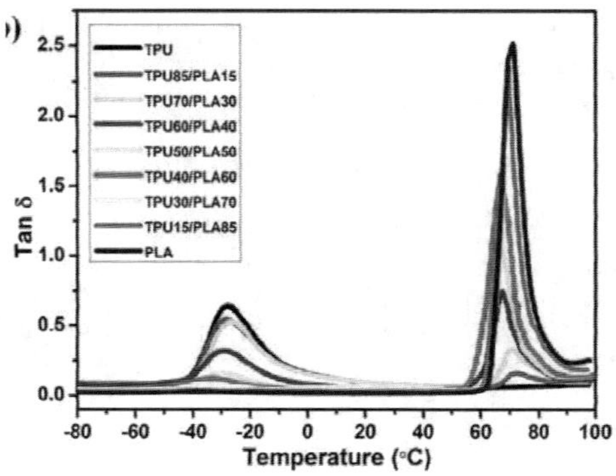

Fig. 7. tan δ versus temperature for TPU/PLA blends [7]

Consequently, TPU/PLA blends exhibited typical phase-separated morphologies controlled by the content of PLA. In the wake of the increment content of PLA, the phase morphology of PLA represented the following variation trend. Firstly; a droplet-like dispersion phase, followed by a continuous phase and finally, the matrix. Qu [8] also suggesting that PLA/TPU blends had much elasticity with increasing TPU content. But the compatibility between PLA and TPU became poor gradually. The phase morphology could provide useful information about the blends. The sea-island structure appeared with the introduction of TPU into PLA matrix. Also the roughness increased evidently, compared with in pure PLA. TPU dispersion phase was spherically dispersed in PLA matrix.

Others, the biodegradable pure PLA were exhibited degraded around 20 days. However it degrades almost 100% after 70 days. Jaso [9] reported that, pure TPU takes early and rapid degradation than PLA. By looking for the rate of degradation, the concentration of PLA in TPU/PLA blends are not depending on this behavior, but the degradation rate are rapidly after 20 days. PLA at higher percentage above 50% showed higher rate. After 60 days, PLA at 80% presented the highest degradation rate among other, which is 20%, 40% 50% and 60%, as refer to Fig. 8.

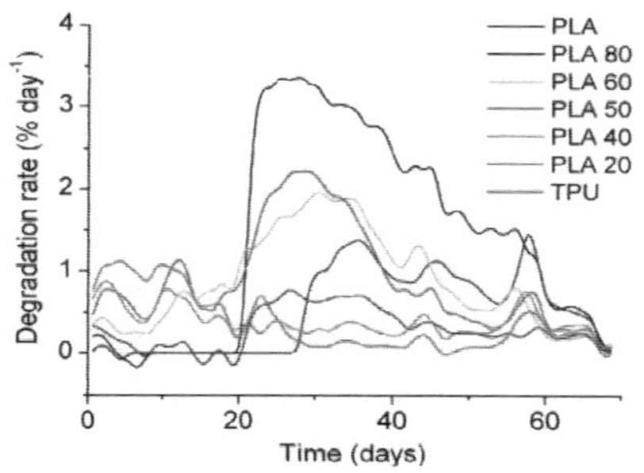

Fig. 8. Rate of degradation for PLA/TPU blends [18]

2.7 Polytetrafluoroethylene (PTFE)

An approach to produce PTFE in TPU has been developed by Huang [10].PTFE is a white powder and under SEM examination, the shape of the PTFE particles was oval or rounded with the individual particles could be separated and had different diameter. The virgin PTFE powder did not have fibril morphology. The TPU/PTFE blends showed the amount of fibrils increased with PTFE concentration. Many of fibrils were entangled and intertwined, as can be seen in Fig. 9.

The formation of the fibrilled PTFE during melt compounding resulted from the low yield strength of PTFE at elevated temperature and plastic deformation. For the mechanical, all TPU/PTFE blends exhibited extremely high tensile strains. Also, because of PTFEs high modulus compared with that of neat TPU thus indicated that PTFE had reinforcing effect.

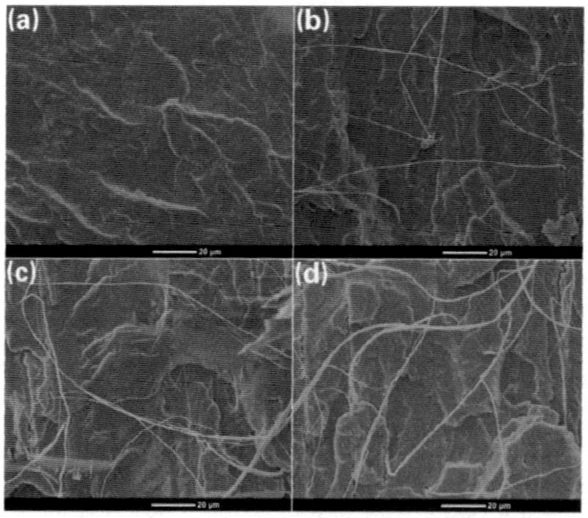

Fig. 9. SEM micrographs of (a) pure TPU (b) – (d) PTFE fibrils for TPU/PTFE with 1%, 3%, 5% of PTFE [10]

The rheological properties of the neat TPU showed liquid-like melt and a typical homopolymer-like terminal behavior at low frequency range. On the other hand, the blends presented a different rheological behavior. The liquid-like changed to solid-like. Huang [10] also investigate the cross section of the blends in foam form. They were found that neat TPU foam exhibited large, elliptical pores, which showed in Fig. 10. The foam exhibited better pore and more uniform structures. The pore size decreased and the density increased after PTFE fibrillation.

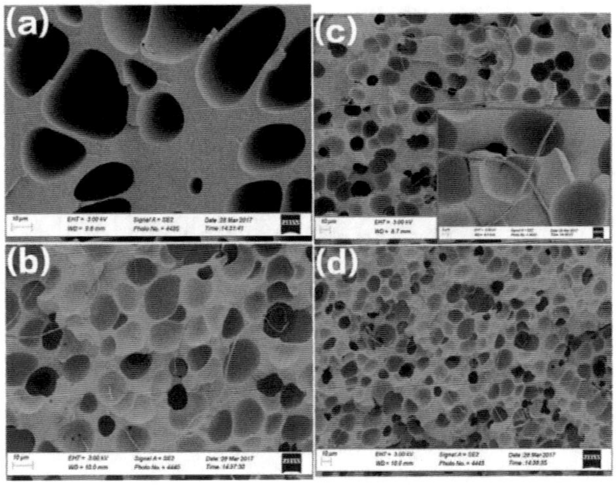

Fig. 10. SEM image of cross section for (a) TPU (foam form) (b) – (d) TPU/PTFE (foam form) blends with 1%, 3 % and 5% PTFE [10].

The presence of PTFE fibrils dramatically improved the foamed structure, resulting in pore density in comparison to neat TPU. The fibrillar of PTFE dramatically enhanced the hydrophobicity, by investigate the contact angle. These behaviors are necessary as competitive materials for shoe sole, food packaging and also automotive interiors [10].

2.8 Polypropylene (PP)

For the study of Ting & research team [11], they broadened their focus on previous PP to impact-resistant PP with TPU. However, the differences in polarity as well as poor interfacial compatibility between these modified PP and TPU provide PP/TPU blends with less comparable functionalities to that of pure PP or TPU. Both are immiscible and incompatible. 30% and 40% of TPU also intrigue the apparent presence of voids, which is evidence that the poor adhesion triggers TPU granule to detach from the PP matrix. The modified PP has improvement in impact resistance and reduced arrangement of crystallinity. The crystallization temperature of the blends is not dependent on the content of TPU, because TPU is non-crystalline polymer. PP has relatively higher thermal stability and helps postpone decomposition of urethane bonds of TPU hard segments and the high crystallization of PPs molecular chains have damaged the molecular chain of TPU's soft segments. Although PP and TPU are incompatible, PP still distinctively improves the thermal stability of TPU. The thermal degradation data showed in Table 2 below, describe the three stage of degradation temperature for pure TPU and PP, also the blends from TPU10, TPU20, TPU30 and TPU40 [11].

Table 2. The 3 stages degradation of TPU/PP [11]

	1st degradation temperature (°C)	**2nd degradation temperature (°C)**	**3rd degradation temperature (°C)**
Pure TPU	302.59	395.29	-
Pure PP	-	-	441.09
TPU10/PP90	338.77	-	456.94
TPU20/PP80	314.28	352.11	459.62
TPU30/PP70	321.08	365.59	467.06
TPU40/PP20	329.05	377.45	470.56

2.9 Polycarbonate (PC)

Surface hydrophilicity and roughness of TPU/PC membranes have been investigate [12]. With addition of TPU content, the hydrophilicity decreased and membranes roughness decreased. SEM image showed the TPU/PC membranes exhibited the fully developed asymmetric membranes including of a dense skin layer, a porous sub-layer and macro-pores with typical sponge-like structure at the bottom. With increasing of TPU content, the sponge-like structure gradually increased at porous sub-layer of asymmetric structure as shown in Fig. 11 below.

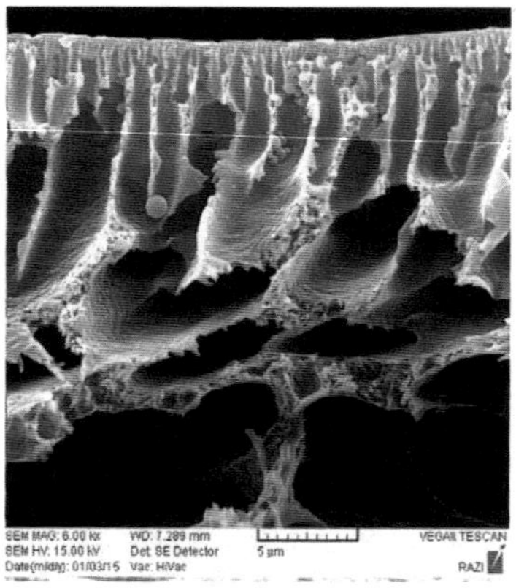

Fig.11. SEM image of cross section for TPU/PC blend membrane [12]

TPU mechanism of pore formation differs since the TPU is a kind of hydrophobic polymer. The more TPU content, the more nascent pore formed in the sub layer. In general, hydrophobic polymer developed sponge-like structure through delayed phase separation. Lastly, the TPU blended in PC matrix improved the pure water flux to be used as membranes in wastewater treatment.

3.0 Conclusions

The blending of various polymers with TPU will provide more advantages in diverse aspects, especially in the desired property modification. The preferred blend properties will meet the requirement of some application such as biomaterial, medical, food packaging, furniture, shape memory foam, transportation, water treatment or others. However, the compatibility of certain polymers with TPU is very important to assess. Incompatible, immiscible and poor interface between them will cause the blends to have diminished in mechanical properties. The blending method also plays an important role. The mixing temperature very important to produce the best blends with good properties.

The authors would like to thank the Ministry of Higher Education Malaysia (MOHE) for the research fund under the Fundamental research Grant Scheme (FRGS) of the Project Code: FRGS/1/2018/TK05/UNIMAP/02/3).

References

1. Pondong W, Kummerlowe C, Vennemann N, Thitithammawong A and Nakason C, Polym. Test. **53** 245-56 (2016)
2. Ahad NA, Ahmad SH and Zain NM, J. Adv. Mater. Res. **879** 107-11, (2014)
3. de Leon AS, Dominguez-Calv A, MolinaSI Mater. Des. **12**1-11 (2019)
4. Tao C, Gao M, Yin B, Li B, Huang Y, Xu G and Bao J Electrochimica Acta **257** 31-9 (2017)
5. Hao Y, Yang H, Zhang H and Mo Z, Fiber Polym 9 **(1)**1-10 (2018)
6. Seegar P, Ratfisch R, Moneke M and Burkhart T Wear **406-407** 68-74 (2018)
7. Liu T, Huang R, Qi X, Dong P and Fu Q Polymer **114** 28-35 (2017)
8. Qu Z, Yin D, Zhou H, Wang X and Zhao S Eur. Polym. J. **116** 291-301 (2019)
9. Jaso V, Glenn G, Klamczynski A and Petrovic ZS Polymer Testing **47** 1-3 (2015)
10. Huang A, Peng X & Turng L Polymer **134** 263-274 (2018)
11. Ting AL, Limin B, Lin M, Lin J, Lou C and Lin JH, J Mater Res Technol **8 (4)** 3389-98 (2019)
12. Saeed N, Mousavi SA and Bastani D, J Water Process Eng **16** 170-182 (2017)

Assessment of Mechanical Properties of Recycled HDPE and LDPE Plastic Wastes

Bibi Intan Suraya Murat[1,*], *Muhammad Syarifuddin* Kamalruzaman[1], *Mohamad Hafiz* Nor Azman[1], and *Muhamad Fazlee* Misroh[1]

[1]Faculty of Mechanical Engineering, Universiti Teknologi MARA, 40450 Shah Alam, Selangor, Malaysia.

Abstract. Most plastic materials can be recycled, and turning them into some useful products can be one of the options in reducing environmental pollutions. In this project, High Density Polyethylene (HDPE) and Low Density Polyethylene (LDPE) were used, shredded into smaller pieces and were hot-pressed to become flat plates. Both materials were then mixed with industrial waste sawdust with an aim to investigate their material properties and their suitability to be used in domestic applications such as bricks, pavement and roof tiles. Three destructive and one non-destructive methods have been conducted: tensile, impact, water absorption and ultrasonic testing. It was found that the tensile properties and impact resistance of the recycled HDPE are better than the recycled LDPE, but 10% lower than their virgin polymers. Higher stiffness and lower impact resistance were observed from specimens with additional sawdust filler in comparison to their original polymers. From the ultrasonic measurement, higher wave velocity and lower attenuation in HDPE specimens were observed, compared to the LDPE specimens, attributed by a higher modulus and density of the HDPE specimens. The ultrasonic velocity, which reflects the elastic modulus, and the attenuation that reflect the microstructure inhomogeneity are respectively decreased and increased when more fillers were added to the original specimens. This study reveals that the mechanical properties of the recycled HDPE and LDPE are closer to their virgin materials, and they have huge potential for plastic recycling.

1 Introduction

In 1950 the world produced plastics only 2 million tonnes per year, but then, the annual production has increased to nearly 8 billion tonnes to date [1]. Very little plastic products are recycled or incinerated in waste-to-energy facilities. In 2015, it was estimated that 55% of global plastic waste was discarded, 25% was incinerated, and 20% recycled [2]. Most of them end up in landfills that may take up to hundreds of years to decompose, piling up the land and finally poluting the soil and environment. In one study, it has been estimated that 1/3 of all plastic wastes end up in soils or freshwater, which later disintegrate and break down

*Corresponding author: bb.intan.suraya@gmail.com

further into nano plastic particles that finally could enter the human food chain [3]. Hence, this plastic waste is such an important issue to be solved quickly.

Apart from cleaning up our oceans or landfills, which is a very significant step, the best way to address this huge plastic pollution is by recycling the plastic wastes [4]. Plastic recycling is the process of collecting plastic wastes and reprocessing the material into useful products. Before reprocessing them, most plastics are sorted according to their resin identification codes. There are represented by Unicode icons and numbered with (1) Polyethylene Terephthalate (PET), (2) High-Density Polyethylene, (3) Polyvinyl Chloride (PVC), (4) Low-Density Polyethylene (LDPE), (5) Polypropylene (PP), (6) Polystyrene (PS), and (7) other plastic types [5]. These resin codes indicate the type of material of each plastic and each of them has a unique material characteristic. Some of them are reusable, some are hazardous after several uses, and some need a more sophisticated recyling process. By understanding these seven codes, it will be easier to choose plastics and to know which plastics to be recycled. Due to their differences in material properties and characteristics, it is recommended to use a singular type of plastic in a production of a new product using recycled materials.

In this project, the HDPE and LDPE thermoplastic wastes were used as they are inexpensive, lightweight and durable. HDPE is a very hard-wearing and does not easily break down under exposure to sunlight or extremes heating or freezing [7]. LDPE is normally used as plastic bags, is considered less toxic than other plastics, but it is not commonly recycled [8]. These two plastics are the easiest plastics to recycle. Their unique properties make them suitable to be moulded into a variety of products for many applications, even after they have been recycled into a new form [6]. One research has demonstrated that for the first time that the plastic molding and shredding process have no property-altering effects on the material over the entire period of reuse [7]. In principle, this result means that this plastic can be reused for a long time without diminishing its properties.

Building from this information, the aim of this study is to investigate the material properties of the newly formed HDPE and LDPE from waste bottles and their suitability to be used in domestic applications such as pavements, bricks or roof tiles. A simple processing route will be used, and explained in the following section.

2 Methodology

2.1 Sample preparation

In this study, two types of plastic wastes were used as the specimens: (1) HDPE, collected from recycled detergent bottles, and (2) LDPE, collected from recycled black plastic bags. These trashes were cleaned and washed before cut into small pieces. Then, they were fed into a shredder machine to obtain the plastic pellets. By using a hot-pressed machine, these plastic pellets were placed in a highly tight square-shaped mold to be pressed to a plate. In a separate study, the HDPE and LDPE pellets were individually mixed with wood sawdust as the filler for the plastic plates. They were mixed at a ratio of 80:20 and 70:30. The HDPE/sawdust and LDPE/sawdust mixture were left to bond under hot press machine at 180°C and 160°C respectively. The pressing duration is 1 hour and 2 hours for the HDPE and LDPE specimens respectively. These four types of specimens are shown in Fig. 1. Then, all specimens were cut into a desired shaped for various material testing. The mass of all specimens was measured for a density calculation.

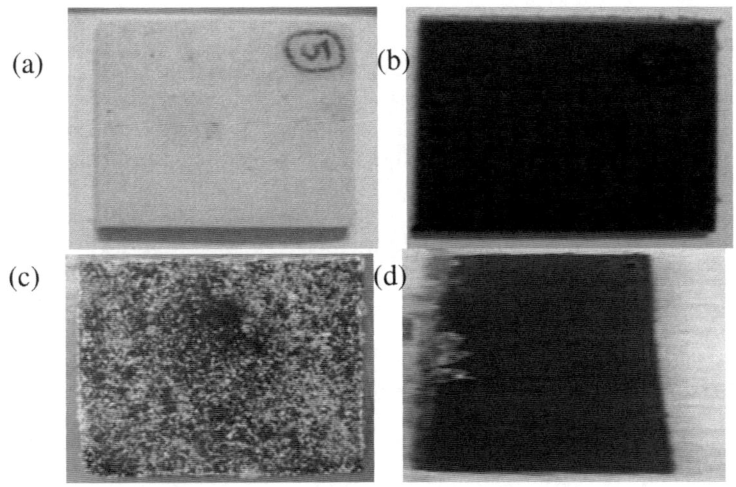

Fig 1. Specimens made of plastic waste materials: (a) HDPE, (b) LDPE, (c) HDPE/sawdust blend, and (d) LDPE/sawdust blend.

2.2 Tensile test

To conduct the tensile test, the specimens were made according to ASTM 3039 standard, with the dimensions of 200 × 25 × 150 mm. The tensile test was conducted at room temperature with test speed of 5 mm/min using an INSTRON 3382 universal test machine. The specimen was clamped on the testing machine with 50 mm length of the specimen both sides. The mechanical properties such as tensile modulus, the ultimate tensile strength and the total elongation of the specimens were taken. The average values were calculated from five runs for each type of specimen.

2.3 Impact test

This test measures the impact energy, or the energy absorbed prior to fracture, using an Izod Impact testing machine. V-notched specimens were prepared according to the ASTM D6110 standard, with a required length of 65 mm mm and width of 13 mm. The notch has a depth of 2 mm, angle of 45°, and a root radius of 0.25 mm. The radius of curvature at the notched specimen is to serve as a stress concentrator. The weight of pendulum is 0.452 kg. Five specimens were used to determine the average impact resistance for the material sample.

2.4 Water absorption test

Water absorption is expressed as an increase in weight gain percentage of a plastic specimen under the following procedure. Referring to the ASTM D570 standard, the specimens were dried in an oven for 24-hour at 105°C. The specimens were immediately weighed after they have been placed to cool in a desiccator. Then, the specimens were soaked in water for 24-hour at room temperature, and the weight of the patted-dried soaked specimens (soaked HDPE and soaked LDPE) were recorded for the calculation of water absorption, as per equation below:

$$\textit{Water absorption \%} = (\textit{wet weight} - \textit{dry weight}) / \textit{dry weight} \times 100 \quad (1)$$

2.5 Ultrasonic measurement

A pulse-echo method was used to measure the properties of longitudinal ultrasonic wave in all specimens. A transducer of 5 MHz center frequency was used, connected to a portable ultrasonic machine GE Krautkramer USM 36. Basic signal parameters such as wave speed and attenuation were recorded. The reflected signals acquired from the measurement setup correspond to the ultrasonic wave velocity that travel twice the thickness of the specimen. The attenuation can be measured by using the reduction in amplitude peaks. Changes in wave speed and energy losses from the interactions with material microstructure are the two key factors in material characterization using non-destructive ultrasonic testing [9]. Both velocity and attenuation can be calculated using Eq. 2 and Eq.3 as below.

$$\textit{wave velocity} = 2(\textit{travelled distance}) / \textit{time} \quad (2)$$

$$\textit{wave attenuation} = (1/ [2(\textit{travelled distance})]) \times \log_{10}(\textit{Peak Amp 1} / \textit{Peak Amp 2}) \quad (3)$$

3 Results and Discussions

3.1 Tensile properties

The stress-strain curves and the bar charts of tensile properties are presented in Fig. 2. From Fig. 2a, the LDPE and HDPE without additional filler shows that there are more ductile and taking a longer time to break than the LDPE- and HDPE- sawdust blend. The HDPE with sawdust shows the most brittleness behavior as they were having less than 5% total elongation during the tensile test (Fig. 2b). As similarly reported in [10], lower elongation at break values was observed with an increase in tensile modulus, which can be seen in HDPE specimens. The effect of the additional filler on the tensile properties can be seen in (Fig 2c and 2d). The elastic moduli of both materials are seen to be increased when they were blended with the sawdust filler, which is in line with the known fact that a stiffer filler particle could enhance the elastic modulus. It is also found that the moduli of both HDPE and LDPE specimens were reduced when they were soaked in water for 24-hrs. The reduction might be due to more water absorption on the interface area between sawdust and polymer, which later caused the reduction in the tensile properties of the composite blend. It is also observed that the tensile properties of all HDPE specimens are higher than the LDPE specimens. Comparing to the literature study [8, 9], there are about 10% differences in the elastic modulus between the recycled products and their virgin materials. Although there is a reduction in the property, the potential of turning these plastic wastes into useful products within the new material property limit is still huge. Overall, the results indicate that recycled HDPE has higher strength, better resistance to deformation, but quite a brittle material when blended with sawdust. In contrast to LDPE, it has a larger elongation rate but weaker in strength compared to HDPE.

3.2 Impact properties

Fig. 3a shows the absorbed impact energy. It can be observed that the HDPE specimens absorbed about 10% more energy compared to the LDPE specimens. This corresponds to the lower elastic modulus (0.38 GPa) and lower density (0.94 g/cm^3) of the LDPE specimen than the HDPE (0.45 GPa and 0.97 g/cm^3 respectively). For the HDPE with a higher modulus (0.45 GPa) and higher density (0.97 g/cm^3), it is expected that the high weight and narrow molecular distribution generally improves the impact resistance. The HDPE- LDPE- without any additional fillers captured a higher amount of energy before failure compared to the

HDPE-LDPE composite blend. It is obvious that the impact resistance is reduced with the increasing of the sawdust filler in both HDPE and LDPE composite blend. This suggests that the sawdust filler is not beneficial to the construction of HDPE- and LDPE- composite blend in term of their impact resistance. It should be noted here that when the recycled specimens were soaked in the water, their impact resistance has shown to be greatly reduced, approximately around 50%. This impact test result confirms the mechanical behaviour of all specimens obtained from the tensile test (Fig. 2a), where it indicates a reduced strain rate and displays more brittle behaviour of the composite blend.

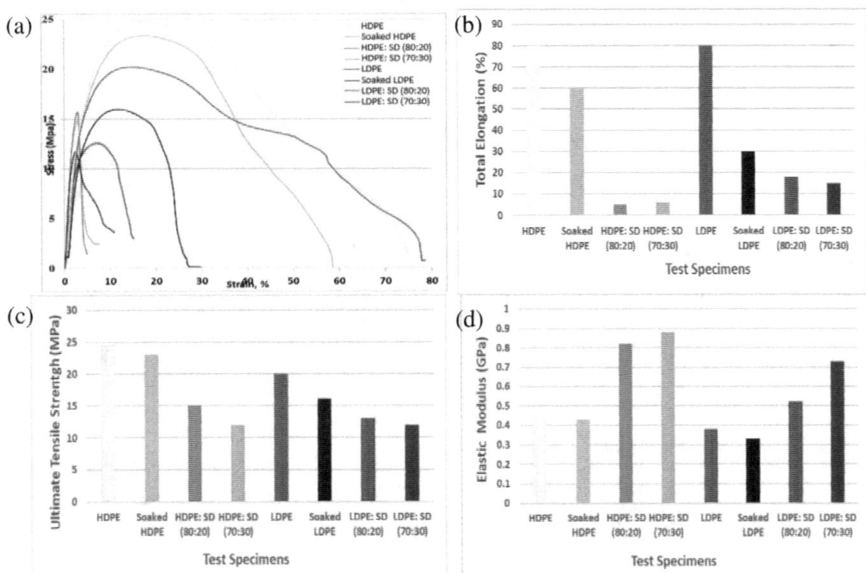

Fig 2. Experimental tensile properties of all specimens: (a) Stress-strain curves, (b) Total elongation (b) Elastic Modulus, (c) Ultimate tensile strength.

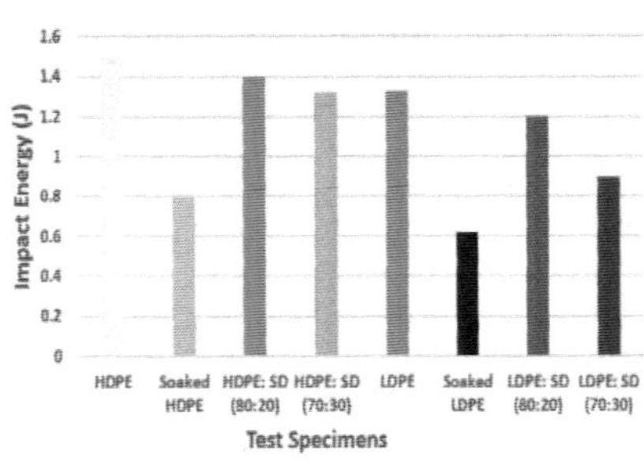

Fig 3. Absorbed impact energy from Izod impact test.

3.3 Water absorption test

From the result, the recycled LDPE specimens absorbed more water compared to HDPE specimens. This could be due to the microstructure of the LDPE, which has more voids and pores [11] that result in water filling the gaps more effectively. Both specimens seem to

IOP Conf. Series: Materials Science and Engineering **957** (2020) 012046 doi:10.1088/1757-899X/957/1/012046

absorb water, but the percentages of the absorption are very small, 0.07% and 0.08% for HDPE and LDPE respectively. Meanwhile, in the composite specimens with the reinforced sawdust, the higher composition of sawdust filler shows higher water absorption capacity. This may be due to the poor mixture between sawdust fibers and the polymer matrix. Furthermore, it is also known that the sawdust particle has a higher ability to absorb water than that the polymer. From a visual inspection, the voids between the fibers and polymers can be clearly seen. If these composites are to be used as roof tiles or bricks, this characteristic is unwanted as they are not waterproof. In addition to that, the composite blend specimens are also floating on the water due to a lighter density compared to their original polymer specimens.

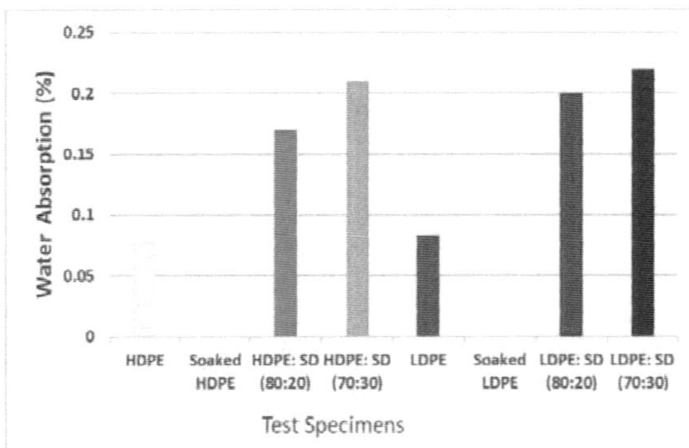

Fig 4. Water absorption capacity of all specimens.

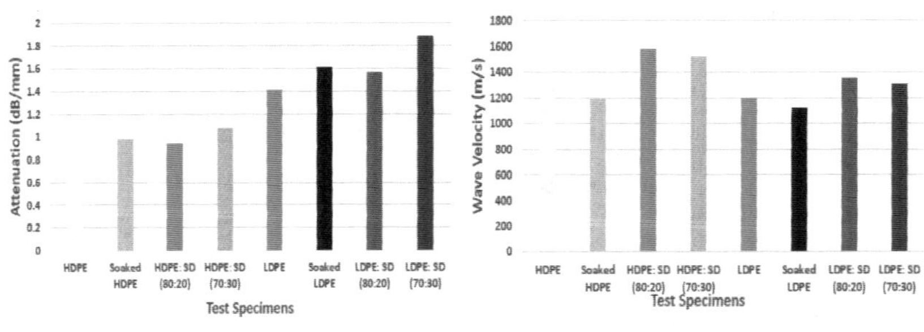

Fig 5. Results from ultrasonic measurement: (a) Wave attenuation, and (b) wave velocity.

3.4 Ultrasonic wave parameters

Two bulk wave parameters, velocity and attenuation coefficient were estimated through pulse-echo measurements. From Fig. 4, the wave attenuation in HDPE specimens are higher compared to the LDPE. Both soaked specimens have higher attenuation than the non-soaked specimens. It can also be seen that adding more filler resulting in higher wave attenuation in both HDPE and LDPE specimens. This attenuation measurement provides valuable information such as material damping property that is directly correlated to the microstructure of the specimens. Higher attenuation means higher inhomogeneity in the grain size distribution, which caused the ultrasonic waves to be absorbed faster and travel slower than homogeneous specimens. This demonstrates that the composites might be good to be used as a soundproofing material or as acoustic absorbers. Meanwhile, the wave velocities of all specimens show the opposite trends. The velocity in LDPE specimens are lower than in the HDPE. Both soaked specimens have lower velocity compared to the non-soaked specimens. The velocity dropped when more filler was added in both types of specimens, due to the increase in the density. Although the ultrasonic velocity is directly proportional to the elastic modulus, this velocity result is not depending on the elastic modulus (Fig 2b). The additional filler has caused the increment in the density of the specimens, which caused to a lower ultrasonic velocity. However, this non-destructive testing has shown its ability to characterize different types of materials because the ultrasonic wave parameters are shown to be a function of the properties of the material.

4 Summary

In general, the tensile properties, impact and water resistance of recycled HDPE specimens are greater than the recycled LDPE. It was found that the sawdust filler in both types of plastic wastes trigger an increase of elastic modulus and lessening the impact resistance. Using the ultrasonic non-destructive method, the higher wave velocity and lower attenuation were obtained in HDPE specimens. This ultrasonic measurement on the specimens without the additional filler can be correlated to the tensile and impact properties: a higher velocity reflects the higher elastic modulus, and a lower attenuation coefficient corresponds to better impact resistance. As a summary, the characterization of the recycled HDPE and LDPE specimens conducted in this study has provided critical insight in finding their recyclability potential and their suitability to be used in domestic applications such as pavements, bricks, wall partition or roof tiles. Consequently, by recycling plastic wastes, water, land, and air pollution can be reduced, which can bring benefits to everyone for a greener and sustainable future.

The authors gratefully acknowledges the support in funding by Faculty of Mechanical Engineering, Universiti Teknologi Mara (UiTM), Shah Alam, Malaysia.

References

1. J. Payne, et al., A circular approach to plastic waste, Polym. Degrad. Stab., **165**, 170-178 (2019)
2. W. L. Filho et al., J. Clean. Prod., **214**, 550–558 (2019)
3. M. Compa et al., Sci. Total Environ., **678**, 188–196 (2019)
4. Joachim Maris et al., Polym. Degrad. Stab., **147**, 245-266 (2018)
5. N. Singh et al., Compos. B. Eng., **115**, 409-422 (2017)
6. G. Faraca, and T. Astrup, Waste Manage., **95**, 388-398 (2019)
7. Retrieved at https://www.plasticstoday.com/packaging/scientific-tests-prove-hdpe-can-be-recycled-least-10-times/68825031358152
8. M. Alzerreca et al., Polym. Test., **46**, 1-8 (2015)
9. M Yamaguchi, S Abe, J. Appl. Polym. Sci., **74**, 3153-3159 (1999)
10. F. L. Savio and M. Bonfanti, Polym. Test., **74**, 235-244 (2019)
11. A. H. Awad and M. H. Abdellatif, Compos. B. Eng., **173**, 106948 (2019)

IOP Conf. Series: Materials Science and Engineering **957** (2020) 012047 doi:10.1088/1757-899X/957/1/012047

Superconducting Lead-free Solder Joint: A Short Review

Rita Mohd Said[1,*], *Mohd Arif Anuar* Mohd Salleh[1] , *Norainiza* Saud[1] and *Mohd Izrul Izwan* Ramli[1]

[1]Centre of Excellence Geopolymer & Green Technology (CeCeoTech), School of Materials Engineering, Universiti Malaysia Perlis (UniMAP), Taman Muhibbah, 02600 Jejawi, Arau, Perlis, Malaysia.

Abstract. Superconducting solders are widely used as joining material to connect superconducting wires such as Niobium-titanium (NbTi) and Niobium–tin (Nb_3Sn) wires in commercial superconducting magnet applications. The physical and superconducting properties of the solder materials are very important since it influence the overall performance of the joint during service. Pb-Bi solder alloy widely used due to their good properties suitable for the superconducting application, however, the restriction on the use of Pb in industry lead to the development of new Pb-free solder material. This paper reviews the superconducting Pb-free solder alternatives, including In-Sn based solder. Besides that, the effect of bismuth (Bi) and antimony (Sb) addition to the binary In-Sn solder alloy were also discussed.

1 Introduction

Solders known as fusible metal or material that become an important element since it will ensure the continuity for the production of electrical and mechanical in the electronic packaging. Previously, Tin-Lead (Sn-Pb) solders have been used for many years due to their good in chemical, physical, thermal and mechanical properties [1]. However, Pb-containing solder has been banned by legislation due to the toxic element of Pb that can cause negative effects on the human health also the environment [2, 3]. Many researches have been done to developed new Pb-free solder to replace the Pb containing solder. Generally, Tin (Sn)-based materials such as Sn-Ag-Cu [4], Sn-Cu-Ni [5], Sn-Ag [6] and Sn-Cu-Bi [7] have been recognized as important alloys for the development of Pb-free solder materials used for the electronics industry.

Eutectic lead-bismuth (Pb-Bi) solder is widely used as a connecting material between the NbTi wire and Nb_3Sn conductors in the superconducting applications such as large engineering research projects for examples poloidal field coils in ITER and in the medical sector (MRI) [8]. These Pb-Bi alloy has low melting point (124 °C for the eutectic 55.5 %

* Corresponding author: rita@unimap.edu.my

Bi composition), relatively high critical temperature, T_c (~8.3 K at about 40 % Bi) and high critical field values [2]. In the future, the legislation may restrict the use of Pb-Bi solder in the superconducting application. Thus, there is also a need to explore the new solder material with suitable superconducting properties to replace the Pb-Bi solder.

The Pb-free superconducting solder candidate should fulfill several requirements such as low in melting point, satisfactory wettability, good thermal, and mechanical properties in addition, the microstructure of the solder materials also crucial due to the fact that overall performance of the solder joint will influence by the microstructure [9-11]. In this short review, the microstructure and also superconducting properties of Pb-free solder that commonly used in the electronics industry were being discussed.

2 Superconducting Joint

Superconducting joining is known as a technique used to connect the superconducting wire. This technique is requirement for the superconducting magnets operated in the persistent current mode, which are usually used in applications such as magnetic resonating imaging (MRI) systems and nuclear magnetic resonance (NMR) spectrometers [12]. NbTi wire has a critical temperature (T_c) of 9.2 K and it is ideal used to fabricate superconducting magnet. There are various technique used for producing the superconducting joints, including solder matrix replacement, ultrasonic welding, diffusion welding, cold-pressing, and spot welding. Among all these techniques, the solder matrix replacement method with lead-bismuth (Pb-Bi) solder become the commonly used technique in industrial [13].

Fig. 1 depicted the fabrication of a superconducting joint using the solder matrix replacing method. This joint was involve in several steps; firstly, the end of the wire was removed using abrasive paper as shown in Fig. 3a. The wires were next immersed in the molten Sn bath (Fig. 3b). After that, the Sn coated wires were then immersed in the Pb-Bi bath can be seen in Fig. 3c. The Pb-Bi solder coated the wires were then being removed from the bath and allowed to cool. Afterwards, the insulated ends of the Pb-Bi coated Nb-Ti wires were cut off. The two wires were aligned, twisted, and tightened with fine Cu wire in open air to enhance their electrical and mechanical properties, as shown in Fig. 3d. Finally, the twisted wires were again immersed in a Cu tube filled with molten Pb-Bi (Fig. 3e) [13].

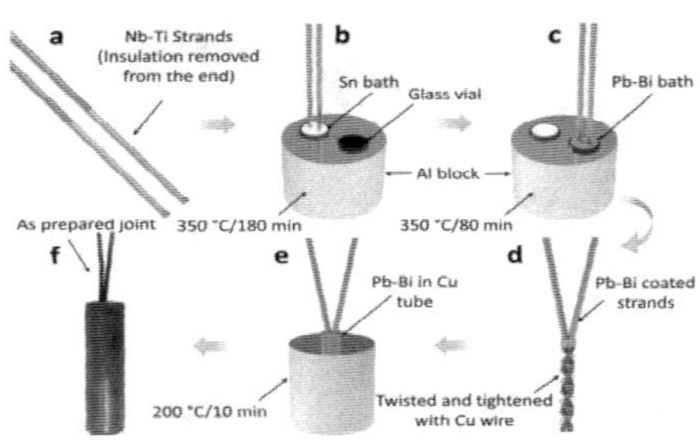

Fig. 1. Multifilamentary Nb-Ti conductors joining process using the solder matrix replacement method [13].

3 In-Sn Solder

The conventional In-Sn lead (Pb) free solder has become the promising candidate for replacing Pb containing solder in the superconducting joint [8, 14]. The composition of $In_{65}Sn_{35}$ solder that contains a matrix of In-rich β (In_3Sn) containing the Sn-rich γ-phase (In_4Sn) has been explore as potential replacement materials. Mousavi et al. have investigated the influence of chemistry and microstructure on the In-Sn solder on the superconducting properties. According to their findings, both β and γ phase was stable over a wide range of chemical composition, with improving superconducting properties for each phase as increasing in solute concentration. The best superconducting properties can be achieved when the alloys have a high volume fraction of β phase [14]. Studied by Tsui et al. found that superconducting properties of $In_{52}Sn_{48}$ solder has low in B_{c2}(T) and critical temperature (T_c) than Pb-containing solder as tabulated in Table 1 [15]. Moreover, more effort has been done to improve the In-Sn solder through microalloying with other elements such as bismuth (Bi) [2, 8, 9, 14], tantalum (Ta) and antimony (Sb) [16].

Table 1. The melting point, resistivity, and superconducting properties of the solders [15].

Solder Alloy	Melting Point °C	ρ (Room Temp.) 10^{-7}Ω·m	ρ (77 K) 10^{-7} Ω·m	T_c K	B_{c2} (0 K) T
$Bi_{49}Pb_{18}In_{21}Sn_{12}$ (Cerrolow 136)	58	9.4	8.1	6.4	3.3
$Pb_{20}Sn_{34}Bi_{46}$	~96	5.5	2.6	8.4	2.3
$Pb_{20}Sn_{60}Bi_{20}$	~170	2.6	1.1	8.5	2.2
$In_{52}Sn_{48}$	118	2.6	1.3	6.4	0.34
$Pb_{38}Sn_{62}$	183	1.5	0.48	7.3	0.30
$Pb_{57}Bi_{36}Sb_7$	200-230	6.1	3.7	8.5	2.5

3.1 In-Sn Solder Microalloyed with Bismuth Addition

The properties of Pb free solder alloy can enhance by adding the third or fourth element. Recently, bismuth (Bi) was added into binary In-Sn solder to improve the performance of the superconducting joint. Mousavi et al. have investigated the influence of the different composition of Sn-In-Bi solder alloy on the properties of the superconducting solder joint. The four ternary alloys that had been studied on the calculated ternary Sn–In–Bi phase diagram as shown in Fig. 2(e). From the finding in Fig. 2 (a-d), ternary Sn-In-Bi solder alloy contains the In-rich β phase, Sn-rich γ phase with more superconducting phases, including BiIn and $BiIn_2$. Based on the Fig. 3, the $Sn_{35}In_{50}Bi_{15}$ solder was the best superconducting properties achieved in Sn–In–Bi phase diagram systems. The solder composition consisting of a majority phase β-phase in which fine fibers of γ and $BiIn_2$ are embedded [14].

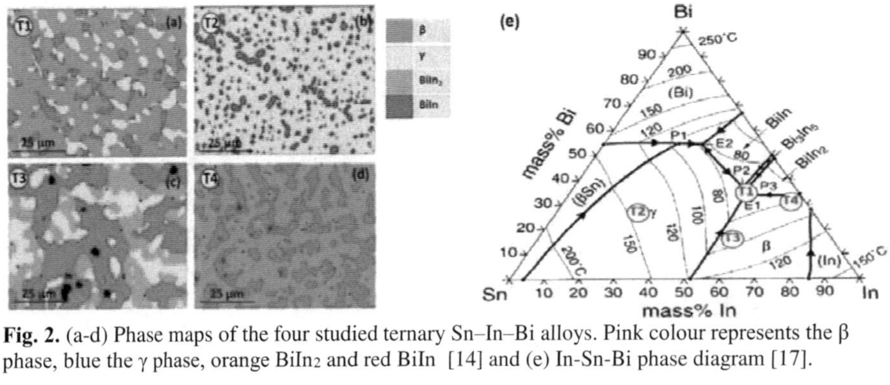

Fig. 2. (a-d) Phase maps of the four studied ternary Sn–In–Bi alloys. Pink colour represents the β phase, blue the γ phase, orange $BiIn_2$ and red BiIn [14] and (e) In-Sn-Bi phase diagram [17].

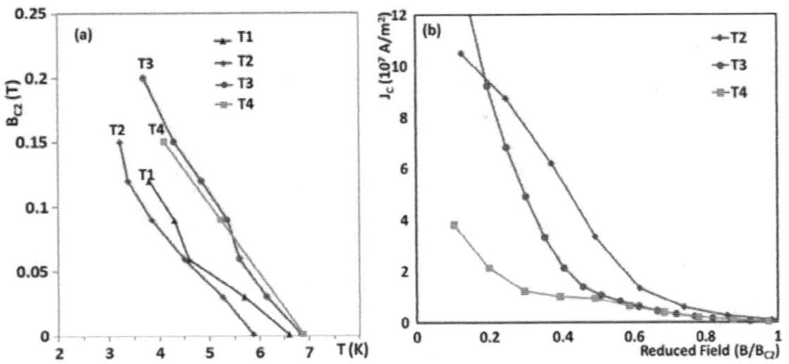

Fig. 3. Superconducting properties of the air-cooled Sn–In–Bi alloys, (a) B_{C2} as a function of temperature, (b) J_C versus reduced field [14].

The interaction of Nb (47wt.%Ti) superconducting filament with liquid ternary (In(In_{50}-Bi_{15})Sn_{35}($In_{54.3}Sn_{36.8}Bi_{8.9}$ at.%) was studied by Santra et al. There are three phase were detected, β, γ, and $BiIn_2$ are existed as shown in Fig. 4. Interaction of NbTi with ternary (In,Bi)Sn results in a lack of any detectable interaction zone. In addition, morphology and chemical composition of the superconducting joints were influenced by the size of NbTi filament [8].

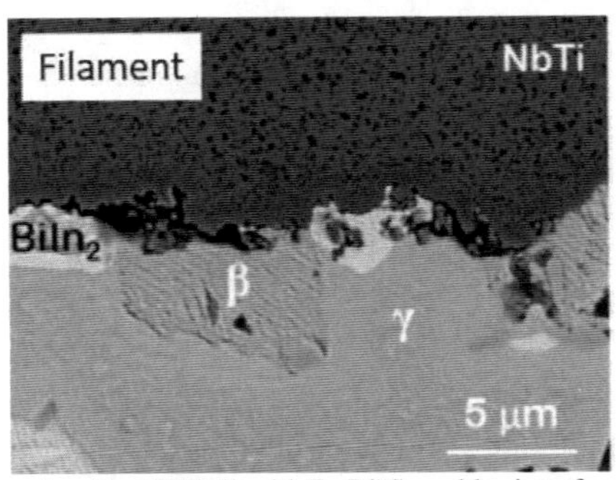

Fig. 4. BSE micrograph showing the NbTi–liquid (In,Bi)Sn solder interface [8].

3.2 In-Sn Solder Microalloyed with Antimony Addition

The microstructure and superconducting properties Sn-In-Sb solder alloy were investigated by Mousavi et al. [16]. As shown in Fig. 5, SnIn-γ is the majority phase with low solubility of Sb in the matrix of γ phase. Moreover, the presence of Sb helps to improve the superconducting properties by increasing the T_c and B_{c2} values.

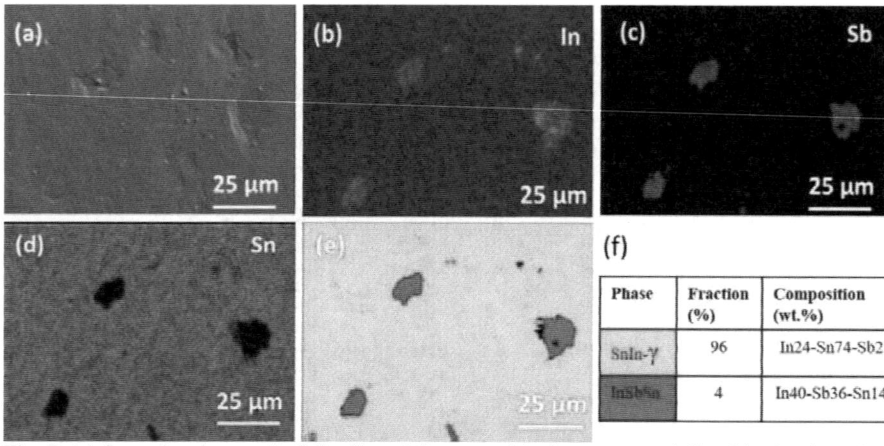

Phase	Fraction (%)	Composition (wt.%)
SnIn-γ	96	In24-Sn74-Sb2
InSbSn	4	In40-Sb36-Sn14

Fig. 5. (a) Secondary image, (b-d) In, Sb and Sn maps, (e) phase map, and (f) table showing phase fraction and the composition of each phase in sample Sn-In-Sb solder [16].

4 Conclusion

The superconducting properties of Pb-free solder such as In-Sn binary solder alloy can be enhanced by dissolving the third element. The addition of alloying elements such as Bi and Sb into the Sn-In phase (Sn-rich γ and In-rich β) can significantly increase the T_c and B_{c2} values which resulting in an improvement in the superconducting properties of the solder. Further research in developing a reliable Pb-free superconducting solder joint was required to search for the solder replacement of Pb-Bi solder.

The authors gratefully acknowledge the Ministry of Higher Education Malaysia, Universiti Malaysia Perlis (UniMAP) and Nihon Superior Co. Ltd. for the financial support throughout the research project.

References

1. S.Cheng, C.-M. Huang, M. Pecht, Microelectron Relia. 75 (2017)
2. C. Aksoy, T. Mousavi, G. Brittles, C.R.M. Grovenor, S.C. Speller, **26 (3)** (2016)
3. H.R. Kotadia, P.D. Howes, S.H. Mannan, Micron. Reliab.54 (2014)
4. M. Aamir, R. Muhammad, N. Ahmed, M. Waqas, Microelectron. Reliab. 78 (2017)
5. M. I.I. Ramli, M.A.A.M. Salleh, M.N. Derman, R.M. Said, N.M. Nasir, N. Saud,. Key Eng. Mater. 700 (2016)
6. L. Zhang, C.W. He, Y.H. Guo, J.G. Han, Y.W. Zhang, X.Y. Wang, Microelectron. Reliab. 52 (2012)
7. M.I.I. Ramli, M.S.S. Yusof, M.A.A.M. Salleh, R.M. Said, K. Nogita, Solid State Phenomena 273 (2018)
8. S. Santra, T. Davies, G. Matthews, J. Liu, C.R.M. Grovenor, S.C. Speller, Mater.Des. 176 (2019)
9. T. Mousavi, C. Aksoy, C. Grovenor, S. Speller, IEEE Transactions on Applied Superconductivity, **26(3)** (2016)
10. M.I.I. Ramli, N. Saud, M.A.A.M. Salleh, M.N. Derman, R.M. Said, N. Nasir, Applied Mechanics Mater. 754-755 (2015)

11. R.M. Said, M. A. A. Mohd Salleh, N. Saud, M. M. Al Bakri Abdullah, A. V. Sandu, AIP Conference Proceedings, 1835 (2017)
12. M. Kodama, K. Okamoto, Y. Koga, T. Yamamoto, H. Watanabe,. Superconductor Sci. Tech. **28(045019)** (2015)
13. D. Patel, S.-H. Kim, W. Qiu, M. Maeda, A. Matsumoto, G. Nishijima, H. Kumakura, S. Choi, J.H. Kim, Sci. Rep. **9(1)** (2019)
14. T. Mousavi, C. Aksoy, C.R.M. Grovenor, S.C. Speller, Superconductor Sci. Tech. **29(015012)** (2016)
15. Y. Tsui, R. Mahmoud, E. Surrey, D. Hampshire, IEEE Transac. Applied Superconductivity **26(3)** (2016)
16. S.C. Speller, T. Mousavi, C.R. Grovenor, W. Darby, C. Aksoy, T.J. Davies, G.D. Brittles, Mater. Res. Society: Advances, **4(1)** (2016)
17. P, V., P. A, and O. H, *Handbook of TernaryAlloy Phase Diagrams*. 1995 Materials Park, OH: ASM International.

IOP Conf. Series: Materials Science and Engineering **957** (2020) 012048 doi:10.1088/1757-899X/957/1/012048

Determination of Protective Potential in the Zero Charge Corrosion Protection System

Shaiful Rizam Shamsudin[1,*], *Rajaselan* Wardan[1], *Azmi* Rahmat[1], *Mohd Nazree* Derman[1], *Muhamad Syazwan* Sanusi[1] and *Mohd Fitri* Mohamad Wahid[1]

[1] Center of Excellence Geopolymer & Green Technology, School of Materials Engineering, University Malaysia Perlis, Jejawi 02600 Arau, Perlis, Malaysia

Abstract. Zero charge corrosion protection, ZCCP, is a new invention in the field of corrosion protection. It applies alternating current, AC, mode while the protective potential to ensure that the metal exposed in the corrosive environment still does not exist in any of the previous works of literature. The study was conducted by the voltage scanning within corrosion potential, E_{corr}, and polarized potential, $E_{impressed}$, using potentiostat. It was carried out by examining the behavior of the current flow in the circuit. The experimental results found that the protective potential for maximum protection in the ZCCP system is 0 mV. By keeping the AC metal potential at 0 mV, the surface will not acquire any charges, hence the corrosion reaction will be inhibited.

1 Introduction

When a piece of metal is exposed in an electrolyte, such as salted or acidic solution, an electrochemical corrosion reaction occurs where the electron conductor (metal electrode) meets the ion conductor (electrolyte) at the electrode-electrolyte interface. Metal in aqueous solutions creates a potential difference due to the adsorption of water molecules on an electrode. Generally, water molecules are seen to be electrically neutral, nevertheless, the water molecule has dipoles moment that described as bipolar molecules that have an excessive positive ($H^{\delta+}$) and negative charges sides ($O_2^{\delta-}$). These polarity differences create a high gradient potential across the metal-solution interface result in a layer of an electric field known as an electric double-layer [1].

The presence of an electric double-layer is due to the separation of the opposite charges in between both metal and solution phases in contact [2]. A potential difference caused by the electric double layer results in the adsorption of ionic charges to metal surfaces then finally brought corrosion reaction. In electrochemistry study, the potential difference between metal and reference electrodes commonly called a half-cell electrode. It is referred to as corrosion potential (E_{corr}), open circuit potential (E_{ocp}), or native potential (E_{nat}). If the electric double-layer can be neutralized, both excessive positive and negative charges on the metal surfaces are eliminated and finally making the metal surface maintained at the potential

* Corresponding author: rizam@unimap.edu.my

of zero charges. The term of zero charge potential (E_{pzc}) is a characteristic value of the electrode potential for any metal at which a clean surface of the metal will not acquire any electrical charge when it comes into contact with an electrolyte [3]. The surface of zero charges occurs when the positive and negative charges on the surface have been neutralized by giving a negative and positive direct current to the metal. In the previous study, it is known that the E_{pzc} is slightly negative than E_{corr} [4].

The application of E_{pzc} concept is manipulated through a method called zero charge corrosion protection (ZCCP). ZCCP technique requires a power supply that capable of giving current/ voltage pulses. When pulses input signals are applied, the current/ voltage must be measured in alternating current (AC) mode. Therefore, in this technique, the protective potential should be measured in different concepts, which is something opposite to the cathodic protection system (CP). It is well known that the protective potential of the CP system for steel in the corrosive environment must be within -850 to -1100 mV vs $Cu/CuSO_4$ [5]. Conversely, to find out a suitable protective voltage in AC mode, a study should be done to determine the presence of electric charges on the metal surface to keep the metal-solution interface at zero charges (or E_{pzc}).

In this study, a potentiostat is manipulated as a power supply to control the net potential of the metal at E_{pzc}. The excessive charges on metal surface is measured based on the current density profile that can be detected by the supersensitive current sensor embedded in the potentiostat. Here, the potentiostat was set in the cyclic mode for micro-voltage scan within E_{corr} and polarized potential ($E_{impressed}$) repeatedly.

2 Methodology

The steel samples were cut into 1.5 cm diameter and 0.4 cm thickness. It was ground by silicon carbide paper up to 600 grit, washed by alcohol, and rinsed by distilled water. The cleaned and dry samples was locked into a flat sample holder and inserted into a corrosion cell. 1 liter of 3.5% NaCl solution was filled into the corrosion cell. Steel sample and saturated calomel reference electrode (SCE) were then connected to the potentiostat (Gill AC, ACM Instruments). The steel samples were initially soaked within 5 days to ensure that E_{corr} is properly in a steady-state condition. The micro-voltage scan was performed at a rate 1 mV/ minute. The micro-voltage scan magnitude ranges (r - s) were ±1 to ±5 mV offset to E_{corr} toward an impressed potential ($E_{impressed}$). The experiment was repeated to the smaller micro-voltage down to ±0.125 mV. The micro-voltage scanning profile is displayed in Fig. 1. All experiments were performed in a Faraday cage and the temperature was controlled at 25°C.

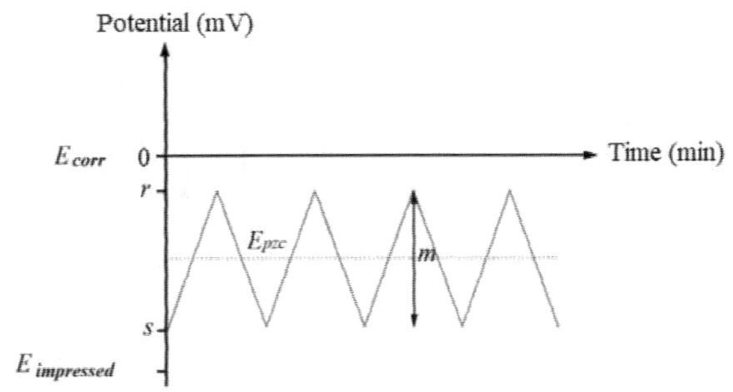

Fig. 1. The micro-voltage scanning profile around, E_{pzc}. (m, is the magnitude, r and s are the maximum and minimum voltage distance respectively).

The micro-voltage scan setup preventing all oxidation & reduction reactions to avoid a reproduction of excessive positive and negative charges on the metal surface. The current density data from the micro-voltage scanning within E_{pzc} to $E_{impressed}$ were analyzed and displayed as a plot of current density against time. To make it easier observation, these data were also converted into a histogram form displaying average current density vs magnitude of micro-voltage scanning.

3 Results and discussion

The voltage scanning test to determine the protective voltage in the ZCCP system was performed by measuring the current density profile (Fig. 2). The current density was recorded 1 reading/s for 24 hours. The reading was taken out as soon as the steady-state E_{corr} achieved within 5 days of immersion period.

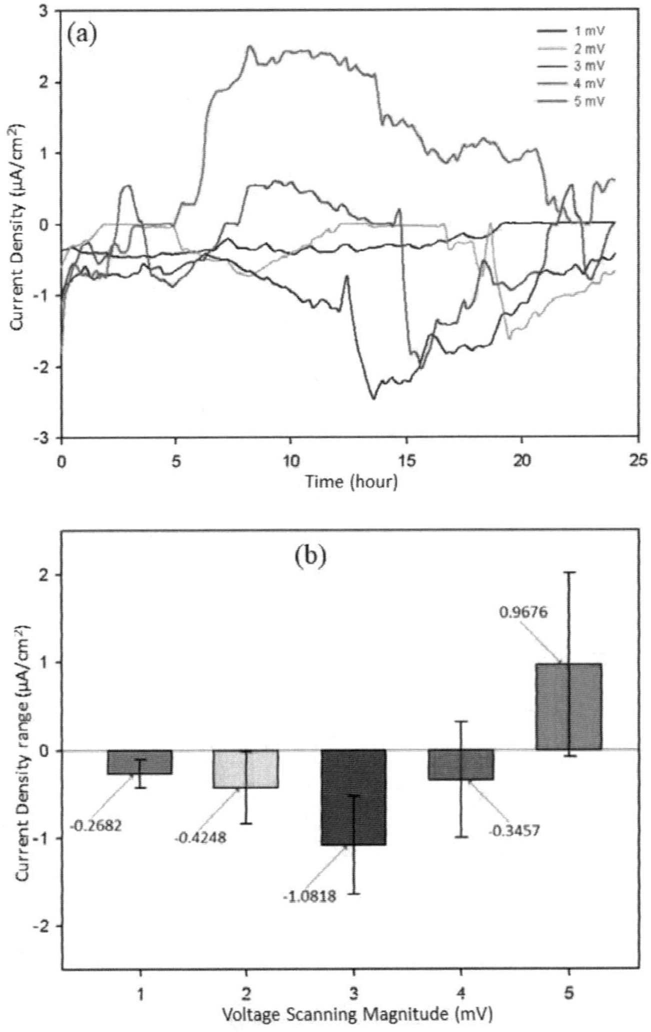

Fig. 2. (a) Plot and (b) histogram for micro-voltage scan from ±1 mV to ±5 mV magnitudes.

Fig. 2(a) shows a plot of current density vs time at the magnitude range i.e. ±1 to ±5 mV within E_{corr} & $E_{impressed}$ as the position E_{pzc} is believed to be within these potentials [4]. There is a wild movement pattern of current flow. The behavior of the electrical current is observed to form an unstable pattern. All current density is still lower as compared to the current requirement for cathodic protection in stagnant seawater (20 to 30 μA/ cm^2) [5]. To make it clear, the average current density is displayed in histogram form (Fig. 2b). The scanned magnitudes, especially for ±5 mV and a part of ±4 mV are found to be pulsating beyond the zero lines of electric current. The negative electric charges were shifted the positive charge region. This means that there are charges transfer at the metal-solution interface, hence promotes the adsorption/ desorption of ionic charge species in the metal-solution interface. Consequently, the metal potential at ±4 and ±5 mV were failed to bring down the metal potential to zero charges line which is automatically failed to act as protective potential in the AC mode of the ZCCP system.

The current densities that do not exceed the zero line occur at ±1 to ±3 mV magnitude of voltage scanned ranges. At this stage, the magnitude of the voltage scan at ±1 mV is considered to be a relatively good tolerance to referred to as the protective potential, while the minimum tolerance in the ZCCP system is about ±3 mV AC. However, the trustworthiness of these three voltage magnitudes is still questionable. It is observed that the flow of electric currents generally tends to be in the region of the negative charges. Although oxidation of metals does not occur in the negative region, the flow of current below the zero line causes the possibility that the metal surface tends to acquire a negative charge. In the long run, these negative charges can potentially accumulate, absorb, and cover the metal surface which eventually makes it possible to enter the immune region as described in the E-pH Pourbaix diagram. This process would make the metal back to the cathode protection system. What is bad about this CP system is that, if the electricity supply is cut off, then the metal reaction rate will soar rapidly, whereas in the ZCCP system, if the metal surface is in a stable state of E_{pzc} in the long period, the corrosion reaction does not occur or takes a long time to suffer failure [4].

To prevent the metal from being covered by the excessive negative charges, a voltage scanning is performed on a smaller magnitude scale. A few similar voltage scan step at ±0.125 to ±1 mV (Fig. 3). Fig. 3(a) shows all voltage magnitude ranges were observed that the current density output does not exceed the zero line. It seems like the current flow, mainly for ± 0.5 and ± 1 mV, moves quite wildly. On the other hand, the scanning at ± 0.125, and ± 0.25 mV scanned magnitude show a very stable situation at the zero lines. Fig. 3 (b) confirms that scans on a scale of magnitude ±1 mV (or less) make the current flow direction would never exceed the zero line position. This indicates that, to ensure that corrosion does not occur on metals which exposed to corrosive environments, protective potential in AC mode or ZCCP systems should reach ±1 mV or below. However, the best protective potential for the ZCCP system can reach a value of 0 mV is confirmed to be more convincing. It is noted that the excessive positive charge on the metal surface is a sign of corrosion reaction. When sufficient electron (or negative current) supplied to the metal just necessary to neutralize the positive charges on the metal surface, the metal will be eventually in a situation that having no charge on the metal surface. Therefore, the water molecules will be no longer can be adsorbed on the metal surface due to the absence of charges. In terms of an electrical double layer model, this is the lowest capacitance value being measured. In the absence of adsorbed water molecules, a sort of higher electrical resistance layer is created between the metal and the solution at the metal-solution interface. In the AC mode of the ZCCP system, 0 mV is the best protection potential where metals and solutions are believed not to be in contact with each other.

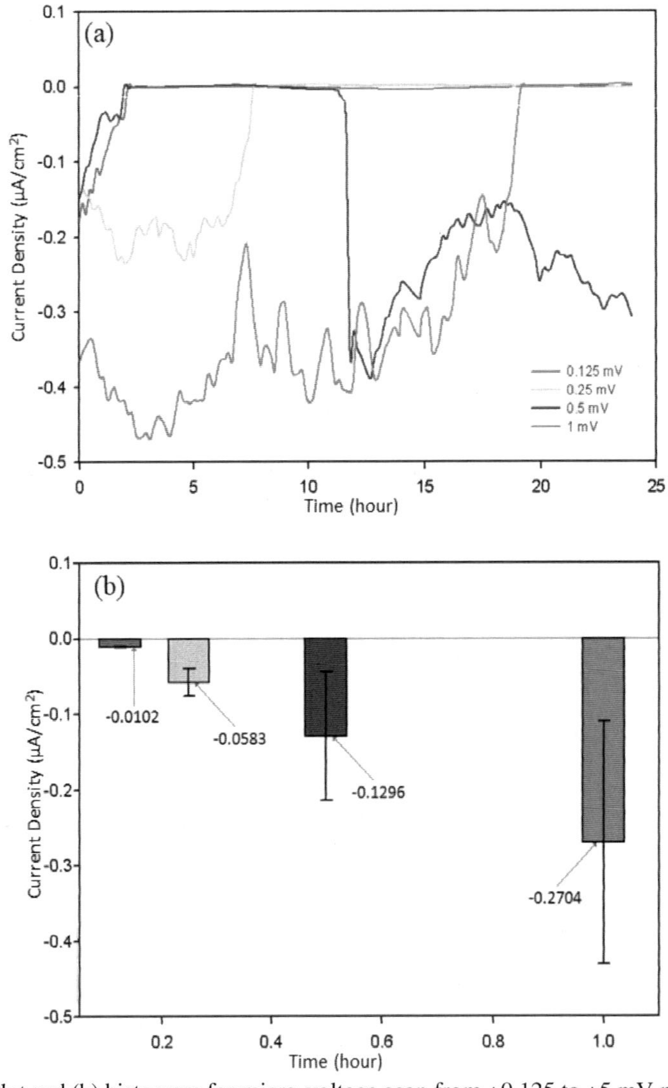

Fig. 3. (a) Plot and (b) histogram for micro-voltage scan from ±0.125 to ±5 mV magnitudes.

4 Conclusion

A corrosion protection method by applying AC mode has been invented and named as zero charge corrosion protection, ZCCP. Micro-voltage scanning has been performed on the steel samples in a corrosive environment shows that the maximum protective potential is at 0 mV. If metal potential could be maintained at 0 mV, its surface will not acquire any charge, hence the corrosion reaction will be inhibited completely.

The author would like to acknowledge the support from the Fundamental Research Grant Scheme (FRGS) under a grant number of FRGS/1/2015/TK10/UNIMAP/02/5 from the Ministry of Education

Malaysia. Thanks to the School of Materials Engineering & Universiti Malaysia Perlis for providing special financial support that enabling the publication of this manuscript.

References

1. Y. L. Wu, D. P. Zhang, G. Y. Cai, X. X. Zhang, Z. H. Dong. Corr. Eng. Sci. & Tech. 1-13 (2020)
2. A. Groß, S. Sakong. Curr Opin Electrochem. **14**. 1-6 (2019)
3. F. El-Hajjaji, M. E. Belghiti, B. Hammouti, S. Jodeh, O. Hamed, H. Lgaz, R. Salghi. Port. Electrochim. Acta, **36**. 197-212 (2018)
4. S. R. Shamsudin. PhD Thesis. Universia Malaysia Perlis (2016)
5. R. Baboian, R. Treseder. NACE Corrosion Engineer's Reference Book. 3 (2002).

A Short Review on Selection of Electrodes Materials for Symmetrical Solid Oxide Fuel Cell

Wan Nor Anasuhah Wan Yusoff[1], *Nurul Akidah* Baharuddin[1,*], *Mahendra* Rao Somalu[1], *Andanastuti* Muchtar[1,2] and *Abdullah* Abdul Samat[3]

[1]Fuel Cell Institute, Universiti Kebangsaan Malaysia, 43600, UKM Bangi, Selangor, Malaysia
[2]Department of Mechanical & Manufacturing Engineering, Faculty of Engineering and Built Environment, Universiti Kebangsaan Malaysia, 43600, UKM Bangi, Selangor, Malaysia
[3]Faculty of Engineering Technology, Universiti Malaysia Perlis (UniMAP), Kampus UniCITI Alam, Sungai Chuchuh, Padang Besar, 02100, Perlis, Malaysia

Abstract. Symmetrical solid oxide fuel cells (SOFCs) operate at high temperatures and consist of a dense electrolyte which is sandwiched between two electrodes. An anode acts as a fuel electrode where fuel is oxidised, while a cathode serves as an air electrode in which oxygen is reduced. However, the use of two different materials as electrodes poses several significant issues concerning durability, reliability and fabrication costs of SOFCs. Furthermore, at least two thermal steps are required for the sintering of both electrodes, thereby increasing energy usage. In overcoming these challenges posed by conventional SOFCs, the concept of symmetrical solid oxide fuel cells (S-SOFCs) has been introduced. S-SOFCs comprise of two identical materials for the anode and cathode. The use of two similar materials decreases the interfacial regions among the electrolyte and electrodes and consequently improves the compatibility among the components. Sulphur poisoning and carbon deposition at the fuel electrode can be solved by merely reversing the flow of oxidant and fuel to oxidise all the species that degrade the performance of SOFCs. Although, the electrode material for S-SOFCs requires some unique properties, such as high electrocatalytic activity for fuel oxidation and oxygen reduction, high electrical conductivity in oxygen and hydrogen/hydrocarbon environments, promising the capability to resist sulphur poisoning and carbon deposition in hydrocarbon fuels and good structural and chemical stability under actual fuel cell conditions. As the works related to S-SOFCs are limited in this respect, this paper reviews the available reports in this field to allow a better understanding of the operational mechanism and the potential of S-SOFCs.

1.0 Introduction

The application of solid oxide fuel cells (SOFC) is one of the most practicable and viable alternatives to alleviate future energy source demands. Offering many advantages with respect to high energy conversion, and with fewer harmful emissions plus greater flexibility on the consumption of fuel makes SOFCs a prominent proposition [1-2]. The configuration of a common single-cell is mostly made of two porous electrodes (anode and cathode) and a densified electrolyte material. The electrode components typically comprise of different material which is able to undertake the specific electrochemical reaction for the SOFC application to operate effectively. The use of two different materials as electrodes has fewer serious issues regarding durability, reliability and the fabrication cost of SOFCs. For example, using two different materials as electrodes leads to the formation of at least two interfacial regions within a single SOFC. Although, these interfaces can cause various thermo mechanical stresses at high operating temperatures primarily due to the difference in their thermal expansion coefficients, less compatibility among components and different compositions. Furthermore, at least two thermal steps are required for the sintering of both electrodes, thereby increasing the fabrication cost of SOFCs [3-4]. Moreover, traditional Ni-based anode materials have a weak ability to resist sulphur poisoning and carbon deposition in hydrocarbon fuels [5]. Therefore, in overcoming and addressing the challenges posed by conventional three-component SOFCs, Badding et al. introduced the concept of symmetrical solid oxide fuel cells (S-SOFCs) in 2001 [6].

The basic configuration of the S-SOFC alongside SOFC is illustrated is compared as shown in Fig. 1 below. The S-SOFC configuration exhibits several advantages over the conventional SOFC such as aiding in the fabrication process while enhancing the interaction between the electrodes and the electrolyte component. Secondly, the S-SOFC fabrication process can minimise production costs, reducing the coking phenomenon in addition to sulphur poisoning [7]. Though,

* Corresponding author: akidah@ukm.edu.my

the exhibition of a unique configuration by the S-SOFC requires both the anode and cathode to have excellent stability (structural and chemical) and good electrical and electrochemical performance. Good electrochemical performance includes the active reaction for an oxidising and reducing condition which is an important characteristic for the electrodes. However, the material that suits the characteristic as the electrodes for the S-SOFC remains limited given the material itself needs to satisfy both the requirement from the anode and cathode while reducing the existence problem that continues to hinder this application to perform at its optimum. By acknowledging the solution to these limitations, it is therefore important to examine the basic properties and characteristic of the recommended material to be utilised, which suits both the anode and cathode components. Thus, a review involving the base material, dopants, material's limitations are among the crucial parts that require detailed understanding for the electrodes to function well during operation.

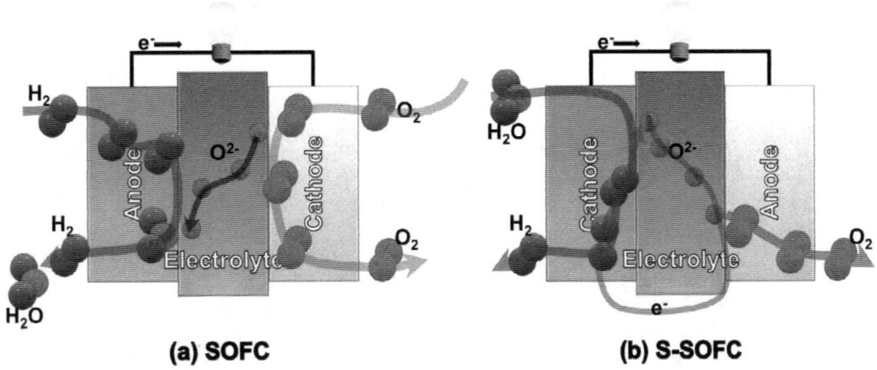

Fig. 1. Basic working mechanism for (a) SOFC and (b) S-SOFC

2.0 Anode

Commonly used anode material in SOFC application, such as Ni-YSZ cermet, is where the utilisation of nickel (Ni) helps in conducting the electron actively while YSZ is charged for ionic conduction [8]. Most Ni-based anodes perform well, especially when the operating system is coupled with hydrogen as a fuel. However, unfortunately, this type of anode experiences unavoidable carbon coking whenever the system is operated using hydrocarbon as fuel. For example methane, is an active catalyst for the reaction involving hydrocarbon cracking [9]. As such, this situation may lead to the deterioration of overall performance since the deposition may be enclosed and reduce the active site for the reaction to occur. Moreover, given the low tolerance towards coking which may limit the development of direct hydrocarbon SOFC, this is one of the significant drawbacks of Ni cermet for anodes [10].

A further reason that has suppressed SOFC advancement to operate using natural gas as the fuel is the susceptibility to sulphur poisoning where this material is present in natural gas due to the formation from the NiS_x compound. Secondly, this problem possibly occurred due to the aggregate fluctuation during the redox reaction and nickel particle agglomeration in the Ni cermet anode [11]. However, despite the drawback presented by Ni cermet, this type of material remains the best option when researchers are working towards developing strategies to enhance coke resistance throughout the modification process. One of the favourable strategies for modification of the anode structure surface is to use other active oxides and to introduce other metals into nickel cermet [10].

3.0 Cathode

The primary function of the cathode is to cause a reaction between the reactant (oxygen or air) and the electrolyte, without itself being consumed or corroded. A cathode plays the role as a catalyst in the oxygen reduction and works in the air or oxygen atmosphere. As such, the problem of non-stability is practically eliminated for the cathodes, which are typically oxide materials [12]. The perovskite-type oxide crystal structure with the general formula of ABO_3 (A is usually a rare earth element and B is usually a transition element) is commonly used as cathode materials. This is due to their stability in geometrical and electrostatic configurations, and many oxides of this type are stable even when a considerable number of defects are introduced [13]. The reality of SOFC operation has mostly evolved around the activation of the oxygen reduction reactions (ORR) occurring on the cathodes [14].

Oxygen activation is normally more laborious as compared to hydrogen activation thereby raising the issue on the effort in reducing the operating temperature which may degrade the cell's performance due to insufficient electrochemical activity driven by the operating temperature [15]. A detailed material design composition, in addition to the microstructure of the cathode material, has a significant consequence on the performance of SOFC operations. For example, controlling the oxygen nonstoichiometric can enhance the ionic and electronic conduction including catalytic activity, which is associated with ORR in the cathode component [16]. Thus, one of the affirmative strategies that can be adopted in idealising the SOFC system is by using new oxides with mixed ionic and electronic conductivity (MIEC). This type of cathode material may elevate the active site for a reaction ahead of the triple-phase boundary (TPB) including the adaptation of nanostructured materials which may increase the electrode specific surface [17].

4.0 Electrode Candidate for S-SOFC

S-SOFCs comprise of two identical compositions/materials for both the anode and cathode. The use of two similar materials decreases the interfacial regions among the electrolyte and electrodes and improves the compatibility among the components. Only one single sintering step is required to fabricate the cells, consequently reducing the fabrication cost. Sulphur poisoning and carbon deposition at the fuel electrode can be solved by merely reversing the flow of oxidant and fuel to oxidise all the species that degrade the performance of the SOFCs [18]. The electrode material for S-SOFCs requires some unique properties, such as having high electrocatalytic activity for fuel oxidation and oxygen reduction, high electrical conductivity in oxygen and hydrogen/hydrocarbon environments, promising the capability to resist sulphur poisoning and carbon deposition in hydrocarbon fuels and good structural and chemical stability under actual fuel cell conditions [19].

Extensive studies have been conducted to develop materials that satisfy the abovementioned requirements. However, all studied materials such as $La_{0.75}Sr_{0.25}Cr_{0.5}Mn_{0.5}O_{3-\delta}$ (LSCM) [20], $Sr_2Fe_{1.5}Mo_{0.5}O_{6-\delta}$ (SFMO) [21], $PrBaFe_2O_{5+\delta}$ (PBFO) [22], $Sm_{0.9}Sr_{0.1}Fe_{0.9}Ru_{0.1}O_{3-\delta}$ (SSFR) [23], and $La_{0.5}Sr_{0.5}Co_{0.5}Ti_{0.5}O_{3-\delta}$ (LSCT) [24] have limitations. LSCM cannot resist sulphur poisoning, PBFO is structurally unstable, and SFMO is chemically unstable [5]. Therefore, innovative materials and novel structures that encounter all the requirements for cathodes and anodes should be developed for the practical application of S-SOFCs under actual fuel cell conditions [18-25].

From this perspective, lithiated nickel oxide has the potential to be used as an electrode for S-SOFCs since this material demonstrates good chemical, physical and thermal stability in oxidation and reduction environments [26]. Although despite the stability of lithiated nickel oxide, its electrical and electrochemical behaviour under actual fuel cell conditions remains a significant challenge. Comprehensive studies have been performed to enhance its electrical and electrocatalytic performance by adding transition metals, such as Zn, Fe, Co, and Al [25–28]. Tan et al. [26] studied lithiated nickel oxide electrodes by adding Cu, Fe and Co and obtaining a power density of 0.38 mW/cm^2 at 560 °C when Co is used as a dopant. Chen et al. [27] co-doped Al and Co into a lithiated nickel oxide electrode and reported a power density of 159.7 mW/cm^2 at 550 °C. However, despite this positive improvement, the values recorded are profoundly different from those of traditional SOFCs that have a power density above 500 mW/cm^2 [29].

Table 1. Summary of previously researched electrode for S-SOFC

Electrode	Electrolyte	Scm^{-1}	ASR	Ref
$BaFe_{0.9}Zr_{0.1}O_{3-\delta}$	$La_{0.9}Sr_{0.1}Ga_{0.8}Mg_{0.2}O_{3-\delta}$	-	0.005 (800 °C) 0.02 (700 °C)	[7]
$La_{0.5}Sr_{0.5}Fe_{0.9}Mo_{0.1}O_{3-\delta}$	$Sm_{0.2}Ce_{0.8}O_{1.95}$	-	0.46 (750 °C)	[30]
$La_{1-x}Sr_xFe_{0.8}Cu_{0.2}O_{3-\delta}$	$La_{0.8}Sr_{0.2}Ga_{0.8}Mg_{0.2}O_{3-\delta}$	-	0.454 (800 °C)	[31]
$La_4BaCu_{5-x}Mn_xO_{13+\delta}$	YSZ	178 (air) 33.3 (Hydrogen)	-	[32]
$Nd_xSr_{2-x}Mn0_{4\pm\delta}$	YSZ	33.4 (air) 0.9 (hydrogen)	-	[33]

Therefore, the structural, electrical and electrochemical properties of doped lithiated nickel oxide should be systematically improved. Table 1 summarises the potential materials reported for S-SOFC. Detailed selection of the based material is not the only the primary consideration in determining the overall performance of the S-SOFC application. The strategy to involve modifications onto the based materials itself plays a vital role to improve the performance. The utilisation of potential dopants possesses suitable characteristics and the ability to reduce drawbacks or to enhance the conductivity. In this context, some reports have shown that ceramics and using metals as dopant, is useful especially when aiming for higher catalytic properties of the electrode, for example, Ni, Ru, Pt and Mo [34].

5.0 Conclusion

Questions about the most suitable material to be used for both electrodes for the S-SOFC remain. However, answering this question depends on several factors, including the selection of materials, the fabrication process, mechanical factors and operating conditions. This study suggests a feasible idea by highlighting the crucial characteristics as well as the drawbacks of electrodes which could help in progressing future studies in this field. However, the practicability surrounding newly developed materials must be first verified for lab-scale before progressing in verification for a prototype.

This research was supported by Universiti Kebangsaan Malaysia under the grant, DIP-2019-011 and the Ministry of Education under the grant, FRGS/1/2019/TK07/UKM/02/1. The authors would like to express their sincere gratitude to the Institute of Fuel Cell, UKM for assisting in this research. They would also like to thank the Centre of Research and Instrumentation Management (CRIM) for providing the analytical facilities and equipment for the completion of this research.

References

1. F. S. da Silva and T. M. de Souza, Int. J. Hydrogen Energy **42**, 26020–26036 (2017)
2. H. Hu, Q. Lin, A. Muhammad, and B. Zhu, J. Power Sources **286**, 388–393 (2015)
3. C. Su, W. Wang, M. Liu, M. O. Tadé, and Z. Shao, Adv. Energy Mater. **5**, 1–19 (2015)
4. B. Niu, F. Jin, L. Zhang, P. Shen, and T. He, **263** (2018)
5. B. Niu, F. Jin, T. Feng, L. Zhang, Y. Zhang, and T. He, Electrochim. Acta **270**, 174–182 (2018)
6. M. E. Badding, J. L. Brown, T. D. Ketcham, and D. J. St. Julien, U.S. Patent 2001/0044043 (2001)
7. W. He, J. Fan, H. Zhang, M. Chen, Z. Sun, and M. Ni, Int. J. Hydrogen Energy **44**, 32164–32169 (2019)
8. J. Xiao *et al.*, Ceram. Int. **45**, 20066–20072 (2019)
9. A. A. Jais, M. R. Somalu, A. Muchtar, and W. N. R. W. Isahak, IOP Conf. Ser. Earth Environ. Sci. **268**, 0–6 (2019)
10. H. Aslannejad, L. Barelli, A. Babaie, and S. Bozorgmehri, Appl. Energy **177**, 179–186 (2016)
11. H. Ding *et al.*, Appl. Energy **233–234**, 37–43 (2019)
12. D. Chen, Q. Zhang, L. Lu, V. Periasamy, M. O. Tade, and Z. Shao, J. Power Sources **303**, 305–316 (2016)
13. G. Chen, Y. Gao, Y. Luo, and R. Guo, Ceram. Int. **43**, 1304–1309 (2017)
14. L. Zhang, J. Yang, and J. Li, J. Power Sources **269**, 723–726 (2014)
15. X. Ding *et al.*, Electrochim. Acta, **236**, 378–383 (2017)
16. R. Scurtu *et al.*, J. Solid State Chem. **210**, 53–59 (2014)
17. D. Garces, A. L. Soldati, H. Troiani, A. Montenegro-Hernandez, A. Caneiro, and L. V. Mogni, Electrochim. Acta **215**, 637–646 (2016)
18. C. Su, W. Wang, M. Liu, M. O. Tadé, and Z. Shao, Adv. Energy Mater. **5**, 1–19 (2015)
19. X. Luo *et al.*, Electrochim. Acta, **260**, 121–128 (2018)
20. D. M. Bastidas, S. Tao, and J. T. S. Irvine, J. Mater. Chem. **16**, 1603–1605 (2006)
21. Q. Liu, X. Dong, G. Xiao, F. Zhao, and F. Chen, Adv. Mater., **22**, 5478–5482 (2010)
22. Y. Chen *et al.*, J. Power Sources **310**, 109–117 (2016)
23. W. Fan, Z. Sun, J. Zhou, K. Wu, and Y. Cheng, J. Power Sources **348**, 94–106 (2017)
24. R. Martínez-Coronado, A. Aguadero, D. Pérez-Coll, L. Troncoso, J. A. Alonso, and M. T. Fernández-Díaz, Int. J. Hydrogen Energy **37**, 18310–18318 (2012)
25. X. Liu *et al.*, Electrochim. Acta **295**, 325–332 (2019)
26. W. Tan, L. Fan, R. Raza, M. Ajmal Khan, and B. Zhu, Int. J. Hydrogen Energy **38**, 370–376 (2013)
27. G. Chen *et al.*, Int. J. Hydrogen Energy **43**, 417–425 (2017)
28. L. Fan and P. C. Su, J. Power Sources **306**, 369–377 (2016)
29. K. Zhao and Y. Du, J. Power Sources **347**, 79–85 (2017)
30. H. Cai, L. Zhang, J. Xu, J. Huang, X. Wei, and L. Wang, Electrochim. Acta **320**, 134642 (2019)
31. F. Zurlo *et al.*, Acta Mater. **112**, 77–83 (2016)
32. S. Durán *et al.*, Solid State Ionics **341**, 115031 (2019)
33. M. V Sandoval *et al.*, Solid State Ionics **317**, 194–200 (2018)

34. N.K. Monteiro, G. A. S. Candido and F. C. Fonseca, J. Electrochem Soc. **78**, 1245–1252 (2017)

IOP Conf. Series: Materials Science and Engineering **957** (2020) 012050 doi:10.1088/1757-899X/957/1/012050

Effect of Different Filler Loading on Flexural Properties and Water Absorption Behavior of Kenaf Core Fiber Reinforced Polypropylene Composite

Nur Rusyidah Mustapa[1,*], *A. R.* Rozyanty[1], and *Mohd Hanif Mohd* Pisal[2]

[1]Center of Excellence Geopolymer and Green Technology (CEGeoGTech), Universiti Malaysia Perlis (UniMAP), 01000 Kangar, Perlis, Malaysia.
[2]School of Material Engineering, Kompleks Pusat Pengajian Jejawi 2, Universiti Malaysia Perlis (UniMAP), Taman Muhibbah, 02600 Jejawi, Arau, Perlis, Malaysia.

Abstract. In this investigation, kenaf core fiber (KCF) reinforced polypropylene (PP) composite was prepared via extrusion and compression moulding. The effect of different ratio of KCF filler with PP on flexural properties of composite had been investigated. PP and KCF filler were prepared with different filler loading of 5, 10, 15, 20 and 25 wt% using twin-screw extruder at temperature 180 °C and rotor speed of 70 rpm, followed by compression moulding at the same temperature. The flexural tests were conducted to evaluate the flexural properties of the composite. The results obtained showed that the incorporation of KCF in PP increased the flexural strength and flexural modulus of the composite. The water absorption behaviour of the composite was also investigated. It was found that water absorption of composites increased with increasing of fiber content.

1.0 Introduction

The utilization of polymer composites over the past few years have shown a remarkable increase in demands due to their advantages such as recyclability, easy fabrication, lightweight, low cost, and their flexibility to be molded into the desired shape and design [1]. Generally, polypropylene (PP), which is known as high volume commodity plastic, has contributed significantly to the development of the plastics industry. Various researches have been conducted to explore and commercialize the incorporation of fillers into a PP matrix to fulfill various requirements in engineering applications.

The interest in natural fiber reinforced polymer composite is rapidly growing both in terms of their industrial applications and fundamental research. Natural fiber, which is derived from natural resources, is often applied as reinforcement of composite. Their availability biodegradability, and low-cost production are the leading causes that attract the interests of researchers toward natural fiber composites. Besides, they are renewable, vast availability, low density and have high specific strength which makes them an attractive alternative to replace the glass and carbon fiber in various applications such as transportation,

*Corresponding author: rusyidahm@gmail.com

military application, buildings and construction industries [2, 3, 4].

It is crucial to study and discuss the use of natural fiber-reinforced composite and their effects on materials' performances, and either they can be considered as favorable reinforcement in a composite system. Presently, one of the most commonly used natural fibers in reinforced plastic composite in Malaysia is kenaf fiber. Kenaf fiber *(Hibiscus cannabinus L.)*, have a low density (1.2 – 1.6 g/cm^3) and offer non-abrasive and high stiffness properties that affect both physical and mechanical properties [5]. Kenaf fiber- reinforced plastic composite, which is used for a wide range of applications, were found comparable to existing synthesis composites. The materials' performances are always presented in terms of their mechanical properties such as flexural properties, tensile properties and impact properties. Their features are connected to engineering performances of materials under critical conditions [6]. For the last decades, numerous studies have been conducted to study the mechanical and water behavior of kenaf fiber-reinforced composite.

Mohammed et al. [7], in his study, stated that the mechanical properties of natural fiber-reinforced composites depend on the type of fiber, its orientation (random or unidirectional), the strength of fiber, physical properties of fiber, and the type of plasticizer used. For example, Subramonian [8] reported in his study of PP-bagasse fiber composite that flexural strength and modulus increased as increasing fiber content. Similar trends, as reported by Fairuz et al. [9] in their study of the effect of fiber loading on mechanical properties of pultruded kenaf fiber reinforced vinyl ester composite. The authors reported that fiber content is having a significant impact on the mechanical properties of composites, as the results obtained from the study showed that the mechanical properties improved by the addition of kenaf fiber to polymer composite.

Besides mechanical properties, water absorption behavior of composite is also considered as one of the essential features in composite studies, to investigate their suitability in outdoor applications. Amoke et al. [10], in their previous studies, reported the relationship between water absorption and fiber loading. They concluded that water absorption increases almost linearly with fiber loading. Besides that, Ismail and Ishak, [11] also reported a similar trend in their studies that the increase in fiber fraction resulted in an overall increase in total water uptake.

Therefore, this study will be focused on the reinforcement of the polypropylene matrix and kenaf core fiber various filler loading. The mechanical properties and water absorption of the composite were investigated.

2.0 Materials and Methods

2.1 Material Selection

In this study, PP, as shown in Fig. 1(a) is used as matrix material was supplied by Lotte Chemical Titan (M) Sdn. Bhd. in the form of homo-polymer pellets. KCF, as presented in Fig. 1(b) was obtained from Rahmatullah Sdn. Bhd. Malaysia. Once collected, KCF was dried, ground, and sieved into 150 μm mesh size to produce powder KCF.

2.2 Sample Preparation

2.2.1 Preparation of PP/KCF composite

Different ratios of KCF fibers (5, 10, 15, 20, and 25 wt%) were added to the PP matrix using formulation shown in Table 1. The sample that contains a pure PP matrix was used as the references and the control of the experiment. The KCF filler was physically mixed with the

PP matrix before extrusion. The mixture was oven-dried at 60 °C for 24 hours to remove the moisture before processing. The mixture then was melt blended in co-rotating twin-screw extruder. Extrusion will be conducted at a temperature around 180 °C and a rotor speed of 70 rpm. After that, the strands were air-dried and pelletized. Then, the pellets were oven-dried at 60°C for 24 hours before they were processed in the hot press compression moulding machine at 180 °C. Once the flat sheet was obtained, the specimens were cut into the following shape and dimension instructed in standard ASTM for flexural test and water absorption test.

Table 1. The formulation of PP/KCF composite.

Samples	PP Content (wt%)	KCF Content (wt%)
PP Neat	100	0
PPKCF5	95	5
PPKCF10	90	10
PPKCF15	85	15
PPKCF20	80	20
PPKCF25	75	25

(a)

(b)

Fig. 1. Raw material before compounding, (a) PP; (b) KCF

Fig. 2. Composite after compression moulding.

2.3 Mechanical Test

The flexural test was carried out by using an Instron Universal Tensile machine in the room atmosphere. The specimens were prepared for the flexural test according to ASTM D-790 with a crosshead speed of 30 mm/min and a support span of 50 mm. The software used in this test calculated the flexural strength and flexural modulus of the specimens.

2.4 Water Absorption Test

The water absorption test was carried out as elucidated in the ASTM D-570 standard. The test specimens were prepared with the dimension of 15.0 mm width x 30.0 mm length x 3.0 mm thickness, with smooth and squarely trimmed edges. The samples were then dried in an oven at 60 °C for 24 hours. After the drying process, all specimens were weighed, and dry weights of specimens were recorded. The samples were horizontally submerged in distilled water at room temperature, and weight change was monitored continuously as a function of time. For each composite, five specimens were tested. Absorption behavior of the composite was studied every 24 hours up to 672 hours (28 days), where the samples were removed from distilled water, wiped dry to remove any surface moisture, and then weighed using a high accuracy analytical balance. The percentage of water absorbency was calculated by using Equation 1.

$$\text{Percentage of Water Absorption, \%} = \frac{W_1 - W_0}{W_0} x\ 100\% \quad (1)$$

Where W_0 = Initial weight of the sample before immersion inside the water W_1 = Weight of wet sample

3.0 Result and discussion

3.1 Flexural properties

The most common purpose of the flexure test is to measure flexural strength and flexural modulus. These two values are often used by researchers to measure the material's ability to resist flexure forces. When flexural forces are applied to the specimens, the upper and lower surface of a sample under a three-point bending load is subjected to tension and compression stresses. In contrast, the axisymmetric plane is subjected to shear stress. Therefore, flexural properties are often considered as the result of combinations of tension and compression. The sample is deemed to fail when the flexural stress reaches a critical point [12, 13].

In this study, the effect of KCF loading on the flexural properties of PP/KCF composite is presented in Table 2. The bar charts for flexural strength and flexural modulus of PP/KCF composite are plotted and presented in Figures 3 and 4, respectively, for better observation. From Fig. 3, the results showed that the flexural strength of composite increases slightly with increasing of KCF filler content. The trend for flexural strength with different filler loading can be shown in Fig. 3. It can be observed that by initial incorporation of 5 wt% of KCF filler, flexural strength showed increment from 44.34 MPa to 46.08 MPa. The flexural strength continuously increased to 48.37 MPa when KCF reached its maximum filler loading at 20 wt% and decreased as more filler was added into the matrix. A significant drop of flexural strength from 48.37 MPa to 40.65 MPa can be observed at 20 wt% to 25 wt% of filler loading.

Based on the results obtained, lower flexural strength recorded at 5 wt% of KCF loading indicates that there were poor dispersion and fiber distribution, thus lead to poor adhesion between fiber and matrix. At the lowest loading of KCF filler, the KCF filler is unable to disperse and distribute well in the polymer matrix, therefore resulting in lower flexural strength. The dispersion and distribution of fiber in the matrix started to improve when more fiber is added to the matrix and give better strength [14]. The flexural strength dropped drastically when the ratio goes beyond 20 wt%. These results suggested that the addition of filler loading up to 20 wt%, the flexural strength decrease due to insufficient filling of the matrix in the composites, which limit the ability to wet and infiltrate the KCF filler. Ilangovan et al. have reported similar trends, in their study of PP composite reinforced raw bagasse. They stated that the flexural and tensile strength of the composite could be improved by adding the number of fibers in composite until it reached its maximum filler loading [15].

The flexural modulus of PP/KCF composite is presented in Fig. 4. The result showed a similar trend with flexural strength, but with significant improvement. It can be observed in Fig. 4 that the flexural modulus of the composites significantly increases from 1.16 GPa to 1.86 GPa with increasing fiber loading. In general, the increment of flexural modulus is heavily influenced by KCF content in composite. Since KCF is made up of lignocellulosic components, it has its stiffness that contributes to overall stiffness of composite [16]. Therefore, the significantly improved flexural modulus of the PP/KCF composite, which can be observed in Fig. 4, is due to high kenaf content combined with the stiffness of the composite. Tarres and the co-workers supported these observations as they stated in their study that good dispersion of reinforcement inside the polymer matrix could be achieved by observing the linear increase in the composite modulus [17].

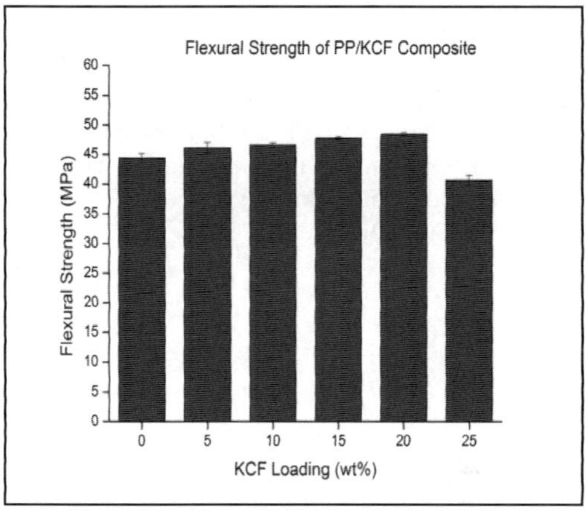

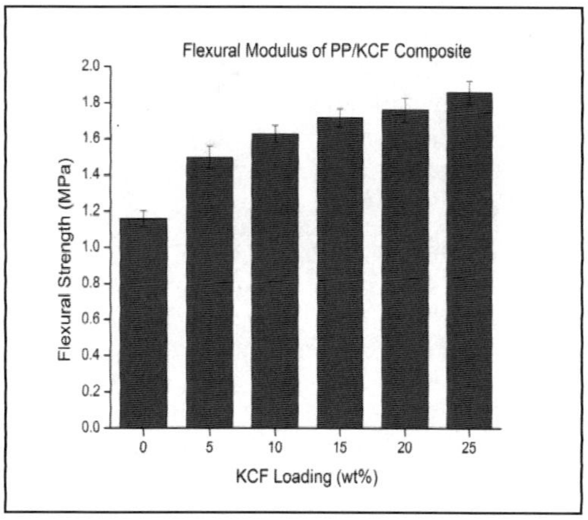

Fig. 3. Flexural strength of PP/KCF composite. **Fig. 4.** Flexural modulus of PP/KCF composite.

Table 2. Flexural properties of PP/KCF composite

KCF Loading (wt%)	Flexural Strength (MPa)	Flexural Modulus (GPa)
0	44.34	1.16
5	46.08	1.49
10	46.53	1.63
15	47.75	1.72
20	48.37	1.76
25	40.65	1.86

3.2 Water absorption behaviour

The water absorption rate of PP/KCF composite with various fibers loading was determined and presented in Fig. 5. The moisture uptake of each composite was identified by the percentage of the weight gain relative to the initial dry weight of the samples. From the result obtained, it can be observed that the water absorption amount of composite increased with increasing fiber content. In the first ten days of immersing the specimens in the distilled water, the results showed a rapid increase of the percentage of water absorption, and then only slightly increased until composite finally reached equilibrium point, where no more water uptake into a composite. Das and Biswas [18] have reported similar trends. In their study. They further explained that increasing water absorption of the composite is due to the hydrophilic nature of fiber compared to the PP matrix and greater interfacial area. Other than that, Hosseinihashemi et al. reported that as the fiber loading increased, the cellulose content also increased, thus increasing the number of hydroxyl (OH) groups in composites. Water absorption of the composite increased as these OH groups come in contact with water to form hydrogen bonding and result in weight gain of the composite [19].

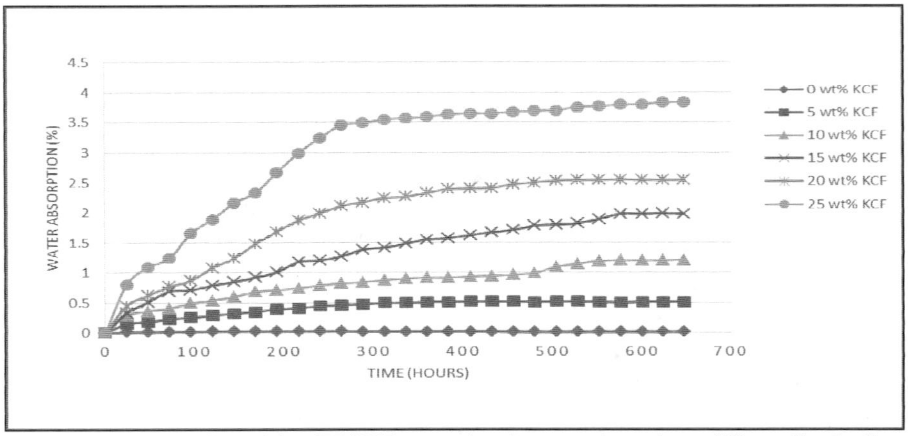

Fig. 5. The percentage by weight of PP/KCF composite after water absorption at different fiber loading.

4.0 Conclusions

The effect of different filler loading on the mechanical properties of PP/KCF composite has been studied. The results of this study showed that the flexural properties of PP/KCF composite increased with increasing of KCF loading until it reached 20 wt%. Further addition of fiber after that decreased the flexural properties due to insufficient filling of the matrix in the composite. Water absorption also shows significantly increased with increasing fiber content.

The authors would like to thank all the staff of Centre of Excellence Geopolymer and Green Technology (CEGeoGTech), School of Material Engineering, Universiti Malaysia Perlis (UniMAP) for all their involvement and support throughout this research. This research received no specific grant from any funding agency, in the public, commercial or not-for- profit sectors.

References

1. M. N. Akanbi, O. Ogbobel, M. U. Obidiegwu, P. C. Uzoma, G. C Onuegbu, P.I. Anynwu, IREJEST. **13**, 11–16, (2015)
2. C. Tezara, J. P. Siregar, H. Y. Lim, F. A. Fauzi, M. H. Yazdi, L. K. Moey, J. W. Lim, JMES, **10**, 12, 2159–2175 (2016)
3. A. Wechsler, S. Hiziroglu, Building and Environment, **42**, 7, 2637–2644, (2007).
4. A. A. Shamsuri, M. N. M. Zolkepli, A. H. M. Ariff, A. K. Sudari, M. A. Zarin, J. Compos., (2015)
5. M. K. F. M. Radzi, N. Muhamad, M. N. Akhtar, Z. Razak & F. M. Foudzi, Sains Malaysiana **47**, 2, 367–376, (2018).
6. M. K. F. M. Radzi, N. Muhamad, M. N. Akhtar, Z. Razak, F. M. Foudzi, Sains Malaysiana, **47**, 2, 367–376, (2018)
7. L. Mohammed, M. N. M. Ansari, G. Pua, M. Jawaid, M. S. Islam, Intl. J. Polym. Sci., **2015**, 1–15, (2015).
8. S. Subramonian, A. Ali, M. Amran, L. D. Sivakumar, S. Salleh, A. Rajaizam, AIME, **8**, 8, 1-5, (2016)
9. A. M. Fairuz, S. M. Sapuan, E. S. Zainudin, C. N. A. Jaafar, JMES, **10**, 1, 1931–

1942, (2016)
10. A. Amoke, O. Ogbobe, G. O. Tenebe, S. I. Ichetaonye, M. D. Ayo, Res. Rev. Polym. **8**, 1, 108, (2017)
11. N. I. Ismail, Z. A. M. Ishak, WOBIC, **368**, (2018)
12. M. Sood, G. Dwivedi, Egypt. J. Pet., **27**, 775–783, (2018)
13. N. M. Nurazzi, A. Khalina, 326–336, (2019)
14. N. I. Ismail, Z. A. M. Ishak, IOP Conf. Series: Materials Science and Engineering, **368** (2018).
15. M. Ilangovan, V. Guna, G. Keshavanarayana, N. Reddy, Sugar Tech, (2017).
16. H. D. Rozman, A. R. Rozyanty, L. Musa, G. S. Tay, J. Wood Chem. Technol., **30**, 152–163, (2010).
17. Q. Tarres, J. Soler, J. I. R. Sola, H. O. Ortega, F. Julian, F. X. Espinach, P. Mutje, M. D. Aguilar, Polymers, **11**, 1244, (2019)
18. G. Das, S. Biswas, Mater. Sci. Eng., **115**, (2016)
19. S. K. Hosseinihashemi, F. Arwinfar, A, Najefi, G. Nemli, N. Ayrilmis, Measurement, **86**, 202–208, (2016)

The effect of aging time on microstructure and hardness value of AZ80 Mg Alloy

Zuraidawani Che Daud[1,*], *Siti Hajar* Azahar[2], *Mohd Nazree* Derman[2], *Nur Maizatul Shima* Adzali [2], *Nur Hidayah* Ahmad Zaidi[2] and *Sinar Arzuria* Adnan[2]

[1, 2]Institute of Engineering Mathematics, Universiti Malaysia Perlis (UniMAP), Kampus Pauh Putra, 02600 Arau, Perlis, MALAYSIA.
[2]Centre of Excellence for Frontier Materials Research, School of Materials Engineering, Universiti Malaysia Perlis (UniMAP), 01000 Kangar, Perlis, MALAYSIA.

Abstract. AZ 80 Magnesium (Mg) alloy (AZ80) is the lightest structural metallic materials with good mechanical properties. However, Mg AZ80 has drawbacks which result in poor ductility and low strength where applications of Mg alloy have been restricted. The AZ80 has high aluminium content can cause the precipitation of ß-phase which is $Mg_{17}Al_{12}$ in Mg-Al alloy. It can affect the mechanical properties such as poor strengthening. This paper was discussed the effect of aging time on microstructure and hardness value of AZ80. The AZ80 samples were cut to 1cm × 1cm. Samples heat treated at 360 °C for one-hour quenching in water. Then, samples aged at 170 °C with different aging times (2 to 8 hours) with same quench. Optical Microscope (OM), Scanning Electron Microscopy (SEM), X-ray Diffraction (XRD) and Vickers Microhardness machine were used to analyse the samples. As the results showed ß-$Mg_{17}Al_{12}$ phase was discontinuously distributed along the grain boundary throughout solid solution treatment. The ß-$Mg_{17}Al_{12}$ phase did not fully dissolve into the α-Mg phase and distributed along the grain boundary. The results showed that sample after 2 hours aging time with highest hardness value 62.5 HV is the optimum sample.

1.0 Introduction

Magnesium (Mg) is the lightest metallic material used in structural applications with a density of 1.738 g/cm^3 compared to aluminium (Al) (2.70 g/cm^3) and iron (Fe) (7.86 g/cm^3). Mg alloys offer an excellent combination of properties that include an excellent strength-to-weight ratio, good impact strengths and fatigue, fairly large thermal and electrical conductivity and excellent biocompatibility [1].

Mg alloy can be classified into two different alloys which are wrought Mg alloy and cast Mg alloy. Mg-Al-Zn series alloys are widely used wrought Mg alloys for AZ31, AZ91, including AZ80 alloy. The wrought Mg alloy performs good mechanical properties compared to cast Mg alloy due to the deformation process that can eliminate pores,

* Corresponding author: zuraidawani@unimap.edu.my

homogenize composition and refine grain [2]. The AZ80 is a light magnesium alloy with excellent corrosion resistance and better forging ability [3]. Besides, a high content of aluminium in alloy such as AZ80 is associated with an appreciable amount of ß-phase, which is $Mg_{17}Al_{12}$. The $Mg_{17}Al_{12}$ is harmful to thc matrix's corrosion behaviour. Thus, by reducing the micro-galvanic effects, corrosion resistance can be enhanced [4]. However, AZ80 have drawbacks which result in poor ductility and low strength where application of Mg alloy has been restricted [2]. As AZ80 has high Al content caused the precipitation of ß-phase in Mg-Al alloy thus effect the mechanical properties such as poor strengthening [3].

Based on the earlier research by Ju-Mei et al, [5], the elongation and tensile strength of the alloy increased significantly. After solid solution treatment, the hardness and yield strength decreased. The durability and strength of the alloy increased with the prolongation of the aging time meanwhile the ductility declined. The Al is the most essential alloying element for magnesium. Commercial AZ80 alloys contain less than 10 wt.% Al (maximum solubility 12.7 wt.% at eutectic temperature 437 °C), and zinc and manganese as secondary alloying elements. The Al contents increases cast ability, tensile strength, compressive strength, and fatigue resistance. In Mg–Al alloys, Al is normally partly in a solid solution and partly precipitated in the form of ß-phase. The $Mg_{17}Al_{12}$, distributed along grain boundaries as a continuous phase as well as part of a lamellar structure [6]. According to Ben-Haroush [4], an appreciable amount of ß-phase, related with a high Al content in alloy such as AZ80. Cast AZ80 alloy typically includes a large fraction of precipitate $Mg_{17}Al_{12}$ along the grain boundaries. The $Mg_{17}Al_{12}$ is cathodic with respect to the matrix and it reveals a passive behaviour over a wider pH range than either of its components of Mg or Al.

Pardo et al., [7] stated that an increase in the relative size of the ß-phase at the expense of the Al rich-area will be increased the cathode to anode area ratio. Consequences, it causes greater localized corrosion. There is no effect on the nucleation of the primary Mg phase by add small amounts of Zn. Most of the Zn is segregated into a secondary phase close to the end of solidification. The morphology of eutectic phase resulting from non-equilibrium solidification of hypoeutectic Mg-Al alloys depends on the Al content of the alloy [8].

The samples will undergo an artificial aging process after past solution heat treated. Due to Ju-Mei et al., [5], the parameter of the samples was artificial aged at 200 °C for 4 to 32 hours. The continuous precipitates start and grow when the aging treatment prolongs to time which limits the nucleation and growth of discontinuous precipitate. Thus, the $ßMg_{17}Al_{12}$ phase precipitates continuous and discontinuous precipitates, which can be identified from their morphologies, in two different ways after aging treatment at 200 °C. Hence, the growth rate of the ß phase is fast at the early stage of nucleation, but the growth will stop as it reaches equilibrium. Reported from Huang et al., [3] the nucleation of the continuous precipitation (CP) is a homogeneous process with the ß-phase forming directly from the matrix. Therefore, after the aging treatment, there will no CP formed without solution treatment [3].

The effect of aging time was carried out on the AZ80. Due to some deficiencies in AZ80 such as the inadequate strength and poor ductility thus, solution and aging treatment is the recommended treatment to improve the mechanical properties of AZ80.

2.0 Methodology

The commercial AZ80 magnesium alloy ingot has composition of Mg - 8.2 wt.% Al and 0.51 wt.% Zn. This ingot cut to be 1cm x 1cm x 1 cm. The samples ground using silicon carbide (SiC) paper and polished to achieve mirror like surface.

Those samples heat treated at 360 °C for 1 hour in a furnace, followed by water quenching. The samples subjected to aging treatment at 170 °C with different aging times (2 hours, 4 hours, 6 hours and 8 hours), followed by water quenching.

The microstructure of the samples observed using optical microscope (OM) and scanning electron microscope (SEM), meanwhile phase analysis characterized using X-ray diffraction (XRD) and hardness using the Vicker's microhardness.

3.0 Results and Discussions

3.1 Microstructure Analysis

The sample revealed two phases which were α-phase and ß-phase. The presence of Al in AZ80 resulted in the formation of the phase ß-$Mg_{17}Al_{12}$ (BCC) structure, forming an incoherent interface with the structure of the α-phase (HCP). The ß-phase is a compositional intermetallic stoichiometric compound. There are two distinct and competitive precipitation modes for continuous and discontinuous precipitation of BCC ß-$Mg_{17}Al_{12}$ from the parent HCP phase [9].

Fig. 1 showed the microstructure of untreated AZ80 and those aged at 170 °C for different aging time. Under all conditions, The AZ80 microstructures consists of which is the light background and ß-$Mg_{17}Al_{12}$ that had the black regions or lines. Most of ß-$Mg_{17}Al_{12}$ formed networks in Fig.1 (a) of untreated sample at the grain boundaries and are discontinuously broken and dispersed after the heat treatment. Few ß phases can be seen in the untreated sample of AZ80. However, the grains of heat treated samples grew bigger.

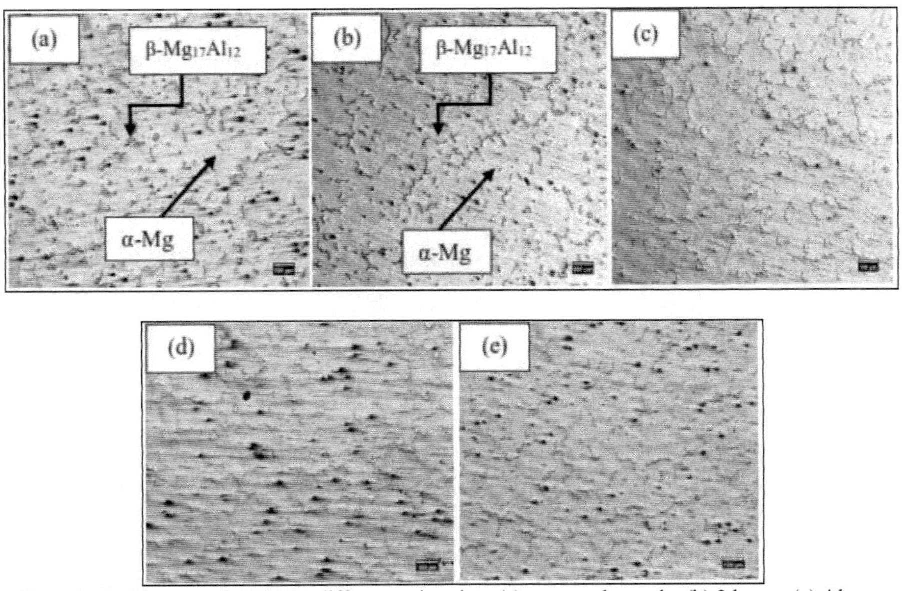

Fig.1. Optical images of AZ80 by different aging time (a) untreated sample; (b) 2 hours; (c) 4 hours; (d) 6 hours and (e) 8 hours.

The main ß-$Mg_{17}Al_{12}$ phase was discontinuously distributed along the grain boundary throughout solid solution treatment and did not fully dissolve into the α-Mg phase. Then, after the aging treatment, the residual ß-$Mg_{17}Al_{12}$ phase distributed along the grain boundary. Fig.1 (b) and (c) showed the amount ß precipitate of each specimen also increases when aging time increases from 2 hours to 4 hours. However, amount of ß precipitate decrease after 4 hours aging as the prolonging holding time. Due to the heat treatment, the amount and distribution of ß phase was significantly changed.

During aging treatment, aluminum (Al) atoms diffuse into grain boundaries to form ß precipitate, reducing the aluminum concentration in the α-phase during aging treatment. As the aging time increasing, the proportions of ß phase $Mg_{17}Al_{12}$ grains also increasing. In spite of that, the ß-$Mg_{17}Al_{12}$ phase as the temperature elevates and the holding time extended. These results were attributable to the solid solution effect of gains in ß-phase $Mg_{17}Al_{12}$, resulting in a reduction in the diffusion phenomenon of ß- $Mg_{17}Al_{12}$ phase grain into the α-phase Mg matrix. This finding also supported by Chye at al.,[10].

Fig.2 represents the SEM images of the untreated and treated AZ80 with different aging time. Few ß phases can be seen in the both untreated and treated samples of AZ80 and also there are large particles along grain boundaries. Fig. 2 (b-e) shows the SEM micrographs of AZ80 aged at 170 °C for different aging time, from which the coarse white ß-$Mg_{17}Al_{12}$ particle can be observed in the specimens where the phase ß-$Mg_{17}Al_{12}$ was distributing discontinuously along the grain boundary after being treated for one hour at 360°C solution treatment followed by water quenching. Due to solution and secondary precipitate, the size of ß-$Mg_{17}Al_{12}$ precipitates in the untreated sample alloys is smaller compared to that in the treated sample alloy.

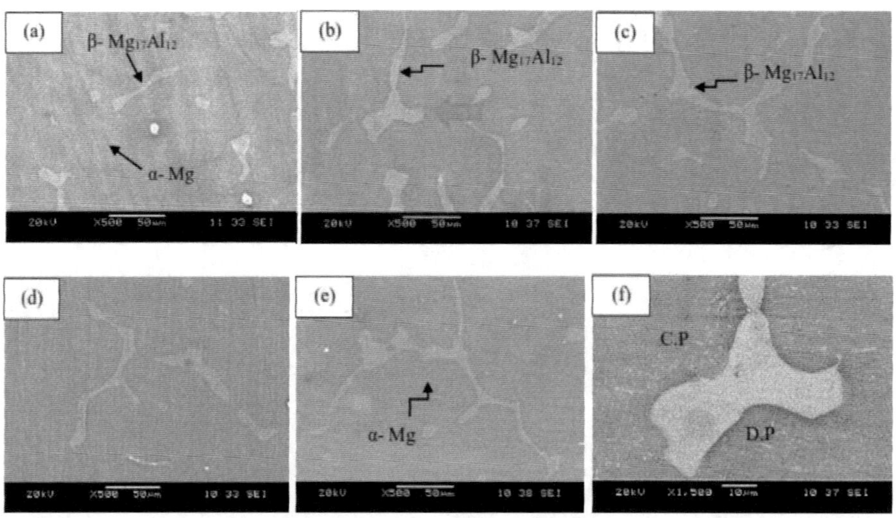

Fig.2 SEM images of AZ80 by different aging time (a) untreated sample; (b) 2 hours; (c) 4 hours; (d) 6 hours and (e) 8 hours; (f) Magnification of blue arrow at **Fig. 3** (b), showing a large amount of lamellar discontinuous precipitation (D.P.) at grain boundaries and granular-shaped continuous precipitates (C.P.) in the matrix

It could be observed that high amount of precipitates existed following aging. Fig.2 (f) shows the high magnification SEM micrograph of discontinuous precipitation at grain boundaries and continuous precipitation within grain boundaries as directed by the blue arrow in Fig.2 (b). The figure shows that the ß- phase was precipitated into two forms from the super-saturated α-Mg solid solution, which is the discontinuous precipitate (DP) and continuous precipitate (CP). Furthermore, DP began to grow into the grains with lamellar growth from the grain boundaries, which is a typical morphology for DP of ß-$Mg_{17}Al_{12}$. In the remaining areas, CP was found to form within the unoccupied grains of DP and the interface in both DP and CP areas was clear. The discontinuous precipitation stops with the rising aging time, and the continuous precipitates formed in the residual grain regions that are not occupied by discontinuous precipitates are continuing to increase.

3.2 X-Rays Diffraction Analysis

Fig.3 shows the XRD pattern for samples aged at 170 °C with different times (2 hours, 4 hours, 6 hours and 8 hours). The XRD result shows that the AZ80 consists of a-Mg, ß-$Mg_{17}Al_{12}$, $Mg_{32}(AlZn)_{49}$, Al_3Mg_2, AlMg and Mg_7Zn_3. However, the α-Mg and ß-$Mg_{17}Al_{12}$ were mainly present. After the aging treatment, the XRD result in Fig.3 (b) revealed that the ß- $Mg_{17}Al_{12}$ phase has the highest diffraction peak between 32°, 34° and 37°. The intensity of ß-$Mg_{17}Al_{12}$ phase becomes higher than α-Mg. Moreover, by comparing the Fig. 3 (a) and (b), the diffraction peak of ß-$Mg_{17}Al_{12}$ phase in Fig.3 (b) is higher than Fig.3 (a). Thus, the higher intensity result in the more phase presents. This result shows positive hardening effect towards the AZ80 alloy that aged at 2 hours.

Fig. 3(c) shows the XRD analysis for AZ80 aged at 170 °C for 4 hours of aging. The α-Mg matrix appears and has the highest intensity at the peak 37°. It's shown five peaks formation of α-Mg at the peak 33°, 37°, 48°, 58°and 69°. While the formation of ß-phase decrease where only three peak formed at 40°, 58° and 63°, this analysis can be conclude that aging for four hours will decreased the hardness compared to aging for two hours. In addition, Fig. 3 (d) and (e) shows the decreasing number of ß-phase on both six and eight hours of aging, the number of ß-$Mg_{17}Al_{12}$ were deceased. Other peaks have full with formation of α-Mg, $Mg_{32}(AlZn)_{49}$, Al_3Mg_2, AlMg and Mg_7Zn_3. Therefore, from the both Fig. 3(d) and (e), hardening effect of AZ80 alloy that aged at 170 °C for six hours and eight hours are poorer than four hours aging.

Ju-mei et al., [5] found that the diffraction peak of the ß- $Mg_{17}Al_{12}$ vanished after solution treatment, but then the ß-phase appeared after the aging treatment. The volume fraction of the ß$Mg_{17}Al_{12}$ phase increases very rapidly during the initial aging treatment period, however when the aging time has been prolonged, the volume fraction increase obviously slows down.

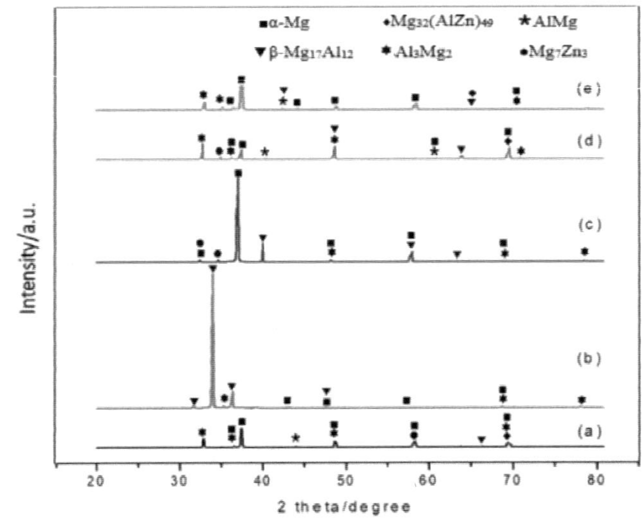

Fig. 3. XRD patterns of the Mg AZ80 aged at 170 °C for, (a) untreated (0 hours); (b) 2 hours; (c) 4 hours; (d) 6 hours; (e) 8 hours.

3.3 Hardness Analysis

Fig.4 is the effect of the aging time on the hardness. After two hours aging process, the hardness of samples was slightly dropped. The average hardness value of the untreated sample is 61.4 HV. During aging treatment, the hardness of all samples increased with aging time until it reached the maximal hardness 62.5 HV at two hours of aging time. However, after the two hours aging time, all treated samples were decreased with the increasing of aging time. The hardness values decrease due to the increasing of aging time. Due to the Chye et al., [10] aluminium (Al) atoms diffuse into grain boundaries during aging treatment to form ß-precipitate and this will reduce Al concentration during aging treatment in a phase. Moreover, the shape and amount of ß precipitate are affecting the hardness. The more the formation of ß precipitate will lead to the higher the aging hardness. The reason is caused by the resistance from the ß-phase formation where the ß-phase movement increased significantly and the structure became more difficult to move. But the prolonging of aging time will cause the material to fracture and reduce the hardness value.

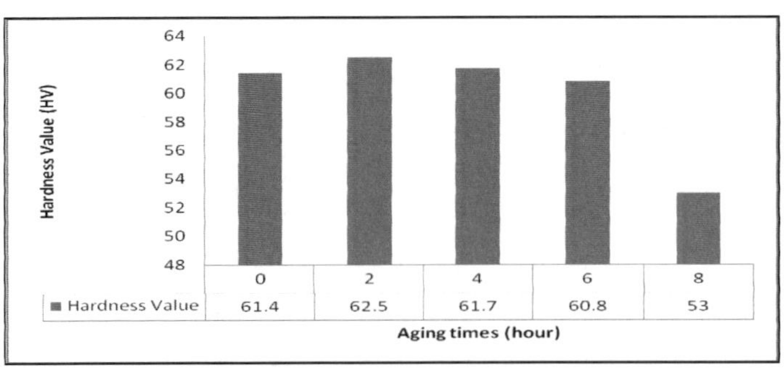

Fig.4. Hardness value (HV) for different aging time.

The hardness after heat treatment affected by the amount of secondary phase, the formation of super saturated grains and the change in average grain size. Reduced amount of secondary phase and grain growth will decrease the hardness while when the formation of super saturated grains decreases, the hardness will increases.

The phenomenon of decreasing the hardness value with increasing sensitization temperature and time can be attributed in part to decreasing the dislocation density during sample recrystallization. The density of dislocations would reduce as other grains are framed in the heat treatment of sensitization. It may also attribute precipitation of the Mg-rich stage and isolation of Mg molecules at grain boundaries, which extract solute Mg particles from the α-matrix. Thus, the remaining Mg particles left behind in a solid solution become ineffective barriers to dislocation movement [11].

4.0 Conclusion

1. The ß-$Mg_{17}Al_{12}$ phase was getting to distribute along the grain boundary discontinuously via aging treatment at 170 °C after solution treatment at 360 °C for 1 hour followed by water quenching.
2. Aging treatment was found as a method of improving mechanical properties at 170 °C after solution treatment. Therefore, the optimal aging time of AZ80 is 2 hours as the highest hardness value is 62.5 HV.

References

1. M. Mezbahul-Islam, A. Mostafa, and M. Medraj, J. Mater., 1–70 (2014)
2. S. Jing, Z. Dingfei, T. Tian, Y. Daliang, X. Junyao, and P. Fusheng, Rare Met. Mater. Eng., **46 (7),** 1768–1774 (2017)
3. X. feng Huang, Y. Zhang, F. Guo, J. chang Yang, Y. Ma, and Y. Hao, China Foundry, **15(2),**103–109 (2018)
4. M. Ben-Haroush, G. Ben-Hamu, D. Eliezer, and L. Wagner, Corros. Sci., **50(6),** 1766–1778 (2008)
5. Z. Ju-mei, J. Bai-ling, W. Zhi-hu, Y. Sen, and N. A. N. Hong-qiang, China Foundary, **4**(4), 3–6 (2007)
6. A. Pardo, M. C. Merino, A. E. Coy, F. Viejo, R. Arrabal, and S. Feliú, Electrochim. Acta, **53,** 7890–7902 (2008)
7. A. Pardo, M. C. Merino, A. E. Coy, F. Viejo, R. Arrabal, and S. Feliú, Electrochim., **53**, 7890–7902 (2008)
8. C. A. Sager *et al.*, Mater. Soc. Annu. Meet.,**12**, 405–410 (2009)
9. Z. Liu, X. Liu, Z. Hou, S. Zhou, and Q. Tian, J. Mater. Eng. Perform.,**25**(9), 3811–3821 (2016)
10. L. Chye, M. Z. M. Zamzuri, S. Norbahiyah, K. A. Ismail, M. N. B. Derman, and S. Illias, Adv. Mater. Res., **685**,102–106(2013)
11. S. C. V, R. Dumpala, A. K. S, K. VV, and R. S. B, J. Magnes. Alloy.,**6**, 52–58, (2018)

Bonding and Phases Analysis of Geopolymer Materials

Nor Shida Dalila Mohd Azhar[1,*], *Farah Farhana* Zainal[1], *Mohd Mustafa Al Bakri* Abdullah[1]

[1]Center of Excellence Geopolymer and Green Technology (CEGeoGTech), School of Materials Engineering, Universiti Malaysia Perlis (UniMAP), P.O. Box 77, D/A Pejabat Pos Besar, 01000 Kangar, Perlis, Malaysia

Abstract. This paper presents the bondings and mineral phases exist in geopolymer material consist of raw fly ash, raw metakaolin and also fly ash and metakaolin blend geopolymer. Fly ash used is a waste materials from Lumut power plant by combusting coal while metakaolin was obtained by calcining the kaolin at 750 °C for 4 hours. Alkaline activators that act as binder were sodium silicate (Na_2SiO_3) solution and sodium hydroxide (NaOH) solution with the ratio of Na_2SiO_3/NaOH is 2.5 to 1. The samples undergo Functional Group Identification (FTIR) and Phase Analysis (XRD). FTIR results show that after geopolymerization process, the band shifted to lower wavenumber (776 cm^{-1}) due to interaction of Al atoms in the silicate geopolymer network because of higher cross-linking. XRD results show mineral phases of quartz and mullite in geopolymer after geopolymerization process and it can be said that not all mineral phases from raw material participating in formation of geopolymer matrix.

1 Introduction

The term geopolymer refers to inorganic aluminosilicate polymers. These inorganic polymers are generally obtained by pozzolanic reaction between aluminosilicate and basic alkaline solution [1]. Pozzolan materials that commonly used for the geopolymer synthesis are fly ash, blast furnace slag or metakaolin [2]. The chemical composition of fly ash and metakaolin are varies usually leads to different physical and mechanical properties from one aluminosilicate material to another [3].

Fly ash is the main product of coal combustion which was vastly used in many industries especially in construction area of interest. This is due to their good pozzolanic and cementitious properties [4,5]. Fly ash is a by-product of coal combustion that contains mainly quartz, hematite, mullite, and amorphous, that have high levels of silicate and aluminate.

On the other hand, metakaolin also have good pozzolanic properties where the calcination condition of kaolin affect the characteristics of resulting inorganic aluminosilicate polymers [6]. Based on previous studies state that geopolymers with optimal characteristics are obtained and found at temperature range 700 °C to 800 °C [1, 5]. Castelein et al investigated

*Corresponding author: norshidadalila@gmail.com

the effect of kaolin to reveal its influence on the formation of mullite also neighboring kaolinite [1, 5]. Metakaolin is prepared by dehydroxylation process of kaolin to remove the chemically bonded hydroxyl ions [6-8].

Because of SiO_2 and Al_2O_3 are the main chemical constituents of fly ash and metakaolin, they are considered as important raw materials for geopolymer [9, 10]. This paper will discussed the bonding and mineral phases between the raw material and the fly ash-metakaolin blend geopolymer by using FTIR and XRD analysis.

2 Methodology

2.1 Materials Selection and Samples Preparation

The materials that used to prepare the geopolymer paste was made from the mixture of fly ash and metakaolin. Fly ash raw used in this study was obtained from Lumut Power Plant, Perak, Malaysia while metakaolins used in this study were obtained from kaolin provided by the from Associated Kaolin Industries Sdn. Bhd. Malaysia. Before using the kaolin, this material was enriched with kaolinite and calcinated at 750 °C in electric muffle furnace [11]. Calcination process lasted 4 hours at the rates of calcination 10 °C/min.

The alkaline solution was prepared by mixing sodium silicate and sodium hydroxide solution (12 M). The sodium hydroxide solution was obtained by dissolving sodium hydroxide pellets with a purity of 99% in distilled water and kept it at room temperature for 24 hours [12].

2.2 Functional Group Identification (FTIR)

FTIR spectroscopy is an analysis method which searches for the major reaction zones of Si-O and Al-O geopolymers. Functional groups shifting of the materials were identified using spectrometer. The specimens was powdered samples scanned from 4000-600 cm^{-1}.

2.3 Phase Analysis (XRD)

The phase analysis was conducted by using Shimadzu X-ray diffractometer. Specimen for analysis also in powder form. X-ray diffraction are very useful to detect structural changes of the material. Characteristics peaks in the XRD pattern of geopolymer materials commonly show quartz, mullite, hematite, magnetite also silliminite. The XRD pattern was analyzed using HighScore Plus software.

3 Results and Discussion

3.1 FTIR Analysis

Fig.1 and Table 1 represent the FTIR spectra of raw fly ash and metakaolin also fly ash and metakaolin blend geopolymers. From Fig.1, the infrared spectrum of fly ash shows the main absorption band occurred at 959-1020 cm^{-1}.These spectrum band are corresponding to symmetric stretching vibration of Si-O-Si and Si-O-Al bonds of SiO_4 and AlO_4 tetrahedrons of the geopolymer network [12]. However, different to fly ash, the FTIR spectra of the metakaolin exhibited absorption bands at 3624 cm^{-1} which are the result of the network of kaolinite.

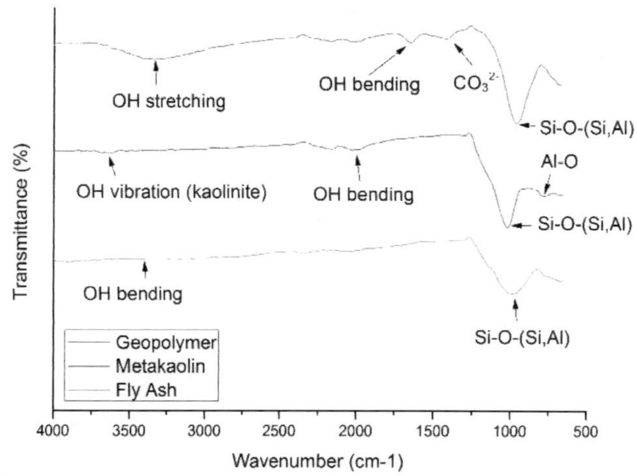

Fig. 1. FTIR spectra of fly ash, metakaolin and Geopolymer.

Table 1. Assignment of main FTIR bands.

Band	Assignment
3624 cm^{-1}	Vibration of OH^- groups of kaolinite
3340-3480 cm^{-1}	OH^- stretching vibration
1648,2020 cm^{-1}	OH^- bending vibration of H-O-H
1390-1420 cm^{-1}	Stretching vibration of CO_3^{2-}
1020 cm^{-1}	Stretching vibration of Si-O-Si and Si-O-Al
959 cm^{-1}	Stretching vibration of Si-O-Si and Si-O-Al
776 cm^{-1}	4-coordinated Al-O bond

In FTIR spectra of geopolymers (Fig. 1) the absorption bands at 3340–3480 ,2020 and 1648 cm^{-1} corresponded to the stretching and bending vibrations of O-H and H-O-H bonds of water molecules [13]. The absorption band at 1390 –1420 cm^{-1} also refer to stretching vibrations of C-O bond of CO_3^{2-} ion which results from the reaction of atmospheric CO_2 with residual sodium content [13]. After geopolymerization process, the band shifted to lower wavenumber (776 cm^{-1}) due to the silicate geopolymer network of Al atoms and it is also indicated formation of higher cross-linking and larger molecular [12].

3.2 XRD Analysis

Fig.2 presents the XRD diffractograms of fly ash, metakaolin and geopolymer. Fly ash and metakaolin was used as primary binder materials. Fly ash shows a range at 2θ between 15°

to 36° indicates the presence of amorphous phase [14]. According to the obtained patterns, the raw fly ash mainly contain of quartz, mullite and hematite. Hematite is the mineral forms of iron oxides, while quartz and mullite indicate the presence of silica and alumina.

Metakaolin showed that the material is rich in kaolinite, quartz, muscovite and hematite peaks. At 2θ,the pattern shows a broad area located in the range 12° to 45°. After the activation process, the crystalline phases were dissolved in the alkaline solution and the aluminosilicate phase was formed in the surface of blend fly ash and metakaolin paste by geopolymerization reaction [15].

However, the differences between raw materials and geopolymer sample show certain changes with the peak representing the amorphous phase. Mineral phases of quartz, mullite were still present in geopolymer but with reduced in intensity and also there were sillimanite phases presence. These results reflected not all mineral phases are reacting in the geopolymerization process.

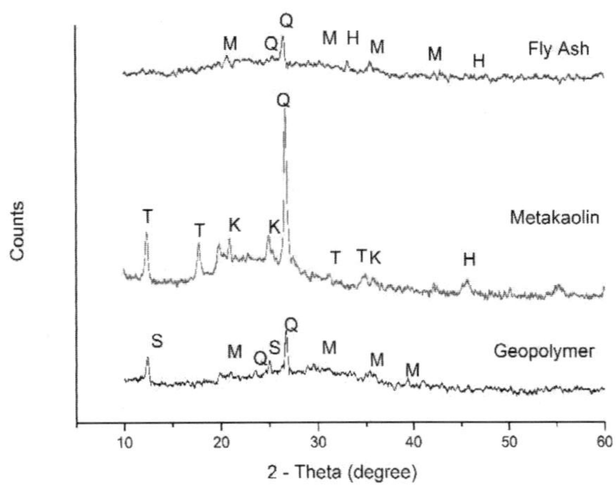

Fig. 2. XRD diffractograms of fly ash, metakaolin and geopolymer (Q – Quartz (PDF no. 01-083-2468), M – Mullite (PDF no. 01-074-2419), S – Sillimanite (PDF no. 01-084-0983), H – Hematite (PDF no. 01-086-0550), T – Muscovite (PDF no. 00-046-1311), K – Kaolinite (PDF no. 00-029-1311)).

4 Conclusions

Thus, as for XRD and FTIR analysis showed that after geopolymerization process, there will be slightly changes in structural phases and the functional group.

The authors would like to thank the staffs of Center of Excellence Geopolymer & Green Technology (CEGeoGTech), School of Materials Engineering, Universiti Malaysia Perlis (UniMAP) for their involvement in the research. This work was supported and funded by the Fundamental Research Grant Scheme (FRGS) under a grant number of **FRGS/1/2018/TK06/UNIMAP/02/2** from the Ministry of Education Malaysia.

References

1. B. K. Diffo, A. Elimbi, M. Cyr, J. D. Manga, and H. T. Kouamo, J Asian Ceram. Soc. **3(1)**, 130–138 (2015)
2. Ding, Y.-C., Chang, W.-H., Lee, W.-H., Wang, S., & Cheng, T.-W. Struct Cooncrere, **19(5),** 1521–1528 (2018)
3. S. N. Fifinatasha *et al.*, Adv. Environ. Biol, **7**(12) 28–29 (2013)
4. D. M. J. Sumajouw, D. Hardjito, S. E. Wallah, and B. V. Rangan, J. Mater. Sci, **42**(9), 3124–3130 (2011)
5. Luna-Galiano, Y., Cornejo, A., Leiva, C., Vilches, L. F., & Fernández-Pereira, C. Mater Construcc, ***65*****(319)** (2015)
6. Boonanunwong, P., Keawpapasson, P., Tippayasam, C., Thavorniti, P., Chindaprasirt, P., & Chaysuwan, D. Key. Eng. Mater. **690,** 179–186 (2016)
7. S.E. Wallah and B.V Rangan, 220-225 (2013)
8. J.M. Khatib, O. Baalbaki, A.A Elkordi, 493–511 (2018)
9. Ma, C.-K., Awang, A. Z., & Omar, W. Constr Build Mater **186,** 90–102 (2018)
10. Zainal, F. F., Hussin, K., Rahmat, A., Abdullah, M. M. A. B., & Sandu, A. V. Appl. Mech. Mater. **754-755**, 892–896 (2015)
11. Diffo, B. K., Elimbi, A., Cyr, M., Manga, J. D., & Kouamo, H. T. J. Asian Ceram. Soc. **3(1),** 130–138 (2015)
12. Cheng-Yong, H., Yun-Ming, L., Abdullah, M. M. A. B., & Hussin, K. Sci. Rep. **7(1)** (2017)
13. N.A. Jaya, L. Y. Ming, H. C. Yong, M. M. A Bakri, " (2018)
14. Kljajević, L. M., Nenadović, S. S., Nenadović, M. T., Bundaleski, N. K., Todorović, B. Ž., Pavlović, V. B., & Rakočević, Z. L. Ceram Int. **43(9)**, 6700–6708 (2017)
15. M. E. Alouani, S. Alehyen, M. E. Achouri, and M. H. Taibi, J. Chem, 1–14, (2019)

IOP Conf. Series: Materials Science and Engineering **957** (2020) 012053 doi:10.1088/1757-899X/957/1/012053

Effect of Pulse Frequencies on Low Carbon Steel in 3.5 wt% NaCl Solution under Zero Charge Corrosion Protection

Rajaselan Wardan[1,*], *Shaiful Rizam* Shamsudin[1], *Mohd Fitri* Mohamad Wahid[1] and *Muhammad* Syazwan Sanusi[1]

[1]Center of Excellence Geopolymer & Green Technology, School of Materials Engineering, University Malaysia Perlis, Jejawi 02600 Arau, Perlis, Malaysia

Abstract. Zero charge corrosion protection (ZCCP) is an alternative to the existing cathodic protection methods. It applies a current/voltage pulse with a certain particular frequency to attain a potential of zero charge (E_{pzc}) on the electrode-electrolyte interface. A study on the pulse frequencies of ZCCP system on low-carbon steel in 3.5 % NaCl Na solution was carried out for 14 days. The pulse frequencies are varied from 4, 20, 50 and 100 Hz. The ability of these frequencies in protecting samples is determined by means of corrosion rate based on weight loss analysis and surface morphology. Experimental observations indicate that low carbon steel corrosion protection is strongly influenced by changes in pulse frequency. Corrosion rates were found to gradually decrease as pulse frequency increased from 4, 20 to 50 Hz, nevertheless at higher frequencies (100 Hz), corrosion protection was observed to be less effective where the corrosion rate dramatically grew up. The surface morphology of the steel surface is smooth and protected at a pulse frequency of 50 Hz. This indicates that the current/ voltage pulse of ZCCP system only requires 50 Hz which is equivalent to a standard AC power frequency applied in most countries including Malaysia.

1 Introduction

Carbon steel is utilized as a significant construction material because of its low generation cost and extraordinary mechanical properties required to be utilized as strengthening bar [1]. However, the challenge is that it has low corrosion resistance especially in seawater environment. A high integrity of corrosion protection is required as it is utilized in numerous destructive conditions, for example, chemical production, marine, and contaminated condition. Corrosion product shaped on the steel surface called rust involves different oxides. Among these oxide, ferrous hydroxide, $Fe(OH)_2$ can be shaped in impartial to fundamental condition, while geothite a-FeOOH and lepidocrocite cFeOOH are the significant constituents in medium to high acidic condition [2]. Moreover, in marine condition, a.k.a.

* Corresponding author: rajclan9143@gmail.com

ganeite, b-FeOOH is present in alpha and gamma variations. Magnetite, Fe_3O_4 and hematite, a-Fe_2O_3 are additionally seen as present in normal just as in faster corrosion state [3].

During corrosion, metals will in general proselyte to all the more thermodynamically stable mixes, for example, oxides, hydroxides, salts, or carbonates. Corrosion is a leading cause of premature failure in metallic structures [4-5].

The zero-charge corrosion protection (ZCCP) is a promising technique to control metal structures harm brought about by oxidation response because of atmosphere changes, ecological variables, and operational conditions. Potential of zero charge (E_{pzc}) is an estimation of the electrode potential for any metal at which a spotless surface of the metal would not procure any electrical charge when it comes into contact with an electrolyte [6-7]. At E_{pzc}, a metal surface is not in contact with a fluid arrangement because of the non-appearance of electrical charge and no electrostatic appeal [8]. ZCCP exploit the frequency of current/voltage pulse waves to the metal that needs to be protected. The frequency of a current is the number of repetitions of one complete cycle of the waveform per second and it is measured in hertz (Hz). The waveform describes the shape of one cycle of the voltage or current. Current can be generated as an alternating current (AC), where the direction of the current flow alternates around zero with positive and negative direction. Most of the countries use 50/60 Hz as the utility frequency because power been supplied is more stable and efficient. Moreover, higher frequency than 50/60 Hz causes some distribution losses [9]. This indicates that frequency change can impart serious effect on the corrosion protection.

The purpose of this study is to investigate the effect of different frequencies of pulse current /voltage of ZCCP system on low carbon steel in 3.5 wt% NaCl solution.

2 Research methodology

The samples of carbon steel was ground and analyzed by optical emission spectrometry, OES (Q8, Magellan, Bruker) for the chemical composition content. The samples were polished to remove the upper oxide layers, rust and corrosion deposits to get a smooth and shiny surface. It was then coated with polymer-based paint. An area of 1 cm^2 was exposed for experimental purposes. The shape and size of the sample are shown in Fig. 1.

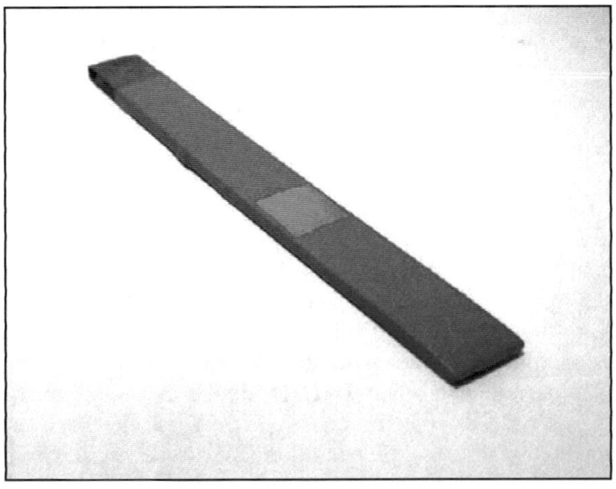

Fig. 1. Shape and size of the sample exposed area.

The painted steel samples were weighted by using analytical balance (GR-200, A&D Weighing) to record the sample weight before the immersion test loss. A custom-made power supply designed for ZCCP experiments is connected to a steel sample. Oscilloscope is used to measure the waveform pulse frequencies generated by the power supply. The experiment was performed by immersing the steel sample in 3.5 % NaCl solution for 14 days at room temperature (25 °C). The ZCCP system was set at different frequencies from 4, 20, 50 and 100 Hz. After immersion testing, exposed steel surface was cleaned by 2 % nital solution and rinsed with the distilled water. Samples were then weighted again to determine their weight loss.The corrosion rates were determined by using Equation (1) [10]

$$\text{Corrosion rate} = \frac{87.6\Delta w}{\rho A t} \tag{1}$$

where A is the exposed surface area (cm^2), t is the exposure immersion time, ρ is the density of the sample (g/cm^3) and w is the weight loss (g).

One unit of steel sample was also immersed without any corrosion protection, which is used as a reference and designated as open circuit potential (E_{ocp}). The surface morphology was then observed using stereomicroscope (SZX16, Olympus) to visualize the presence of corrosion attack on the steel samples. The purpose of this study is to investigate the effect of different frequencies of pulse current /voltage of ZCCP system on low carbon steel in 3.5 wt% NaCl solution.

3 Results and discussion

The carbon steel used in this study is in compliance with AISI 1016 (Table 1). Thus, it is a thermodynamically unstable material in aqueous environment, which tends to corrode, moreover in the presence of saltwater to give a product of oxyhydroxides of iron, reddish color, FeOOH. [11].

Table 1. Chemical composition of samples

Element	Sample	AISI 1016
Iron, Fe	98.55	98.13 – 99.58
Manganese, Mn	0.78	0.60 – 0.90
Carbon, C	0.13	0.12 – 0.18
Sulfur, S	0.038	≤0.050
Phosphorous, P	0.020	≤ 0.040

Fig. 2 shows the corrosion rate versus immersion time for different pulse frequencies of ZCCP system. The corrosion rate of E_{ocp} shows the highest value 3.018 µm/year. It is the actual thermodynamically stable potential of AISI 1016 steel in 3.5 % NaCl solution under experimental condition. A high corrosion rates in E_{ocp} samples are indeed expected. It is noted that in literature, the E_{ocp} value of mild steel is within the range of 60 - 70 µm/year [12].

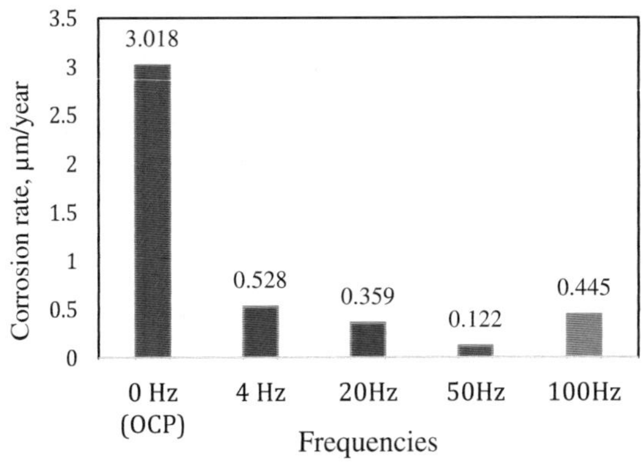

Fig. 2. Corrosion rate of different frequency of zero charge corrosion protection.

In ZCCP system, lower pulse frequency (4 Hz) recorded 0.528 µm/year and it was gradually decrease to 0.359 µm/year at 20 Hz frequency. The lowest corrosion rate, 0.122 µm/year was obtained at 50 Hz, while at 100 Hz frequency, the corrosion rate was dramatically increase to 0.445 µm/year. The changes in the corrosion rate is due to the pulse frequency factor. In this case, frequency is a function of the pulse energy [13]. Frequency plays an important role in pushing the flow of electrons out of the metal center into the steel "skin" layer. This mechanism is called "skin effect". This skin effect reduces the ability of a metal to act as an electrical conductor. The amount and energy of electrons in the skin's effect prevent the conversion of Fe atoms to Fe^{2+} ions. Taking into account the characteristics of charges on the electrode-electrolyte interface, the effect of the skin is equivalent to the metal character at the zero charge potential, E_{pzc}. At the low frequencies, 4 Hz, the pushing energy of the electrons in the metal center is insufficient to develop a skin effect. Therefore, a corrosion reaction occurs at high rates. The corrosion rate decreases as the frequency of the pulse energies to form a skin effect gradually reaching the maximum effect to 50 Hz where all the Fe atoms exchange to Fe^{2+} ions are inhibited. Coincidently, this 50 Hz frequency is the same as the alternating current in the household electrical outlets in most countries including Malaysia [14]. It is possible that 60 Hz frequency can be studied where it could have an effect that is almost identical to the corrosion rate at 50 Hz. However, at high frequencies (more than 50 or 60 Hz) the current/voltage pulse produces high energy. The flow of the electrons in the metal center is pushed out beyond the metal skin or surface. This mechanism causes the metal surface to have an excessive negative charge. This in turn causes the metal with excess negative charge to react with the ionic charges of opposite polarity.

The results have been strengthened by the evidence of the surface morphology shown at Fig. 3 as the 4 Hz has the worst corrosion attack as the surface represented by crater-like morphology. Flat but rough and uneven surface was produced at 20 Hz. At the 50 Hz, smooth and much even surface was produced indicating the samples were well protected. Meanwhile, sample of 100 Hz shows identically morphology as the lower frequency. However, smaller craters were produced.

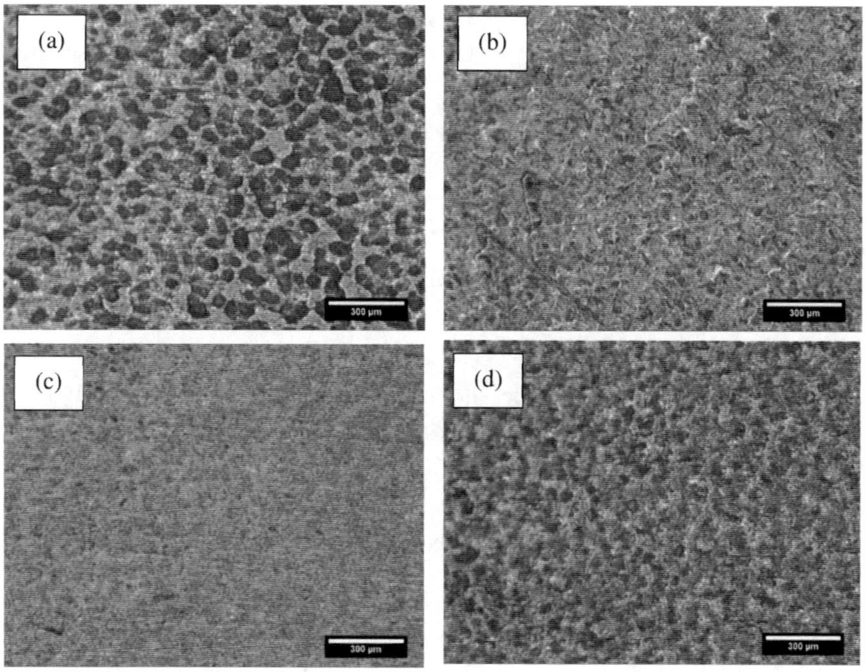

Fig. 3. Surface morphology of immersed samples, (a) 4 Hz, (b) 20 Hz, (c) 50 Hz and (d) 100 Hz.

Low frequencies of pulse current/voltage (e.g. 4 Hz) were insufficient to restrain the adsorption of the excessive negative charges from oxygen water dipole. It also failed to prevent the conversion of Fe atoms to Fe^{2+} ionic charges. Therefore, at low frequencies, the ZCCP mechanism is unable to act as a carbon steel corrosion protection. However, this corrosion protection capability gradually increases as the frequency were amplified from 20 Hz to 50 Hz, With the removal of the charges at the interface until the magnitude of the charge reaches zero, the charge transfer does not occur and no corrosion response occurs. At zero charge, the double-layer no longer exists at the metal-solution interface. To protect against corrosion, this zero-charge situation must be maintained to prevent the charge transfer. This shows that at this frequency range, ionic charges are depleted or removed from the surface of the samples and it is sufficient enough to maintain the zero-charges layer and promote the capability of ZCCP towards corrosion protection. But at 100 Hz, the current become overload than the current needed. At the higher pulse frequency, the charging and discharging processes vigorously applied to the ZCCP system, resulting the changes of water molecule orientation to be more energetic and faster. It provides sufficient energy to reinduced the adsorption and charges transfer at the electrode-electrolyte interface [15]. On the metal side, it also made the surface loses the capability to produce a metastable protective layer to hinder further corrosion reaction to occur.

4 Conclusion

The effect of corrosion on low carbon steel was significantly influenced by different pulse-current frequency of current/voltage in the ZCCP system. Corrosion rate at E_{ocp} indicates the highest value, 3.018 μm/year and 50 Hz the lowest value, 0.122 μm/year. Increasing the frequency from 0 to 50 Hz showed changes in both corrosion rate and the surface morphology. At high frequency of 100 Hz, it has high corrosion rate compared to 20 to 50 Hz, and the surface are rougher, with small crater like surface. 50 Hz has much smooth and flat surface demonstrate the sample was well protected by ZCCP technique. Therefore, the ZCCP for the carbon steel (AISI 1016) in 3.5 % NaCl solution only required 50 Hz, which was equivalent to the standard AC power frequency applied in most countries.

The author would like to acknowledge the support from the Fundamental Research Grant Scheme (FRGS) under a grant number of FRGS/1/2015/TK10/UNIMAP/02/5 from the Ministry of Education Malaysia. Thanks to the School of Materials Engineering & Universiti Malaysia Perlis for providing a special financial support for enabling the publication of this manuscript.

References

1. J. Edington, The new world of steel. *In The Hatfield Memorial Lectures*. (101-122). (2005)
2. Z. B. Wang, N. R. Tao, S. S. Li, W. Wang, G. Liu, J. Lu, & K. Lu,. Mater. Sci. Eng. A, **352(1-2)**, 144-149 (2003)
3. B. K. Panigrahi,. Bull. Mater.Sci. **24(4)**, 361-371 (2001)
4. M. G. Fontana & N. D. Greene, Corros. Eng. (2018)
5. D. De la Fuente, I. Díaz, J. Simancas, B. Chico, & M. Morcillo. Corros. Sci. **53(2)**, 604-617 (2011)
6. D. E. Talbot, & J. D. Talbot, Corros. Sci Technol. (2018)
7. J. Kaur, R. Kant, J. Phys. Chem. C, **121(4)** (2017)
8. S. R. Shamsudin, PhD Thesis. UniMAP, (2016)
9. Y. Yang, & Y. F. Cheng,. Surf Coat Tech **216**, 282-288. (2013)
10. M. Barbalat, L. Lanarde, D. Caron, M. Meyer, J. Vittonato, F. Castillon, & P. Refait, Corros. Sci, **55**, 246-253 (2012)
11. ASTM Standard. Standard Practice for Laboratory Immersion Corrosion Testing of Metals.Annual Book of ASTM Standard. ASTM G31-72, ASTM International (1999)
12. A. Ismail and N. H. Adan, Am. J. Eng Res. **03**, 64-67 (2014)
13. M. A. Sobhan, M. Ams, M. J. Withford, & E. M. Goldys, J. Nanopart Res. **12(8)**, 2831-2842. (2010)
14. Y. He, G. Tian, H. Zhang, M. Alamin, A. Simm, & P. Jackson, IEEE Sens. J, **12(6),** 2113-2120. (2012)
15. A. P. Jamale, C. H. Bhosale, & L. D. Jadhav, J. Alloys Compd. **623**, 136-139. (2015)

IOP Conf. Series: Materials Science and Engineering **957** (2020) 012054 doi:10.1088/1757-899X/957/1/012054

Review on the effect of Gallium in solder alloy

Ervina Efzan M.N[1,*] and *Nur Faziera* M.N[2]

[1]Faculty of Engineering and Technology, Multimedia University, 75450 Ayer Keroh, Malacca, Malaysia

Abstract. Soldering is a well-known metallurgical joining method that uses a filler metal, the solder, with a melting point below 425°C. The trends toward to miniaturization in microelectronic packaging industries have motivated the development of high performance lead-free solders. However, the properties of the lead free solder cannot match the properties of lead solder. To overcome this problem, a dopant is added to the based lead free solder. In this paper, review on the effect of Gallium as doped was discussed.

1 Lead free solder

Recently, the regulation of certain hazardous substances (RoHS), and waste electrical and electronic equipment (WEEE) has resulted in extensive researches on lead-free solders due to their harmful effects of Pb on the environment and human health [1].

The elimination of Pb-containing components in the microelectronics industry is a worldwide significant target. A number of available Pb-free solder alloys such as Sn–Bi, Sn–Ag, Sn-Au, Sn-In, Sn-Sb, Sn-Pd and Sn–Zn are considered as potential lead-free alternatives to the classical Sn–Pb alloy. These alloys have low melting temperatures and good wettability when compared with the traditional Sn-Pb alloy [2].

2 The effect of dopant in lead free solders

In the last few decades, researchers have been focused on developing lead-free solder for the electronic industry [3][4][5]. In parallel of developed lead-free solder, many researchers have studied the addition of different micro/nanoparticles into the solder matrix to form a solder composite, aiming to improve the microstructure and control the growth rate of IMCs, as well as to enhance the mechanical properties [5][6]. Hence, producing better solder alloy especially for electronic packaging, this is to produce better solder alloy especially for electronic packaging [7]. The effect of dopant can be classified into three which are microstructure, mechanical properties and physical properties.

* Corresponding author: ervinaefzan@gmail.com

2.1 The effect of Gallium as dopant in lead free solders.

The element of Gallium (Ga) has several interesting physical properties. One of the interesting is it has melting point of 29.78°C [8][9][10]. Besides that, its boiling point is 1983°C and it expands 3.4% in volume during solidification [12].

Ga has been used as an additive in lead free solder [11]. It is applied to increase the tensile strength, enhance fatigue life, increase the ductility, increase the wettability and lower the melting temperature of solders [13][14][15]. In addition, its inherent characteristic of wetting most metals and oxides without the use of flux becomes a very attractive feature for microelectronic packaging industry [16][17]. It is founded that Ga may form solid solutions with Sn and Zn [18][19][20].

Mohanty and Lin [18] studied the effect of alloying element gallium in the polarization characteristics of Pb-free Sn-Zn-Ag-Al-xGa solders in NaCl solution. Based on their studies, they concluded that low content of gallium (wt.%) shifts the corrosion potential towards more noble values but contents of gallium more than 0.25 wt.% shifts the corrosion potential towards active values. The passivation behavior of the solder is improved by increasing the Ga content up to 0.25 wt.%. However, the film formed is not completely protective. The maximum corrosion of the solder is observed at 5.0 wt.%. This is reflected in the increase of the maximum passivation current density.

Another study conducted by Zhang et al, reported that with the addition of Ga to the Sn-Zn, the melting properties and wetting abilities of the solder alloy improved. To examine the melting point of the solder alloy, differential scanning calorimetry (DSC) method is used. Sn-Zn-6Ga solder has a melting point of 174.4°C and wetting rate of 48.9% compared with Sn-Zn solder that has melting point of 198.5°C and wetting rate is about 20%. As the conclusion, the addition of Ga can decrease the melting point of Sn-Zn eutectic system. As the content of Ga increases, the melting point will decrease. It also can decrease the wetting angle and increase the wetting rate.

Addition of Ga element on Sn-Ag-Cu solder can obviously decrease the melting point [21]. Each percentage addition of Ga element can decrease the melting point of Sn-Ag-Cu eutectic solder alloy by 4°C. The Ga element forms Cu2Ga phase around the joint interface during the soldering process. This can leads to the decreasing of growth rate of the interfacial IMCs layer and also decrease the ductility of solder. Compared to Sn-Ag-Cu/Cu, less Cu3Sn forms at the Sn-Ag-Cu-Ga/Cu.

Increasing the Ga concentration will effects on microstructures, thermal and mechanical properties of the Sn-9Zn solder alloys, the melting temperatures of the alloys will decrease and the pasty ranges of the alloys are simultaneously enlarged [22-25]. They showed that addition of Ga element causes the Zn-rich phase precipitates to grow and disperse in the β-sn matrix. Moreover, it also decreases the fraction of the Sn/Zn eutectic region, increase the Sn-matrix region and make the needle-like Zn-rich phase becomes coarser and longer. Small additions of 0~0.75wt.% Ga decrease the melting point of the solders alloy but maintain the same strength and ductility as Sn-9Zn solder. The effect Gallium in lead free solder is summarized in the Table 1 based on the properties [25-36].

Table 1. The effect of Gallium as dopant in lead free solders.

Properties	Solder alloy	Description of Ga	Effects
Polarization	Sn-Zn-Ag-Al	Low (<0.25wt.%)	Corrosion potential towards noble values
		High (>0.25wt.%)	Corrosion potential towards active values
		High (5.0wt.%)	Maximum corrosion occur
Passivation	Sn-Zn-Ag-Al	Increase	Improved
Melting temperature	Sn-Zn	Increase (6wt.%)	Decrease about 24.1°C
	Sn-Ag-Cu	Increase (1wt.%)	Decrease about 4 °C
Wetting rate	Sn-Zn	Increase (6wt.%)	Increase about 28.9 %
Wetting angle		Increase	Decrease
Growth rate	Sn-Ag-Cu	Increase (0.25wt.%)	Decrease Cu_2Ga phase formed rather than Cu_3Sn
	Sn-Bi	Increase (0.25wt.%)	Various IMCs formed
Ductility	Sn-Ag-Cu	Increase	Decrease
Microstructure	Sn-Zn	Increase	•Phase precipitation grow and disperse •Zn-rich phase becomes coarser and longer •Zn-rich phase becomes blunted, enlarged and less densely
Tensile strength	Sn-Zn	Increase	Increase
Fracture	Sn-Zn	Increase (0.75wt.%)	Larger cup and cone fracture seen
Oxidation resistance		Increase	Much better
Concentration on surface		Increase	40 times higher in bulk solder

3 Conclusion

Even though the usage of lead solder has been banned, the lead free solder can be developed and match the properties of lead solder. Furthermore, the lead free solder can give better properties than lead solder if adding dopant such as Gallium Ga to solder alloy, but more research need to be develop. In addition, the temperature of the lead free solder is lower than lead solder where the temperature for lead solder is 183°C.

References

1. R. W. Messler, *Principles of Welding: processes, Physics, Chemistry, and Metallurgy* (John Wiley & Sons, 2008)
2. M. L. Minges, *Electronic Materials Handbook: Packaging: Volume 1 of Electronic Materials Handbook*, (ASM International, 1989)
3. J. Norish, *Advanced Welding Processes: Woodhead Publisshing Series in Welding and Other Joining Technologie,* (Elsevier, 2006)
4. G. Humpston, D. M. Jacobson, *Principles of soldering*, (ASM International, 2004)
5. D. M. Jacobson, *Principles of Brazing*, (ASM International, 2005)
6. F. C. Campbell, Joining; Understanding the Basics. Engineering Pro Collection (Understanding the Basics, (ASM International, 2011)
7. S. Ratchev, *Precision Assembly Technologies and Systems: 7th IFIP WG 5.5 International Precision Assembly Seminar, IPAS 2014, Chamonix, France, February 16-18, Revised Selected papers*, **435** (2014)
8. A. Jayendran, English for Maschinenbauer. Aus den program Grudlagen Maschinenbau. Vieweg + Teubner Verlag (2000)
9. Messler, W. Savage, Joining of Advanced Materials, (Elsevier, 2013)
10. G. Humpston, D. M. Jacobson, *Principles of soldering*, (ASM International, 2005)
11. H. h. Manko, *Solder and Soldering: 4th Edition. Electronic Packaging,* (McGraw-Hill Professional, 2001)
12. J. Bath, *Fundamental Properties of Pb-Free Solder Alloys. Lead-Free Solder*, 21-74, (Springer US, 2007)
13. Chipscale Review, Lead Free Solders for Surface mount Technology Applications (Part 1), (1998)
14. Europian Commission, Austria: Europian Union, (2003)
15. K. N. Subramaniam, *Lead-Free Solders: A special Issue of the Journal of Materials Science: Materials in Electronics*, (Springer US, 2007)
16. N. C. Lee, *Reflow Soldering Processes*, (Elsevier Science, 2002)
17. H. Ipser, H. Flandorfer, C. Luef, C. Schmetterer, U. Saeed, *Thermodynamics and Phase Diagrams of Lead-free Solder Material*, Lead-Free Electronic Solders, 3-17, (2006)
18. K.J. Puttlitz, K.A. Stalter, *Handbook of Lead-free Solder Technology for Microelectronic Assemblies*, Marcel Dekker, New York, (2004)
19. F.Q. Lang, H Tanaka, O Munegata, T Taguchi, T Narita, Materials characterization, **54**, 223-229, (2005)
20. F.X. Che, W.H. Zhu, E.S.W. Poh, X.W. Zhang, X.R. Zhang, J. Alloys Compd, **507**, 215-224, (2010)
21. Y. Li, K.-S. Moon, C. P. Wong, J. Sci., **308**, 1419–1420, (2005)
22. K. I. Chen, K. L. Lin, J Electron. Mater., **31**, 861–867, (2002)
23. J. S. Hwang, Z. Guo, H. Koenigsmann, Solder Surf Mt Tech, **13/2,** 7–13, (2001)
24. T. Shimizu, H. Ishikawa, I. Ohnuma and K. Ishida, J. Electron. Mater. **28**, 1172–1175, (1999)
25. K.I. Chen, K.L. Lin, J. Electron. Mater., **31**, 861, (2002)

26. J. G. Swallow , W. H. Woodford, Y. Chen, Q. Lu, J. J. Kim, D. Chen, Y. M. Chiang , W. C. Carter, B. Yildiz, H. L. Tuller, K. J. Van Vliet, *Chemomechanics of ionically conductive ceramics for electrical energy conversion and strorage*, J. Electroceramics, (2014)
27. H. S. Stoker, *General, Organic, and Biological Chemistry: 6th Edition*. Cengage Learning, (2012)
28. Y. Zhu. Mechanical Properties of Materials (Chapter 2): Strength of Mechanical Components.
29. H. Czichos, T. Saito, L. R. Smith, Springer Handbook of Materials Measurement Methods. Springer Science &Business Media, (2006)
30. C. E. Ophardt: Virtual Chembook. Elmhurst College, (2003)
31. Sattiraju, S.V.; Dang, B.; Johnson, R.W.; Yali Li; Smith, J.S.; Bozack, M.J., IEEE Transactions, **25**, 168,184, (2002)
32. Bagrets, N., Barth, C., & Weiss, K.-P, IEEE Trans. Appl. Supercond., **24**, 1–3, (2014)
33. Billah, M. M., Shorowordi, K. M., & Sharif, A, J. Alloys Compd., **585**, 32–39, (
34. Bracco, G., & Holst, B., Surface science techniques. Springer Series in Surface Sciences, **51**, (2013)
35. M. N. Ervina Efzan, A. Aisyah Mariani, Int. J. Appl. Sci. Eng., **1**, 1–10, (2012)
36. E. E. M. Noor, N. M. Shariff, C. K. Yew, T. Ariga, A. B. Ismail, Z. Hussain, J. Alloys Compd., **507**, 290–296, (2010)

IOP Conf. Series: Materials Science and Engineering **957** (2020) 012055 doi:10.1088/1757-899X/957/1/012055

Phase formation and microstructure analyses of $La_{0.6}Sr_{0.4}CoO_{3-\delta}$-$BaCe_{0.54}Zr_{0.36}Y_{0.1}O_{2.95}$ composite material prepared by solid state mixing method

Abdullah Abdul Samat[1,2,*], *Muhammad Syafiq Amsyar* Ahmad Pisor[1], *Meor Ahmad Faris* Meor Ahmad Tajudin[1], *Mohd Al-Hafiz* Mohd Nawi[1], *Murizam* Darus[3], *Nafisah* Osman[4] and *Nurul Akidah* Baharuddin[5]

[1]Faculty of Engineering Technology, Universiti Malaysia Perlis (UniMAP), Kampus UniCITI Alam, Sungai Chuchuh, Padang Besar 02100, Perlis, Malaysia.
[2]Centre of Excellence for Unmanned Aerial Systems (CEOUAS), Universiti Malaysia Perlis (UniMAP), 01000 Kangar, Perlis, Malaysia.
[3]School of Materials Engineering, Universiti Malaysia Perlis (UniMAP), 02600 Arau, Perlis, Malaysia.
[4]Faculty of Applied Sciences, Universiti Teknologi MARA, 02600 Arau, Perlis, Malaysia.
[5]Fuel Cell Institute, Universiti Kebangsaan Malaysia, 43600 UKM Bangi, Selangor, Malaysia.

Abstract. A composite material of $La_{0.6}Sr_{0.4}CoO_{3-\delta}$-$BaCe_{0.54}Zr_{0.36}Y_{0.1}O_{2.95}$ (LSC-BCZY) was prepared by mixing sol-gel derived LSC and BCZY powders in different weight percentage (wt%) ratio of LSC to BCZY (LSC:BCZY). The prepared composite powders were denoted as S1 (30:70), S2 (50:50) and S3 (70:30). The powders were characterized by an X-ray diffractometer (XRD), a Brunauer-Emmett-Teller (BET) surface area and porosity analyzer and a scanning electron microscope (SEM) equipped with an energy dispersive X-ray (EDX) spectrometer. XRD analysis confirmed that all of the powders were not pure enough due to the presence of impurity phases such as barium carbonate ($BaCO_3$) and strontium carbonate ($SrCO_3$) and unknown phases. S1 powder has the highest amount of impurity phases (81.19 %) and the largest BET surface area (4.82 m^2 g^{-1}). All of the powders formed typical clump-like network structure as proven by SEM analysis. EDX analysis revealed that the elemental compositions of La, Sr and Zr were deviated from their nominal mole fractions in all powders due to the Zr-rich clusters formation. The results indicate that the formation of pure and homogenous LSC-BCZY composite powder prepared by solid state mixing method requires further modification and improvement of the preparation method.

1 Introduction

Solid oxide fuel cell (SOFC) is an electrochemical device that converts chemical energy directly into electricity via electrochemical processes, producing water molecule and high quality of heat as by-products. Generally, an SOFC consists of three major components

* Corresponding author: abdullahabdul@unimap.edu.my

which are electrolyte, electrodes (anode and cathode) and interconnect. Nowadays, SOFC with proton-conducting electrolyte (H^+-SOFC) has received greater attention than SOFC with oxide ion-conducting electrolyte (O^{2-}-SOFC) due to its advantages including higher ionic conductivity and electrical efficiency at SOFC operating temperature range (500 °C - 800 °C) [1,2]. Performance of an SOFC either H^+-SOFC or O^{2-}-SOFC is strictly determined by the properties of its components such as purity, porosity, homogeneity and conductivity [3–5].

One of the most crucial components of an SOFC is the cathode as the low performance of an SOFC is closely related to the cathode polarization resistance (R_p) and degradation [6–8]. To minimize the cathode R_p issue, cathode materials for an SOFC must have high electrical conductivity (> 100 S cm^{-1}) and adequate porosity (20 - 40 %), stable in working environment as well as chemically and thermally compatible with other components. In addition, the most important criterion is the value of area specific resistance (ASR) must be $\leq$ 0.30 Ω cm^2 at intermediate temperature of 500 °C - 800 °C [9]. Mixed ionic and electronic conductor (MIEC) materials such as lanthanum strontium cobaltite, $La_{0.6}Sr_{0.4}CoO_{3-\delta}$ (LSC) is a common cathode material for O^{2-}-SOFC application [8,10, 11]. The LSC material has also been used for H^+-SOFC application which uses doped-barium cerate-zirconate proton conductor materials such as $BaZr_{0.1}Ce_{0.7}Y_{0.1}Yb_{0.1}O_{3-\delta}$ (BCZYYb) and $BaCe_{0.54}Zr_{0.36}Y_{0.1}O_{2.95}$ (BCZY) as reported by Wang et al. [12] and Abdul Samat et al. [13], respectively. The ASR values of the LSC cathode at 600 °C working with the BCZYYb was 1.78 Ω cm^2 and the BCZY was 0.70 Ω cm^2.

To achieve the targeted ASR value ($\leq$ 0.30 Ω cm^2), Wang et al. [12] have developed a composite material of LSC-BCZYYb. The developed composite material recorded lower ASR value than the LSC at 600 °C which is 0.34 Ω cm^2. The noteworthy result indicates that the composite material has better electrochemical properties than the MIEC material. Therefore, the development of LSC-BCZY composite material is expected to significantly reduce the ASR value of LSC cathode working with BCZY electrolyte material for intermediate temperature H^+-SOFC application. However, the properties of LSC-BCZY composite material such as phase purity and homogeneity need to be optimized first before any electrochemical performance testing can be done. The effect of different weight ratio percentage of LSC to BCZY on the properties of LSC-BCZY is systematically presented and discussed in this present work.

2 Materials and Methods

2.1 Powder preparation

In-house developed $La_{0.6}Sr_{0.4}CoO_{3-\delta}$ (LSC) and $BaCe_{0.54}Zr_{0.36}Y_{0.1}O_{2.95}$ (BCZY) powders were used to prepare the LSC-BCZY composite powders. Based on citrate-EDTA sol gel method [14], LSC powder was synthesized using $La(NO_3)_3.6H_2O$, $Sr(NO_3)_2$ and $Co(NO_3)_2.6H_2O$ as starting materials. A stoichiometric amount of the starting materials was dissolved in deionized water to form a metal nitrate salt precursor solution. Then, a stoichiometric amount of a combined chelating agent consisting of citric acid and ethylenediaminetetraacetic acid (EDTA) was added into the precursor solution. Then, the pH of the solution was adjusted to 0.5 by adding an appropriate amount of ammonium hydroxide solution and followed by adding a stoichiometric amount of ethylene glycol as polymerizing agent. The solution was heated at 150 °C for several hours to form a viscous gel the gel was dried at 250 °C for 5 hours to obtain as-synthesized black ash powder. Subsequently, the powder was calcined at 900 °C for 5 hours to obtain black fine powder.

In the similar method, BCZY powder was synthesized using $Ba(NO_3)_2$, $Ce(NO_3)_3.6H_2O$, $Zr(NO_3)_2O.xH_2O$ and $Y(NO_3)_3.6H_2O$ as precursor materials. After adding the combined chelating agent, the pH of the solution was adjusted to 7. The obtained viscous gel was dried at 325 °C for 12 hours and the produced black ash powder was further calcined at 1100 °C for 10 hours to obtain yellowish fine powder. The details for the synthesis of BCZY powder was described elsewhere [15].

The LSC-BCZY composite powders were prepared via solid state mixing method. The calcined LSC and BCZY powders in different weight ratio percentage (wt%) of LSC to BCZY (LSC:BCZY) were grounded in an agate mortar for an hour. Then, the grounded mixture was dried in a drying oven for overnight. The produced LSC-BCZY composite powders were denoted as S1 (30:70), S2 (50:50) and S3 (70:30).

2.2 Powder characterization

To confirm the formation of LSC and BCZY perovskite phases in the prepared LSC-BCZY composite powders (S1, S2 and S3), the powders were subjected to X-ray diffractometry (XRD) analysis by using an XRD machine (Bruker, D8-Advance, Germany) within the 2θ range of 20° to 70°. The XRD analysis was done at room temperature by using Cu-K_α radiation source (λ = 0.1540558 nm). A Brunauer-Emmett-Teller (BET) surface area and porosity analyzer (Micromeritics, ASAP 2020, USA) was used to analyze the BET-specific surface area. A Carl Zeiss SMT Supra-40VP scanning electron microscope (SEM) equipped with an energy dispersive X-ray (EDX) spectrometer was used to capture the morphology image of the prepared composite powder. The image was captured in secondary electron mode using an accelerating voltage of 10 kV. The elemental composition analysis was performed by using the captured image of the prepared composite powder.

3 Results and discussion

Fig. 1 shows the XRD plots of the prepared LSC-BCZY composite powders (S1, S2 and S3). The crystalline peaks of LSC perovskite phase were matched with Joint Committee of Powder Diffraction Standards (JCPDS) file no. 48-0121 and the crystalline peaks of BCZY perovskite phase were matched with JCPDS file no. 89-2485. The results are in-line with the XRD plots of the pure in-house developed LSC and BCZY powders as reported elsewhere, respectively [14,15]. However, the XRD plots confirmed that all of the prepared composite powders did not form a pure composite phase of LSC-BCZY. It is due to the presence of impurity phases such as barium carbonate ($BaCO_3$, JCPDS file no. 52-1528) and strontium carbonate ($SrCO_3$, JCPDS file no. 05-0418) and the formation of peaks of other unknown impurity phases. The percentage of impurity phases (by considering the known peak only) in each of the prepared composite powder is calculated by using the following equation:

$$\% I_m = I_m/(I_m + I_p)*100\% \quad (1)$$

where I_m represents the maximum intensity of the impurity phase and I_p represents the maximum intensity of the perovskite phase. The results are tabulated in Table 1. S3 has the lowest percentage of impurity phases as compared to S1 and S2. From the table, it can be seen that the percentage of impurity phases decreased as wt% of LSC and BCZY in the prepared LSC-BCZY composite powders increased and decreased, respectively. The reason for this observation is still unclear since some of the peaks in the XRD plots could not be identified yet. Further analysis and investigation on this observation is in progress and the results will be reported elsewhere.

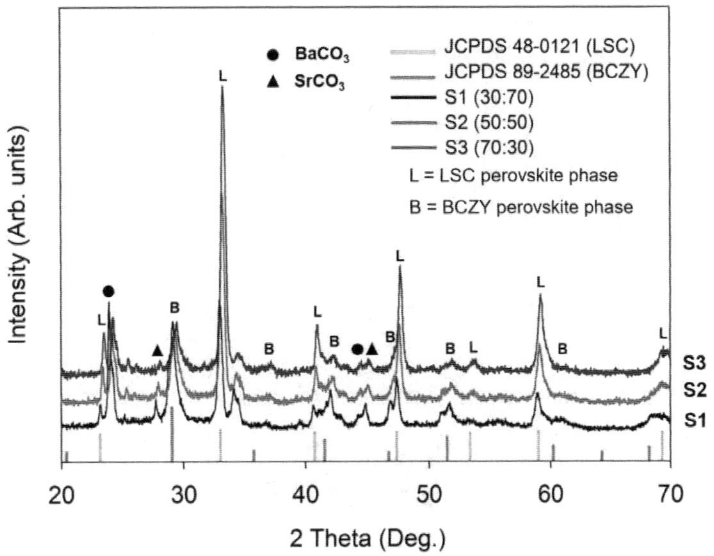

Fig. 1. XRD plots of the prepared LSC-BCZY composite powders

Table 1. Percentage of perovskite and impurity phases in the prepared LSC-BCZY composite powder

Powder	LSC		BCZY		Total impurity phase (%)
	Perovskite phase (%)	Impurity phase (%)	Perovskite phase (%)	Impurity phase (%)	
S1 (30:70)	72.11	27.89	46.70	53.30	81.19
S2 (50:50)	78.96	21.04	48.13	51.87	72.91
S3 (70:30)	81.16	18.84	49.44	50.56	69.40

Ba and Sr are the two elements that easily react with carbon dioxide (CO_2) to form stable carbonate compounds. Formation of impurity phases of $BaCO_3$ and $SrCO_3$ is common in the preparation of powder through solid state mixing method or other methods including sol-gel, co-precipitation and hydrothermal [16–20]. The stable carbonate compounds usually require high calcination temperature (> 1300 °C) to decompose [19], [21,22]. However, high heat treatment lead to poor quality of the produced powders [23–25]. From the Table 1, it can be also clearly seen that the percentage of $BaCO_3$ is higher than the $SrCO_3$ in all composite powder. It might be contributed to a larger ionic radius of Ba than Sr, thus Ba reacts more towards CO_2 to form carbonate compound than the Sr. Further investigation on this observation is also in progress and the results will be reported elsewhere.

Typical SEM micrographs of the prepared composite powders (S1, S2 and S3) and their EDX spectra are shown in Fig. 2 and Fig. 3, respectively. All of the composite powders showed similar clump-like network structure. It can be clearly seen that the S1 is slightly more porous than the S2 and S3. It is supported by a larger value of BET-specific surface area for S1 (4.82 $m^2 g^{-1}$) than the S2 (3.58 $m^2 g^{-1}$) and S3 (3.48 $m^2 g^{-1}$). Comparing the three SEM micrographs, it was evident that the particle size of the produced composite powders prepared by solid state mixing method was small, but the particle size distribution was not even due to the formation of agglomerated particles. The same observation was also reported by Chen et al. [26] in producing Pr_2NiO_4-$Pr_{0.2}Ce_{0.8}O_{1.9}$ composite powder for intermediate temperature SOFC application.

Fig. 2. SEM micrograph of the prepared LSC-BCZY composite powders

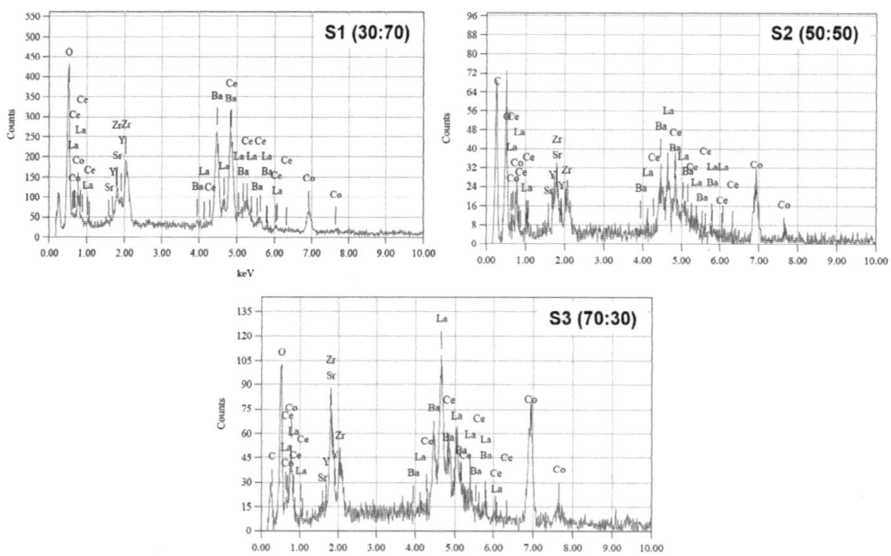

Fig. 3. EDX spectra of the prepared LSC-BCZY composite powders

The non-homogenous phase and particle size led to the non-homogeneous elemental composition in the prepared LSC-BCZY composite powders as shown in Table 2. Non-homogeneous elemental composition is another drawback of powders prepared by solid state mixing method [27,28]. From the table (Table 2), it can be seen that the calculated elemental mole fractions of Co/Ba/Ce/Y in the prepared LSC-BCZY composites are close to their nominal mole fractions. However, the calculated elemental mole fraction of Zr in all of the prepared composite powders is deviated from its nominal mole fraction although it is decreased as the wt% of BCZY decreased from 70% (S1) to 30% (S3) in the prepared LSC-BCZY composite powders. It is due to the fact that Zr tends to form Zr-rich clusters as previously reported by Osman et al. [29]. Consequently, the non-homogenize distribution of Zr element in the prepared powders gave rise to the fluctuations of chemical composition. The formation of Zr-rich clusters might have contributed to the deviations of La/Sr mole fractions from their nominal mole fractions in all of the prepared composite powders. The Zr-rich clusters might have covered these two elements from being fully detected during EDX analysis as the analysis is done at a specific area or spot and elemental distribution at an area is different from other areas [30].

Table 2. Elemental composition of the prepared LSC-BCZY composite powders

Element	Nominal mole fraction	Calculated mole fraction		
		S1	S2	S3
La	0.60	0.15	0.13	0.17
Sr	0.40	0.14	0.10	0.19
Co	1.00	1.00	1.00	1.00
Ba	1.00	1.00	1.00	1.00
Ce	0.54	0.55	0.57	0.56
Zr	0.36	1.16	0.83	0.77
Y	0.10	0.55	0.63	0.10

4 Conclusions

Composite powders of $La_{0.6}Sr_{0.4}CoO_{3-\delta}$-$BaCe_{0.54}Zr_{0.36}Y_{0.1}O_{2.95}$ (LSC-BCZY) with different weight percentage (wt%) ratios of LSC to BCZY (LSC: BCZY = 30:70, 50:50 and 70:30) were successfully prepared through solid state mixing method. However, the purity of the prepared composite powders is quite low due to the formation of carbonate compounds impurity including $BaCO_3$ and $SrCO_3$ and the presence of other unknown impurity phases. All of the prepared composite powders have porous and clump-like network structures. Furthermore, the powders tend to form Zr-rich clusters, resulting to the non-homogeneous elemental composition distribution. Overall, S3 is better than the S1 and S2 since it has the lowest percentage of impurity phases and the closest calculated mole fraction to the nominal mole fraction of elemental composition. In order to yield a higher purity and better elemental composition distribution of LSC-BCZY composite powder, further modification of the preparation method is in progress and the outcomes will be reported elsewhere. Later, a high purity and a homogeneous LSC-BCZY composite powder will be subjected to the electrochemical performance testing for the application as cathode material in intermediate temperature H^+-SOFC.

The authors would like to thank the School of Materials Engineering, Universiti Malaysia Perlis (UniMAP), Universiti Teknologi MARA (UiTM) Perlis Branch and Centre for Research and Instrumentation Management (CRIM) of Universiti Kebangsaan Malaysia (UKM) for the facility support.

References

1. S. Hossain, A.M. Abdalla, S.N.B. Jamain, J.H. Zaini, A.K. Azad, Renew. Sustain. Energy Rev. **79**, 750 (2017)
2. N.L.R.M. Rashid et al., Ceram. Int. **45**, 6605 (2019)
3. N.A. Baharuddin, N.A. Mohd Nazrul Aman, A. Muchtar, M.R. Somalu, A. Abdul Samat, M.I. Aznam, Ceram. Int. **45**, 12903 (2019)
4. A. Abdul Samat, W.N.A. Wan Yusoff, N.A. Baharuddin, M.R. Somalu, A. Muchtar, N. Osman, Process. Appl. Ceram. **12**, 277 (2018)
5. M. Jafari, H. Salamati, M. Zhiani, E. Shahsavari, Int. J. Hydrog. Energy, **44**, 1953 (2019)
6. R.A. Budiman et al., Solid State Ion. **332**, 70–76 (2019)
7. D. Xie, K. Li, J. Yang, D. Yan, L. Jia, B. chi, J. Pu, J. Li, Int. J. Hydrog. Energy, **In Press** (2020) doi: 10.1016/j.ijhydene.2020.01.014.
8. O. Gwon, S. Yoo, J. Shin, G. Kim, Int. J. Hydrog. Energy, **39**, 20806 (2014)
9. N.P. Brandon, S. Skinner, B.C.H. Steele, Annu. Rev. Mater. Res. **33**, 183 (2003)
10. C. Benel, A.J. Darbandi, R. Djenadic, A. Evans, R. Tolke, M. Prestat, H. Hahn, J. Power Sources, **229**, 258 (2013)
11. J. Hayd, L. Dieterle, U. Guntow, D. Gerthsen, E. Ivers-Tiffée, J. Power Sources, **196**, 7263 (2011)
12. M.S. Wang, J.X. Wang, C.R. He, Y.J. Xue, H. Miao, Q. Wang, W.G. Wang, Ceram. Int. **41**, 5017 (2015)
13. A. Abdul Samat, M.R. Somalu, A. Muchtar, N. Osman, IOP Conf. Ser. Earth Environ. Sci. **268**, 012136 (2019)
14. A. Abdul Samat, M.R. Somalu, A. Muchtar, O.H. Hassan, N. Osman, J. Sol-Gel Sci. Technol. **78**, 382 (2016)
15. N.A. Abdullah, S. Hasan, N. Osman, J. Chem. **2013**, 1 (2013)
16. S.V. Seyed-Vakili, C.R. Graves, A. Babaei, S. Heshmati-Manesh, M.B. Mogensen, Fuel Cells, 17, 108 (2017)
17. S. Zhuang, Y. Liu, S. Zeng, J. Lv, X. Chen, J. Zhang, J. Sol-Gel Sci. Technol. **77**, 109 (2016)
18. R. Gawel, K. Przybylski, M. Viviani, Mater. Chem. Phys. **147**, 804 (2014)
19. W. Acuña, J.F. Tellez, M.A. Macías, P. Roussel, S. Ricote, G.H. Gauthier, Solid State Sci. **71**, 61 (2017)
20. A. Abdul Samat, W.N.A. Wan Yusoff, N.W. Norman, M.R. Somalu, N. Osman, J. Kejuruter. S1, **1**, 49 (2018)
21. A. Chakroborty, A. Das Sharma, B. Maiti, H.S. Maiti, Mater. Lett. **57**, 862 (2002)
22. N. Osman, I.A. Talib, H.A. Hamid, Sains Malays. **38**, 103 (2009)
23. S.K. Durrani, S. Naz, M. Mehmood, M. Nadeem, M. Siddique, J. Saudi Chem. Soc. **21**, 899 (2017)
24. X.-Z. Fu, J.-L. Luo, A.R. Sanger, N. Luo, K.T. Chuang, J. Power Sources, **195**, 2659 (2010)
25. A. Abdul Samat, S. Mohd Senari, M.R. Somalu, A. Muchtar, O.H. Hassan, N. Osman, Sains Malays. **47**, 589 (2018)
26. X. Chen, J. Wang, Q. Liang, X. Sun, X. Zhu, D. Zhou, J. Meng, Solid State Sci. **100**, 106108 (2020)
27. Z. Shao, W. Zhou, Z. Zhu, Prog. Mater. Sci. **57**, 804 (2012)
28. A. VahidMohammadi, Z. Cheng, J. Electrochem. Soc. **162**, F803 (2015)
29. N. Osman, I.A. Talib, H.A. Hamid, Sains Malays. **39**, 479 (2010)
30. B. Choudhury, A. Choudhury, Curr. Appl. Phys. **13**, 1025 (2013)

IOP Conf. Series: Materials Science and Engineering **957** (2020) 012056 doi:10.1088/1757-899X/957/1/012056

Cryorolling of SAC305 solder : Microstructure analysis and shear strength of solder joint

Nurulakmal Mohd Sharif[1,*] and *Wan Asilah* Wan Azalan[1]

[1]School of Materials and Mineral Resources Engineering, USM Engineering Campus, 14300 Nibong Tebal, Penang, Malaysia

Abstract. Miniaturization of devices along with demand of higher performance in electronic packaging industries led to development of higher reliability solder alloys. Solder material must meet the requirements of high solder joint strength, better wettability, high fatigue and creep resistance in order to provide highly reliable solder joint. This work attempts to evaluate effectiveness of cryorolling in providing ultrafine grains which could lead to higher solder strength. The microstructure of bulk solder and IMC layer at interface between solder and Cu substrate, and lap joint shear strength of cryorolled commercial SAC305 solder alloy are presented here. Solder alloys were dipped in liquid nitrogen at different dipping times (0, 10, 30 and 60 minutes) prior to rolling. Cryorolling was observed to encourage grain refinement in solder with 30 min dipping gave lowest grain size of 5.88 μm, compared to 11 μm for as-cast solder. Longer dipping up to 60 minutes resulted in larger grain size and tend to be brittle with small cracks started to appear during rolling. Shear strength test showed an improved and stronger joint for SAC305 dipped in liquid nitrogen for 30 minutes compared to sample without dipping in liquid nitrogen. As a conclusion, cryorolled solder alloys had smaller grain size and this led to an improved solder joint shear strength.

1.0 Introduction

The mechanical properties and reliability of solder joints are affected by microstructures, movement of dislocations and grains reconfiguration. As solders typically have relatively low melting point, the deformation mechanisms at either room temperature or service temperature are considered as high-temperature mechanisms. These mechanisms are dependent on microstructure, which is related to composition and soldering process. Thus, understanding the relationship between structure - property and structure – process are important to produce desired microstructure which give high reliability products [1]. For instance, larger grain size microstructure will become more resistant towards creep when grain-boundary sliding dominates the total creep strain. On the other hand, fine grain solder joint have larger fatigue life compared to coarse grain solder joint.

* Corresponding author: srnurul@usm.my

Grain refinement in solder alloy is typically because of alloying [2, 3] or via severe deformation process (SPD) such as equal channel angular pressing (ECAP) [4, 5]. Although alloying is more common in refinement of solder alloy, SPD technique has been reported to be effective in production of materials with submicron or nano-sized grains [6]. One of the main advantages of SPD is that ultra-high plastic deformation (with strain in order of 10 and higher) is achieved without a substantial change in dimensions of the worked sample, thus allowing very strong grain refinement effect [7]. By combining cryogenic treatment and deformation process, the tensile strength and yield strength can be considerably improved as dynamic recovery is suppressed during cryorolling [8]. The cryogenic treatment is able to suppress the recrystallization along the process and thus preserving the grain refinement effect. Researches based on cryogenic treatment had been performed on magnesium alloy, titanium alloy and some steels, and the properties of these materials have been reported to improve to various extent [8]. Therefore, this study was developed to investigate the effect of various dipping time in liquid nitrogen prior to rolling on the microstructure and solder joint strength of commercial SAC305 solder alloy.

2.0 Methodology

A commercial Sn-3.0Ag-0.5Cu (SAC 305) solder alloys was used in this project, provided by RedRing Solder. The solder was initially prepared by melting the ingot bar and casted into stainless steel mold to get flat sheet with approximately 4 mm thickness. Several thickness reduction percentages were tried to observe deformation behavior after nitrogen dipping and it was found that 75% thickness reduction was maximum before the solders formed severe cracks and break. Dipping time was set to be 10, 30 and 60 minutes prior to rolling process.

The solder alloys after cryorolling were ground with SiC abrasive paper grit 360 until 2000 before polished using 1 and 3 μm diamond suspension. Once mirror-like surface was obtained, the samples were chemically etched with etchant of 2% hydrochloric acid (HCl), 5% nitric acid (HNO3) and 93% ethanol (CH3OH) for 15 seconds to reveal the microstructure of the alloy using ZEISS SUPRA 35VP scanning electron microscope (SEM) equipped with energy dispersive X-ray (EDX). For evaluation of shear strength of solder joint, solder samples were reflowed in T200C+ lead-free reflow oven at 280°C for 15 seconds on bare copper substrate. For all reflowed solder joints, activated rosin (RA) flux was used. Single lap joint shear test was carried out following ASTM D 1002 standard, and sample dimension is shown in figure 1. The test is conducted on Instron Universal Testing Machine at displacement rate of 13 mm/min. The maximum load by each sample was then divided by the overlap area, which is 9 mm2 to get the strength value, and the strength value presented was an average of 5 tests.

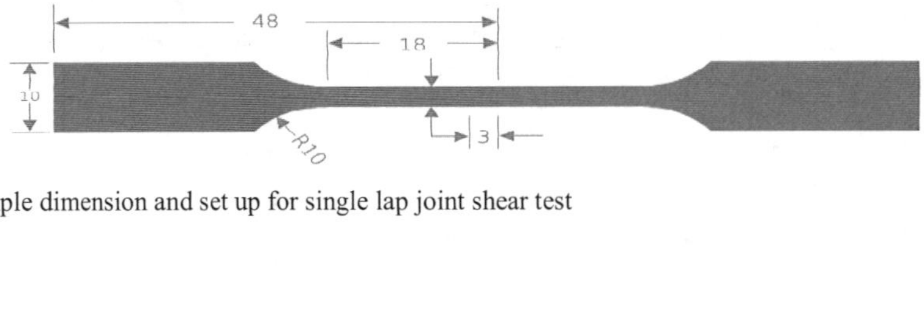

Fig.1. Sample dimension and set up for single lap joint shear test

3.0 Results and Discussion

3.1 Microstructure

Fig. 2 shows the evolution of microstructure as dipping time in nitrogen is varied, from as-cast solder to 60 minutes dipping time. All samples show typical SAC305 features which consists of a matrix of β-Sn and eutectic phase of Cu_6Sn_5 and Ag_3Sn IMC particles [9]. EDX analysis (not presented here) was done to confirm elemental composition of the IMCs. It can be observed that dipping in nitrogen encouraged grain refinement of the β-Sn matrix, and the eutectic network was also more evenly distributed within the matrix. The average grain size, measured using lineal intercept method following ASTM 112, is presented in Table 1 for all samples.

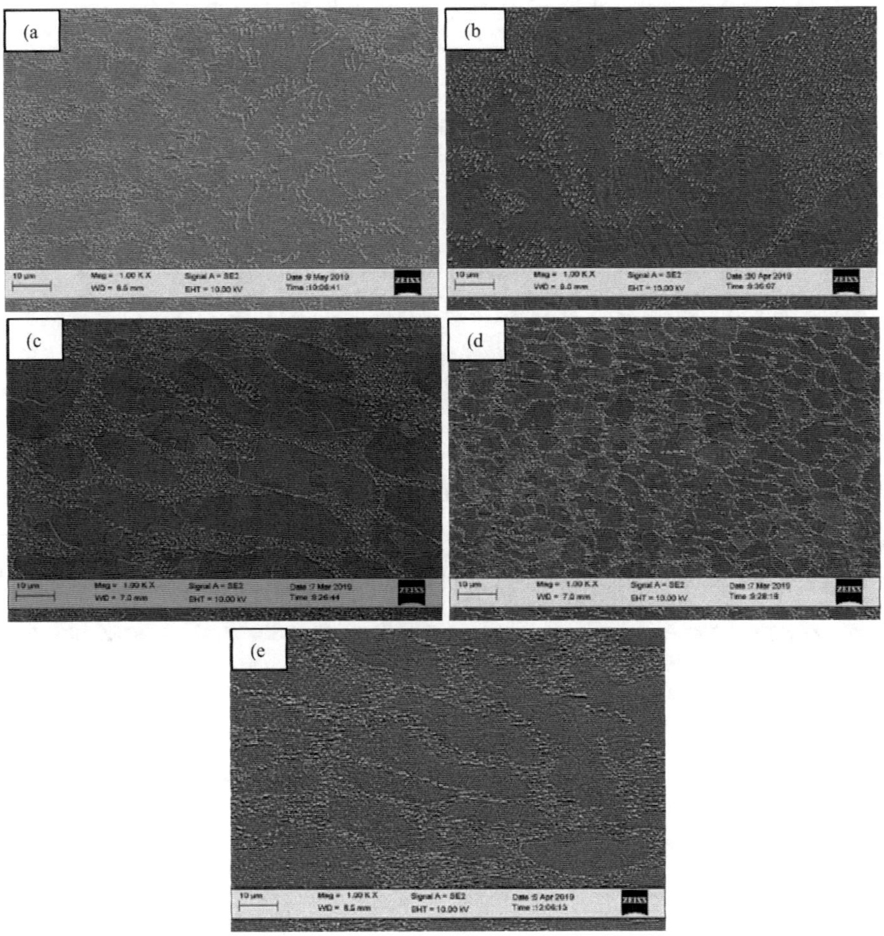

Fig. 2. SEM micrograph of cryorolled SAC305 (a) as-cast, (b) 0 dipping, (c) 10 min, (d) 30 min, (e) 60 min dipping, 1000x

IOP Conf. Series: Materials Science and Engineering **957** (2020) 012056 doi:10.1088/1757-899X/957/1/012056

From Table 1, SAC305 solder sample with dipping in liquid nitrogen for 60 minutes shows the largest grain size which is 12.50 μm. On the other hand, sample dipped for 30 minutes prior to cryorolled gave the smallest average grain size which is 5.88 μm. The 10 minutes dipping sample and sample without dipping exhibited average grain size value of 6.90 and 7.14 μm respectively. Meanwhile, average grain size of as cast sample is 11.00 μm which is 87% larger than the smallest grain size (5.88 μm) and this shows great potential of cryorolling for grain refinement of solder alloys. Longer dipping time of 60 minutes however, resulted in increment of grain size and it is also observed that the longer dipping time tend to increase brittleness of sample with small cracks started to appear at the edges during rolling. At this point of time, we are still unclear as to the exact reason why longer duration of dipping led to brittleness.

Table 1. Average grain size of all samples

SAC305 samples	Average grain size (μm)
As-cast	11.00
0	7.14
10	6.90
30	5.88
60	12.50

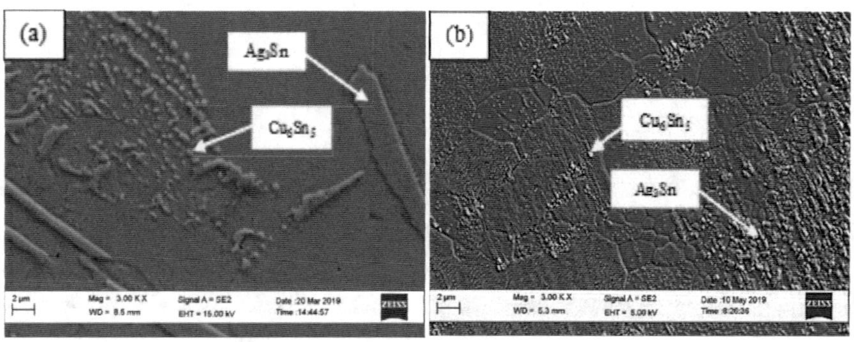

Fig. 3. Higher magnification showing changes in IMC particle sizes with cryorolling, (a) as-cast sample, (b) sample dipped 30 minutes

Cryorolling was also observed to result in break-up of IMC particles resulting in smaller particles distributed more evenly throughout the β-Sn matrix. Figure 3 shows higher magnification of as-cast and 30 minutes dipping samples and it is clear that cryorolled sample had smaller Ag_3Sn particles compared to the as-cast sample. On the other hand, Cu_6Sn_5 particles might not have any changes in size as they are originally in small rounded shape in the as-cast solder. Smaller β-Sn grains were also more prominent in figure 3(b) compared to that of the as-cast sample. This shows another advantage of SPD, in that it is able to break up compound or particles resulting in more evenly distributed fine particles within the matrix. Finer particles, as has been commonly reported in literature, are more efficient in blocking movement of dislocation and thus, providing excellent improvement in mechanical properties.

3.2 Lap Joint Shear Strength

Solder samples were reflowed on bare pure copper substrate in the form of a single lap joint for shear strength test. The shear strength of all solder joints are given in Table 2. As can be seen, the strength increased from sample with 0 dipping to 30 minutes dipping before the strength decreased for the 60 minutes dipping sample.

Table 2. Average shear strength for all solder joints

Dipping time	Ave. shear strength, MPa
0	51.9
10	50.6
30	55.7
60	48.8

Cryogenic treatment has been reported to induce a large internal stress in the matrix of the material, which is the incentive of amounts of dislocations generation. This means that the strength and toughness of alloy will be improved because of the entanglement and pinning of these dislocations. With larger dislocation density, the mechanical properties of solder material will be enhanced [8, 10]. At the same time, cryogenic treatment suppresses dynamic recovery process during deformation, and thus effectively increase dislocation density without being erased by re-crystallization especially in low melting temperature alloy such as solder. Cross slip and dislocation annihilation processes are prevented because of the limited mobility of defect at such low temperature. Therefore, cryogenic treatment can be seen as a potential technique in increasing mechanical properties of solder.

The combination of cryogenic treatment and rolling also was observed to encourage grain refinement in the solder samples. Therefore, based on the Hall-Petch relation, this would help to improve the dislocation motion resistance, and thus, the increased dislocation density would result in improved strength in the solder. As grain size reduces, the yield strength will increase. From the microstructure observed under SEM and based on the average size for SAC305 solder samples in Table 1, sample with dipping in liquid nitrogen for 30 minutes has the smallest grain size. As listed in Table 2, this particular sample gave highest shear strength compared to sample rolled without prior dipping in nitrogen. In addition to that, as cryorolling also resulted in breaking up of large Ag3Sn particles, the solder could potentially has better toughness and possibly creep resistance. This comes from the more uniform distribution of small IMC particles which could act as pinning mechanism and improve the creep resistance. Work on creep behavior of cryorolled solder alloys is currently underway to evaluate how cryorolling influence the creep resistance of SAC305 alloy.

4.0 Conclusion

Several conclusions can be made in this work :

1. Cryorolling was observed to encourage grain refinement in SAC305, up to dipping for 30 minutes. Longer dipping time of 60 minutes resulted in larger grain size.

2. Cryorolling process also resulted in breaking up of particularly the Ag3Sn particles into smaller size. Smaller IMCs and more evenly distributed eutectic network was observed in cryorolled samples.

3. Shear strength of the solder joint improved in cryorolled samples due to smaller β-Sn grains, smaller IMC particles and more evenly distributed eutectic network. These would be more efficient in blocking dislocation movement leading to higher strength.

The authors would like to acknowledge the funding provided by FRGS grant (grant no : 1001.PBAHAN.6071378).

References

1. N. C. Lee, (2002) Reflow Soldering Processes, Elsevier Science.
2. H.Shang, Z.L.Ma, S.A. Belyakov, C.M. Gourlay, J. Alloy Compd, **715** , 471 – 485 (2017)
3. B.X Vuong., N.S.H.Vu, T.D.Manh, M.Vaka, D.X. Du, N.D.Nam Mater. Lett. **228** 309 313 (2018)
4. 4.H.Vafaeenezhad, S.H.Seyedein, M.R.Aboutalebi, A.R. Eivani ,Comput. Mater. Sci. **163**, 196-208 (2019)
5. Q.S. Zhu, Z.G. Wang, S.D. Wu, J.K. Shang, Mater. Sci. Eng. A **502** , 153–158 (2009)
6. Y. Zhao, H. Guo, Y. Ning & Z. Yao, Mater. Des. **46**, 889-894 (2013)
7. G. E. Totten, M. Tiryakioglu & O. Kessler, *Encyclopedia of Aluminum and Its Alloys*, 1054 (2018)
8. Y. Yao, X. Li & X. He, J. Mater. Sci. **29**, 4517-4525 92018)
9. T. Lu, D. Yi, H. Wang, X. Tu, B. Wang ,J. Alloys Compd. **781** p. 633 – 643 (2019)
10. G.R. Li, T. Qin, A.G. Fei, H.M. Wang, Y.T. Zhao, G. Chen, X.Z. Kai ,J. Alloys Compd. **802** ,50 – 69 (2019)

IOP Conf. Series: Materials Science and Engineering **957** (2020) 012057 doi:10.1088/1757-899X/957/1/012057

Tin and germanium substitution in lead free perovskite solar cell: current status and future trends

Ayu Wazira Azhari[1,2*], *Faith Shi Xin* Then[1], *Dewi Suriyani* Che Halin[2], *Suhaila* Sepeai[3] and *Norasikin* Ahmad Ludin[3]

[1]Water Research Group (WAREG), School of Environmental Engineering, Universiti Malaysia Perlis 02600 Arau, Perlis, Malaysia.
[2]Center of Excellence Geopolymer & Green Technology (CEGeoGTech), School of Material Engineering, Universiti Malaysia Perlis 02600 Arau, Perlis, Malaysia.
[3]Solar Energy Research Institute, Universiti Kebangsaan Malaysia, 43600 UKM Bangi, Selangor, Malaysia

Abstract. Tin and germanium-based perovskite solar cell is gaining interest in lead-free perovskite solar cells as it is less toxic as compared to lead but possess almost all the characteristics of a perfect solar cell materials. Within 5 years, the reported efficiency of tin-based solar cells has increased from 6.4% to 9%. Although facing with stability issues as it is easily oxidised in ambient air, several studies have proven that the stability issues can be reduced. One of it is by using cesium as the 'A' cation. On the other hand, although studies on germanium-based perovskite solar cells are rarely conducted, promising results are shown when it is alloyed with tin producing narrower bandgaps and better stability owing to the protection of the GeO_2 surface layer.

1 Introduction

Perovskite solar cell (PSC) has shown notable progress since its first debut in 2009 [1]. Ever since then, the PCS has shown remarkable breakthrough and advances rapidly as the next generation solar cells. PSC consist of a perovskite compound with chemical formula of ABX_3 where A is an organic cation (methyl-ammonium 'MA' and/or formamidinium 'FA'), B a divalent metal ion (Pb^{2+}, Bi^{2+}, Sb^{2+}, Sn^{2+}, Ge^{2+} etc.) and X a halide ion (F^-, I^-, Br^- or Cl^-). The typical perovskite structure is depicted in Fig. ***1*** [2]. Since the fabrication of PSC is solution based, wide range of alteration is possible that include band-gap and conductivity-type modification through doping and/or alloying resulting in various perovskite structures [3, 4]. Direct substitution of the cations or anions will subsequently lead to distortion to the perovskite structure as the formability of this structure is based on the ratio between the cations and anions (A:X and B:X) [5, 6]. The Goldschmidt tolerance factor (GTF) is used to determine the crystal structure of the perovskite materials and is given as:

* Corresponding author: ayuwazira@unimap.edu.my

$$t = \frac{r_A + r_X}{\sqrt{2(r_B + r_X)}} \quad (1)$$

where r is the ionic radius of A-cation, B-cation and X-anion. An ideal perovskite material should have a perfect cubic structure with t of 0.8 to 1 [7-9]. When the tolerance factor is lower than 0.8 or higher than 1, distorted octahedral structure is form that will results in loss in symmetry thus affecting the electronic properties of the material [10].

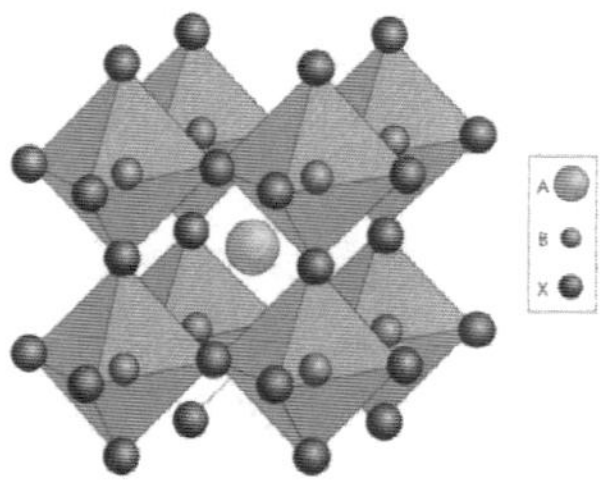

Fig. 1. Typical crystal structure of a perovskite with A, B and X ions [2]

With lead (Pb) as the active material, the power conversion efficiency (PCE) of PCS has escalated from merely 3.81% [1] to 23.7% [11] in less than a decade. However, the major obstacle in commercialization of Pb-based PCS is the toxicity of Pb that possess treat to the environment as well as human health [12]. Numerous studies have been conducted by various researchers in order to find the prospective substitute for Pb while maintaining the high efficiency and stability.

2 Approaches to lead-free perovskite solar sell

An ideal PSC material to replace Pb should be less toxic while still maintaining its high stability and excellent electrical properties. Besides that, these materials should also possess all the necessary characteristics of a perfect PCS i.e. having narrow direct bandgap, high light absorption coefficients, high charge carrier mobilities, low exciton-binding energies, long charge-carrier lifetime and long exciton diffusion length [13, 14]. A number of potential metals have been identified based on the Goldschmidt's model that include the group 14 elements, alkaline earth metals, p-block elements, transition metals and lanthanides (Fig. *2*, all of those elements with stable oxidation state of +2. However, those elements with large bandgap (Be^{2+}, Ca^{2+}, Sr^{2+}, Ba^{2+}), high toxicity (Cd^{2+} and Hg^{2+}), high radioactivity and low stability are considered less suitable [15].

Amongst these candidates, the group 14 elements (Sn^{2+}, Ge^{2+} and Si^{2+}) are most favourable due to their similar electronic configuration to Pb [16, 17]. However, the stability of the +2 oxidation state is decreasing moving up the Periodic Table due to the reduction in inert electron pair effects while the reduction of bandgap is also observed owing to increased ionic characteristics [14].

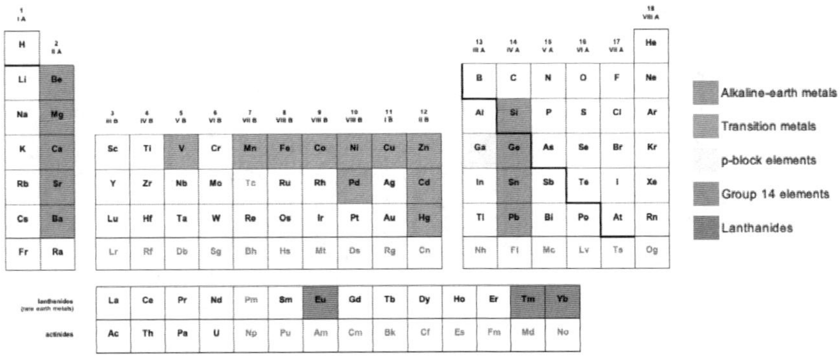

Fig. 2. Elements to replace Pb in perovskite in the periodic table

3 Tin as substitute for lead in perovskite solar cell

To attain Pb-free PSC, Sn^{2+} was among the first divalent metal considered as an alternative candidate to substitute Pb^{2+} due to the comparable electronic configuration and close effective ionic radius (1.35 Å) as of Pb^{2+} (1.49 Å) [18, 19]. Besides that, Sn-based perovskites also possess lower direct optical bandgap (1.2 – 1.4 eV) and higher charge carrier mobility ($10^2 – 10^3$ $cm^2/V·s$) compared to Pb [20-22]. The first reported study on Sn-based PSC was conducted by [23] with PCE of 6.4%. In this study, a mesoporous structure was fabricated. Following the success of this study, several attempts have been conducted to further improve the efficiency of the Sn-based PSC. This include integrating different 'A' cation such as MA [24-26], FA [27, 28] and cesium (Cs) [29-31] resulting in a symmetrical 3D structures [22]. Besides that, mixed 'A' cation is also employed [32] as it combine the best characteristics of individual cation and boost resistance in normal ambience [14]. A study by [33] reported that addition of "hollow" ethylenediammonium (en) cations enhance the performance and stability of a PSC by improving the film morphology, increasing the bandgap, unit cell volume and carrier lifetime while lowering the trap-state density. The highest efficiency realize by this PSC is 6.63%.

Besides varying the 'A' cation, numerous studies were also conducted employing several halide ions 'X' namely I^-, Br^- and Cl^-. Reported studies employing I^- as the halide ion has shown promising results with highest PCE of 9% using $FASnI_3$ [34]. A study by [35] has indicated that the for Sn-based PSC, the bandgap increases from I_3, Br_3 and Cl_3. Although replacement of I^- with Br^- or Cl^- shows improvement in certain composition, the formation of cubic perovskite is difficult either by using Br^- or Cl^- owing to the smaller ionic radius [36].

The major shortcoming in Sn-based PSC is the stability issue where Sn^{2+} is easily oxidize in ambient conditions into Sn^{4+} that can be observed by the discoloration of the perovskite solution in the presence of oxygen leading to reduction in its photovoltaic properties [37-39]. Although Cs cation is most stable for Sn-based perovskite, the major drawback lies in the low mobility, high minority carrier and the disability to form a fully 3D perovskite structure [13]. Table 1 depicted some of the major studies on Sn-based PSC to date.

Table 1. Major achievement in Sn-based perovskite solar cells

Perovskite type	Bandgap (eV)	J_{sc} ($mA.cm^{-2}$)	V_{oc} (V)	FF	PCE (%)	References
$CsSnI_3$	1.30	4.80	0.42	0.22	0.90	[18]
$MASnI_3$	1.23	16.80	0.88	0.42	6.40	[23]

3. J.H. Noh, S.H. Im, J.H. Heo, T.N. Mandal and S.I. Seok, Nano Letters, **13**, 1764 (2013)
4. Y. Zhou, Z. Zhou, M. Chen, Y. Zong, J. Huang, S. Pang and N.P. Padture, Journal of Materials Chemistry A, **4**, 17623 (2016)
5. M. Kouhnavard, S. Ikeda, N.A. Ludin, N.B. Ahmad Khairudin, B.V. Ghaffari, M.A. Mat-Teridi, M.A. Ibrahim, S. Sepeai and K. Sopian, Renewable and Sustainable Energy Reviews, **37**, 397 (2014)
6. L. Yang, A.T. Barrows, D.G. Lidzey and T. Wang, Reports on Progress in Physics, **79**, 026501 (2016)
7. A.E. Fedorovskiy, N.A. Drigo and M.K. Nazeeruddin, Small Methods, **0**, 1900426 (2019)
8. Z. Li, M. Yang, J.-S. Park, S.-H. Wei, J.J. Berry and K. Zhu, Chemistry of Materials, **28**, 284 (2016)
9. W. Travis, E.N.K. Glover, H. Bronstein, D.O. Scanlon and R.G. Palgrave, Chemical Science, **7**, 4548 (2016)
10. M. Abd Mutalib, N. Ahmad Ludin, N.a.A. Nik Ruzalman, V. Barrioz, S. Sepeai, M.A. Mat Teridi, M.S. Su'ait, M.A. Ibrahim and K. Sopian, Materials for Renewable and Sustainable Energy, **7**, 7 (2018)
11. National Renewable Energy Laboratory, Best Research-Cell Efficiencies, (2018) https://www.nrel.gov/pv/assets/pdfs/pv-efficiency-chart.20181221.pdf
12. F. Gagan, G. Deepesh and T. Archana, Interdisciplinary Toxicology, **5**, 47 (2012)
13. W. Ke and M.G. Kanatzidis, Nature Communications, **10**, 965 (2019)
14. M.I.H. Ansari, A. Qurashi and M.K. Nazeeruddin, Journal of Photochemistry and Photobiology C: Photochemistry Reviews, **35**, 1 (2018)
15. S. Chatterjee and A.J. Pal, Journal of Materials Chemistry A, **6**, 3793 (2018)
16. R. Ali, G.-J. Hou, Z.-G. Zhu, Q.-B. Yan, Q.-R. Zheng and G. Su, Chemistry of Materials, **30**, 718 (2018)
17. T. Nakajima and K. Sawada, The Journal of Physical Chemistry Letters, **8**, 4826 (2017)
18. Z. Chen, J.J. Wang, Y. Ren, C. Yu and K. Shum, Applied Physics Letters, **101**, 093901 (2012)
19. A.K. Jena, A. Kulkarni and T. Miyasaka, Chem Rev, **119**, 3036 (2019)
20. K.P. Marshall, M. Walker, R.I. Walton and R.A. Hatton, Nature Energy, **1**, 16178 (2016)
21. L.M. Herz, ACS Energy Letters, **2**, 1539 (2017)
22. C.C. Stoumpos, C.D. Malliakas and M.G. Kanatzidis, Inorganic Chemistry, **52**, 9019 (2013)
23. N.K. Noel, S.D. Stranks, A. Abate, C. Wehrenfennig, S. Guarnera, A.-A. Haghighirad, A. Sadhanala, G.E. Eperon, S.K. Pathak, M.B. Johnston, A. Petrozza, L.M. Herz and H.J. Snaith, Energy & Environmental Science, **7**, 3061 (2014)
24. F. Hao, C.C. Stoumpos, D.H. Cao, R.P.H. Chang and M.G. Kanatzidis, Nature Photonics, **8**, 489 (2014)
25. F. Hao, C.C. Stoumpos, P. Guo, N. Zhou, T.J. Marks, R.P.H. Chang and M.G. Kanatzidis, Journal of the American Chemical Society, **137**, 11445 (2015)
26. H. Do Kim, Y. Miyamoto, H. Kubota, T. Yamanari and H. Ohkita, Chemistry Letters, **46**, 253 (2017)
27. W. Liao, D. Zhao, Y. Yu, C.R. Grice, C. Wang, A.J. Cimaroli, P. Schulz, W. Meng, K. Zhu, R.G. Xiong and Y. Yan, Advanced Materials, **28**, 9333 (2016)

28. M. Zhang, M. Lyu, J.-H. Yun, M. Noori, X. Zhou, N.A. Cooling, Q. Wang, H. Yu, P.C. Dastoor and L. Wang, Nano Research, **9**, 1570 (2016)
29. W. Li, J. Li, J. Li, J. Fan, Y. Mai and L. Wang, Journal of Materials Chemistry A, **4**, 17104 (2016)
30. A. Kaltzoglou, D. Perganti, M. Antoniadou, A.G. Kontos and P. Falaras, *Stress Tests on Dye-Sensitized Solar Cells with the Cs2sni6 Defect Perovskite as Hole-Transporting Material*, 102, 49 (2016)
31. S. Gupta, T. Bendikov, G. Hodes and D. Cahen, ACS Energy Letters, **1**, 1028 (2016)
32. Z. Zhao, F. Gu, Y. Li, W. Sun, S. Ye, H. Rao, Z. Liu, Z. Bian and C. Huang, Advanced Science, **4**, 1700204 (2017)
33. W. Ke, C.C. Stoumpos, I. Spanopoulos, L. Mao, M. Chen, M.R. Wasielewski and M.G. Kanatzidis, J Am Chem Soc, **139**, 14800 (2017)
34. S. Shao, J. Liu, G. Portale, H.-H. Fang, G.R. Blake, G.H. Ten Brink, L.J.A. Koster and M.A. Loi, Advanced Energy Materials, **8**, 1702019 (2018)
35. I.E. Castelli, J.M. García-Lastra, K.S. Thygesen and K.W. Jacobsen, APL Materials, **2**, 081514 (2014)
36. P.P. Boix, S. Agarwala, T.M. Koh, N. Mathews and S.G. Mhaisalkar, The Journal of Physical Chemistry Letters, **6**, 898 (2015)
37. M.a.A. Wadi, T.H. Chowdhury, I.M. Bedja, J.-J. Lee, N. Amin, M. Aktharuzzaman and A. Islam, Electronic Materials Letters, **15**, 525 (2019)
38. A. Babayigit, A. Ethirajan, M. Muller and B. Conings, Nature Materials, **15**, 247 (2016)
39. F. Giustino and H.J. Snaith, ACS Energy Letters, **1**, 1233 (2016)
40. T.-B. Song, T. Yokoyama, S. Aramaki and M.G. Kanatzidis, ACS Energy Letters, **2**, 897 (2017)
41. L. Liang and P. Gao, Adv Sci (Weinh), **5**, 1700331 (2018)
42. T. Krishnamoorthy, H. Ding, C. Yan, W.L. Leong, T. Baikie, Z. Zhang, M. Sherburne, S. Li, M. Asta, N. Mathews and S.G. Mhaisalkar, Journal of Materials Chemistry A, **3**, 23829 (2015)
43. C.C. Stoumpos, L. Frazer, D.J. Clark, Y.S. Kim, S.H. Rhim, A.J. Freeman, J.B. Ketterson, J.I. Jang and M.G. Kanatzidis, Journal of the American Chemical Society, **137**, 6804 (2015)
44. L.-J. Chen, RSC Advances, **8**, 18396 (2018)
45. M. Roknuzzaman, K. Ostrikov, H. Wang, A. Du and T. Tesfamichael, Scientific Reports, **7**, 14025 (2017)
46. P.-P. Sun, Q.-S. Li, L.-N. Yang and Z.-S. Li, Nanoscale, **8**, 1503 (2016)
47. L.-C. Tang, Y.-C. Chang, J.-Y. Huang, M.-H. Lee and C.-S. Chang, Japanese Journal of Applied Physics, **48**, 112402 (2009)
48. I. Kopacic, B. Friesenbichler, S.F. Hoefler, B. Kunert, H. Plank, T. Rath and G. Trimmel, ACS Applied Energy Materials, **1**, 343 (2018)
49. M.-G. Ju, J. Dai, L. Ma and X.C. Zeng, Journal of the American Chemical Society, **139**, 8038 (2017)
50. N. Ito, M.A. Kamarudin, D. Hirotani, Y. Zhang, Q. Shen, Y. Ogomi, S. Iikubo, T. Minemoto, K. Yoshino and S. Hayase, J Phys Chem Lett, **9**, 1682 (2018)
51. C.H. Ng, K. Nishimura, N. Ito, K. Hamada, D. Hirotani, Z. Wang, F. Yang, S. Likubo, Q. Shen, K. Yoshino, T. Minemoto and S. Hayase, Nano Energy, **58**, 130 (2019)

IOP Conf. Series: Materials Science and Engineering **957** (2020) 012058 doi:10.1088/1757-899X/957/1/012058

The Effect of Sodium Hydroxide (NaOH) Solution Concentration on Properties of Geopolymer Paste

Farah Farhana Zainal[1,*], *Nurqistina* Sulotoha[1], *Yusrina* Mat Daud[1], *Mohammad Firdaus* Abu Hashim[1], *Hasri*[2] and *Hartati*[2]

[1]Center of Excellence Geopolymer and Green Technology (CEGeoGTech), School of Materials Engineering, Universiti Malaysia Perlis (UniMAP), P.O. Box 77, D/A Pejabat Pos Besar, 01000 Kangar, Perlis, Malaysia
[2]Fakultas Matematika & Ilmu Pengetahuan Alam, Universitas Negeri Makassar, Kampus FMIPA UNM, Parang Tambung, JL. Mallengkeri, Makassar, 90224, Indonesia

Abstract. This research analyzes the effect of sodium hydroxide (NaOH) solution concentration ranging from 8M to 12M as alkaline activator on the properties of geopolymer paste. Alkaline activator is essentially a mixture of sodium hydroxide and sodium silicate solution. In this research, the main component used was raw kaolin. In order to produce kaolin based geopolymer paste, the alkaline activator solution was prepared with 0.24:1.00 ratio of Na_2SiO_3/NaOH and this alkaline activator solution was prepared for 24 hours prior before used in another process. The solid-to-liquid ratio which is kaolin-to alkaline activator solution ratio was 0.80:1.00. The mixture of kaolin based geopolymer paste were placed in cube moulds with a size of 50x50x50 mm, and left for 24 hours until it hardened. Then, the samples were cured at 80°C in the oven for 24 hours. The samples of kaolin based geopolymer paste were tested based on compressive strength, morphology analysis, water absorption and porosity after 28 days. In this project, 8M concentration of NaOH solution was the best concentration in order to study the synthesis of kaolin based geopolymer paste as the sample had the highest amount of compressive strength with 0.992 MPa and had the lowest water absorption and porosity with 1.246% and 24.08% respectively. Scanning Electron Microscope (SEM) was used to observe the morphological structure of the kaolin based geopolymer paste. The sample with 8M concentration of NaOH solution shows that least amount of unreacted particles. The structure of kaolin based geopolymer paste was more dense at 8M concentration as the size of pore decreases. The kaolin based geopolymer paste is suitable for use in the construction building industry as a finishing product such as coating.

*Corresponding author: farahfarhana@unimap.edu.my

1 Introduction

Geopolymer is an inorganic material and inorganic polymers formed by activating silica-aluminum rich minerals such as kaolinitic clay, metakaolin and fly ash mixed with the alkaline hydroxide solution and alkaline silicate solution at ambient conditions or higher temperature level [1-4]. Geopolymer is a new generation of the inorganic substances that differ in microstructure and chemical composition compared to Portland cement [5]. In this study, kaolin was used as the sources of alumino-silicate. Many geopolymer synthesis studies have been conducted using fly ash as the source materials, so in previous research there is lack of information on the use of kaolin in geopolymer synthesis. Kaolin was chosen to see how much kaolin in geopolymer synthesis can stand alone. Concentration of sodium hydroxide (NaOH) solution and curing regime are the important parameters to consider when designing for a particular application of kaolin-based geopolymer [6]. The major problem with this type of study is that not many researchers have conducted studies on the effects of NaOH solution concentration on metakaolin or calcined kaolin, so little information is available regarding the effect of NaOH solution concentration on kaolin-based geopolymer paste [7-9].

Other than that, it is important to investigate the various variables affecting the strength and durability of the kaolin geopolymer paste. It is necessary to identify the variables that have the significant effect on the sample so that the proper solution can be made to avoid the limitations. The two variables that can affect the strength and durability is the concentration of NaOH and curing temperature [10-12]. The solution of NaOH mixed as an alkaline activator with Na_2SiO_3 solution will affect the properties of the kaolin-based geopolymer. The porosity and water absorption of kaolin-based geopolymer paste is very important to measure, as the highest percentage of porosity and water absorption will affect the strength of the sample. Several studies have produced estimates of porosity on geopolymer [5, 6], but there are still insufficient data on kaolin-based geopolymer paste for porosity and water. Different NaOH concentrations was used in this project to investigate the effect of NaOH solution concentration in term of physical and mechanical properties of kaolin-based geopolymer paste.

2 Experimental Procedures

2.1 Materials and Sample Preparation

Kaolin which is an industrial waste acts as a source of aluminosilicate in this study. The kaolin was obtained from a local station. Table 1 shows the elemental composition of raw kaolin examined using Xray Fluorescence (XRF) analysis. The alkaline activator solution was prepared by mixing NaOH solution with Na_2SiO_3 solution.

Table 1. Chemical composition of kaolin.

Compound	wt (%)	Compound	wt (%)
SiO_2	56.5	MnO	0.042
Al_2O_3	37.3	ZrO_2	0.035
K_2O	2.50	CuO	0.031
Fe_2O_3	2.06	V_2O_5	0.028
TiO_2	0.764	PbO	0.019
CaO	0.284	Ga_2O_3	0.015
RuO_2	0.185	Cr_2O_3	0.012
Rb_2O_3	0.0762	ZnO	0.011
BaO	0.065	Y_2O_3	0.0099
Eu_2O_3	0.063	Ra_2O_7	0.007

In this study, NaOH solution was prepared by dissolving specified amounts of sodium hydroxide (NaOH) pellets with distilled water to obtain NaOH concentration in the range of 8M to 12M and cooled down to room temperature for 24 hours. Then, the NaOH solution was mixed with sodium silicate (Na_2SiO_3) solution to form alkaline activator with a Na_2SiO_3/NaOH ratio of 0.24:1.00 [13]. For the preparation of kaolin based geopolymer paste, the kaolin powder was mixed with alkaline activator until homogenous using solid-to-liquid ratio of 0.80:1.00. Heah et al., stated that the ratio of Na_2SiO_3/NaOH and S/L ratio with 0.24:1.00 and 0.80:1.00 gives the optimum compressive strength after 28 days [1, 14]. After the homogenous paste was obtained, the mixture of kaolin based geopolymer paste was placed into 50x50x50 mm cube moulds and cured in the oven at 80°C for 24 hours [15]. All the samples were sealed using a thin plastic layer after the curing process in order to prevent the excess of Na^+.

2.2 Sample Characterization

2.2.1 Compressive Strength Test

Compressive strength tests of all specimens were evaluated according to the ASTM C109/ C109M-08 by using the Shimadzu Universal Testing machine. A minimum of 3 samples for each concentration were examined after 28 days in order to evaluate the early strength gain for each concentration.

2.2.2 Morphology Analysis

Morphology analysis was performed by using Scanning Electron Microscope (SEM) JSM-6460 LA JEOL Japan in order to study the morphology of the raw kaolin and kaolin based geopolymer paste at different concentration of NaOH solution. The samples were cut into small pieces and coated using Auto Fine Coater before the SEM analysis.

2.2.3 Water Absorption and Porosity Test

Water absorption test was conducted by immersing a minimum of 3 samples for 24 hours and calculating the average data by using equation (1) :

$$\text{Percentage of water absorption} = \frac{\text{Wet weight - Dry weight}}{\text{Dry weight}} \times 100\% \qquad (1)$$

In order to identify the value of porosity of the sample, pycnometer was used to analyze the data. The readings were taken by calculating the average value of the total pore within 3 samples and the data is usually presented in percentage.

3 Results and Discussion

3.1 Compressive Strength

The mechanical properties of kaolin based geopolymer paste were identified by compressive strength testing. The compressive strength testing was conducted after 28 days ensure the optimization of geopolymerization process. Fig. 1 shows the compressive strength of different concentration NaOH solution (8M, 10M and 12M) that was cured at

80°C after 28 days. It can be seen that the compressive strength slightly decreased from 8M to 12M. Sample 8M shows the highest strength with 0.992 MPa, followed by Sample 10M with 0.946 MPa and Sample 12M with 0.754 MPa.

Sample 8M shows the highest value compared to the other samples due to the water content. At lower NaOH solution concentration, the sample mix has excess mixing water. This water content gives benefits to the geopolymerization process where it helps in the transportation of dissolved ions [6]. Other than that, due to the early age strength at 28 days, sample 8M achieved optimum strength. This is because at 28 days, the reaction is mostly fully reacted as kaolin based geopolymer paste.

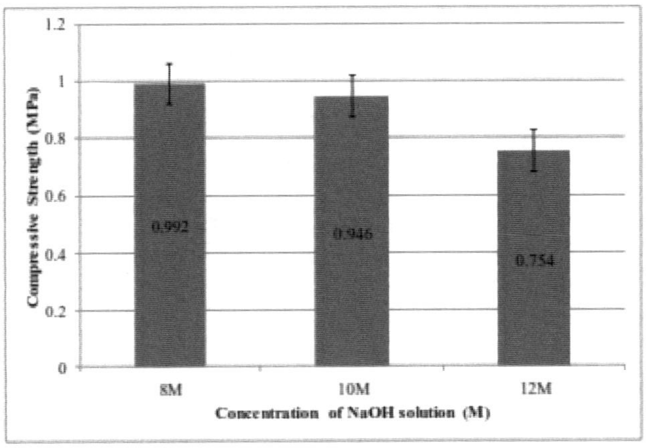

Fig. 1. Compressive strength with various concentration of NaOH solution.

3.2 Morphology of Geopolymers

Analysis of morphology was observed after 28 days and curing at 80 °C for 24 hours. Fig. 2 shows the changes in morphologies of kaolin based geopolymer paste for various NaOH solution concentrations. From Fig. 2 (a), (b), (c) and (d), it can be seen that the kaolin based geopolymer paste particles changed in shapes after 28 days when it changes from plate-like structure to spherical sponge-like structure. Raw kaolin was activated by the alkaline activator with various NaOH solution concentrations and the geopolymerization reaction occurred after mixing the raw kaolin with alkaline activator solution [16, 17].

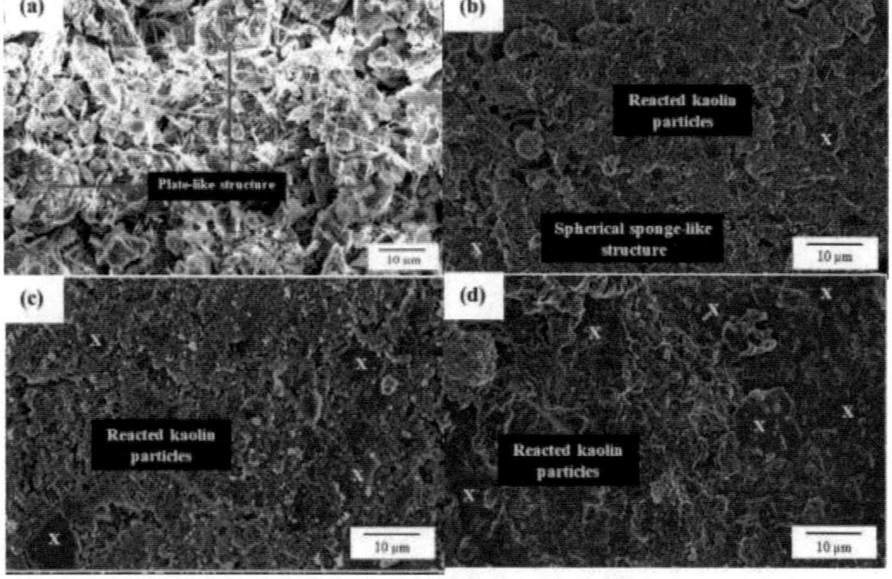

Fig. 2. SEM image of kaolin (a), and geopolymer synthesized using different NaOH concentration of 8M (b), 10M (c) and 12M (d) at 28 days. (X shows unreacted kaolin zones).

Fig. 2 displays the morphology of kaolin based geopolymer paste with different concentrations of NaOH solution at constant Na_2SiO_3/NaOH ratio and S/L ratio with 0.24:1.00 and 0.8:1.00 respectively. Other than that, Fig. 2 also shows that sample 8M contained less amount of unreacted particles compared to sample 10M and sample 12M. Sample 12M shows the largest amount of unreacted particles of raw kaolin. Sample 8M contributes the highest strength with 0.992 MPa and this can be related to the morphology of the sample where sample 8M is more homogeneous compared to other samples as shown in Fig. 2. The morphology of sample 8M has more geopolymeric gel than sample 10M and sample 12M due to the water content that contributes to optimum strength of the sample. This morphology test revealed that sample 12M with the highest concentration of NaOH solution has a larger amount of unreacted kaolin based geopolymer paste as shown in Fig. 2 (d). It can also be observed that sample 12M has less geopolymeric gel that is likely to be caused by the excess Na^+. Sample 10M and sample 12M have the lowest strength due to the incomplete chemical reaction during geopolymerization process between raw kaolin and alkaline activator solution presented by the plate-like structure of kaolin. However, all the samples displayed partially reacted particles and showed large amounts of unreacted particles at 28 days. This is due to the characteristics of the kaolin itself. As the plate-like and layered structure provides low reactivity of the kaolin, the dissolution of kaolin and alkaline activator solution was extremely slow [18]. Kaolin based geopolymer paste does not undergo complete geopolymerization process and is not fully incorporated into the structure.

3.3 Water Absorption and Porosity Test

Fig. 3 presents the relationship between water absorption and porosity for different concentration of NaOH solution in terms of molarity. As illustrated in Fig. 3, the trend of the graph shows that the amount of water absorption and porosity are slightly increased

$MASnI_3$	1.30	16.30	0.68	0.48	5.23	[24]
$MASnIBr_2$	1.75	12.30	0.82	0.57	5.73	[24]
$MASnI_3$	1.34	21.40	0.32	0.46	3.15	[25]
$FASnI_3$	1.40	22.07	0.465	0.60	6.22	[27]
$CsSnIBr_2$	-	17.40	0.31	0.57	3.20	[29]
Cs_2SnI_6	-	13.97	0.58	0.42	3.44	[30]
$FASnI_2Br$	1.68	6.82	0.467	0.54	1.72	[28]
$CsSnBr_3$	1.75	9.10	0.41	0.58	2.17	[31]
$MASnI_3$	1.23	26.00	0.232	0.386	2.33	[26]
$CsSnI_3$	-	25.71	0.38	0.49	4.81	[40]
$enMASnI_3$	1.40	24.28	0.429	0.64	6.67	[33]
$(FA)_{0.75}(MA)_{0.25}SnI_3$	1.33	21.20	0.61	0.63	8.12	[32]
$FASnI_3$	-	24.1	0.525	0.71	9.0	[34]

4 Germanium as substitute for lead in perovskite solar cell

Germanium (Ge) is another potential candidate in lead free PSC as it is in the same group 14 as Pb and Sn. Studies have shown that although Ge possesses similar transport and optical properties as Pb and Sn analogues due to its divalent nature they are hardly been investigated due to its mercurial character in +2 oxidation state [14]. With smaller ionic radius, poor solubility in polar solvent, and relatively wide bandgap (>1.6eV), the PCE for Ge-based PSC to date is still less than 5% [41-43]. Lowest bandgap for PSC is observed when Cs is used as the 'A' cation as compared to MA and FA, hence it is anticipated that CsGeX perovskite is more suitable for solar cell application [44, 45]. On the other hand, changing the 'X' anion will also increase the bandgap of the PSC with I^- showing the lowest bandgap followed by Br^- and Cl^- [46, 47]. Most studies suggest that in line with the Sn-based PSC, the stability issue is the major concern in Ge-based PSC. Hence further studies are to be carried out either by using tandem structures [15, 19], the introduction of additive [48] or alloying.

5 Mixed Sn-Ge perovskites for solar cell

Mixed 'M' cation has been carried out in mixed Pb:Sn based PSC and has shown promising results [22]. Varying the Pb:Sn ratios will enable tuning in of the bandgap and further increase the stability. The same trend is observed in mixed Sn:Ge based PSC where lower bandgap and better stability can be achieved. A theoretical study by [49] predicted that the preparation of a stable Sn:Ge PSC is possible with higher carrier mobility and better resistance to water penetration. Another study by [50] employing mix Sn:Ge PSC resulted in higher performance of 4.48% PCE compared to 3.31% without Ge. It is believed that passivation of Ge has effectively reduce defects on the surface of the Sn-based perovskite hence reducing recombination and increasing the efficiency. The highest PCE (7.9%) of Sn:Ge-based PSC by far is realised in another study by [51] where addition of Ge is believed to increase the stability of the perovskite structure while reducing the trap density. Hence it is believed that more promising results is anticipated in this area of studies.

References

1. A. Kojima, K. Teshima, Y. Shirai and T. Miyasaka, Journal of the American Chemical Society, **131**, 6050 (2009)
2. M.A. Green, A. Ho-Baillie and H.J. Snaith, Nature Photonics, **8**, 506 (2014)

with increasing NaOH solution concentration. The samples with 8M concentration of NaOH solution shows the lowest amount of water absorption and porosity with 1.246% and 0.422% respectively compared to other samples. As expected, the porosity affected the water absorption of kaolin based geopolymer paste. More water will enter the sample that has higher amount of porosity and will cause damage to the sample such as crack. As the amount of porosity increases, the size of the pore also increases, causing the water absorbed to also increase.

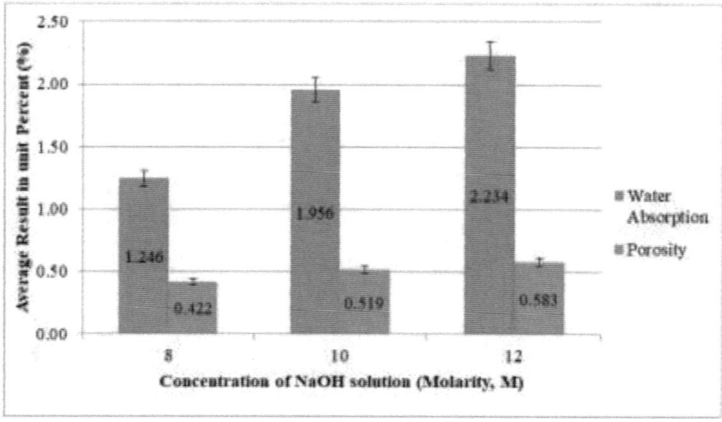

Fig. 3. Water absorption and porosity with various concentration of NaOH solution at 28 days.

Conclusions

Detailed work has been done to achieve the objectives of this study. Referring to the experimental investigation, this paper has highlighted the effect of Sodium Hydroxide (NaOH) solution concentration on the properties of geopolymer paste. Results from this investigation can be concluded as following:

1) The concentration of NaOH solution has significant effect on the compressive strength of kaolin based geopolymer paste.
2) Kaolin based geopolymer paste showed the optimum strength at 0.992 MPa for the samples prepared with 8M concentration of NaOH solution cured at 80°C for 24 hours.
3) The relationship of compressive strength with water absorption and porosity of kaolin based geopolymer paste showed the similar trends.
4) Samples with lower water absorption and porosity yielded optimum strength after 28 days.
5) Sample with 8 M concentration of NaOH solution showed the least amount of porosity and displayed less portion of unreacted particles of kaolin based geopolymer paste. This can be proved by morphology study using Scanning Electron Microscope (SEM).

The authors would like to thank the staffs of Center of Excellence Geopolymer & Green Technology (CEGeoGTech), School of Materials Engineering, Universiti Malaysia Perlis (UniMAP) for their involvement in the research. This work was supported and funded by the Universitas Negeri Makassar Kampus FMIPA UNM, Makassar, Indonesia. The authors also would like to acknowledge the support from the Fundamental Research Grant Scheme (FRGS) under a grant number of **FRGS/1/2018/TK06/UNIMAP/02/2** from the Ministry of Education Malaysia.

References

1. A. Nmiri, N. Hamdi, M. Duc, and E. Srasra, "Synthesis and characterization of kaolinite-based geopolymer: Alkaline activation effect on calcined kaolinitic clay at different temperatures," vol. 8, no. 2, pp. 276–290, 2017.
2. H.D. Rozman, A.R.Rozyanty, L. Musa, G.S. Tay. J Wood Chem Technol **30(2)**, 152-163 (2010)
3. A.R. Rozyanty, H.D. Rozman, G.S. Tay. Adv. Mater. Res. **264-265**, 712-718 (2011)
4. H.D. Rozman, A.R. Rozyanty, G.S Tay, R.N. Kumar. J. Appl. Polym. Sci. **115(5)**, 2677-2682 (2010)
5. M. T. S. Alehyen, M. EL Achouri, J. Mater. Environ. Sci., 8 (5), 1783–1796 (2017)
6. C. Y. Heah et al., Int. J. Miner. Metall. Mater. **20(3)**, 313–322 (2013)
7. Ahmad Faris, M., Mustafa Al Bakri Abdullah, M., Nizar Ismail, K., Aida Mohd Mortar, N., Firdaus Abu Hashim, M., Hadi, A. IOP Conf. Ser. Mater. Sci. Eng. **551 (1)**, 012080 (2019)
8. Hashim, M.F.A., Abdullah, M.M.A.B., Sandu, A.V., Puskas, A., Daud, Y.M., Zainal, F.F., Faris, M.A., Hasri, Hartati, IOP Conf. Ser. Mater. Sci. Eng. **572 (1)**, 012037 (2019)
9. Abu Hashim, M.F., Abdullah, M.M.A.B., Ghazali, C.M.R., Hussin, K., Binhussain, M., Omar, M.F. Effect of Glass Reinforced Epoxy (GRE) pipe filled with Geopolymer Materials for Piping Application: Compression Properties (2016) MATEC Web of Conferences, 78, art. no. 01066.
10. Daud, Y.M., Kamarudin, H., Ruzaidi, C.M., Osman, A.F., Al-Bakri, M. (2015) Mater. Sci. Forum **803**, 58-62 (2015)
11. Daud, Y.M., Hussin, K., Osman, A.F., Ghazali, C.M.R., Al Bakri Abdullah, M.M., Sandu, A.V. Mater. Plast. 54 (3), 543-545 (2015)
12. Mat Daud, Y., Hussin, K., Ruzaidi, C.M., Osman, A.F., Al-Bakri, M., Binhussain, M. Mater.Sci. Forum **819**,290-294 (2015)
13. S. Ramasamy et al., Rev. Adv. Mater. Sci. **42(1)**, 83–91(2015).
14. Farah Farhana Zainal, Sahrul Abd Hakim, Yusrina Mat Daud, Mohammad Firdaus Abu Hashim, Mohd Mustafa Al Bakri Abdullah, Hasri and Hartati IOP Conf. Ser. Mater. Sci. Eng. 701 012002 IOP Publishing (2019)
15. S. Pangdaeng, " Int. J. GEOMATE, 14(46) (2018)
16. H. Kamarudin, a M. M. Al Bakri, M. Binhussain, C. M. Ruzaidi, M. Luqman, and C. Y. Heah,, **10**,18–24 (2011)
17. Azhar, N.S.D.M., Zainal, F.F., Abdullah, M.M.A.B. AIP Conf. Proc. 2129, art. no. 020046 (2019)
18. C. Y. Heah et al., Constr. Build. Mater., **35**, pp. 912–922 (2012)

A short review: Properties of superconducting solder

Chang M.S.[1,*], *Mohd Salleh* M.A.A.[1], and *Halin* D.S.C.[1]

[1]Centre of Excellence Geopolymer and Green Technology, School of Materials Engineering, Universiti Malaysia Perlis (UniMAP), Taman Muhibbah 02600, Jejawi, Arau, Perlis, Malaysia.

Abstract. Development of electronic devices were moving towards miniaturization, multifunction and increasingly specific service environment of solder interconnects, where this drive to the higher requirements on the solder properties. Without solder, it would be impossible to produce the infinite electronic devices that define the 21st century. Hence, solder is important in the design and engineering process. This review paper was summarized on the literature of the low-temperature solder systems for superconducting solder materials, which provide further theoretical basis for the study of superconducting solder of electronic and aerospace applications. Pb-Bi system is the most satisfactory solder as a superconducting solder, but due to the restrictions of the lead usage, so new Pb-free superconducting solder need to be invented. Of those Pb-free superconducting materials had been studied, the Sn-In-Bi ternary solder system show the greatest superconducting properties as compared to other Pb-free superconducting solder where $T_C = 6.9$ K, H_{C2} (4.2 K) = 0.18 T and J_C (4.2 K, 0.01 T) = 1.3×10^8 A/m^2 respectively.

1 Introduction

In the electronic assemblies, solder plays an important role to hold the electronic components together while permitted various components to expand and contract, to transfer heat [1], and to conduct electrical signals [2]. Superconducting solder is crucial in the industry as standard joints between superconductors. The overall behavior of the solder interconnects are highly dependent on the performances of the solder materials and the microstructure such as residual strains, precipitates, and imperfect sintering [3]. Superconducting solders should have the properties of low melting temperature to avoid degradation of the superconducting filaments when interconnecting, good wetting properties, suitable liquid phase viscosity, and sufficient superconductivity to allow sufficient current to penetrate through the joint [4].

It has been reported that most of the superconducting solders systems are binary or ternary eutectic alloys. The dominant phase of the eutectic superconducting solder alloy act as continuous matrix holding the other disconnected phases, while the secondary phase work as the flux pinning sites to improve the magnetic irreversibility and critical current

* Corresponding author: cmshin@ymail.com

densities in the magnetic field [5]. Besides, at least one phase of the solder alloy must being superconducting element such as In (T_C = 3.4 K), Pb (T_C = 7.2 K), Bi (T_C = 8.7 K) or Sn (T_C = 3.7 K) [4]. In addition, Levy et al [3] also conclude the concept of flux pinning in the solder system where solder alloys with combination of superconducting and non-superconducting phase (e.g: Pb-Bi) have greater structure sensitivity compared to the solder alloys with two superconducting phases because of the interphase boundary separating a area of radically distinct electronic properties would be expected to impede the motion of flux lines.

Nevertheless, in certain condition, solute contents [5], values of critical temperature (T_C), critical field (H_C), and critical currents (J_C) of the constituent phases in the eutectic alloys will be influenced by the composition, thermal and mechanical history of the material [3, 5, 6]. For instance, Levy et al [3] reported that Sn-Zn and Sn-Cd eutectics solder alloys have H_C of ~0.1 T, but other researchers discovered much inferior values for these materials [7]. This paper reviews the research progress of the different compositions of Pb-free superconducting solder system. The effect of the solder composition to the solder properties such as Tc, H_C, and J_C of the solder system were discussed.

2 Lead-based binary superconducting solders.

The lead-bismuth alloys (Pb-Bi) is the typical superconducting solder has been practiced for many decades as a metallic superconductor. These solder interconnects were able to carry 1000 A current at 4.2 K and 1 T [8]. Fig. 1 shows the Pb-Bi phase diagram, which indicated there are ε-Pb-rich and α-Pb-rich superconducting phases in this system, but there is an extensive doubt in the ε-phase composition at the Bi-rich region due to the solidus steep gradient. These alloys have low melting temperature of 124°C (eutectic $Pb_{44.5}Bi_{55.5}$), high T_C = 8.4 K ($Pb_{60}Bi_{40}$), H_{C2} (4.2 K) = 1.77 T ($Pb_{60}Bi_{40}$) [6, 8], and $J_C = 2.0 \times 10^8 A/m^2$ [9].

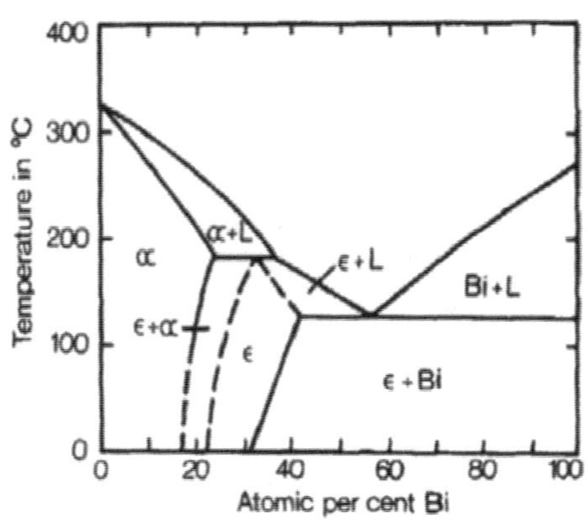

Fig. 1. Pb-Bi binary phase diagram [6].

However, solid state diffusion of the phases in this system happened rapidly even at 300 K, due to the equilibrium solubility with temperature alter the structures and thus superconducting behavior is significantly alter by time at room temperature [6]. For instance, the eutectics $Pb_{44.5}Bi_{55.5}$ H_{C2} values decline notably with room-temperature ageing, apparently due to the precipitation of Bi from the superconducting phase [5]. Moreover, Thornton [10] revealed that increasing of Bi content in the Pb-Bi solders joints in the range 30-55.5 wt% Bi could increase the J_C values due to the flux pinning at finely dispersed Bi precipitates. The H_{C2} values of Pb-Bi joints also can be reached to function at fields lower than 2 T at 4.2 K [11, 12].

Apart from Pb-Bi, the lead-tin (Pb-Sn) solder is considered most suitable mixture for soldering because it has the eutectic composition of $Sn_{63}Pb_{37}$ combination with melting point of 183 °C [13], relatively high T_C = ~7.5 K, good wettability, strength, and low cost. Pb permitted the solder melted at low temperature and Sn improved the wetting properties of the alloy by lending higher strength joining. The compositions usually used for superconducting solders are $Sn_{50}Pb_{50}$ (T_C = 7.75 K) [4], $Sn_{30}Pb_{70}$ (T_C = 7.45 K) [4] and $Sn_{60}Pb_{40}$ (T_C = 7.05 K) [14]. Among all these solder systems, $Sn_{60}Pb_{40}$ solder system is more applicable for the assembly of the superconducting joint which has a deformation range of 5 °C. Moreover, high Sn content can improve the H_C value, wettability consequently reduce the liquidus temperature of the solder alloys [15]. Thus, they account 80-90% of all solders in electronics production.

The Sn equilibrium solubility in Pb is depending on temperature which differ from 29 at.% till ~2 at.% (from eutectic point till room condition). For instance, the microstructures of as-cast Pb-Bi are instable even though at room condition and the Pb phase still consist of adequate Sn in solution for post-solidification sample that aged for over one month in room condition to increase H_C to 0.12 T (at 4.2 K) [5]. Consequent annealing at 170 °C then rapid cooling to cryogenic temperatures could raise both Sn content and H_C to 0.19 T. Yet, room-temperature precipitation of excess Sn happened quickly in rapid cooling sample, and H_C decline quickly to 0.05 T. In the Sn-Pb system, cold work also can diminish H_C value mainly caused by the increasing of precipitation.

In summary, the composition of $Pb_{60}Bi_{40}$ is the most compromised leaded superconducting solder used as the joining for the superconducting metallic in terms of the superconducting properties. However, future research on new Pb-free superconducting solder must be invented due to the restrictions of Pb usage in the electronics application.

3 Lead-free binary superconducting solders.

Since the legislation of restriction Pb in industries with a few exceptions, it is important to invent alternative Pb-free solders that meet the elementary properties for superconducting solders [16, 17]. The superconducting solder systems are required to have low melting points, generally eutectic alloys wherein the majority phase is essential to have good super-conductivity [3, 6]. Among all the Pb-free binary superconducting solders that has been studied up to date, the tin-indium (Sn-In) systems are a capable replacement for Pb-free superconducting solders according to the limited available reports which has the advantages of low melting point of 120 °C for the eutectic alloy, relatively high T_C = 5.4 ± 0.2 K ($Sn_{48}In_{52}$) [9], higher ductility, better wetting properties, and longer fatigue life compared to the other competing Pb-free solder system [15]. Additionally, the In-rich phase in the solder system is desirable for low melting temperature and good ductility which can reduce the coefficient of thermal expansion (CTE) mismatches between component and substrate.

The binary phase diagram of Sn-In as shown in Fig. 2 (a) shows two superconducting phases, β and γ in this system. The β-phase has better superconductivity compared to the γ-phase where the T_C, H_{C2} and J_C values of the β-phase are 5.9 K, 0.066 T and 6.1 $\times 10^7$A/m^2, and γ-phase are 4.5 K, 0.025 T and 1.22 × 10^7A/m^2 respectively [18], whereas Warren Jr and Bader [14] state that β-phase has T_C of 6.5 K and γ-phase has T_C of 4.7 K. This shows that the T_C values for the intermetallic phases is highly dependence to the composition of the solder system [19]. Furthermore, Levy et al [3] proposed that strong flux pinning at the interfaces boundary in the Sn-In system by discovered a linear variation among the critical current density and the interphase boundaries density for randomly and highly oriented specimens.

Sn-In rapid cooling from the liquid state samples attain greater transition temperature (~2-3 K) compared to those of annealed samples as shown in Fig. 2 (b). There is an argument arise and state that the raise of transition temperatures is derive from the samples internal strain and not caused by short-range disorder of the structure [19], nevertheless it may cause by the β-phase variances in the composition thus result in the higher gradient to lower Sn contents of the solvus line among β and β + γ phase region. However, even optimized the binary Sn-In alloys, the superconductivity is still much lower than those required in industry.

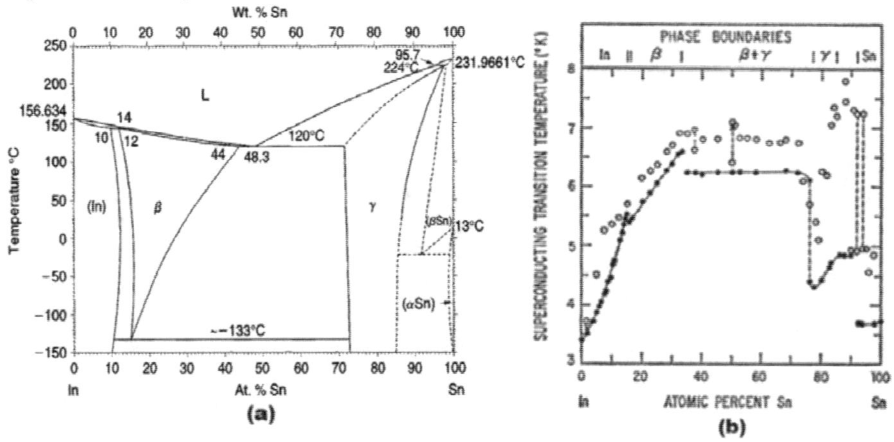

Fig. 2. (a) Sn-In binary phase diagram [20]. (b) T_C values versus Sn-In system composition: ○quenched samples; ●annealed samples [19].

In short, Sn-In is the best Pb-free binary superconducting solder that had been studied up to date. However, the superconductivity of the solder is highly structure sensitive. The cooling rate of the solder system could significantly affect the solder transition temperature, where quenching can improve the transition temperature.

4 Ternary superconducting solders.

Some modification has been done by previous researchers on the Pb-free superconducting solder by incorporated ternary elements into the solder to enhance the properties. Tayebeh Mousavi [18] studied that incorporation of Bi into Sn-In solders have better superconductivity, where the optimum composition for this solder system is $Sn_{35}In_{50}Bi_{15}$ which show T_C = 6.9 K, H_{C2} (4.2 K) = 0.18 T and J_C (4.2 K, 0.01 T) = 1.3×10^8 A/m^2 respectively. the improvement of the superconductivity is due to the system contain superconducting phases such as Bi-In (T_C = 4.8 K) and $BiIn_2$ (T_C = 5.5K) [21]. The superconductivity of the Sn-In-Bi shows good superconducting properties even the microstructures are relatively coarser, because the β-phase with best superconductivity is the major phase in this alloy. This can be concluded that the enhancement of J_C of the solder system is mainly attributed by the intrinsically higher values of T_C and B_{C2} of the phase as compared to the interphase boundaries act as the active flux pinning sites.

Susannah Speller et al [4] discovered the addition of 5 wt% Sb into the Sn-In alloy, can improved the overall Sn-In system superconductivity where the T_C and H_{C2} values can be greatly improved, primarily the T_C value improved from 4.4 K to 5.5 K. However, the low solubility (2-3wt% Sb) in the Sn-In is due to the excess Sb reacted with In alloy and result in the development of ternary In-Sb-rich phase islands and cause diminish of In content within the matrix and cause the Sn-rich γ-phase development only [4] which has lower superconductivity than In-rich β-phase [18]. Additionally, Susannah Speller et al [4] has compared the addition of Bi and 5 wt.% Sb into Sn-In and found that the T_C and H_{C2} values of the Sn-In-Bi is higher than the Sn-In-Sb and conclude that the addition of Bi into the Sn-In alloy is better than Sb where it shows better superconducting properties.

In a word, the addition of ternary elements can enhance the superconducting properties and improve the reliability of the Pb-free superconducting solder. However, the addition have a critical value, where excessive addition can cause negative effect to the characteristics of the solder material. For example, excessive of Sb addition into Sn-In can lower the superconductivity of the solder. Thus, determining the optimum content is very important.

5 Conclusion

In spite of Pb-Bi solder satisfied all the criteria as the superconducting solder, but the prohibitions of the Pb usage in the industry has driven towards the exploration of the Pb-free superconducting solder materials. Sn-In is the most promising Pb-free binary solder system. By incorporation of Sb or Bi to the Sn-In phases, the T_C and B_{C2} values can be increased significantly. Sb shows lower solubility (up to 2-3wt% Sb) in the Sn-In phases than Bi, and the excess Sb in the system will cause the formation of Sb-In-rich phase islands. Bi element has greater solubility in β and γ phases, and can enhance higher superconductivity of Sn-In phases. However, even the greatest T_C Sn-In-Bi alloy are still far away under the Pb-Bi alloys (8.4 K) and indicating that the suitable replacement for the Pb-Bi solder have not yet found. The undiscovered ternary Pb-free systems are a fertile field to be explored for the novel of high-performance superconducting solders.

The authors would like to acknowledge the support from Nihon Superior Co. Ltd., Fundamental Research Grant Scheme (FRGS) under a grant number of FRGS/1/2017/TK07/UNIMAP/02/6 from the Ministry of Education Malaysia, and Tin Solder Technology Research Grant (TSTRG) under grant number of 9002-00082 from Tin Industry (Research and Development) Board. The authors wish to thank the Center of Excellence Geopolymer & Green Technology (CEGeoGTech), School of Materials Engineering, Universiti Malaysia Perlis, UniMAP for their partial support.

References

1. Lee, H., et al., Microelectron Reliab **87**,75-80 (2018)
2. Jung, D., A. Sharma, and J. Jung, J. Mater. Sci. **53**,1-19 (2018)
3. Levy, S.A., Y.B. Kim, and R.W. Kraft, J.Appl Phys **37**(10),3659-3665 (1966)
4. Speller, S.C., et al., Mater. Res. Soc **4**(1),3483-3488 (2016)
5. Livingston, J.D., J. Appl Phys **38**(5),2408-2410 (1967)
6. Evetts, J.E. and J.M.A. J Phys Chem Solids **31**(5), 973-982 (1970)
7. Lutes, O.S. and D.A. Clayton, Phys. Rev. **145**(1), 218-224 (1966)
8. Brittles, G.D., et al.,. Supercond. Sci. Technol.**28**(9) (2015)
9. Aksoy, C., et al., IEEE Trans. Appl. Supercond. **26**(3), 1-3 (2016)
10. Roy F. Thornton, S., N.Y., U.S. Patent, Editor. 1990, General Electric Company, Schenectady, N.Y.
11. Swenson, C.A. and W.D. Markiewicz, IEEE Trans. Appl. Supercond. **9**(2),185-188 (1999)
12. Brittles, G.D., et al., Supercond. Sci. Technol. **27**(12), (2014)

13. Barnak, J.P., A.F. Sprecher, and H. Conrad, Scr Metall Mater **32**(6), 879-884 (1995)
14. Warren, W.H. and W.G. Bader, Rev. Sci. Instrum. **40**(1), 180-182 (1969)
15. Puttlitz, K.J. and K.A. Stalter, *Handbook of Lead-Free Solder Technology for Microelectronic Assemblies*.Taylor & Francis (2004)
16. Suganuma, K., Curr Opin Solid State Mater. Sci. **5**(1), 55-64 (2001)
17. Ogunseitan, O.A., JOM, **59**(7), 12-17 (2007).
18. Mousavi, T., et al., Supercond Sci Technol **29**(1), (2016)
19. Merriam, M.F. and M. Von Herzen,. Phys. Rev.,**131**(2), 637-643 (1963)
20. *Magnetic materials and their properties*, in *Smithells Metals Reference Book (Eighth Edition)*, W.F. Gale and T.C. Totemeier, Editors. 2004, Butterworth-Heinemann: Oxford. p. 20-1-20-24.
21. Berger, L. and B. Roberts, *Properties of superconductors.* CRC handbook of chemistry and physics 78th edn. CRC Press, Boca Raton, FL, 12-60 (1997)

The Effect of Germanium Addition on the Lead-free Solder Alloys: A Short Review

Chi Ying Tan[1,*], *Mohd Arif Anuar* Mohd Salleh[1] and *Norainiza* Saud[1]

[1]Center of Excellence Geopolymer and Green Technology (CEGeoGTech),
School of Materials Engineering, University Malaysia Perlis (UniMAP),
Kompleks Pusat Pengajian Jejawi 2, Taman Muhibbah, 02600 Jejawi, Arau, Perlis, Malaysia

Abstract. The Restriction of Hazardous Substances (RoHS) has been enforced a law to restrict the use of hazardous materials in electronic and electrical industries. Hence, it leads to the development of lead-free solder among the electronic industry. SnAg, SnCu, SnAgCu, and SnZnBi solders are found to be alternatives to replace SnPb solder alloys. However, they still have many problems, such as large undercooling and large intermetallic compounds are still present in the solder alloys. Later, researchers come up with the idea of adding alloying elements to lead-free solders to further enhance the properties of lead-free solders. This review paper is aimed to analyze and summaries the effects of germanium (Ge) addition to lead-free solders focusing on its microstructure and thermal properties. The Ge has an almost similar crystal structure as pure tin (Sn), so it is expected that the properties of the lead-free solder could be enhanced by adding an appropriate amount of Ge. Nevertheless, Ge has a unique characteristic as it could act as an antioxidant agent in the lead-free solders.

1 Introduction

The Restriction of Hazardous Substances (RoHS) has enforced the ban on using hazardous materials for electrical and electronic equipment since the year of 2006. Lead is one of the banned substances by the RoHS, yet, it is still used for solder alloy in certain electronic applications [1]. Therefore, electronic industries have raised extensive concerns about the development of environmental friendly lead-free solders. As a result, many Sn-based solders are developed as alternatives to leaded solder alloys. For instance, SnAgCu (SAC), SnAg, SnBi, SnCu, and SnZnBi are the candidates to take over SnPb solder in the current electronic industries [2-4]. The most dominant lead-free solders in current industries are made up of Sn-based materials, this is because most of the properties of Sn-rich solders, such as chemical and mechanical properties are influenced by the properties of pure Sn [5].

Although there are suitable candidates to replace SnPb, they still have many unsolved problems and unknown characteristics of the lead-free solder alloy. For instance, the precise composition of the solder, large intermetallic compounds that still present in the

* Corresponding author: chiying95@outlook.com

IOP Conf. Series: Materials Science and Engineering **957** (2020) 012060 doi:10.1088/1757-899X/957/1/012060

microstructure and the problems of large undercooling. Consequently, some researchers have added some metal alloys or rare-earth elements into lead-free solder. For example, silver, copper, nickel, aluminum, and bismuth are added to Sn-based solder to improve the solder properties [6-9]. Based on the literature, the addition of alloying elements can strongly influence the microstructure of Sn-rich eutectic solders. With a small amount of alloying elements can enhance the properties of solder alloys, by reducing the undercooling, refine the grain structure and improve the solderability of the solder alloys [6, 8-10].

Some authors have driven further development of germanium (Ge) addition into lead-free solder. The Ge has a similar crystal structure to Sn element and they are belonging to group 14 elements in the periodic table. Therefore, researchers believe that adding an appropriate amount of Ge into the Sn-rich lead-free solder could enhance effectively the microstructure and leads to improvement in their properties. Furthermore, Ge shows a great property of anti-oxidant and dross reduction during the soldering process. Formation of dross during oxidation will influence the solderability and wetting properties of the solder. Yet, it also adds the cost of the process as maintenance is required to clean the mechanical parts damaged by the dross. Thus, it causes value loss of the solder [11]. Nevertheless, oxidation of high-temperature solder alloy also important as it might give rise to crack propagation. In the previous studies, it is proven that the growth of oxides can be inhibited by adding Ge [12, 13]. Therefore, this paper is aimed to review the effects of Ge addition on the microstructure and thermal properties of lead-free solders.

2 Effect of Ge Addition to Lead-free Solders

2.1 Microstructure

Fig. 1 (a, b) depicts the microstructure of Ge added SnCu and SnCu solder alloys. According to M. Hasnine *et al.* [7], the microstructure of SnCu solder alloy consists of rich beta-tin phases and Cu_6Sn_5 particles. It is noted that the Ge rich precipitation was not present in the microstructure. This may due to the content of Ge in the solder alloy was very low, which was only 0.01%. These results were in agreement with T.H. Chuang *et al.* [14]. The authors further noted that as it can be seen that after the addition of 0.5Ge into Sn-3Ag-0.5Cu-0.5Ce solder alloy, new phases were formed, which are $AgSn_3$ and $CeSn_3$. Furthermore, G. Meng *et al.* [15] stated that the microstructure of intermetallic compounds, $AgSn_3$ and Cu_6Sn_5 have been refined and their dispersion tends to be well spread.

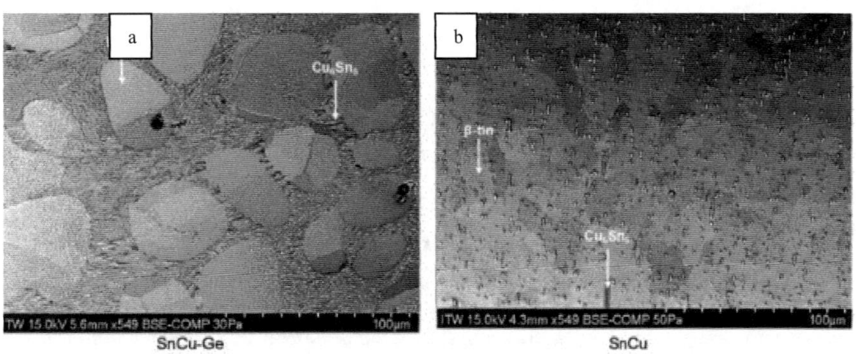

Fig. 1. Microstructure of (a) SnCu-Ge and (b) SnCu solder alloys [7].

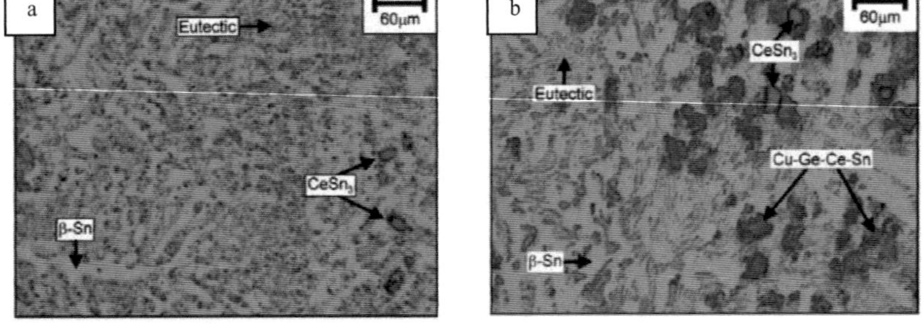

Fig. 2. Microstructure of (a) SnAgCu-Ce and (b) SnAgCu-CeGe solder alloy [14].

2.2 Thermal properties

From the results of S.H. Wong et al. [16] and his team, the melting points of Ge added solder is almost similar with and without Ge to the SnZnBi solder alloy. Referring to Fig. 3, it can be seen that the addition of a small amount of Ge does not significantly affect the melting point of the SnZnBi solder. While X.D. Zhang *et al.* [17] added 0.01-0.10wt% of Ge into Sn0.7Cu and the melting points were found that slightly increase with an increase in the Ge content.

In contrast, the undercooling of Sn0.7CuGe solder alloy decreased as the amount of Ge increase. As explained by M. Hasnine *et al.* [7] the undercooling of the solder alloy can be described as the difficulties of a solder alloy to be solidified. This is because undercooling is a period when liquid solder turns to solid, thus, the microstructure and mechanical properties can be influenced [18]. Furthermore, a large amount of undercooling can give an impact on the reliability performance of solder joints [20-21]. In view of the fact that stress concentration might occur in an uneven solidified solder and could lead to mechanical failure. Therefore, with the addition of Ge into lead-free solders is expected to enhance the mechanical properties as they have lower undercooling value.

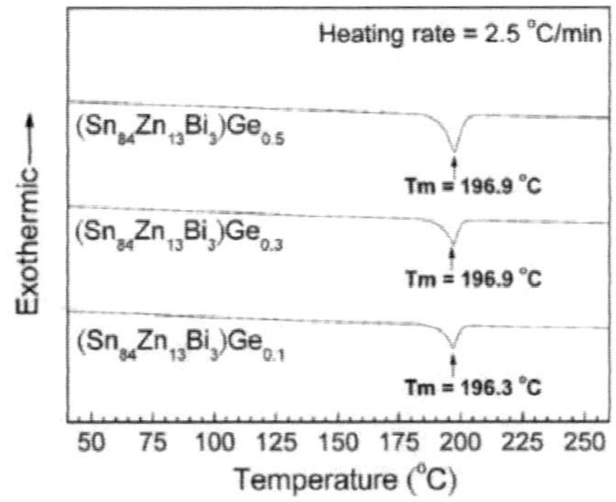

Fig. 3. Differential Thermal Analaysis (DTA) curves of SnZnBiXGe ; x = 0.1, 0.3, and 0.5 respectively [16].

It is well acknowledged that the CTE values decrease with an increase in the amount of Ge [7]. The coefficient of thermal expansion of a solder alloy can be defined as the change in the size of the solder per degree change in temperature at the constant pressure [19]. Based on the results published by S.H. Wong et al. [16] in Fig. 4, the thermal deformation of $(Sn_{84}Zn_{13}Bi_3)_{99.9}Ge_{0.1}$ is almost the same as Cu, which is 17 ppm/◦C. Therefore, $(Sn_{84}Zn_{13}Bi_3)_{99.9}Ge_{0.1}$ / Cu joints will have the least CTE mismatch. CTE mismatch will cause residual stress and stress concentration to happen in between the solder and Cu substrate. Hence, it will lead to crack propagation to happen in the interfacial [17].

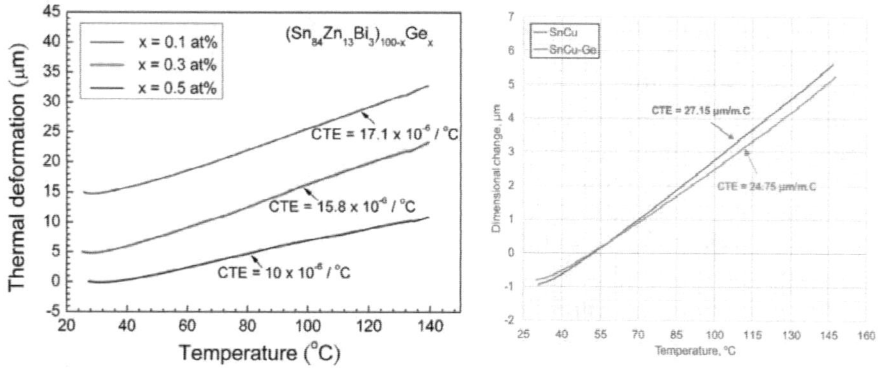

Fig. 4. Thermomechanical Analysis (TMA) measurement of Ge added into (a) SnZnBi and (b) SnCu solder alloys [7, 16].

3 Summary

The effects of Ge addition to lead-free solders were reviewed in this paper and be summarised as below:

- The dominant microstructure that presents in Ge-added solder alloy is rich beta-Sn phases and Cu_6Sn_5. However, there are no Ge-rich phases as the amount of Ge added into solder alloys is very little. Also, it would seem that the microstructure of lead-free solder alloy was significantly refined with a proper amount of Ge addition.
- The addition of Ge does not significantly affect the melting points of SnZnBi and SnCu solder alloys. While the degree of undercooling decreased with an increase in Ge content.
- CTE of the Ge-added lead-free solder alloy was found to decrease as an increase in Ge content. In the literature, a small amount of Ge content is sufficient to reduce the difference of CTE of solder and Cu substrate.

This work was supported by an Institutional Links grant, ID 332397914, under the Newton-Ungku Omar Fund partnership and Nihon Superior Co. Ltd.. The grant is funded by the UK Department of Business, Energy and Industrial Strategy (BEIS) and Malaysia and delivered by the British Council.

References

1. Tae-Kyu Lee, et al., *Fundamentals of Lead-Free Solder Interconnect Technology*. 1 ed. 2015: Springer US.
2. Chidambaram, V., J. Hattel, and J. Hald, Mater. Des. **31**, 4638-4645 (2010)
3. Sharif, A., *Harsh Environment Electronics: Interconnect Materials and Performance Assessment*. 2019: Wiley.
4. Chidambaram, V., J. Hattel, and J. Hald, Microelectron. Eng. **88**, 981-989.(2011)
5. Kang, S.K., et al., J. Res. Develop. **49**, 607-620. (2005)
6. Gao, L., et al., Microelectron. Eng. **87**, 2025-2034 (2010)
7. Hasnine, M., B. Tolla, and M. Karasawa, J. Mater. Sci. **28**, 16106-16119. (2017)
8. Niu, X. and K.-L. Lin, Mater. Sci Eng. A **677**, 384-392 (2016)
9. Tay, S.L., et al., Intermetallics. **33**, 8-15 (2013)
10. Wang, Y., et al., J. Alloys. Compd. **478,** 121-127. (2009)
11. Evans, J.W., *A guide to lead-free solders: physical metallurgy and reliability*. 2007: Springer Science & Business Media.
12. Leidheiser Jr, H., *The corrosion of copper, tin and their alloys*. 1971.
13. Takemoto, T., et al.*Proceedings Second International Symposium on Environmentally Conscious Design and Inverse Manufacturing*. 2001.
14. Chuang, T.-H. and C.-C. Chi, J. Alloys. Compd **480**, 974-980. (2009)
15. Meng, G., T. Yang, and L. Chen, TRANSACTIONS-CHINA WELDING INSTITUTION. **28**, 65. (2007)
16. Wang, S.-H., et al.,J. Alloys Compd **497**, 428-431 (2010)
17. Zhang, X., et al., Mater. Res. Express **6**, 016556 (2018)
18. Cho, M.G., et al., Appl. Phys. Lett. **95**, 021905 (2009)
19. Tipler, P.A. and G. Mosca, *Physics for Scientists and Engineers*. 2007: W. H. Freeman.
20. R.M Said, M.A.A.M Salleh, M.I.I Ramli, N. Saud, Solid State Phenom. 280, 163-16 (2018)
21. R. M. Said, M. A.A. M. Salleh, M. N. Derman, M.I.I. Ramli Key Eng. Mater. **700,** 123-131 (2016)

Thermal behaviour and microstructural analysis of Sn-40Pb alloy and Sn-40Pb soldered on electroless nickel/immersion gold

Ai Ling Teoh[1,*], *Mohd Arif Anuar* Mohd Salleh[1] , *Dewi Suriyani* Che Halin[1] , and *Kai Loong* Foo[2]

[1]School of Materials Engineering, Universiti Malaysia Perlis (UniMAP), Taman Muhibbah, 02600 Jejawi, Arau, Perlis, Malaysia
[2]Nano Biochip Research Group, Institute of Nano Electronic Engineering (INEE), Universiti Malaysia Perlis (UniMAP), 01000 Kangar, Perlis, Malaysia

Abstract. Due to the toxicity of lead solder, the lead-free solder has been developing and improved in order to replace the lead solder. However, the behaviour and properties of lead solder is still better than the lead-free solder, especially in application when soldering on substrate. Hence, a common lead solder, Sn-40Pb is studied in this research by comparing with Sn-40Pb soldering on electroless nickel/immersion gold (ENIG) substrate. The thermal behaviour by differential scanning calorimeter (DSC) and microstructure formation with elemental analysis by scanning electron microscope (SEM) were carried out for Sn-40Pb solder alloy and Sn-40Pb soldered on ENIG substrate in this study. The result showed that the pasty range of Sn-40Pb was lower than Sn-40Pb/ENIG while the undercooling was higher. The diffusion and dissolution of Ni and Cu elements from ENIG substrate into the solder, forming the lead-rich phase with Ni elements and interfacial $(Cu,Ni)_6Sn_5$ with Pb elements in Sn-40Pb/ENIG. The diffusion and dissolution of elements from substrate into the solder affects the thermal behaviour and microstructural of solder.

1 Introduction

Due to the legislation by the Waste of Electrical and Electronic Equipment (WEEE) and the directives on the restriction of hazardous substance (RoHS), the lead-free solders that is green and friendly to the environment is focused for developing since lead solder is a toxic material which is harmful to human health and also the environment. However, the lead solder is still having better eutectic composition properties which are good in mechanical properties, good wettability, low cost and low melting point. Hence, the further understanding of lead solder is important for enhancing the properties of lead-free solder [1-4].

As different formation of microstructure resulted in various solder joint properties mechanically which causes different reliability of electronic devices, the thermal behaviour of solder plays a main role for application. Chu et al. has compared and studied the thermal

* Corresponding author: akirateoh@gmail.com

analysis by differential scanning calorimeter method and microstructures for different composition of tin-lead solders in past [5]. On the other hand, Cruz et al. has carried out the thermal analysis for different composition of tin-lead solders by cooling curve method [6].

Solder is always used for the bonding and interconnect purpose in application, and hence it is direct contacting the substrate in electronic devices during soldering process. The degree of undercooling during solidification for solder on substrate affects the solder joint's microstructure forming which resulted in different electronic device's performance. However, most of the researches were focused on the thermal analysis of solder itself or soldering above the substrate of copper. Hence, the aim of this research is to understand the effects of ENIG substrate to the Sn-40Pb solder through the behaviour of thermal and its changing on microstructures.

2 Experimental Procedures

2.1 Sample preparation

The Sn-40Pb solder alloy was melted at 350 °C for 1 hour in the electric resistance furnace and then poured into the mould of stainless-steel for casting. Some casted metal sheets were rolled into the thickness of 50 μm as metal foils and next punched into round shape with 3 mm diameter puncher, while other metal sheets were kept for the solder analysis. After applying the rosin mildly activated (RMA) flux onto the 3 mm punched metal foil, it was then put on the Pyrex sheet and reflowed in the oven forming sphere shape as the 900 μm solder ball due to the surface tension effect. Next, some solder balls were reflowed on the ENIG substrate once the RMA flux was applied, while other solder balls were kept for another testing.

2.2 Thermal analysis

The differential scanning calorimeter (DSC) model Q10 was used to analyse the solder's thermal behaviour of 5 mg of Sn-40Pb metal sheet. To understand the Sn-40Pb/ENIG behaviour of thermal, the RMA flux was applied on the 3.2 mg solder ball and then this solder ball was put on the ENIG substrate. By using the rate of 10 °C/min, heating up both test samples to 270 °C and then following by cooling down to 30 °C inside the inert nitrogen gas chamber to obtain the exothermic and endothermic curves for thermal analysis.

2.3 Microstructural analysis

Scanning electron microscope (SEM) model JEOL JSM 6460LA equipped with Energy Dispersive X-ray (EDX) was used to observe the microstructure and analyse the element of Sn-40Pb solder and the Sn-40Pb/ENIG after sample grinding, polishing and etching.

IOP Conf. Series: Materials Science and Engineering **957** (2020) 012061 doi:10.1088/1757-899X/957/1/012061

3 Result and Discussions

3.1 Thermal behaviour

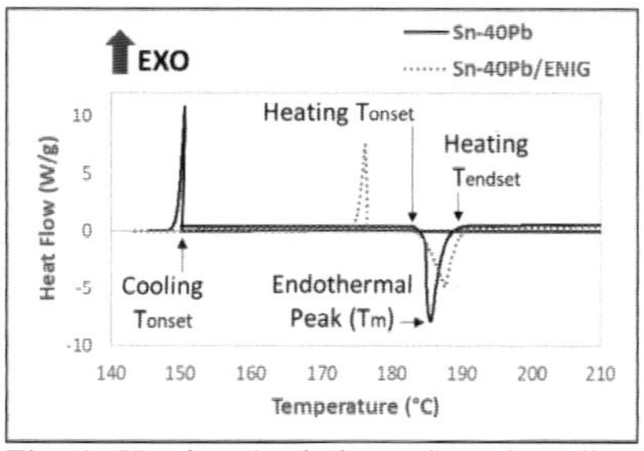

Fig. 1. Heating (endothermal) and cooling (exothermal) DSC results of Sn-40Pb and Sn-40Pb /ENIG.

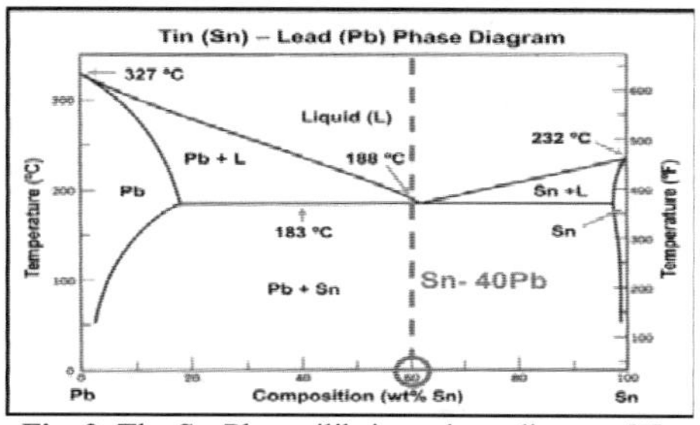

Fig. 2. The Sn-Pb equilibrium phase diagram [7].

In order to obtain a good performance of electronic device, the solder joint properties need to be studied and understood through the solder characterization of cooling and melting. To mimic the ball-grid array (BGA) application [8], DSC analysis was carried out to study the samples' thermal behaviour at 10 °C/min of cooling and heating rate. Fig. 1, Table 1 and Table 2 show the results of DSC. Both thermal behaviour of Sn-40Pb solder and Sn-40Pb/ENIG were tested and analysed. The heating, T_{onset} of Sn-40Pb and Sn-40Pb/ENIG where the endothermic reaction during heating up was begun at 184.71 and 182.76 °C respectively. The heating result plot is then achieving the peaks at 185.57 and 187.57 °C for Sn-40Pb and Sn-40Pb/ENIG respectively and labelled as Tm. Fig. 2 displays the Sn-Pb binary phase diagram. Based on the equilibrium phase diagram for Sn-Pb, the phase changing of (Pb+Sn) into the phase of (Pb+L) was occurred at the temperature of 183 °C which is the T_{onset}. Accordingly to the research by Cho et al., it has showed that the findings are similar to above DSC results [9].

On top of that, El-Daly et al. and Sayyadi et al. have reported that the pasty range is one of the thermal behaviour of material and can be formulated as:

$$\text{Pasty range} = \text{Heating } T_{endset} - \text{Heating } T_{onset} \quad (1)$$

where heating T_{onset} is the solidus temperature and T_{endset} is the liquidus temperature during heating [8, 10]. By calculation, Table 1 displays the pasty range of Sn-40Pb and Sn-40Pb/ENIG respectively 6.80 and 10.01 °C. With the existing of heat, soldering process has been taking part between the solder and substrate causing an increasing of pasty range of Sn-40Pb/ENIG, compared to Sn-40Pb solder itself. Furthermore, the degree of undercooling is another important parameter that affects the reliability of solder joint which determines the solder solidification behaviour ; and this parameter can be obtained from the DSC curve. Elmer et al. and Kang et al. have defined the undercooling as the deduction of heating T_{onset}, which is the starting point of melting in heating curve; to cooling T_{onset}, which is the starting point of solidification in cooling curve [11, 12]; and it can be formulated as:

$$\text{Undercooling, } \Delta T = \text{Heating } T_{onset} - \text{Cooling } T_{onset} \quad (2)$$

The undercooling degree of Sn-40Pb and Sn-40Pb/ENIG is 34.43 and 6.25 °C respectively ; showing that with the existing of substrate, the undercooling degree for Sn-40Pb/ENIG is decreasing compared to Sn-40Pb. Cho et al. has been reported the similar results where the tin-rich solder alloys have lower undercooling when the solder reacts with the substrate. The wettable surface of substrate facilitates the reaction between both substrate and solder during the solidification [9]. In this research, the substrate with ENIG surface finish has given a wettable surface to the Sn-40Pb solder during the soldering process; enabling the substrate elements for example Ni to diffuse and dissolve into the solder during solidification, resulted in the decreasing of undercooling degree.

Table 1. Heating peak (T_m), solidus temperature (heating T_{onset}), liquidus temperature (heating T_{endset}) and pasty range for Sn-40Pb and Sn-40Pb/ENIG.

Sample	Heating Peak T_m (°C)	Heating T_{onset} (°C)	Heating T_{endset} (°C)	Pasty Range $T_{onset} - T_{endset}$ (°C)
Sn-40Pb	185.57	184.71	191.51	6.80
Sn-40Pb/ENIG	187.57	182.76	192.77	10.01

Table 2. Starting point of solidification (cooling T_{onset}) and degree of undercooling (ΔT) for Sn-40Pb and Sn-40Pb/ENIG.

Sample	Cooling T_{onset} (°C)	Undercooling ΔT (°C)
Sn-40Pb	150.28	34.43
Sn-40Pb/ENIG	176.50	6.25

3.2 Microstructure observation

Fig. 3 shows the SEM image of Sn-40Pb solder alloy, where it consists of the primary lead-rich phase and secondary tin-rich phase. The research by Cruz et al. has stated that the Sn-12Pb, Sn-22.5Pb and Sn-28.5Pb formed the primary lead-rich phase and secondary tin-rich phase after the solders were solidified [6].

Fig. 4 displays no significant difference of the Sn-40Pb solder microstructure when it is soldered on a substrate, except an interfacial layer was formed between the solder and substrate. The formation of crystals in white color and its grey color matrix in Fig. 4 shows the similar structures as in Fig. 3. EDX analysis was done to study the element formed in the microstructure, especially the interfacial layer between the substrate and solder.

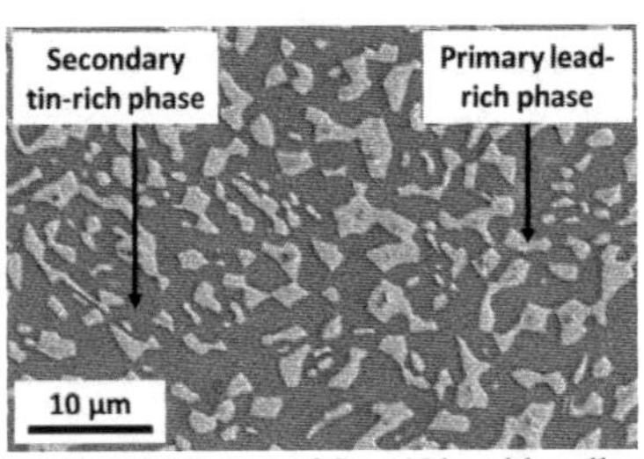

Fig. 3. SEM image of Sn-40Pb solder alloy.

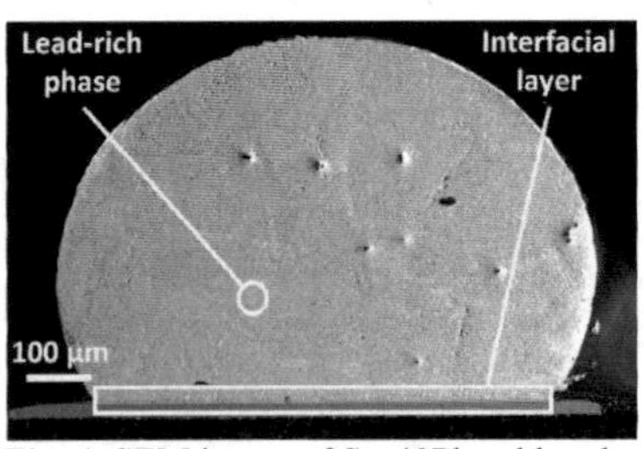

Fig. 4. SEM image of Sn-40Pb soldered on ENIG substrate

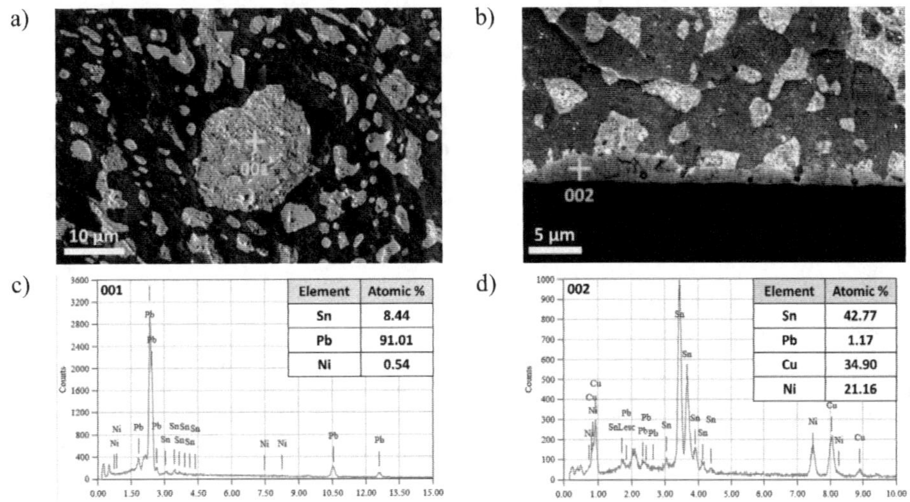

Fig. 5. SEM images and EDX results of Sn-40Pb soldered on ENIG substrate. (a) Lead-rich phase (b) Interfacial layer (c) EDX on lead-rich phase (d) EDX on interfacial layer.

From the EDX result of the lead-rich phase (Fig. 5a) inside the solder ball in Fig. 5c, the phase is having the elements of Sn, Pb, Ni; showing the atomic percentage that forming the highest quantity of lead. Chu et al. has reported in the research that Sn-Pb composition with more than 25 weight percent of lead consists of two nucleation phases which are the primary lead-rich phase and the secondary tin-rich phase as the matrix [5]. On the other hand, Cho et al. has found in the study that the Ni elements are diffused and dissolved from the substrate into the solder during the interfacial reaction when heat is applied [9]. Thus, the crystal nucleated in the solder ball after soldering in Fig. 5a is the lead-rich phase with Ni element since the elements from substrate have diffused and dissolved into the solder.

Moreover, at the contacting area of substrate and solder, there is a growing of a crystal layer as illustrated in Fig. 5b. The forming of this crystal layer is because of the interfacial reaction between the substrate and solder during soldering process. According to the EDX result in Fig. 5d, it exhibits that the interfacial crystals are formed by the Sn, Pb, Cu and Ni elements. Based on the atomic percentage results, the interfacial crystal layer formed was the interfacial intermetallic compound of $(Cu,Ni)_6Sn_5$ with a little amount of Pb element. Salleh et al. has reported the similar intermetallic compound formation of $(Cu,Ni)_6Sn_5$ when Ni element is added into the Sn-0.7Cu alloy [13]. Cu and Ni elements are originated from the substrate where both elements are diffused and dissolved into the solder with the existing of heat during the interfacial reactions [9]. The existence of other elements in the solder caused the decreasing of the undercooling degree since the nucleation of intermetallic compound was begun during solidification process [14-18].

4 Conclusion

The diffusion and dissolution of Ni and Cu elements from ENIG substrate into the solder during solidification resulted in higher pasty range for Sn-40Pb/ENIG compared Sn-40Pb while it reduced the undercooling degree. Both microstructures of Sn-40Pb and Sn-40Pb/ENIG consisted of the primary lead-rich phase and the secondary tin-rich phase.

However, Ni elements dissolved in the lead-rich phase of Sn-40Pb/ENIG while there is no Ni elements in Sn-40Pb solder itself. Furthermore, a layer of intermetallic compound of $(Cu,Ni)_6Sn_5$ with a little amount of Pb element was formed at the interfacial between the substrate and solder.

It is thankfully to the research supporter, Nihon Superior and; the support of materials and financial from an Institutional Links grant, ID 332397914, under the Newton-Ungku Omar Fund partnership. The grant is funded by the UK Department of Business, Energy and Industrial Strategy (BEIS) and Malaysia and delivered by the British Council. It is to acknowledge School of Materials Engineering from Universiti Malaysia Perlis (UniMAP) for the facilities supporting too.

References

1. W.-T. Chen, C. Ho, and C. Kao, J. Mater. Res. **17,** 263-266 (2002)
2. A. El-Daly, A. El-Taher, and S. Gouda, Mater. Des. **65**,796-805 (2015)
3. M. Fazal, N. Liyana, S. Rubaiee, and A. Anas, Measurement (2018)
4. H. A. Jaffery, M. F. M. Sabri, S. M. Said, S. W. Hasan, I. H. Sajid, N. I. M. Nordin, M. M. I. M. Hasnan, D. A. Shnawah, and C. V. Moorthy, J. Alloys Compd. 151925 (2019)
5. M. Chu, Y. Shiohara, and M. C. Flemings, Metall Trans. A **15**, 1303-1310 (1984)
6. H. Cruz, C. Gonzalez, A. Juárez, M. Herrera, and J. Juarez, J. Mater. Process. Technol.**178**, 128-134 (2006)
7. S. Xu, A. H. Habib, A. D. Pickel, and M. E. McHenry, Prog. Mater. Sci. **67**,95-160 (2015)
8. A. El-Daly, W. Desoky, A. Saad, N. Mansor, E. Lotfy, H. Abd-Elmoniem, and H. Hashem Mater. Des. **80**, 152-162 (2015)
9. M. G. Cho, S. K. Kang, and H. M. Lee, J. Mater. Res. **23**, 1147-1154 (2008)
10. R. Sayyadi and H. Naffakh-Moosavy, Mater. Sci. Eng. A **735**, 367-377 (2018)
11. J. W. Elmer, E. D. Specht, and M. Kumar, J. Electron. Mater.**39**, 273-282 (2010)
12. S. K. Kang, M. G. Cho, P. Lauro, and D.-Y. Shih, in 2007 Proceedings 57th Electronic Components and Technology Conference,1597-1603 (2007).
13. M. M. Salleh, S. McDonald, C. Gourlay, S. Belyakov, H. Yasuda, and K. Nogita, J. Electron. Mater. **45**, 154-163 (2016)
14. T. Lu, D. Yi, H. Wang, X. Tu, and B. Wang, J. Alloys. Compd **781**, 633-643 (2019)
15. H. Shang, Z. Ma, S. Belyakov, and C. Gourlay, J. Alloys Compd **715**, 471-485 (2017)
16. X. Tu, D. Yi, J. Wu, and B. Wang, J. Alloys Compd **698**, 317-328 (2017)
17. R. M. Said, M. A.A. M. Salleh, M. N. Derman, M.I.I. Ramli Key Eng. Mater. **700,** 123-131 (2016)
18. M.I.I Ramli, N. Saud, M.A.A Mohd Salleh, M. Nazree Derman, R.Mohd Said, Mater. Sci.Forum 803, 228-232 (2015)

IOP Conf. Series: Materials Science and Engineering **957** (2020) 012062 doi:10.1088/1757-899X/957/1/012062

Preliminary study on the effect of Ni addition on tin (Sn) whisker growth from lead-free solder coating

Aimi Noorliyana Hashim[1,*], *Mohd Arif Anuar* Mohd Salleh[1], *Muhammad Mahyiddin* Ramli[2], *Khor Chu Yee*[3] and *Noor Zaimah* Mohd Mokhtar[1]

[1]Centre of Excellence Geopolymer and Green Technology, School of Materials Engineering, Universiti Malaysia Perlis (UniMAP), Taman Muhibah 02600, Jejawi, Arau, Perlis, Malaysia
[2]School of Microelectronic Engineering, Pauh Putra Campus, University Malaysia Perlis (UniMAP), 02600 Arau, Perlis, Malaysia
[3]Faculty of Engineering Technology (FETech), Universiti Malaysia Perlis (UniMAP), Level 1, Block S2, UniCITI Alam Campus, Sungai Chuchuh, 02100 Padang Besar, Perlis, Malaysia

Abstract. It has been widely confirmed that alloying significantly can mitigate the formation of tin (Sn) whiskers and consequently became a possible candidate as a lead-free alloys coating material in the micro-electronics industry. In this study, the effect of 0.05 wt.% Ni addition on the formation and growth of Sn whisker in Sn-0.7Cu solder coatings have been investigated under continuous mechanically stress induced at room temperature. It is clearly found that Ni addition have significantly enhanced stress relaxation by reducing the growth rate of formation Sn whiskers in Sn-0.7Cu solder coatings. The morphology of $(Cu,Ni)_6Sn_5$ interfacial IMC was more refine and thinner with a fine scallop-shaped interfacial intermetallic layer purposely to lower the compressive stress of Sn coating and extend the nucleation period, thus mitigating the formation and growth of Sn whiskers.

1 Introduction

The lead (Pb) free solder has attracted more attention worldwide since the implementation of the Restriction of Hazardous Substances Directive (RoHS) in 2006 prohibiting the use of lead in electronics as it is harmful to both environment and human health. However, the requirement of lead free solder in electronic industry has revived the problem of Sn whiskers that leads to various of interconnect system failures [1-4]. Previous studies have been found that a fundamental factor accounting for Sn whisker formation is compressive stress generated by the diffusion of copper (Cu) atoms into tin (Sn) film, followed by the formation of Cu-Sn intermetallic compound (IMC) [3, 5, 6]. As the Cu-Sn IMC particles grow, their volume expansion creates a stressed region that extends into the Sn layer. High local stresses near the Sn/IMC interface can cause the emission of dislocations and creation of a plastically deformed region in the vicinity of the particles [7]. The compressive stress

* Corresponding author: aimiliyana@unimap.edu.my

induced by IMCs growth is the most likely driving force leading to the prolonged growth of Sn whiskers from coatings because the IMC continues to grow under room temperature (RT) exposure conditions [4].

It has been widely established that alloying significantly can mitigate the formation of Sn whiskers and thus became a possible candidate as a lead-free alloys coating material in the micro-electronics industry [2, 8]. The addition of Ni to Sn–Cu solder have been one of the most promising candidates for their superior solderability and reliability in industrial applications [9, 10]. It has also been shown that Ni addition can create stabilization of the hexagonal high-temperature phase of Cu_6Sn_5 [10] and may assist to enhance stress relaxation in the layer through the formation of fine, needle-like $(Cu,Ni)_6Sn_5$ at the solder–substrate interface. However, the influence of Ni microalloying on the Sn whisker growth is still not clear, and needs further research.

In this study, the effect of 0.05 wt.% Ni addition on the formation and growth of tin (Sn) whisker in Sn-0.7Cu solder coatings have been investigated under continuous mechanically stress induced at room temperature. Morphology analysis of Sn whisker growth was done using scanning electron microscope (SEM). The quantitative analysis of intermetallic compound (IMC) interface layer on whisker growth was performed using atomic force microscopy and Profilm software. These results outcomes can be used as a basis to elucidate systematically in mitigating Sn whisker kinetic growth of tin solder joints.

2 Experimental Procedures

Solders of Sn-0.7Cu solder and Sn-0.7Cu-0.05 Ni were ready in the form of ingots supplied by Nihon Superior Co. Ltd., Japan. The Cu (>99.9%) substrate with dimensions of 1.5 cm x 1.5 cm squares of 1.0 cm in thickness was immersed into acid liquid bath that contains 5 g (35%) of hydrochloric acid with 95 g of deionized water (1.75%) for 5 s to remove surface oxides, rinsed with acetone followed by distilled water and dried. The Cu substrate was initially dipped in a flux solution for 2 s, then the flux was drained by placing its vertically on a filter paper for 2 s. The flux used was a standard B type (JIS Z3198- 4), which consists mixtures of rosin, 2-propanol and diethylamine hydrochloride to remove the oxidation from the surface of the substrates. The elemental composition analysis of solders and substrate using arc spark spectrometer is shown in Table 1.

Table 1. Metal Composition Analysis Results of Sn-based solders

Element (%)	Sn0.7Cu	Sn0.7Cu0.05Ni	Cu Substrate
Sn	99.23	99.2	-
Cu	0.768	0.656	99.94
Ni	-	0.052	-
Sb	0.002	0.006	-
Pb	0.0013	0.016	0.0053
Zn	<0.0002	<0.0002	0.014
Fe	0.004	0.004	-
Al	0.0004	0.0004	0.01
In	0.0013	0.003	-

The Cu substrate was then mounted on the clip holder of a solder-dip machine followed by immersed in the molten solder bath with withdrawal speeds of 10 mm/s and an immersion dipping time of 5 s. Table 2 shows the temperature profile parameters of the solder-dip process. The solder bath is held at a peak temperature of 265 °C. After the solder-dip, the samples were cooled down in air before the water rinse in ultrasonic bath.

Table 2. Solder-dip temperature profile

Solder-Dip Profile Data	Value
Preheat Temperature (°C)	200
Preheat Time (s)	60
Preheat Rate (°C/s)	2.5
Peak Temperature (°C)	265
Immersion Speed (mm/s)	10
Immersion Dipping Time (s)	5

Micro-indentation test was used to induce mechanical stress through a constant compressive force to accelerate the growth of Sn whiskers. As is shown in Fig.1, the device consists of a 200 g weight, a 1 mm ball and a quartz holder that can fix the weight in a horizontal position and the sample is on the bottom of the holder. The devices placed in the clean room where the temperature and relative humidity is 20–25°C and 40–50% respectively.

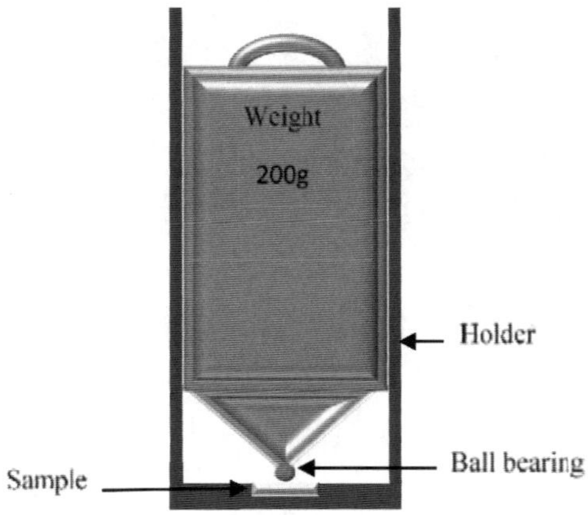

Fig. 1. Schematic of micro-indentation test apparatus for accelerated Sn whisker growth

The formation and growth of whisker were periodically observed for each specimen throughout of 30 days under continuous stress induced using scanning electron microscope (SEM). The samples were then completely deep-etched using a mixed solution of NaoH and ortho-nitrophenol in order to observe the three-dimensional morphology of the interfacial IMC of the solder coating. The shape and root mean square (RMS) roughness of IMC interfacial layer was analysed by atomic force microscope (AFM) and Profilm software.

3 Results and Discussion

Fig. 2 shows the surface morphology of Sn whisker distribution for 10 days and 30 days at room temperature on Sn-0.7Cu and Sn-0.7Cu-0.05Ni solder coatings. Sn whiskers growth rate were observed greater with various type of Sn whiskers on the Sn-0.7Cu solder compared to Sn-0.7Cu-0.05Ni solder. Moreover, the growth of long filament-type Sn whisker were observed on the Sn-0.7Cu solder as shown in Fig. 2a and Fig. 2b compared to Sn-0.7Cu-0.05Ni solder with the growth of nodule and hillock type of whiskers. According to Fig. 2c and Fig. 2d, it is found that Ni addition have significantly enhanced stress relaxation by reducing the growth rate of formation Sn whiskers in Sn-0.7Cu solder coatings.

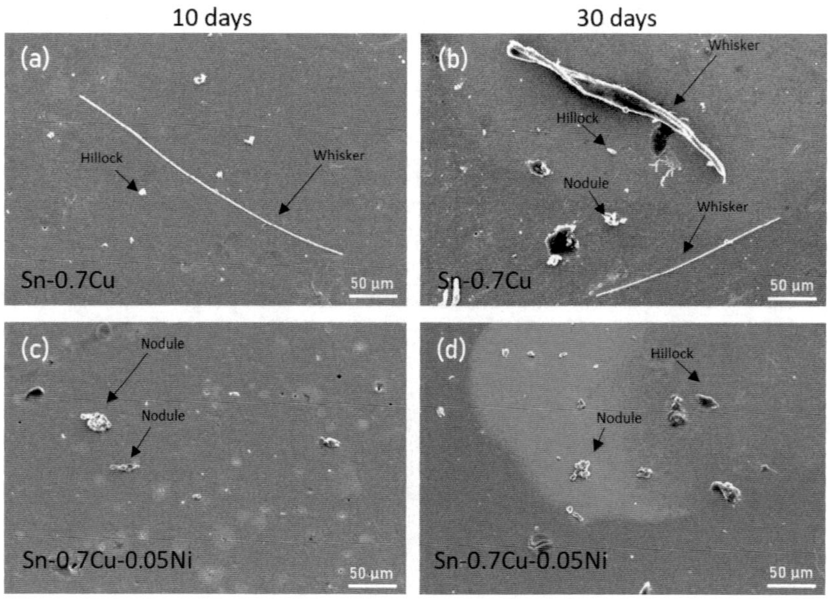

Fig. 2. Surface morphology of Sn whisker distribution for 10 days and 30 days at room temperature on solder coatings (a-b) Sn-0.7Cu (c-d) Sn-0.7Cu-0.05Ni

IMC induced compressive stress is the most widely accepted driving force leading to the prolonged growth of whiskers from solder coatings [4]. To investigate the relationship of the Sn whiskers formation and the IMC interfacial growth, a quantitative analysis of the AFM images for IMC interfacial layer formed on Sn–0.7Cu and Sn–0.7Cu-0.05Ni solder coatings are considered in Fig 3. The morphology of the Cu_6Sn_5 IMC interfacial layer on Sn–0.7Cu (Fig. 3a) is approximately thick scalloped than the $(Cu,Ni)_6Sn_5$ IMC layer on Sn–0.7Cu-0.05Ni (Fig. 3b). It is seen that the highest peak of Sn-0.7Cu IMC interface layer is ~1.328 µm. However, Sn-0.7Cu-0.05Ni IMC interface layer are thinner at ~1.142 µm. It is found that the thickness of $(Cu,Ni)_6Sn_5$ IMC interfacial layer is thinner than Cu_6Sn_5 IMC interfacial layer which agrees well with that reported by Salleh et. al. [10]. The thinner IMC interface layers lead to lower compressive stress of Sn coating, thus mitigating the formation and growth of Sn whiskers [4].

IOP Conf. Series: Materials Science and Engineering **957** (2020) 012062 doi:10.1088/1757-899X/957/1/012062

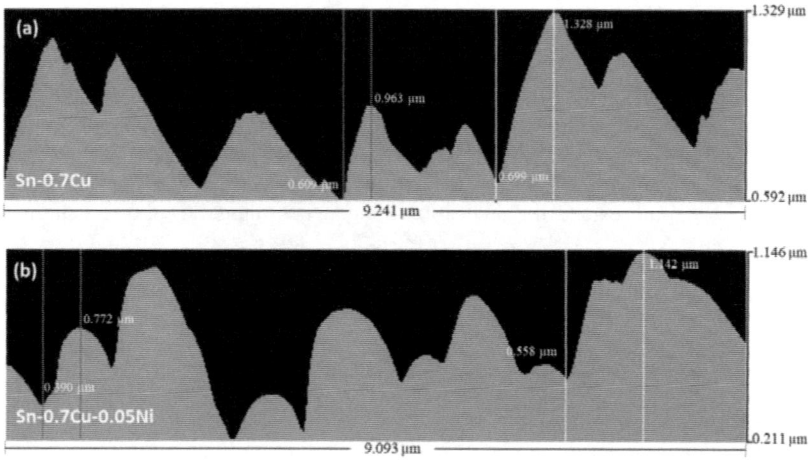

Fig. 3. AFM images of cross section IMC interfacial layer of (a) Cu_6Sn_5 (b) $(CuNi)_6Sn_5$

Fig. 4 shows the AFM topographies of IMC interfacial layer of deep etched Sn-0.7Cu and Sn-0.7Cu-0.05Ni solder. Surface roughness (RMS), a commonly used parameter in defining the surface characteristics, is represented in terms of the statistical deviation from average height. The RMS roughness values of IMC interfacial layer of Cu_6Sn_5 and $(Cu,Ni)_6Sn_5$ were 1.59 µm and 1.39 µm respectively. Furthermore, a thick scalloped of Cu_6Sn_5 IMC apparently is higher compared to a scalloped of $(Cu,Ni)_6Sn_5$ IMC. The morphology of $(Cu,Ni)_6Sn_5$ interfacial IMC was more refined with a fine scallop-shaped interfacial intermetallic layer pointedly extend the incubation period and reduced Sn whiskers kinetic growth. This is consistent with a previous study by Nogita et al. [11] on the stabilization of the hexagonal $(Cu,Ni)_6Sn_5$ phase occurs when Ni-containing solder alloys are used. This phase stabilization may prevent volume changes that could contribute to the stress relaxation in the solder layer.

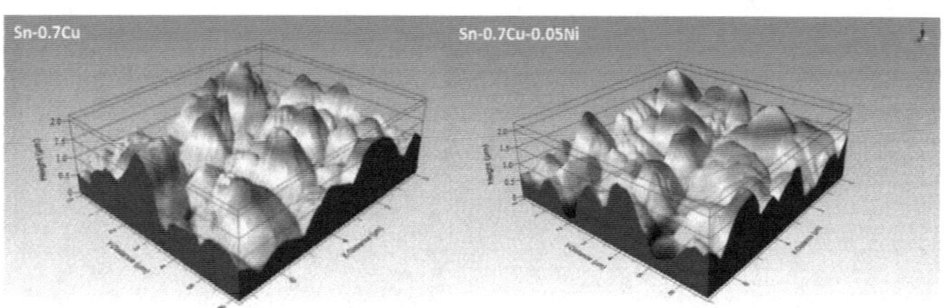

Fig. 4. AFM 3-D topographies images of IMC interfacial layer of (a) Sn-0.7Cu (b)Sn-0.7Cu-0.05Ni

4 Conclusion

There is a robust indication that the Ni addition have significantly enhanced stress relaxation by reducing the growth rate of formation Sn whiskers in Sn-0.7Cu solder coatings. The morphology of $(Cu,Ni)_6Sn_5$ interfacial IMC was more refine and thinner with a fine scallop-shaped interfacial intermetallic layer purposely to lower the compressive stress of Sn coating and extend the nucleation period, thus mitigating the formation and growth of Sn whiskers.

The authors gratefully acknowledge Nihon Superior (Grant No. 2016/10/0001) and Tin Solder Technology Research Group (Grant No. 9002-00084) for the materials and finance support.

References

1. N. Jadhav, M. Williams, F. Pei, G. Stafford, E. Chason, J. Electron. Mater., **42** (2013) 312-318.
2. N.Z.M. Mokhtar, M.A.A.M. Salleh, A.N. Hashim, S.F. Nazri, IOP Conference Series: Mater. Sci. Eng. 701 (2019) 012001.
3. E. Chason, N. Jadhav, F. Pei, JOM, 63 (2011) 62.
4. J.S. Chen, C.H. Ye, J.M. Chen, J.J. Xu, C. Yu, H. Lu, Materials Letters, 161 (2015) 201-204.
5. A. Baated, K.-S. Kim, K. Suganuma, J. Mater. Sci.: Mater. Electron., 22 (2011) 1685-1693.
6. J. Cheng, P.T. Vianco, B. Zhang, J.C. Li, Appl. Phys. Lett., 98 (2011) 241910.
7. E. Chason, N. Jadhav, W. Chan, L. Reinbold, K. Kumar, Appl. Phys. Lett., 92 (2008) 171901.
8. A.N. Hashim, M.A.A.M. Salleh, Solid State Phenomena, 280 (2018) 151-156.
9. H. Tsukamoto, Z. Dong, H. Huang, T. Nishimura, K. Nogita, Mater. Sci. Eng. : B, 164 (2009) 44-50.
10. M.M. Salleh, S. McDonald, C. Gourlay, S. Belyakov, H. Yasuda, K. Nogita, J. Electron Mater, 45 (2016) 154-163.
11. K. Nogita, Intermetallics, 18 (2010) 145-149.

Impact of Thermal Ageing and Multiple Reflow on Lead Free Composite Solder : A Short Review

Nur Syahirah Mohamad Zaimi[1,*], *Mohd Arif Anuar* Mohd Salleh[1] , *Mohd Mustafa Al Bakri* Abdullah[1], *Marliza* Mostapha[1] and *Romisuhani* Ahmad[2]

[1]Centre of Excellence Geopolymer and Green Technology, School of Materials Engineering, Universiti Malaysia Perlis (UniMAP), Taman Muhibbah 02600, Jejawi, Arau, Perlis, Malaysia
[2]Faculty of Engineering Technology, Uniciti Alam Campus, Universiti Malaysia Perlis (UniMAP), Perlis, Malaysia

Abstract. This paper reviews the impact of the thermal ageing and multiple reflow to the intermetallic compound (IMC) layer and mechanical performance of lead free solder reinforced with various reinforcements. Based on available literatures it had been proved that, the IMC layer and mechanical properties can be affected with the exposure of different temperatures and time. Besides that, the incorporation of the reinforcement in the matrix of solder alloys could improve the thickness of IMC layer and strength of the solder joints although it had been subjected to thermal ageing and multiple reflow cycles as compared to unreinforced solder. Therefore, the composite solder system could be possibly applied and adapted in current electronic devices even with the exposure of high temperatures.

1 Introduction

Soldering technologies have been prominent electronic interconnections since the early age of electronics and the technology will continually grow further in the future [1]. The performance and the reliability of the solder interconnections in the electronic devices are undeniably crucial in ensuring the longer lifetime of the electronic devices. In addition, current trend in the electronic industry are associated with the high speed, lightweight, miniaturization and multifunction of electronic devices. Therefore, a very robust interconnected materials need to be developed to ensure good performance of the interconnections prior to current trends. As more than 50 years ago, tin lead (Sn-Pb) solder alloys were dominantly used in the electronic industry as one of an outstanding interconnected material [2]. However, the concern regarding the toxicity of lead to the environment led to the emergence of many lead-free solder alloys that were expected to perform equivalently or better than lead solders. Examples of lead free solders were tin-copper (Sn-Cu), tin-copper-silver(SAC) and tin-copper-nickel (Sn-Cu-Ni) [3-5]. To date, researchers also introduced composite solders, solder that intentionally reinforced with

* Corresponding author: syahirahzaimi25@gmail.com

reinforcements either it was intrinsic or extrinsic reinforcements [6]. The incorporation of the reinforcements in the solder matrix were aimed to improve the service temperature capability, altered and stabilized the fined-grain microstructure as in returns affect the mechanical performance of the solder joints. Reinforcements reportedly used by previous researchers are aluminium (Al) [7], iron (Fe) [8], silicon carbide (SiC) [9], titanium oxide (TiO_2) [6, 10], cerium oxide (Ce_2O) [11] and kaolin geopolymer ceramics [2]. It was reported in terms of microstructures, thermal properties, mechanical properties and solderability were seen with the additions of reinforcements [1, 10, 12-14].

In a semiconductor devices, the solder joints between the die and substrate, and the substrate with the base plate were exposed to the thermal and mechanical stresses either during the processing and also subsequent operations [15]. In additions, during the service of the electronic components, solder joints were extremely exposed to the changes in the cycles of temperatures which in return would induce the thermal fatigue failures. Therefore, it was extremely important to investigate the performance of the new developed composite solders as it will subjected to different thermal conditions, so as to improve the reliability of the solder joints. Moreover, the most important parts that might be affected with different thermal exposure during the operations and services were the growth of intermetallic compound layer (IMC) which then affects the mechanical properties of the solder [16].

Therefore, this paper discussed on the impacts of different thermal treatments to the growth of intermetallic compound layer and mechanical properties of composite solders with the incorporation of various types of reinforcements.

2 Thermal effects in composite solders

Temperature is one of the important factors that lead to changes in the microstructure of the solder and induce internal stress in the solder joints which later promotes to the occurrence of the failure at the solder joints. The effects of temperature to the solder joint investigate as the solder joints were subjected to thermal ageing, thermal cycle and multiple reflow conditions. Thermal ageing is a condition where the solder joints are exposed to the various constant temperatures for some period of times [17]. Table 1 shows some of the temperatures and time that had been used by previous researchers in order to investigates the influence of thermal ageing to the composite solders. While, thermal cycle involves exposing the solder joints to the alternate extremely low and high temperatures at distinct temperature cycles such as from – 40 to 125 °C, - 40 to 150 °C and – 55 to 125 °C [18, 19] . Multiple reflow is a term referring to the process experienced by the solder joints in the electronic assemblies starting from the wafer bumping to the board level assembly [20]. Solder joints are commonly the weakest points as it was subjected to different kind of thermal treatments and it was the most likely part that would initiate the failure in the electronic devices. Therefore, the reliability and performance of the solder joints would dictate the reliability and also overall performance of electronic devices. Moreover, the intermetallic compound (IMC) layer directly linked with the mechanical properties of the solder. Thus, failure at the IMC layer of the solder joints would affects the mechanical performance of the solder. In next section, effects of the thermal ageing and multiple reflow on the intermetallic compound (IMC) layer and mechanical properties of the composite solder.

Table 1: Thermal ageing of different solder compositions at different temperatures and times.

Solder compositions	Temperatures (°C)	Time (hours)
Sn-3.0Ag-0.5Cu + TiC [21]	150	169, 324, 484
Sn-3.5Ag-0.5Cu + ZnO [22]	120	2.5, 5, 10, 50
Sn-3.5Ag-0.5Cu + TiO_2 [23]	125	168
Sn-58wt.%Bi + multiwalled carbon nanotubes [24]	85, 95, 105 and 115	100, 300, 500 and 1000
Sn-0.7Cu + TiO_2 [25] Sn-Cu-Ni + TiO_2 [25]	150	500, 1000, 1500 and 2000
Sn-3.5Ag-0.5Cu+ TiO_2 [26]	100, 125, 150, 175	168
Sn-Ag-Cu + Sm_2O_3 [1]	125 and 175	120, 240 and 720
Sn-Cu-Ni + TiO_2 [27]	75, 125 and 150	24 and 240

3 Effects of thermal ageing and multiple reflow to the intermetallic compound layer of composite solders

Solder joints in the electronic devices were often exposed to the elevated temperature at a longer period of time. Thus, this condition could affect the reliability of the solder joints since the intermetallic compound (IMC) layer and mechanical performance could be affected with the temperature. According to Chen et. al [21], the addition of titanium carbide (TiC) in the matrix of Sn-3.0Ag-0.5Cu could change the morphology of the interfacial layer of Cu_6Sn_5 from circular granular to short rod like with the rate of growth of the interfacial layer were suppressed. Fig. 1 shows the morphology of the interfacial layer which subjected to isothermal ageing at 150 °C for 169 h, 324h and 484 h. He also reported that, Cu_3Sn interfacial layer was significantly suppressed with the incorporation of TiC which could be attributed by the adsorption of reinforcement particles on the surface of interfacial layer that hindered the diffusion of copper to the solder matrix [21]. The formation of Kirkendal voids were associated with the thermal ageing time and temperature. It was interestingly proved that, the formation of Kirkendal voids at the interfacial layer of Cu_3Sn were reduced in the composite solders of Sn-3.0Ag-0.5Cu/TiO2 as compared to pure Sn-3.0Ag-0.5Cu [25].

During multiple reflow, the thickness of interfacial IMC layer were decreased with the addition of titanium oxide (TiO_2) with the maximum thickness of interfacial IMC layer for the composite solder Sn-3.0Ag-0.5Cu/ TiO_2 with multiple reflow cycle was 8 μm [25]. However, in pure Sn-3.0Ag-0.5Cu the thickness of interfacial IMC would grow until 13 μm. In addition, the morphology of Ag_3Sn in the composite solders of 96.4Sn-3.6Ag/Al_2O_3 transformed from the coarser to finer spheroidal shape with multiple reflow cycles [28]. Therefore, the reinforcements particles added shows the ability to suppressed the formation of thicker IMC layer although it was subjected to different ageing time and temperature and also multiple reflow cycles [29-30].

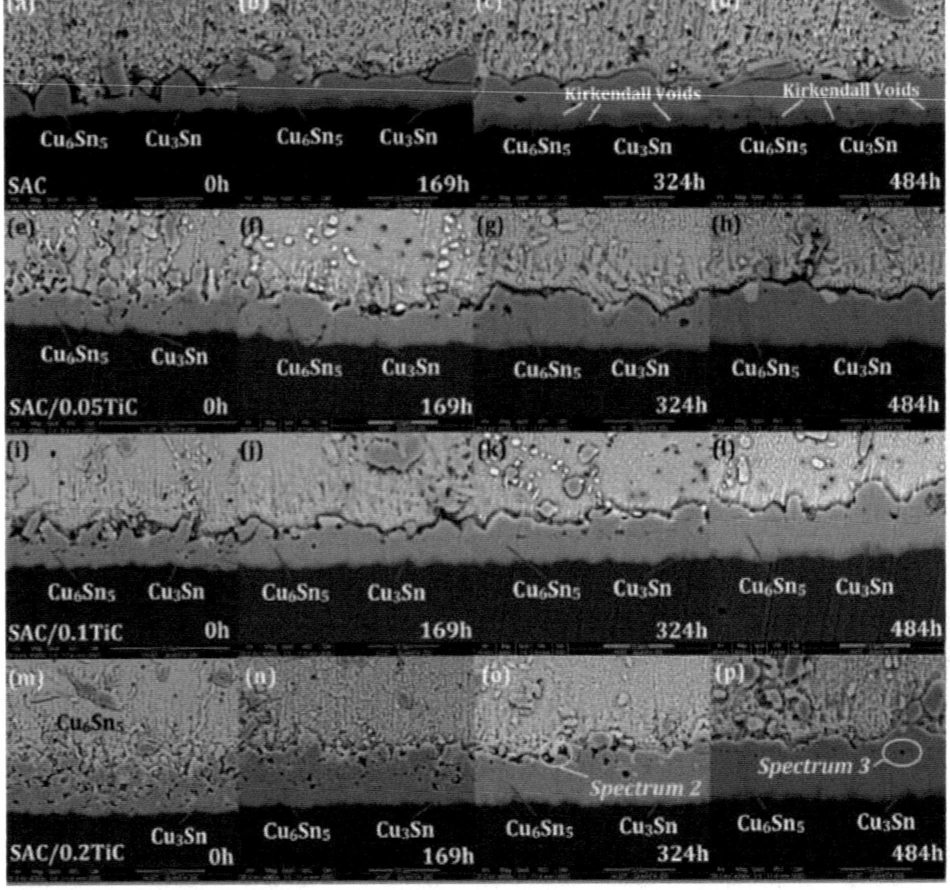

Fig.1. Morphology of Cu_6Sn_5 interfacial IMC subjected to thermal ageing at 150 °C with different ageing time (a-d) SAC, (e-h) SAC/0.05 TiC, (i-j) SAC/0.1 TiC, (m-p) SAC/0.2 TiC [21].

4 Effects of thermal ageing and multiple reflow to the mechanical properties of composite solders

In previous sections, it was proved that the thickness and morphology of interfacial IMC changed with the exposure of different temperature at different times. Thus, the strength of the solder joint could also affect which would dictates the subsequent mechanical performance of the solder. Fig.2 shows the graph of stress strain curve for Sn-3.5Ag-0.5Cu (SAC355) composite solders. As reported by [22], the tensile strength of SAC355 lead free solder and SAC355/ZnO were affected with the increasing in the ageing temperature. According to stress strain curve in Fig.2, the stress strain curves were shifted to the lower stress level as compared to the lower ageing temperature. Moreover, the ultimate tensile stress, yield stress and fracture stress were reduced with an increase in the ageing temperature for SAC355 with ZnO and SAC355 without addition of ZnO. However, the reduction in the ultimate tensile stress, yield stress and fracture stress in the SAC355 with ZnO was not as higher as in SAC355 without ZnO, which proved the effectiveness of the ZnO as the reinforcement particles. The reason in the improvement of solder strength in SAC355/ZnO was due to the formation of finely dispersed IMCs and β-Sn dendrite size which help in suppressing the plastic deformation.

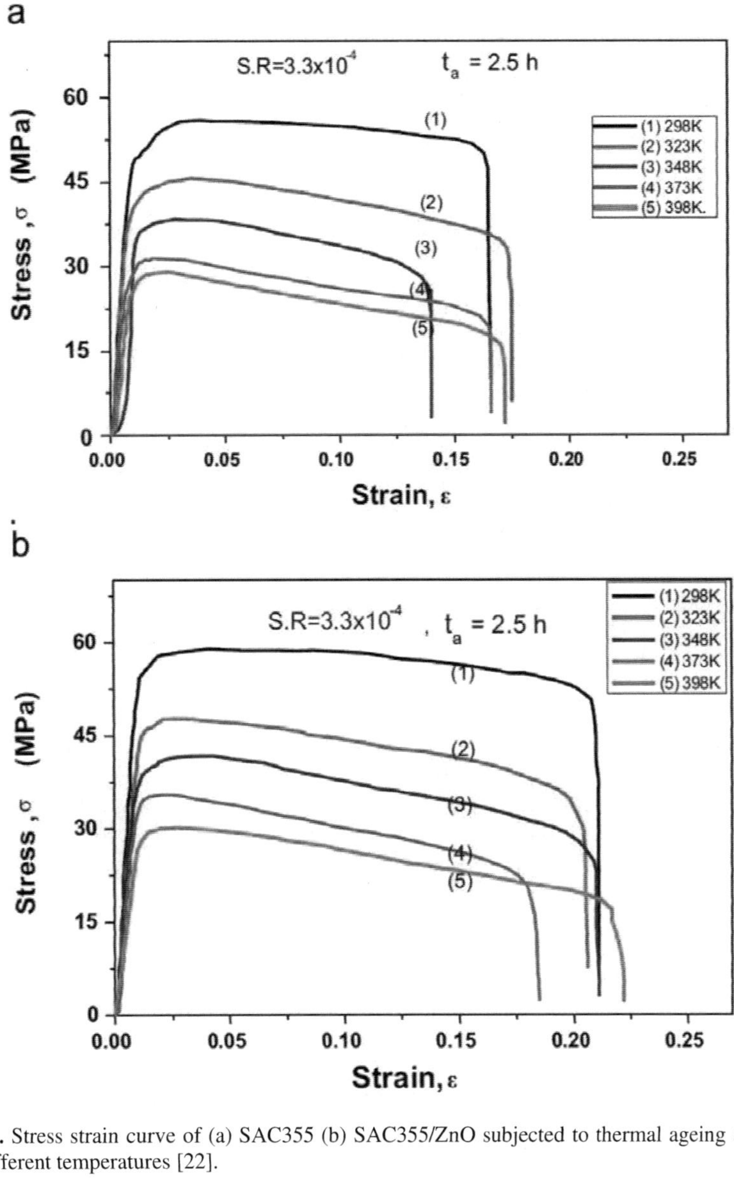

Fig. 2. Stress strain curve of (a) SAC355 (b) SAC355/ZnO subjected to thermal ageing at 2.5 hours for different temperatures [22].

Besides that, addition of strontium titanate ($SrTiO_3$), to Sn-Ag-Cu reduced the shear strength of the solder joints with increasing in ageing times. Although the shear strength was reduced with respected to ageing time, but the strength in Sn-Ag-Cu reinforced with $SrTiO_3$ were still higher than pure Sn-Ag-Cu.

In terms of multiple reflow, the strength of the solder joint was decreased with the increasing of reflow cycle. Tikale et. al [28], reported that the strength of Sn-Ag-Cu solder was gradually decreased start after the second reflow.The decreament in the strength of the solder alloys was due to the formation of thick Cu_6Sn_5 and the formation of Cu_3Sn layer at the interfacial IMC. Formation of Cu3Sn IMC could lead to the mismatch of thermal coefficient of thermal expansion between IMCs. Thus, promotes the existence of micro-cracks along the IMC layer as it was subjected to multiple reflow cycle.

5 Summary

This reviews had documented the changes in the intermetallic compound layer and mechanical properties of solder under thermal ageing and multiple reflow cycle. It can be concluded that,the addition of reinforcement particles into existing lead free solder alloy could enhanced the properties of the solder eventhough it was subjected to thermal ageing and multtiple reflow cycle. The strength of the solder were reduced with ageing time and temperatures, but the value still higher than the solder alloys without the addition of reinforcement particles. Therefore, in developing new solder composite, the effects of different temperatures exposure to the intermetallic layer and mechanical properties should of composite solder should be carefully examined in order to integrate and applied composite solder so as to adapt with the modern electronic devices.

The authors acknowledge the financial support through the Ministry of Higher Education, Malaysia, under the Fundamental Research Grant Scheme (FRGS/1/2017/TK10/UNIMAP/03/2) (9003-00612). The authors also acknowledge Center of Excellence Geopolymer & Green Technology (CEGeoGTech) and School of Materials Engineering, Universiti Malaysia Perlis.

References

1. A.K. Gain, L. Zhang, Mater. Res. Express. **6,** 066526 (2019)
2. N.S.M. Zaimi, M.A.A.M. Salleh, M.M.A.B. Abdullah, M.Mostapha, R.Ahmad, IOP Conf Ser Mater Sci Eng. **701,** 012033 (2019)
3. M.I.I. Ramli, M.A.A.M. Salleh, R. Mohd Said, N. Saud, AIP Conf Proc. **1835,** 020027 (2017)
4. N.S.M. Zaimi, M.A.A.M. Salleh, M.M.A.B. Abdullah, M.Mostapha, R.Ahmad, IOP Conf Ser Mater Sci Eng. **551,** 012081 (2019)
5. S. Cheng, C. Huang,M. Pecht, Microelectron. Reliab. **75**, 77-95 (2017)
6. M.A.A.M. Salleh, S.D. McDonald, Y. Terada, H. Yasuda, K. Nogita, Mater. Des. **82**, 136-147 (2015)
7. N. Razliana, N. Adli, N. Saud, S.A. Musa Appl. Mech Mater. **754-755**, 166-170 (2015)
8. P. Zhang, J. Wang, S. Xue, P. Xue Appl. Sci. **9**, 2044 (2019)
9. A.A. El-Daly, W.M. Desoky, T.A. Elmosalami, M.G. El- Shaarawy, A.M. Abdraboh, Mater. Des. **65,** 1196-1204 (2015)
10. Z. Liu, H. Ma, S. Shang, Y. Wang J. Mater. Sci. **30,**18828-18837 (2019)
11. Z.H. Li, Y. Tang, Q.W. Guo, G.Y. Li J. Alloys Compd. **789,** 150-162 (2019)
12. Z.H. Li, Y. Tang, Q.W. Guo, G.Y. Li J. Alloys Compd. **818,** 152893 (2019)

13. A. Wattanakornphaiboon, R. Canyook, K. Fakpan, Mater. Today **5,** 9213-9219 (2018)
14. A.K. Gain, L. Zhang, Materialia, **3**, 64-73 (2018)
15. G. Khatibi, A. Betzwar Kotas, M. Lederer, Microelectron. Reliab. **85,** 1-11(2018)
16. J. Shen, M. Zhao, H. Peipei. J. Alloy. Compd. **574,** 451-458 (2013)
17. N. Jiang, L. Zhang, Z.Q. Liu, L. Sun, W. Long, P. He, M. Xiong, M. Zhao Sci. Technol. Adv. Mat. 20, 876-901 (2019)
18. R.J. Coyle, K. Sweatman, B. Arfaei JOM **67**, 2394-2415 (2015)
19. L. Zhang, S.B. Xue, G. Zeng, L. Gao, H. Ye J. Alloy. Compd. **510**, 38-45, (2012)
20. S.R.A. Idris, S. Zuleikha, Z.A. Abd Malek, MATEC **74,** 00038 (2016)
21. G. Chen, H. Peng, V. Silberschmidt, Y.C. Chan, C. Liu, F. Wu J. Alloy. Compd. **685,** 680-689 (2016)
22. M. Sobhy, A.M. El- Rafei, M.M. Mousa, G. Saad Mater. Sci. Eng. **646**, 82-89 (2015)
23. T.H. Chuang, L.C Tsao, C. Chung, S.Y. Chang Mater. Des. **39**, 475-483 (2012)
24. C. J. Lee, K.D. Min, H. J. Park, S.B. Jung J. Alloy Compd. **820,** 153077 (2019)
25. M.A.A.M Salleh, S.D. McDonald, K. Nogita,. J. Mater Process Technol. **242,** 235-245 (2017)
26. L.C. Tsao J Alloy Compd. **509**, 8441-8448 (2011)
27. R. M. Said, M. A.A. M. Salleh, M. N. Derman, M.I.I. Ramli Key Eng. Mater. **700,** 123-131 (2016)
28. S. Tikale, K. Prabhu, T Indian I Metals, **71,** 2855-2859 (2018)
29. R.M Said, M.A.A.M Salleh, M.I.I Ramli, N. Saud, Solid State Phenom. 280, 163-16 (2018)
30. R. M. Said, M. A.A. M. Salleh, M. N. Derman, M.I.I. Ramli Key Eng. Mater. **700,** 123-131 (2016)

Effect of infill pattern on mechanical properties of 3D printed PLA and cPLA

Cheow Keat Yeoh[1,*], *Chee Son* Cheah[1], *Raguvaran* Pushpanathan[1], *Cheah Chie* Song[1], *Mei Ai* Tan[1], and *Pei Leng* Teh[1]

[1]School of Materials Engineering, Universiti Malaysia Perlis

Abstract. PLA are the most sustainable alternatives and can fit in a wide-range of applications of electronics, nonwoven fabrics and food packaging. With these PLA's adaptability and suitability in many techniques of production such as injection moulding, extrusion, and blow moulding, PLA has become high interest in the production process. Besides, PLA as a thermoplastic polyester that mostly obtained from renewable materials [1]. Infill patterns can affect the mechanical properties of 3D printed PLA and cPLA. PLA with zig zag infill pattern has higher tensile strength of 23.409 MPa compared to PLA with grid and concentric infill pattern. Meanwhile, cPLA with grid infill pattern has higher tensile strength of 30.5638 MPa compared to cPLA with concentric and zig zag infill pattern. By using the suitable infill pattern parameter, the 3D printed PLA and cPLA can have good mechanical properties and can be applied in packaging, pharmaceuticals, textiles, automotive, biomedical and tissue engineering. It has been widely investigated for biomedical applications due to its biodegradability and biocompatibility.

1 Introduction

3D printers can be used to fabricate more complex designed objects, allow flexible large design and also improve the performance of the fabricated objects [5]. In FDM, the thermoplastics polymer filament heated above its glass transition temperature, Tg and extruded through a movable nozzle in the x-y plane to form a 3D structure [5].

PLA as a thermoplastic polyester is versatile and mostly obtained from materials that are annually renewable. The materials are normally fermented into agriculturally originated lactic acid. PLA possesses high strength, modulus of elasticity, stiffness, brittleness and is a completely biodegradable matrix [1]. Additionally, PLA has a low glass transition temperature (Tg = 60–65 °C) and melting temperature (Tm = 173–178 °C), making it very useful for 3D printing since it does not require a heated surface for the objects being printed. While there is a good amount of evidence that PLA materials supplied by various vendors produces final prints which vary considerably in terms of mechanical properties, interlayer adhesion, and appearance, very few published studies exist which discuss the

* Corresponding author: ckyeoh@unimap.edu.my

relationship between source material chemistry and the properties of the structures created by desktop FDM printers, despite their use in research applications [6].

Although PLA is an eco-friendly bioplastic with excellent biocompatibility, processibility, and less energy dependence, it still has drawbacks as well, which limit PLA use in certain applications [7]. The main drawback for PLA is they are very brittle material with less than 10% elongation at break. Although its tensile strength and elastic modulus are comparable to polyethyleneterephthalate (PET), the poor toughness limits its use in the applications that need plastic deformation at higher stress levels. Therefore, incorporation of plasticizers and impact modifiers into the composites of PLA for the improvement of the toughening and impact properties of the material becomes necessary.

The different parameter of 3D printing will affect the mechanical properties of produced objects. Thus, the mechanical properties of different 3D printing parameter of the 3D printed PLA and cPLA are the focus of this study.

2 Methodology

2.1 Preparation of PLA and cPLA by RepRap 3D printer

The design of PLA and cPLA structure are drawn according to ASTM D638. Appropriate composition of raw materials, poly lactic acids (PLA) and cPLA are fed in the RepRap 3D printer to produce PLA and cPLA products. The printing parameters are shown in Table 2.1.1.

Table 2.1.1 Printing parameters of 3D printer.

Printing Parameters	Values
Printing temperature	200 °C
Layer height	0.2 mm
Printing speed	14 mm/s
Layer thickness	0.2 mm
Wall thickness	0.8 mm

The infill pattern of the printed PLA and cPLA are varied during the fabrication of the PLAand cPLA products. The infill density is set at 80% while the infill pattern is varied with zigzag, grid and concentric pattern. At the end of preparation of products, the different density with different infill patterns of the PLA and cPLA are produced.

2.2 Mechanical properties lab testing for different parameters of PLA and cPLA

There are several mechanical properties tests that used to identify the mechanical properties of PLA and cPLA. The mechanical properties tests are:

2.2.1 Tensile test

The tensile test is used to identify the stress, strain and Young's modulus of the samples under a specific tension. The test is conducted according to ASTM D638 and the Universal Testing Machine with its model Instron 5569 is used. The crosshead speed or strain rate that used is 5mm/mins with surrounding temperature at around 25±2oC. Before the tensile test, the dimension of the sample is measured by the Vernier calliper for the use of calculate the stress, strain and the Young's modulus of the samples. To make sure the accuracy of the results, 5 specimens for each different parameter were used to obtain the average tensile properties.

2.2.2 Microhardness Vickers Test

To determine the microhardness of the samples, ASTM E384 will be the standard for the whole microhardness Vickers testing. A square-based diamond indenter with 136o between the face of the samples is used or equipped on the microhardness Vickers machine. During the testing, the indenter with a load of 9.807N is used and a dwelling time around 10 seconds is apply to make sure the accurate microhardness of the samples. Besides that, each sample is tested by the indenter at 5 different points on the surface of the sample to obtain average values.

3 Results and Discussion

3.1 Tensile test results for different infill pattern of PLA

The value of the stress and strain of PLA with three different infill pattern had been used to compare their tensile properties. Figure 3.1 show the stress-strain graph for three different infill patterns of PLA. Figure 3.2 and 3.3 show the bar chart for the mean of tensile strength and elongation at break of three infill pattern of PLA.

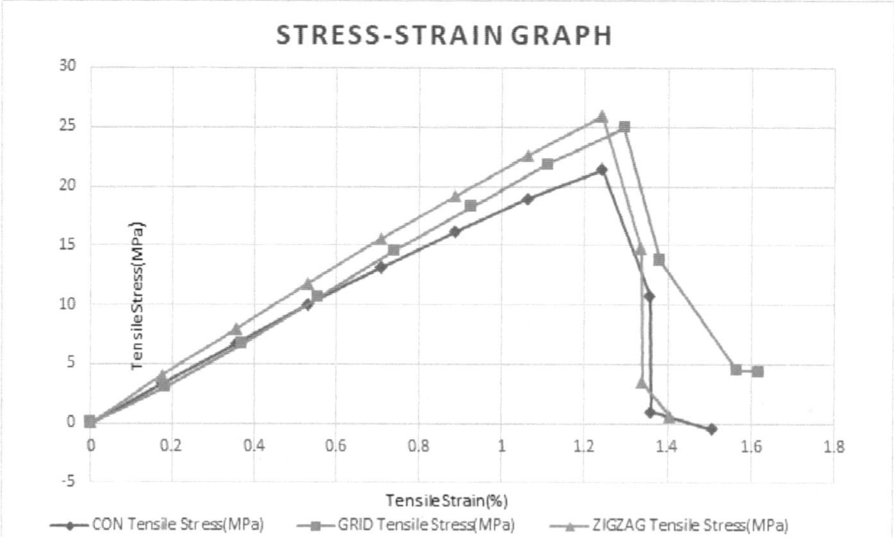

Fig. 3.1. Stress-strain graph for three different infill patterns of PLA.

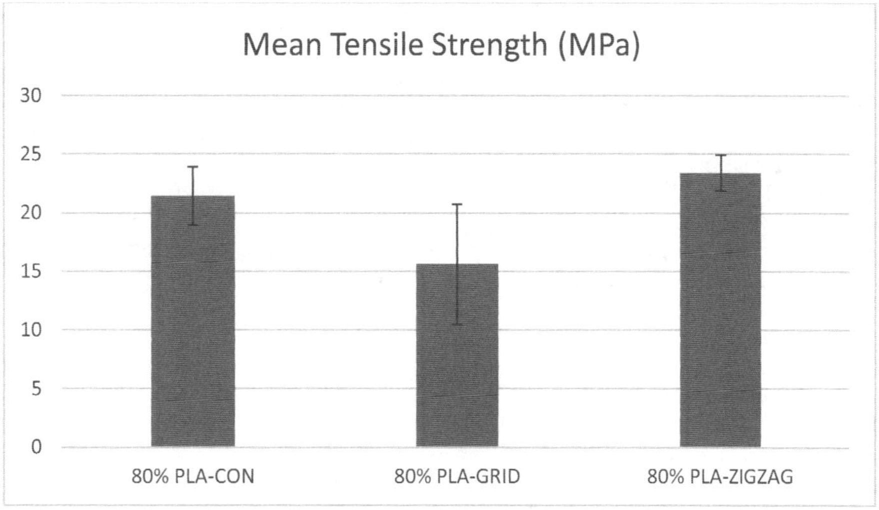

Fig. 3.2. Tensile strength of three infill pattern specimens

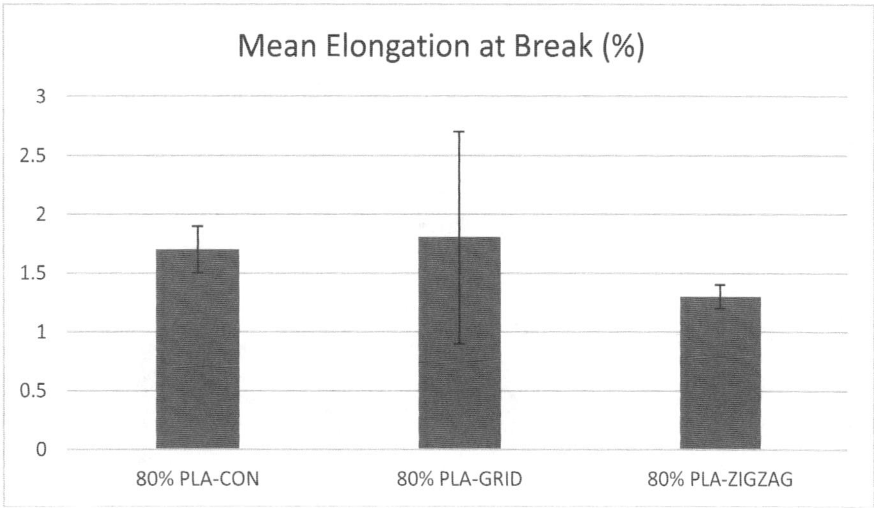

Fig. 3.3. Elongation at break of three infill pattern specimens

Fig. 3.4. Zigzag pattern

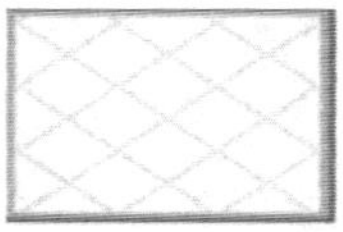

Fig. 3.5. Grid pattern

Fig. 3.6. Concentric pattern

Based on the result, PLA with zigzag infill pattern have the highest tensile strength if compared with PLA with concentric and grid infill pattern. Zigzag infill pattern printing continuously in one diagonal direction and cause stronger inter raster bonding with tiny raster gap and allow more consistent adhesion between the layers [2]. So, the consistent adhesion between the layers and the inter raster bonding cause the zigzag infill pattern experience highest tensile strength at 23.409 MPa but also cause the zigzag infill pattern to experience the lowest elongation at break,1.3%, compare to concentric and grid infill pattern of specimens.

The grid infill pattern specimen (Figure 3.5) has lines in both diagonal directions on each layer. This causes the 3D printed specimens to have weak inter raster bonding with large raster gap So, the grid infill pattern specimen experiences the lowest tensile strength at 15.620 MPa compared with other two types of infill pattern but due to weak inter raster bonding and low adhesion between the layers, the grid infill pattern specimen to expeirence the highest elongation at break of 1.8%.

For concentric infill pattern specimens, they have the moderate tensile strength, 21.442 MPa, and elongation at break, 1.7%. This is because the infill pattern prints from the outside towards the centre causing the 3D printed specimens to have moderate inter raster bonding with moderate raster gap.

As conclusion, the infill pattern of the 3D printing will affect the raster gap, inter raster bonding of the 3D printed specimens and the consistency of the extruder motion that will affect the adhesion between the layers during 3D printing. Hence, infill pattern is the important parameter for 3D printing.

3.2 Microhardness Vickers Test Results for PLA

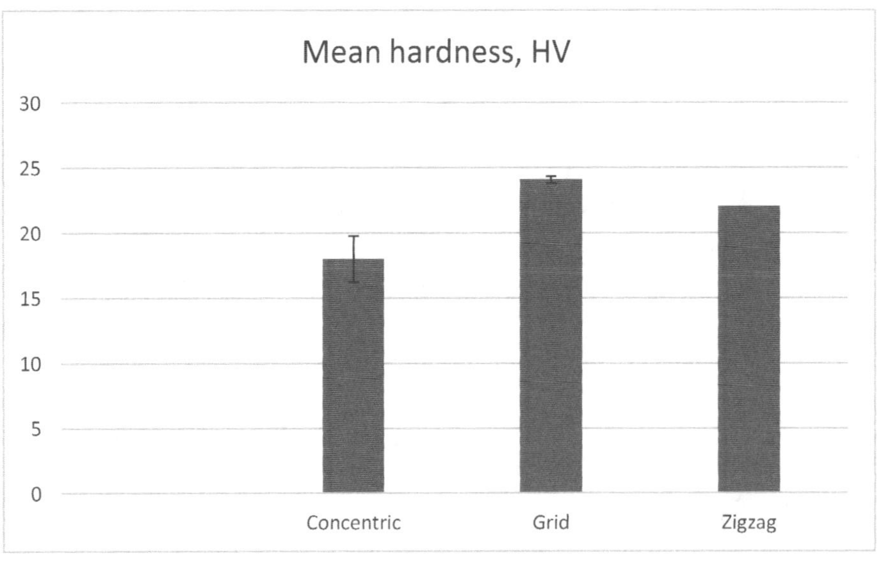

Fig. 3.7. Hardness values.

Based on Figure 3.7, grid infill pattern 3D printed PLA has the highest hardness values followed by zigzag infilled pattern and concentric infill pattern. This is due to inter raster bonding in the grid infill pattern are more resistant to deformation [3]. Besides that, the concentric infill pattern has the lowest hardness value, this is due to the inter raster bonding in concentric infill pattern are less resistance to deformation. While the zigzag infill pattern 3D printed PLA has the moderate resistance to deformation.

3.3 Effect of different infill pattern on stress and strain properties of CPLA

Figure 4.1 shows the effect of the three infill patterns, concentric, grid and zigzag on the stress and strain properties of CPLA. Figure 4.2 and Figure 4.3 shows the effect of the three infill patterns, concentric, grid and zigzag on the mean tensile strength (MPa) and mean break elongation (%) of CPLA.

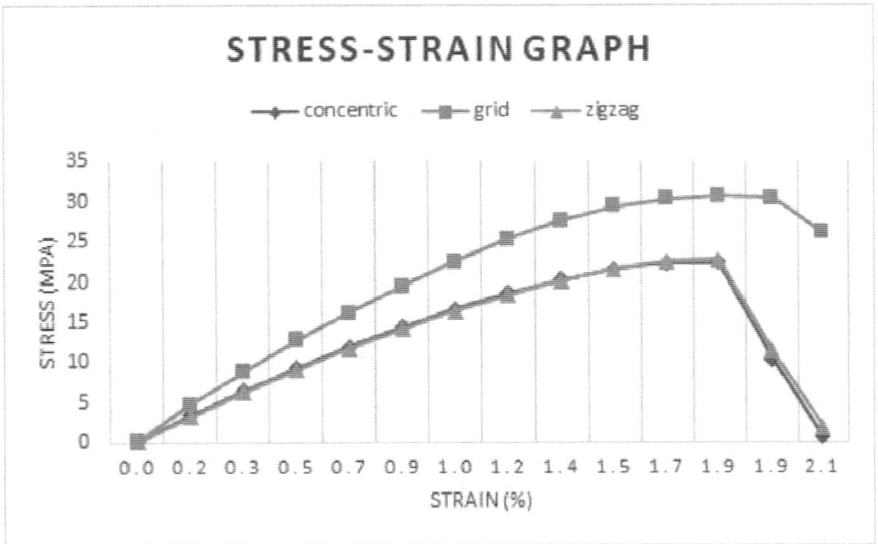

Fig. 4.1. Effect of different infill pattern on stress and strain properties of CPLA.

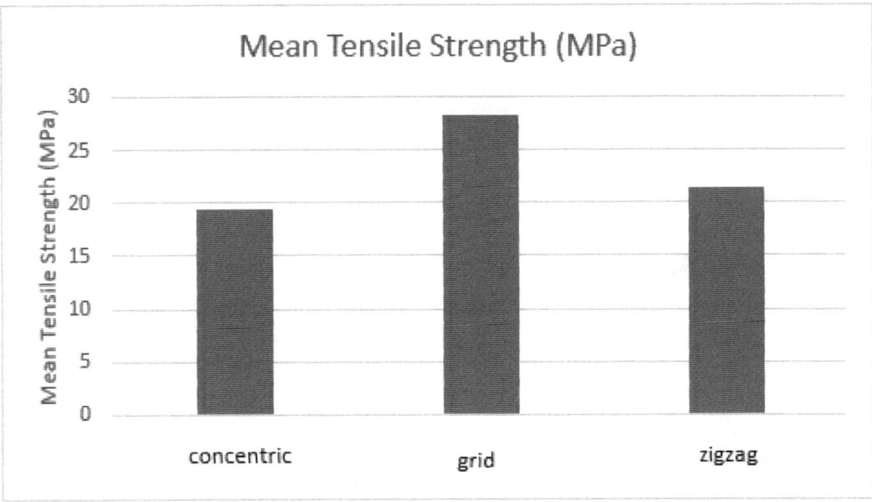

Fig. 4.2. Effect of different infill pattern on mean tensile strength (MPa) of CPLA.

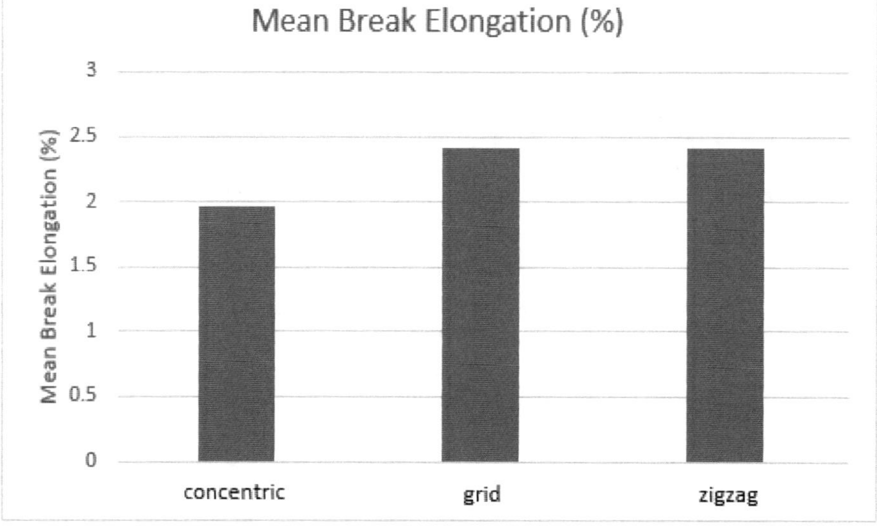

Fig. 4.3. Effect of different infill pattern on mean break elongation (%) of CPLA.

Based on Figure 4.1, CPLA with grid infill pattern has achieved the highest stress value of 30.5638 MPa at a strain percentage of 1.871%. A grid shaped infill has lines in both diagonal directions. Since the lines forms in both directions, therefore it increases the adhesion between print raster reducing the formation of pores or air gap in the print. When the print or specimen has lower pores, it has higher resistance towards external force which eventually resulting in high ultimate tensile strength.

According to Figure 4.2, the specimen which is printed with grid infill pattern has a mean tensile strength of 28.328 MPa. Specimen of grid infill pattern has a higher mean tensile strength. This shows that the specimen with grid infill pattern has stronger tensile properties than the specimen with concentric and zigzag infill pattern. This is due to formation of closely packed raster which increases the adhesion between print layers. Therefore, the bonding formed between the raster becomes stronger and higher force is required to break these bonds resulting in high mean tensile strength.

Figure 4.3 shows that grid infill pattern and zigzag infill pattern has higher break elongation value of 2.42%. Concentric infill pattern has lower break elongation due to print layers which are too compact. This will restrict polymeric chain mobility in the specimen which causes crack propagation [8]. Meanwhile, the grid infill pattern is not too compact which allows for the mobility of polymeric chain.

3.4 Effect of different infill pattern on micro hardness (HV) of CPLA

Figure 4.4 shows the mean micro hardness of specimens with concentric infill pattern is higher than specimens with grid and zigzag infill pattern. Concentric infill pattern is much more compact compared to grid and zigzag pattern since the infill prints from the outside towards the center of the model, making the infill lines to be too close to each other. This would increase the consistency of print layers [9]. Thus, higher load indentation is needed for surface penetration, therefore increases the micro hardness of the specimen.

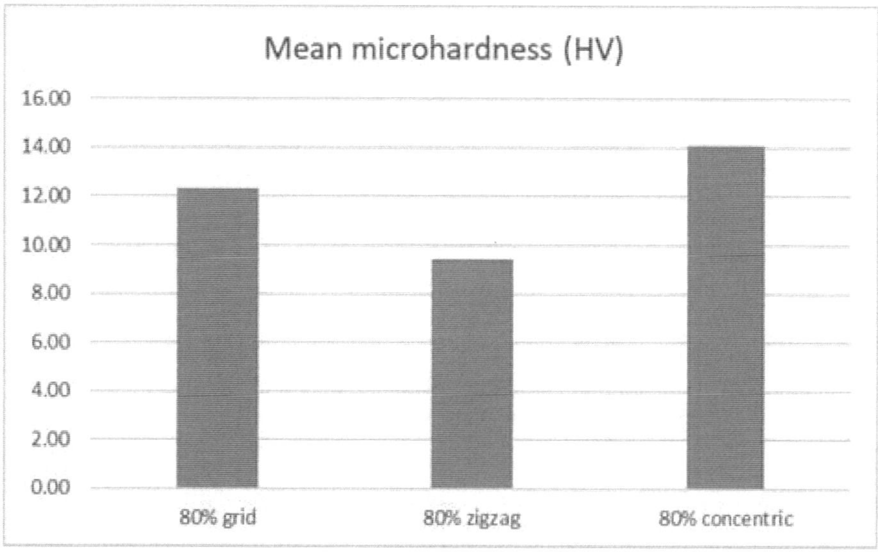

Fig. 4.4. Effect of different infill pattern on Mean micro hardness (HV) of CPLA with 80% infill density.

4 Conclusion

Based on the results, the infill pattern can affect the mechanical properties of 3D printed PLA and cPLA. PLA with zig zag infill pattern has the highest tensile strength among the three different infill patterns. Meanwhile, PLA with grid pattern has the highest hardness value compared to concentric and zig zag infill pattern. cPLA with grid pattern has the highest tensile strength among the three different infill patterns and cPLA with concentric pattern has the highest hardness value compared to grid and zig zag pattern. As for application, 3D printed PLA and cPLA can be applied in packaging, pharmaceuticals, textiles, automotive, biomedical and tissue engineering.

This work was supported by the Fundamental Research Grant Scheme (FRGS) under a grant number of FRGS/1/2018/TK05/UNIMAP/02/14 from the Ministry of Education Malaysia.

References

1. D. M. Abdalhadi, Z. Abbas, A. F. Ahmad, K. A. Matori, and F. Esa, Fibers Polym., **19**, 7, (2018).
2. D. M. Patel, Int. J. Eng. Dev. Res., **5**, 3 (2017).
3. K. A. Hamzah et al., J. Mater. Eng. Perform., **28**, 9 (2019).
4. K. Bryll, E. Piesowicz, P. Szymański, W. Slaczka, and M. Pijanowski, *MATEC Web of Conferences*, **237** (2018)
5. K. Gnanasekaran et al., Appl. Mater. Today, **9** (2017).
6. M. A. Cuiffo, J. Snyder, A. M. Elliott, N. Romero, S. Kannan, and G. P. Halada, Appl. Sci., **7**, 6 (2017)

7. R. M. Rasal, A. V. Janorkar, and D. E. Hirt, Prog. Polym. Sci., **35**, 3 (2010)
8. V. Kumar, J. RamKumar, S. Aravindan, S. K. Malhotra, K. Vijai, and M. Shukla, Int. J. Plast. Technol., **13**, 2 (2009)
9. Y. Y. Aw, C. K. Yeoh, M. A. Idris, P. L. Teh, K. A. Hamzah, and S. A. Sazali, Materials (Basel)., **11**, 4 (2018)

Influence of squeegee impact on stencil printing process: CFD approach

M.H.H. Ishak[1], *M.S.* Abdul Aziz[1,2,*], *M.H.S.* Abdul Samad[2], *Farzad* Ismail[1] and *M.A.A* Mohd Salleh[3]

[1]School of Aerospace Engineering, Universiti Sains Malaysia, 14300 Nibong Tebal, Penang, Malaysia.
[2]School of Mechanical Engineering, Engineering Campus, Universiti Sains Malaysia, 14300 Nibong Tebal, Penang, Malaysia.
[3]Center of Excellence Geopolymer and Green Technology (CEGeoGTech), School of Materials Engineering, Universiti Malaysia Perlis (UniMAP), Perlis, Malaysia

Abstract. This work has been conducted to predict the real time observation of solder paste Sn96.5Ag3.0Cu0.5 (SAC305) filling process into stencil apertures as well as print quality in stencil printing by using Computational Fluid Dynamics (CFD) approach. A 3-Dimensional stencil printing model was developed and simulated in FLUENT by using different squeegee parameters which are the angle and printing speed. An experimental work was performed to be compared with part of the simulation results in term of print quality for validation purpose. It is found that squeegee angle 60° to 80° has potential to obtain good print quality of solder paste. In addition, print speed range between 35 mm/s to 95 mm/s also can be the good print speed option to achieve good print quality in stencil printing process. Finally, the maximum pressure distribution of solder paste also changes substantially as the squeegee travel further with respect to different values of tested parameters.

1 Introduction

Stencil printing is a process of depositing a viscous fluid which is solder paste onto a bare substrate such as printed circuit board (PCB) through the aperture openings of a stencil. It is originally adapted from the screen-printing process. This process is one of the widely used methods of soldering in Surface Mount Technology (SMT) today as a solution for higher pin count and fine pitch size. This is due to the high requirement of smaller size, lighter weight, and high-performance PCB in electronic packaging. The use of stencil printing for the deposition of conductive interconnects was developed during the late 1960s as companies looked to increase the densities of the products, improve assembly time and drive down production costs. This process was termed in surface mount assembly (SMA) and is currently used today in at least 70% of electronic packaging. The stencil printing is commonly known in SMA as the largest contributor to soldering defect by 60% averagely [1].

* Corresponding author: msharizal@usm.my

In stencil printing, a thin layer of stainless-steel foil is placed onto a PCB. It has laser cut configuration of apertures that will determine the basic layout of the solder paste deposits on the board. The stencil will be aligned on the top surface of the board and brought in close to or in direct contact with the surface. A squeegee is used to transfer solder paste inside the stencil apertures to deposit a required amount of solder paste on pad of the substrate [2]. For metal blade squeegee, the angle of 60° is the most suitable angle to handle the ultra-fine and larger pitch size because it offers consistency of print and the closest print volume to the designed volume of solder paste. It is also mentioned that the volume of solder paste deposits is proportional to the paste pressure and the highest region of the pressure is at the edge of squeegee [3]. The squeegee provides hydrodynamic pressures in the roll of solder paste that assists the filling of paste inside the apertures by shearing as it moves over the stencil [4]. Once the squeegee is fully swiped the solder paste through the apertures, the stencil is removed out from the board leaving the small volumes of solder paste in the filled apertures on the board as deposits. Fig. 1 below shows the schematic diagram of stencil printing process [10].

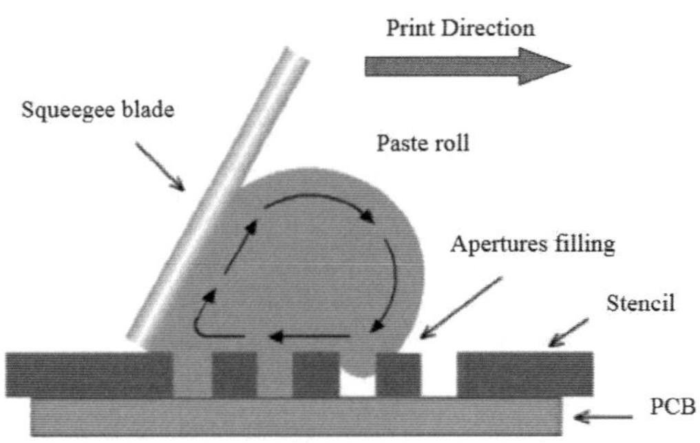

Fig. 1. Schematic diagram of stencil printing process.

One of the remarkable capabilities of stencil printing in SMT packaging is it can place millions of solder paste deposits onto a substrate such as PCB by one print stroke only. In addition, the process can be repeated thousands of times with the same stencil onto the further substrates with good consistency of soldering performance. Thus, it reduces soldering time as well as offers higher process output. More advantages about this process are it does not require large capital investment or highly skilled operators. The combination of the cost advantage with the capability to handle wide range of materials makes it in one of the top list of best option for high volume applications. Furthermore, the compatibility of the method with pre-existing equipment of printing and capability to produce solder bumps within all range of compositional spectrum such as lead-free solders have increased its reputational in SMT packaging [2].

2 Literature Review

Stencil printing is one of the best method for high volume application where the time and process output are concerned. The need of high-performance PCB with smaller size, lighter weight, and higher pin count has contribute to the development of fine pitch and ultra-fine pitch for stencil printing applications [3]. There is various study has been conducted through the years in order to improve the performance of stencil printing in SMT technology numerically and experimentally which is the printing quality. Numerical studies have been performed to simulate stencil printing process by using Computational Fluid Dynamics (CFD) methods at macroscopic and microscopic scales by using 2-D model that involves Finite Volume Method (FVM), Lattice-Boltzmann Method (LBM), and Discrete Element Method (EDM). It is found that the volume fraction of solder paste is far lower than normal due to the unrealistic representation of spherical particles by 2-D cylinder that cause artificial blockage of fluid passage at greater volume fraction [12].

Neural network approach model has been proposed to solve the fine pitch printing quality issue of the non-linear behavior stencil printing through the volume prediction of solder paste deposits. The proposed approach is determined to be effective to predict and control the printing quality in surface mount assembly such as paste volume deposited on PCB pad with small prediction errors which is less than 7% [13]. In stencil printing, the stencil experiences pressure exerted by the squeegee on its surface that might affect the behavior of the structure. The deformation behavior of stencil which can affect printing quality has been investigated by using Finite Element Method (FEM) associated with the uneven surface of Printed Wiring Circuit (PWB). Greater surface level differences between the stencil and PWB affect the stencil bending behavior that can lead to excessive solder paste deposits and solder bridges issue [5].

Furthermore, stencil printing process also has been modeled by using Finite Volume Method (FVM) to study the effect of two different characteristics of solder paste on the simulation results. The flow field and pressure distribution of the solder paste along the stencil surface are determined to be compared with non-Newtonian and Newtonian cases. The pressure distribution of the non-Newtonian case is found to be differed from the Newtonian case substantially which means modeling the stencil printing process with Newtonian solder paste properties causes serious mistakes since solder paste is known to be a non-Newtonian fluid [14]. Moreover, the effect of squeegee speed and solder paste density variation on stencil printing process has been studied by using CFD approach through the flow characteristic of solder paste simulation. Based on the study, increment of squeegee speed can cause the shear stress to increase proportional to the increment of solder paste's shear strain [15].

3 Methodology

A stencil printing model is created in 3-Dimensional by using ANSYS Design Modeler that consists of two significant components for the simulation which are aperture bodies and printing body as explained below. In this case, stencil is neglected because the stencil thickness is too thin. However, the thickness of the stencil is counted in the model preparation as the thickness of aperture bodies. Aperture bodies are drawn based on the exact dimension measured by using Alicona Infinite Focus microscope to form a complete stencil printing model as shown in Fig. 2 and Fig. 3. Fig. 4 shows the modeling geometry of the aperture body.

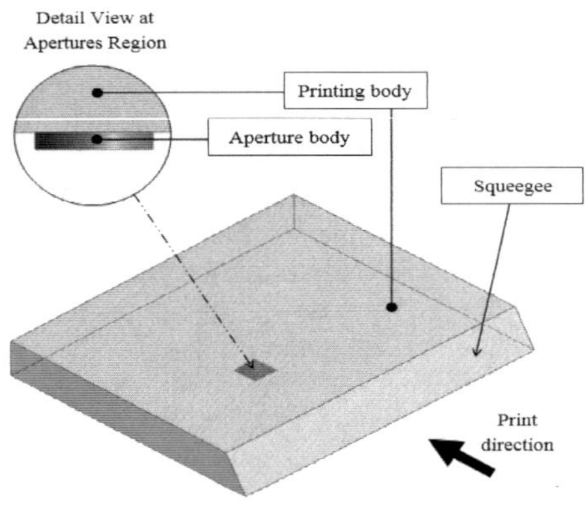

Fig. 2. 3-Dimensional stencil printing model.

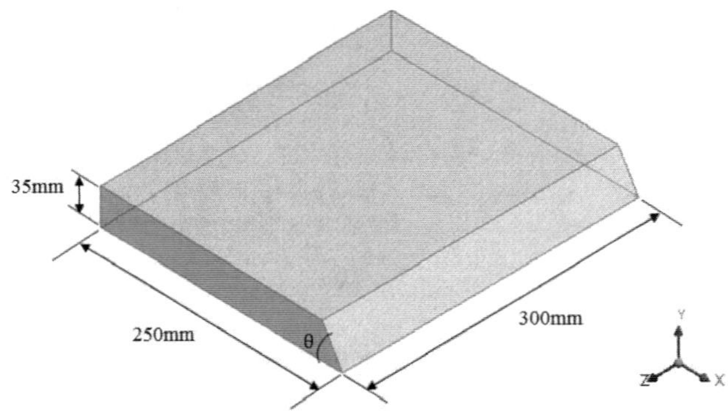

Fig. 3. Printing body dimension of stencil printing model.

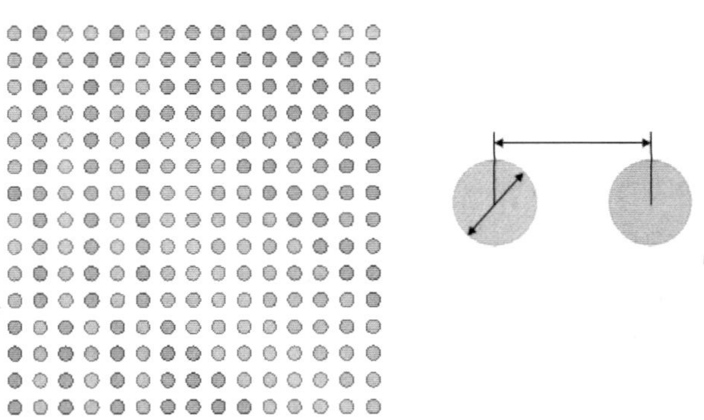

Fig. 4. Design of aperture bodies and dimension.

The thickness of the aperture bodies is the same due to the adoption of stencil thickness that are assumed as constant and perfect (thickness = 0.1 mm). The printing body is made to have a slanted part as the squeegee where the angle is the variable parameter to be tested in the simulation. The angles tested in the simulation are 50°, 60°, 70°, and 80°. The specification of the selected apertures design is 225 I/O; 1.5 mm PITCH. The design has 225 individual aperture bodies by pitch distance 1.5mm to each other. All the fluid bodies are meshed with structured elements consists of hexahedral and wedges for both aperture and printing bodies as shown in Fig. 5. For the printing body, the bottom surface is declared as fluid interface while others are declared as wall. The walls of the printing body are further used in dynamic mesh setting. The interior of the printing body is the zone where solder paste and air are placed together inside it. Furthermore, the top surface of the aperture bodies is declared as fluid interface while others are set as wall. The aperture bodies also have interior zone where fluid from printing body is filled in it. The surface contact of fluid interface between printing and aperture bodies formed fluid connection between the bodies in simulation.

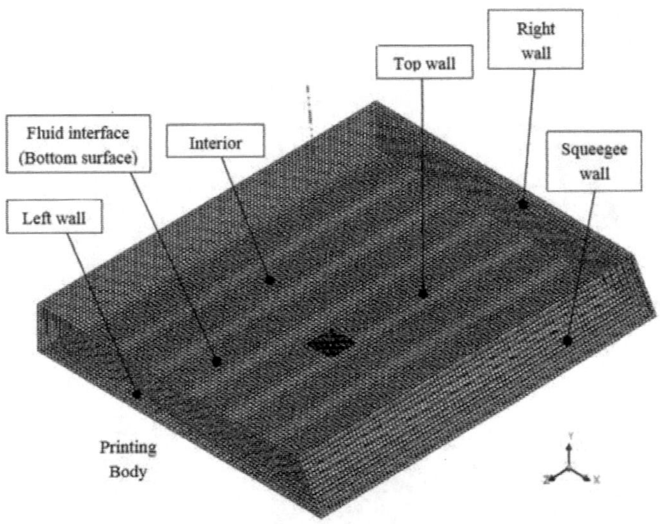

Fig. 5. FLUENT structured meshed model and boundary conditions.

4 Results and Discussions

From Fig. 6, it is shown that the experimental solder deposits coverage area at all comparison locations are greater than simulation at 100% filling stage; squeegee print speed 35mm/s and angle 60°. This indicates the area of paste deposition on the copper plate is bigger than the size of the apertures. In addition, solder deposits coverage area in simulation are similar at all locations which are fully covered the apertures. The highest percentage difference of coverage area is at location 3 by 7.62% while the lowest is at location 1, 2.65% as shown in Fig. 16. This is due to the experimental procedure that has uncontrolled parameters such as printing speed due to manual printing process compared to the simulation. However, the percentage differences for all locations are within 10%. Thus, simulation results generated by FLUENT in the prediction of print quality in stencil printing it is fairly acceptable.

IOP Conf. Series: Materials Science and Engineering **957** (2020) 012065 doi:10.1088/1757-899X/957/1/012065

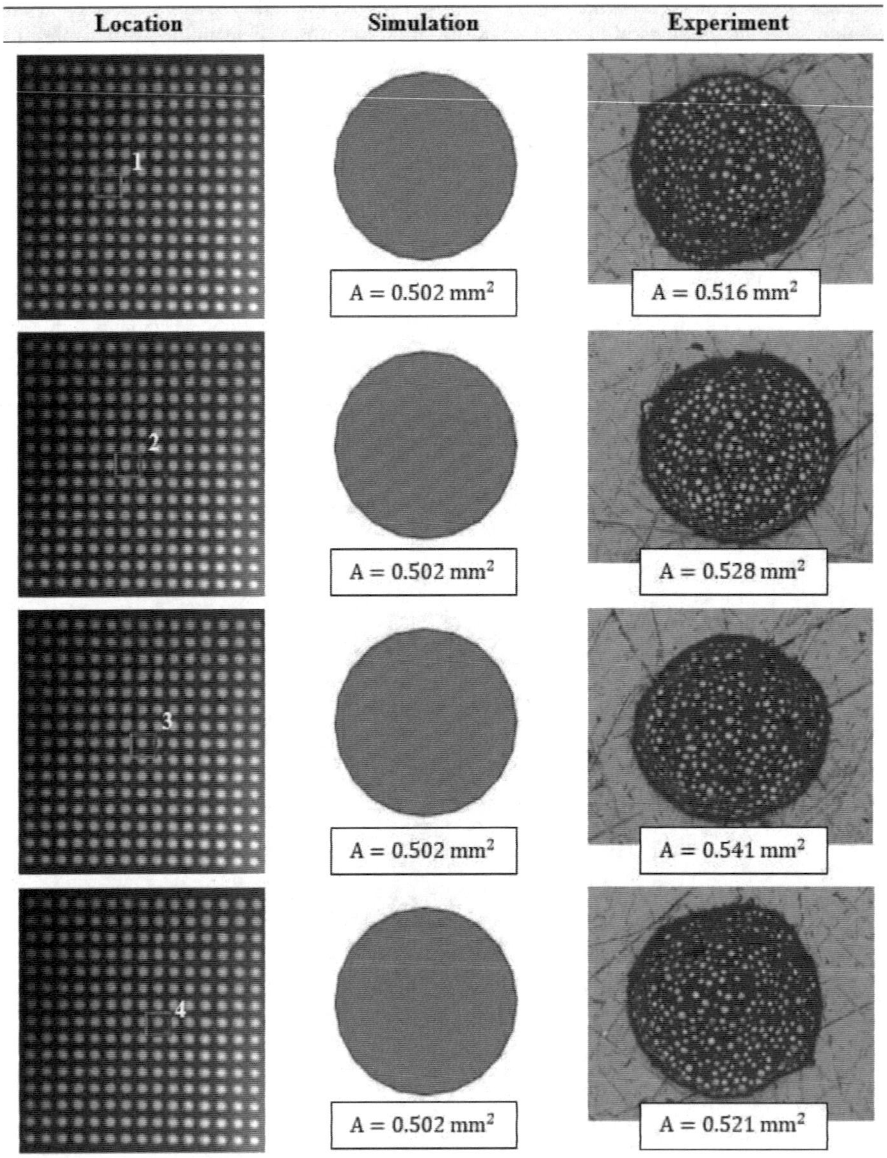

Fig. 6. Observation of solder deposits coverage area, A at different locations.

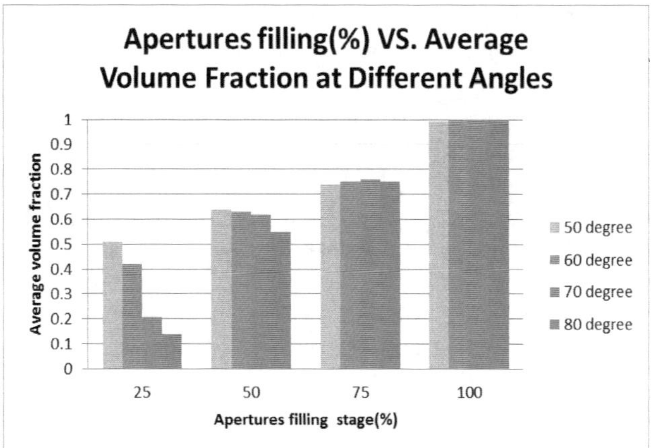

Fig. 7. Average volume fraction (volume-weighted average based) over apertures filling stages at different squeegee angles, θ.

The filling pattern of the solder paste into the aperture is observed by stages with respect to the different angle of squeegee at constant squeegee print speed, 35 mm/s. The average volume fraction(volume-weighted average based) of solder paste inside the aperture through filling stages is shown in Fig. 7. At 25 % filling stage of apertures, the filling pattern between the squeegee angles shows significant differences. 50° squeegee angle has the highest volume fraction while 80° squeegee angle has the lowest average volume fraction. The trend of the filling continues to 50% filling stage but started to have approximately the same average volume fraction at 75%. All squeegee angles achieved average volume fraction at 1 which mean all apertures were fully filled with solder paste at 100% filling stage except for 50° squeegee angle that has a slight drop of average volume fraction due to presence of unfilling aperture holes.

5 Conclusion

The 3-Dimensional stencil printing modeling by using Computational Fluid Dynamics (CFD) approach shows good prediction of stencil printing performance. The approach gives real time observation of printing quality which is difficult to obtain by experimental work during printing process regarding to squeegee print speed and angle. The filling process of solder paste into the apertures shows significant difference filling pattern across the apertures at all filling stages for whole squeegee angles at constant speed 35 mm/s. However, all squeegee angles were successfully filled all the apertures with solder paste except 50° squeegee angle. Thus, operating stencil printing process by using squeegee angle 60° to 80° is possible to achieve good solder paste print quality.

The authors would like to extend their appreciation to Universiti Sains Malaysia (Grant No: 1001/ PMEKANIK/ 8014072) for providing financial and technical support for this research.

References

1. T. N. Tsai, "Modeling and optimization of stencil printing operations: A comparison study," Comput. Ind. Eng., **54**, 3 (2008).

2. D. Manessis, R. Patzelt, A. Ostmann, R. Aschenbrenner, and H. Reichl, "Technical challenges of stencil printing technology for ultra fine pitch flip chip bumping," Microelectron. Reliab, **44**, 5 (2004).
3. J. Pan and G. L. Tonkay, "A study of the aperture filling process in solder paste stencil printing," will appear in the Proceedings of ASME 1999 International Mechanical Engineering Congress and Exposition. (1999).
4. R. Durairaj, S. Ramesh, S. Mallik, A. Seman, and N. Ekere, "Rheological characterisation and printing performance of Sn/Ag/Cu solder pastes," Mater. Des., **30**, 9 (2009).
5. O. Krammer, L. M. Molnár, L. Jakab, and A. Szabó, "Modelling the effect of uneven PWB surface on stencil bending during stencil printing process," Microelectron. Reliab., **52**, 1(2012).
6. M. Whitmore, J. Schake and C. Ashmore, "The impact of stencil aperture design for next generation ultra-fine pitch printing," 2012 35th IEEE/CPMT International Electronics Manufacturing Technology Conference (IEMT), Ipoh, (2012).
7. J. Pan, G. L. Tonkay, R. H. Storer, R. M. Sallade, and D. J. Leandri, "Critical variables of solder paste stencil printing for micro-BGA and fine-pitch QFP," IEEE Trans. Electron. Packag. Manuf., **27**, 2 (2004).
8. T. A. Nguty, "The rheological properties of solder and solar pastes and the effect on stencil printing,"(2000).
9. B. An, Z. Xu, H. Chen, F. Wu, and Y. Wu, "Solder Paste Printability Tester and Method," 2006 7th International Conference on Electronic Packaging Technology, Shanghai, (2006)
10. B. An and Y. Wu, "Evaluating the printability of solder paste from paste roll characteristics," 2010 11th International Conference on Electronic Packaging Technology & High Density Packaging, Xi'an, (2010)
11. R. Mohanty, F. Ma, R. L. Iyer, and D. Santos, "Effect of Board Clamping System on Solder Paste Print Quality," IPC APEX EXPO Proceedings (2012)
12. G. P. Glinski, C. Bailey and K. Pericleous, "Simulation of the stencil printing process [solder pastes]," International Symposium on Electronic Materials and Packaging (EMAP2000) (Cat. No.00EX458), Hong Kong, (2000)
13. T. Yang, T. N. Tsai, and J. Yeh, "A neural network-based prediction model for fine pitch stencil-printing quality in surface mount assembly," Eng. Appl. Artif. Intell., **18**, 3,(2005).
14. O. Krammer, "Finite volume modeling of stencil printing process," 2014 IEEE 20th International Symposium for Design and Technology in Electronic Packaging (SIITME), Bucharest, (2014)
15. V. Thakur, Dr. S Mallik, and V. Vuppala, "CFD Simulation of Solder Paste Flow and Deformation Behaviours during Stencil Printing Process," International Journal of Recent advances in Mechanical Eng.(IJMECH), **4**, 1, (2015).

Development of Integrated Solar Water Heater Using Phase Change Material

Yamuna Munusamy[1,*] *and Ong* Kok Seng[1] ,

[1]Faculty of Engineering and Green Technology, Jalan Universiti, Bandar Barat, 31900 Kampar, Perak.

Abstract. Main barriers for installation of conventional solar water heating system (SWHS) are requirement of large installation area and heavy weight of the heat storage tank consisting water. A storage tank with minimum capacity of 230 liters are needed for household usage and this tank when empty can weigh up to 100 kg while a filled tank could weigh up to 400 kg. In reality, installation of this conventional system require more space and heavy structural support on the roof, which is not available in most houses. As a solution, a form stable light harvesting phase change material (FSLHPCM) integrated compact system is proposed. This new SWHS consists of glass – topped collector cum storage tank filled with FSLHPCM. This new FSLHPCM is form stable, does not leak and exhibit stable latent heat when tested to 1000 thermal cycles. It is capable to not only store heat but can improve the collector efficiency by harvesting the visible light and converting it to heat. The additional light to heat conversion efficiency was 22%. In the new system a maximum 75 liters' glass-topped collector tank is sufficient to substitute a 230-liter tank. This will reduce installation, renovation and maintenance costs for the SWHS.

1 Introduction

According to statistics residential energy usage for hot water supply accounts for 26% of total residential energy usage in Australia [1], 18.3 % in USA [2] and 28.1% in Japan [3]. In most countries electrical water heaters and stove heating remain as main sources of hot water supply. Both of these system relies heavily in burning of fossil fuels which leads to CO^2 emissions. Application of solar water heater system (SWHS) can reduce the reliance on fusil fuel based water heating and results in lower emission of CO_2 into the environment. The current commercially available SWHS is proven to be able to reduce residential energy consumption for water heating between 50-85% [4]. Despite all these economic and environmental benefits, in reality the installation of SWHS even in well developed nations are far below expectation. In 2104 only 13% of households in Australia installed SWHS [5], 22% in Turkey as for 2016 [9] and 4% in Taiwan as for 2011 [10]. This is due to long payback period. A payback period of 7 years in Jordan [13], 9.2 years in Hong Kong for centralized SWHS in high rise building [6], 6 years for galvanized steel SWHS in Pakistan

* Corresponding author: yamunam@utar.edu.my

[7] and 8 years in Brazil after 5.5 % discount in installation cost by government had been reported [8]. The long payback period are mainly contributed by the initial capital cost of SWHS installation related to the building architecture required to support this system. To overcome this problem over the past two decades latent heat storage (LHS) using phase change material (PCM) had been extensively studied. LHS is used in SWHS to reduce the size of storage tank and enhance the performance and reliability of thermal storage system by extending its operational time [9]. A shorter payback period, less initial setup cost and more efficiency is expected from application of PCM.

2 Methodology

2.1 Materials

PMMA (molecular weight = 120, 000 kg/mol) was purchased from Sigma-Aldrich (M) Sdn. Bhd. (Subang Jaya, Malaysia). MA (purity ≥ 98%) and chloroform (stabilized with 0.6-1.0% ethanol) were supplied by Evergreen Engineering & Resources Sdn. Bhd. (Semenyih, Malaysia). NBR (total solid content of 44.70%) was obtained from Synthomer Sdn. Bhd. (Kluang, Malaysia). PA (total solid content of 36.35%) was synthesized in Universiti Tunku Abdul Rahman (UTAR) (Kampar, Malaysia). RGO was synthesized by oxidizing graphite nanofiber via conventional Hummer's method [10] followed by reduction using hydrazine hydrate [11].

2.2 Preparation of RGO-PARB/MA tablet

RGO was mixed with Triton X-100 (dispersing agent) before blending with the PA coating solution. The solution was stirred for 1 h to obtain a homogenous blend. MA powder was compressed into a tablet using hydraulic press model E-Z Press™ supplied by International Crystal Laboratories, USA at the pressure of 4000 psi. The pressed MA tablet was then coated with NBR followed by a PA coating solution blended with RGO. For each layer of coating, the MA tablet was immersed in the coating solution for five seconds. After that, MA tablet was taken out and dried at room temperature for 4 days. The steps were then repeated with the same tablet being flipped upside down. RGO-PARB/MA tablets were prepared in 0, 0.5, 1.0, 1.5 and 2.0 wt% loading of RGO.

2.3 Testings

For outdoor solar energy conversion and storage efficiency measurement, RGO-PARB/MA tablets with different RGO loading were placed on a polystyrene sheet and covered with an acrylic transparent box as illustrated in Fig. 1. A pyranometer, model CMP3 supplied by Kipp & Zonen, Netherlands was used to measure the solar irradiance. Temperatures of the composite samples were recorded using thermocouples connected to the data logger, model GL820 supplied by Graphtec, Tokyo, Japan. The obtained solar energy conversion efficiency was used to calculate the size of the intergrated system.

Fig.1. Outdoor Experiment Setup for Solar Energy Conversion Efficiency Measurement

Thermal cyclic test was carried out using a thermal cyclic system designed by the University of Malaya to determine the thermal reliability of FSPCMs over a period. In each thermal cycle, the tablet samples were put into a sample holder and heated from 30 to 80 °C and then the samples were let to cool to room temperature. This procedure was repeated for 1000 cycles. Each cycle represents a day of heat gain and heat release.

3 Results and Discussions

Compared with pure PANBR/MA pellet (0.0 wt% RGO), the temperature of all RGO-PANBR/MA pellet rapidly increased upon sunlight irradiation. This is because RGO in PANBR coating absorb sunlight and transform it into heat through non-radiation thermal decay. Similar finding was reported by Tang et al. who used dye as light absorptive material on PCM (Tang et al., 2016). Fig. 2 shows that 1.5 wt% RGO-PANBR/MA pellet had higher temperature (61.1 °C) than pure PANBR/MA pellet (55.2 °C) after exposing to sunlight for 8000 seconds. An increment of 10.69% in temperature was recorded.

The total heat energy stored in RGO-PANBR/MA pellet with different RGO loading is shown in Fig.3. The total stored heat energy is calculated over the period of 3 sunny days. The RGO-PANBR/MA pellet with 1.5 wt% RGO loading stored the highest amount of heat energy among the pellets: 49 wt% higher than the PANBR/MA pellet without RGO. The total stored heat energy was lower when the RGO loading is further increased to 2.0 wt%. This is because RGO is agglomerate at 2.0 wt% loading and thus reduce the effective surface area for sunlight absorption.

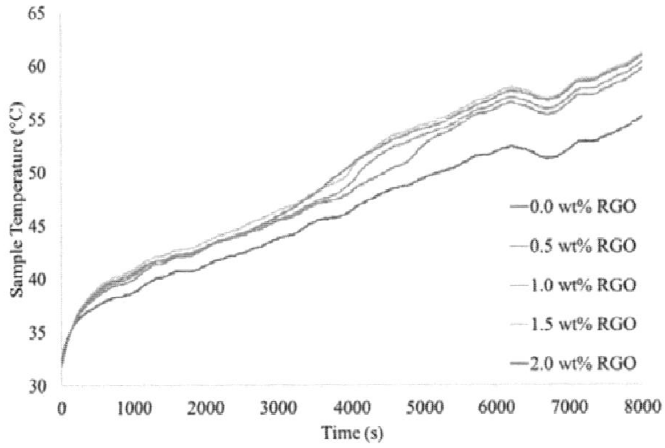

Fig.2. Temperature of RGO-PANBR/MA Pellets with Various Loading after Exposing to Sunlight for 8000 seconds.

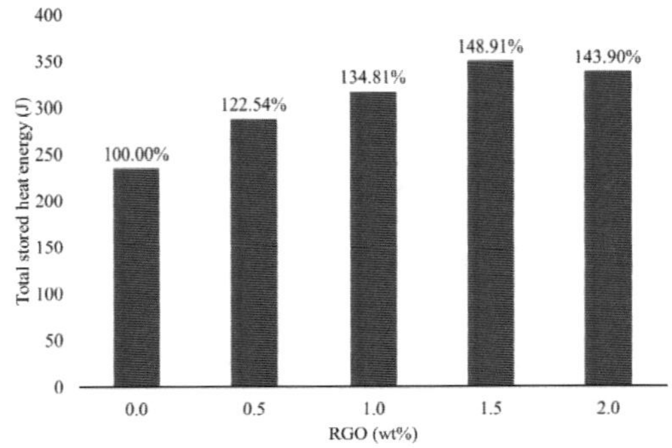

Fig.3. Total Stored Heat Energy in RGO-PANBR/MA Pellets with Various Loading.

Solar energy conversion efficiency for RGO-PANBR/MA pellets with different loading is shown in Fig.4. The efficiency is dependent on the solar energy absorbed by the top surface as well as heat loss to the environment which is mainly affected by the difference between the pellet temperature and the temperature inside the PMMA container. All pellets absorb the same amount of solar energy on their top surfaces. However, the heat loss is the same at the beginning of the experiment on each day, but the hotter pellets has significantly higher heat losses than the colder pellets after a short period of time. Therefore, to make a fair comparison of the solar energy conversion efficiency of all the pellets, only the results from the first 20 minutes are used in the calculation when the temperature difference is less than 5°C. The result shows that the 1.5 wt% RGO-PANBR/MA pellet had the highest solar energy conversion efficiency which was 21%. This is because 1.5 wt% RGO-PANBR/MA has the highest loading of RGO before RGO start to aggregate.

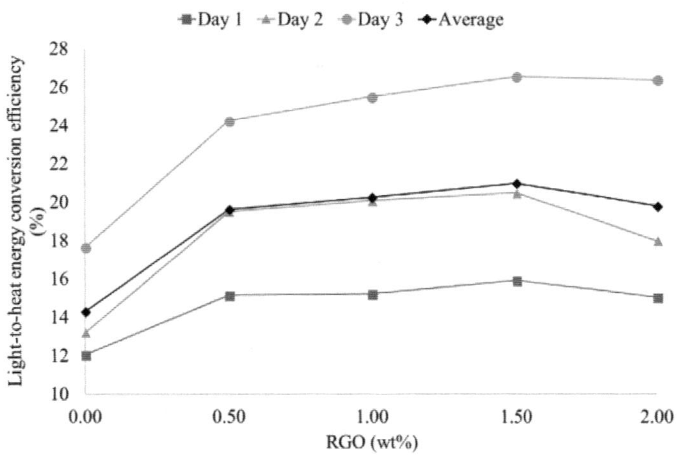

Fig.4: Solar Energy Conversion Efficiency of RGO-PANBR/MA Pellets with Various Loading

Based on the energy conversion efficiency the tank size was calculated as following:

Total solar irradiation is;

$$4.5\ \frac{kwh}{m^2} = 16200\ \frac{kJ}{m^2} \tag{1}$$

Sunlight to heat average conversion efficiency obtained = 22% from Fig.4
Daily hot water consumption for 4 person in a household is estimated as

$$V_w = 4 \times 5 \left(\frac{l}{min}\right) \times 10\ (min)$$
$$= 200\ l \tag{2}$$

$$C_p \text{ of water} = 4.2\ \frac{kJ}{kg°C} \tag{3}$$

Required hot water temperature = 40°C
Inlet/cold water temperature = 22°C
Latent heat of FSLHPCM = 125 $\frac{kJ}{kg}$
Energy required to heat the water

$$Q_w = 200(kg) \times 4.2 \left(\frac{kJ}{kg}\right) \times 18°C$$
$$= 15120\ kJ \tag{4}$$

Surface area of the proposed prototype based on the conversion efficiency:

$$A_c = \frac{15120\ kJ}{16200\ kJ \times 0.22} \tag{5}$$

$$= 4.2\ m^2$$

Thus it will be fixed to 4 m^2

From the energy point of view, the required amount of FSLHPCM will be:

$$Mass\ of\ FSVLHPCM = \frac{0.22\ x\ 16200\ kJm^{-2}}{125\ kJkg^{-1}} \times 4\ m^2 \tag{6}$$

$$= 114\ kg$$

From the point of view of design of the system as shown in Fig. 5.

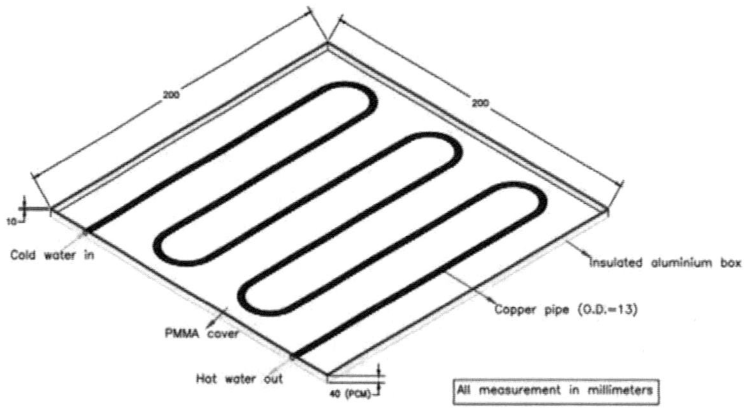

Fig.5: Design of the integrated system

The tank will have array of copper pipes (10 coils) = 13 mm OD x 10 m long
The volume of the pipes will be 0.0013 m3
Height of the tank is fixed to 50 mm (PCM filled up to 40 mm and 10 mm allowance)
l x w x h of the tank will then be: 2 m x 2 m x 0.04 m

The volume of FSLHPCM will then be: (volume of the filled tank) –(volume of pipe) (7)
= (0.16 m^3 – 0.0013m^3)
= 0.1587 m^3

Density of FSLHPCM: 860 $\frac{kg}{m^3}$
Mass of FSLHPCM required;

$$0.1587\ m^3 \times 860\ \frac{kg}{m^3} \tag{8}$$

= 136 kg

Thus the mass based on the design is sufficient to store the energy required. The minimum mass of FSLHPCM required is only 114 kg.

Total weight of the conventional system for a household of four:

Sensible heat storage water tank (270 liter tank) = 400 kg
Collector panel: 50 kg
Total weight: 450 kg

Total weight for the integrated system:
Weight of insulated aluminum tank: 20 kg
Weight of FSLHPCM: 136 kg
Weight of PMMA panel: 5 kg

Total weight: 161 kg (latent heat storage thus there are no water storage and no collector panel)

Weight reduction percentage:

$$= \frac{450 - 161}{450} \times 100 \tag{9}$$

= **64%**

4 Conclusion

FSLHPCM filled system was found to be effective in reducing the weight of SWHS by 64%. The reduction in term of weight of the system is caused by high latent heat storage capacity of the FSLHPCM compared to water and additional light to heat conversion efficiency by 22%. The capability to reduce the weight of the SWHS will encourage the installation of SWHS in many households.

References

1. T. Umee, E. Walker, P.A. Bahri, G. Baverstock, S. Rezvani, W. Saman. Sustain. Energy Techn., **30** (2018).
2. E. Kabir, P. Kumar, A.A. Adelodun, K.H.Kim. Renew. Sust. Energ. Rev., **82** (2018).
3. H. Farabi-Asl, A. Chapman, K. Itaoka, F. Taghizadeh-Hesary. J. Sol. Energy. **158** (2019).
4. M.J. Booysen, J.A.A. Engelbrecht, M.J. Ritchie, M. Apperley, A.H. Cloete. Energy. Sustain. Dev. **51** (2019).
5. Clean Energy Council, Clean Energy Australia Report 2016-2017. Clean Energy Council: Victoria Australia.
6. K.C. Chang, W.M Lin, K. M. Chung. Sustain. City. Soc. **41** (2019).

7. O.L. Jing, M.J. Bashir, J.J. Kao. Renew. Sust. Energ. Rev. **48** (2015).
8. Energy Rating. Types of Water Heaters. 2014[cited 2019 May 19]. Available from : http://www.energyrating.gov.au/products/water-heaters.
9. H. Benil. Renew. Sust. Energ. Rev.**54** (2016).
10. W. S. Hummers, R.E. Offeman. J. AM. Chem. Soc. **80** (1958).
11. S. N. Tripathi, P. Saini, D. Gupta, V. Choudhary. J. Mater. Sci. **48** (2013).

Influence of electron beam irradiation on crosslink behaviour of reclaimed tire rubber/EVA blend in the presence of radiation sensitizers

Suganti Ramarad[1,*], *Chantara Thevy* Ratnam[2], *Mohammad* Khalid[3], and *Luqman* Chuah[4]

[1]School of Engineering and Physical Sciences, Heriot-Watt University Malaysia, 62200, Putrajaya, Wilayah Persekutuan Putrajaya, Malaysia
[2]Radiation Processing Technology Division, Malaysia Nuclear Agency (Nuclear Malaysia), Bangi, 43000 Kajang, Malaysia
[3]Graphene & Advanced 2D Materials Research group (GAMRG), Research Centre for Nano-materials and Energy Technology (RCNMET), School of Science and Technology, Sunway University, Bandar Sunway, 47500 Subang Jaya, Selangor, Malaysia
[4]Department of Chemical & Environmental Engineering, Faculty of Engineering, Universiti Putra Malaysia (UPM), 43400 Serdang, Selangor, Malaysia

Abstract. Non-degradable waste tire generation around the world is growing at an alarming rate. Diversifying the recycling route of these waste tires is essential to solve the problem. To address the issue, this study utilized radiation sensitizers and electron beam irradiation to enhance the poor properties of reclaimed waste tire rubber/poly(ethylene vinyl acetate) (RTR/EVA) blend. The RTR, EVA and radiation sensitizers were mixed in the internal mixer followed by electron beam (EB) irradiation with doses ranging from 50 to 200 kGy. Radiation sensitizers loading was fixed at 4 wt%. The degradation and stability in RTR were studied in term of crosslink behaviour upon irradiation. Electron beam irradiation revealed the presence of radical stabilizing and scavenging additives within RTR which retards the crosslinking process in RTR and the blends. RTR, EVA and the RTR/EVA blend suffered from further oxidative degradation from irradiation in air. Radiation sensitizers, trimethylol propane triacrylate (TMPTA) and tripropylene glycol diacrylate (TPGDA), were used to accelerate the irradiation induced crosslinking in RTR and the RTR/EVA blend. Presence of radiation sensitizers leads to enhanced crosslink formation within RTR and the RTR/EVA blend. These findings validated the feasibility of using ionizing radiation in the presence of radiation sensitizers for recycling inferior waste polymers into higher quality material.

* Corresponding author: r.suganti@hw.ac.uk, suga4u2c@gmail.com

1 Introduction

Ionizing radiation is an upcoming powerful technology in addressing polymeric waste issues. The extent of irradiation can be controlled to introduce chain crosslinking or scission to tweak a recycle material's properties as desired. Previous publications [1,2] have addressed the lack of efficiency in radiation induced crosslink formation in reclaimed tire rubber and ethylene vinyl acetate (RTR/EVA) blends due to readily present radical stabilizing and scavenging additives in RTR. This study is dedicated to the use of different radiation sensitizers to accelerate radiation induced crosslinks in RTR/EVA blends. The three radiation sensitizers used are N,N-1,3 Phenylene Bismaleimide (HVA2), trimethylol propane triacrylate (TMPTA) and tripropylene glycol diacrylate (TPGDA) with two, three and two functional sites, respectively. The chemical structure of all three radiation sensitizers are shown in Figure 1. The loading of the radiation sensitizers has been set to 4 phr based on previous studies [3-5]. Presence of radiation sensitizers would allow for faster reach to optimal irradiation dose and reduce the undesired effect of oxidative degradation. The influence of electron beam irradiation and radiation sensitizers on crosslink network and the gel content properties are reported in this study.

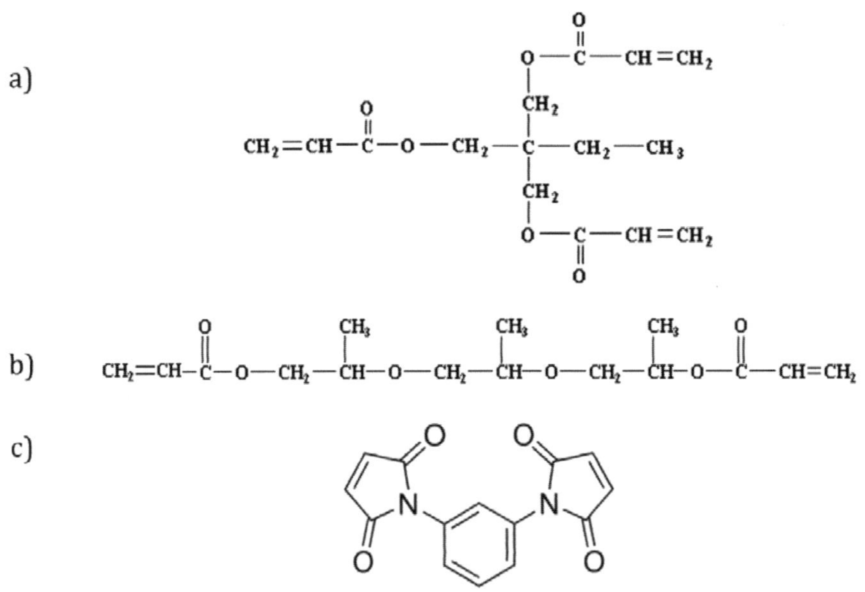

Fig. 1. Chemical structure of a) TMPTA, b) TPGDA and c) HVA2

2 Materials and Methods

2.1 Materials

Poly(ethylene-co-vinyl acetate) (Grade EVA N8045), EVA, having 18% vinyl acetate content with melt flow index, MFI, value of 2.3 g/10 min and a density of 0.947 g/cm^3 was purchased from the TPI POLENA Public Limited Company, Thailand. Reclaimed tire rubber (RECLAIM Rubberplas C), RTR, from waste, heavy duty tires used in this study was supplied by Rubplast Sdn. Bhd., Malaysia. General properties of the RTR are 48% rubber hydrocarbon, 5% ash content, 15% acetone extract, 25% carbon black fillers and density of 1.3 g/cm^3. (3-Aminopropyl)triethoxy silane (APS) was used as compatibilizers in RTR/EVA blends. Multifunctional acrylates (MFA); trimethylol propane triacrylate (TMPTA) and tripropylene glycol diacrylate (TPGDA) were used as irradiation sensitized crosslinking agents. Whereas N,N-1,3 Phenylene Bismaleimide (HVA2) was used as a conventional irradiation sensitized crosslinking agent.

2.2 Melt compounding

RTR and EVA were melt blended in an internal mixer (Brabender Plasticoder PL2000-6 equipped with co-rotating blades and a mixing head with a volumetric capacity of 69 cm^3). The rotor speed was set at 50 rpm while blending temperature was set at 120 °C. The processing parameters were determined from a preliminary work.

Melt compounding was done according to the designation of the prepared blends as shown in Table 1. APS was added based on rubber weight percentage while radiation sensitizers were added based on total matrix weight percentage. EVA was fed into the internal mixer chamber and allowed to melt for two minutes, followed by the addition of RTR. Both EVA and RTR were allowed to mix for 8 minutes before collecting the blends from the internal mixer. Compatibilizer (when used) was added simultaneously with RTR at the second minute. Multifunctional acrylates and bismaleimide crosslinking agents (when used) were added at third and seventh minute respectively. Bismaleimide crosslinking agent was added later as it induces crosslinking process in polymers beyond 70 °C; unlike multifunctional acrylates which would only crosslink with exposure to irradiation energy. This ensures the compounded materials can still be processed (or flowing) during subsequent compression moulding step. Total mixing time was kept constant to 10 minutes for EVA, RTR and all the blends to ensure similar thermal history. The collected materials were immediately cut into smaller pieces and kept in sealed plastic bags for compression moulding process.

2.3 Compression moulding and electron beam irradiation

Materials obtained from internal mixer were compression moulded to obtain test specimens. The compounded materials were placed into a steel frame mould covered by aluminium plates at both sides. The materials were pressed at 130 °C into sheets of 1 mm thickness. The moulding cycles involve 3 minutes of preheating without pressure, 20 seconds of venting and 3 minutes of compression under 14.7 MPa pressure using hot and cold pressing machine (LP-S-50 Scientific Hot and Cold Press). Cooling was done immediately between two platen of cold press at 20 °C for 2 minutes.

The moulded sheets were irradiated using 3 MeV electron beam accelerator (model NHV-EPS-3000) at dose ranging between 0 – 200 kGy. The acceleration energy, beam current and dose rate were 2 MeV, 2 mA, and 50 kGy per pass, respectively

2.4 Gel content analysis

The samples gel content was determined according to ASTM D2765. Samples were placed in a stainless steel wire mesh of 120 mesh size and extracted in boiling Toluene using Soxhlet apparatus for 24 hours. Samples were then collected and dried in an oven at 70 °C until constant weight is obtained. Gel content was calculated as per Equation 1 below. Where W_0 and W_1 are the dried weight of sample before extraction and after extraction, respectively

$$Gel\ content\ (\%) = (W_1 / W_0) \times (100) \quad (1)$$

Table 1. Designation of prepared samples

Designation	RTR (wt%)	EVA (wt%)	APS (wt%)	Radiation sensitizers (wt%)
RTR	100	0	0	0
RTR/TPMPA	100	0	0	4
RTR/TPGDA	100	0	0	4
RTR/HVA2	100	0	0	4
EVA	0	100	0	0
EVA/TMPTA	0	100	0	4
EVA/TPGDA	0	100	0	4
EVA/HVA2	0	100	0	4
50RTR	50	50	0	0
50RTR/5APS	50	50	5	0
50RTR/5APS /TPMPA	50	50	5	4
50RTR/5APS /TPGDA	50	50	5	4
50RTR/5APS /HVA2	50	50	5	4

3 Results and Discussion

Figure 2 shows the influence of TMPTA, TPGDA and HVA2 on radiation induced gel formation in RTR, EVA and 50RTR/5APS blend as a function of irradiation dose. Neat RTR shows 68% of the gel content prior to irradiation (0kGy), affirming the presence of readily existing crosslinks within its matrix. The gel content of neat RTR increased only marginally with the increase in irradiation dose. Similar findings by Ratnam *et. al.* [6] explained the stabilization of the rubber and the radical scavenging effects by the additives causing a marginal increment in the gel content upon irradiation. Presence of radiation sensitizers helped to enable more crosslink formation, yielding higher gel content values. RTR/HVA2 shows about 20% increase in gel content compared to neat RTR at equivalent irradiation dose, netting a maximum value of 91% gel content at 200 kGy irradiation dose. Addition of HVA2 proves to be efficiently introducing crosslinking in RTR sample. Similar observation has also been reported in EVA/NR study [7].

However, dynamically vulcanized RTR/HVA2 samples were found to undergo degradation during blending process. Thus, these made it difficult for RTR/HVA2 formulation to be compression moulded into testing specimens. Hence, influence of HVA2 on RTR properties could not be verified. TMPTA and TPGDA containing RTR sample also displayed higher gel content than neat RTR. This indicates that presence of TMPTA and TPGDA accelerates irradiation induced crosslinking in RTR [8]. RTR/TMPTA records higher gel content values as compared to RTR/TPGDA. This is due to the higher number of functionalities in TMPTA (trifunctional) contrast to TPGDA (difunctional) [9-10]. TMPTA is capable of forming more crosslink bridges due to higher functionality resulting in higher gel content values.

Figure 2b shows the influence of TMPTA, TPGDA and HVA2 on gel formation of EVA matrix. TMPTA was found to be the most efficient radiation sensitizer in accelerating formation of the crosslinks on EVA matrix compared to TPGDA, HVA2 and neat EVA. TPGDA and HVA2 only displayed slight increase in gel content values compared to neat EVA (inset of Figure 2b). These finding suggest the efficiency of crosslink formation in EVA matrix are in the order of TMPTA>TPGDA=HVA2. This is again due to difference in the functionality of radiation sensitizers as mentioned earlier. Unlike RTR/HVA2, no crosslink formation was observed in EVA/HVA2 composition before irradiation. HVA2 with multi-radical accepting capabilities interacts with radicals to stabilize the overall reaction. This suggests the presence or formations of radicals are more likely happening in RTR aided by the heat energy available during blending and compression moulding process. Whereas EVA is accounted as more stable at the processing temperature, leading to lower possibility for crosslink formation in EVA/HVA2 composition before irradiation. Similar observation has also been reported by [7]. Moreover, previous work has shown that HVA2 have higher tendency to interact with rubber component than the thermoplastic matrix used in the study [7,11].

Figure 2c shows the influence of TMPTA, TPGDA and HVA2 on gel formation of 50RTR/5APS blend. Gel content yield of neat 50RTR blend was also shown on the figure for comparison. It is observed that the 50RTR blends require an irradiation dose above 50kGy in order to achieve a significant increase in the gel content. This could be due to the presence of additives in the RTR retarding the crosslinking process in these blends as discussed earlier. At above 100kGy irradiation dose, the gel content of the 50RTR blend increases slowly and exhibits a marginal difference. 50RTR/5APS blend, the APS compatibilized blend, showed an interesting result where the gel content before irradiation (0 kGy) recorded lower values compared to the control, 50RTR blend. Gel content observed before irradiation in a blend was a contribution from the partially devulcanized structure of RTR. An obvious reduction in the gel values of APS compatibilized blends suggests that APS plays major role in further devulcanizing or reclaiming the RTR. This is supported by the fact that amines have been long used as reclaiming agent in rubber materials due to its strong nucleophilic nature [12-14].

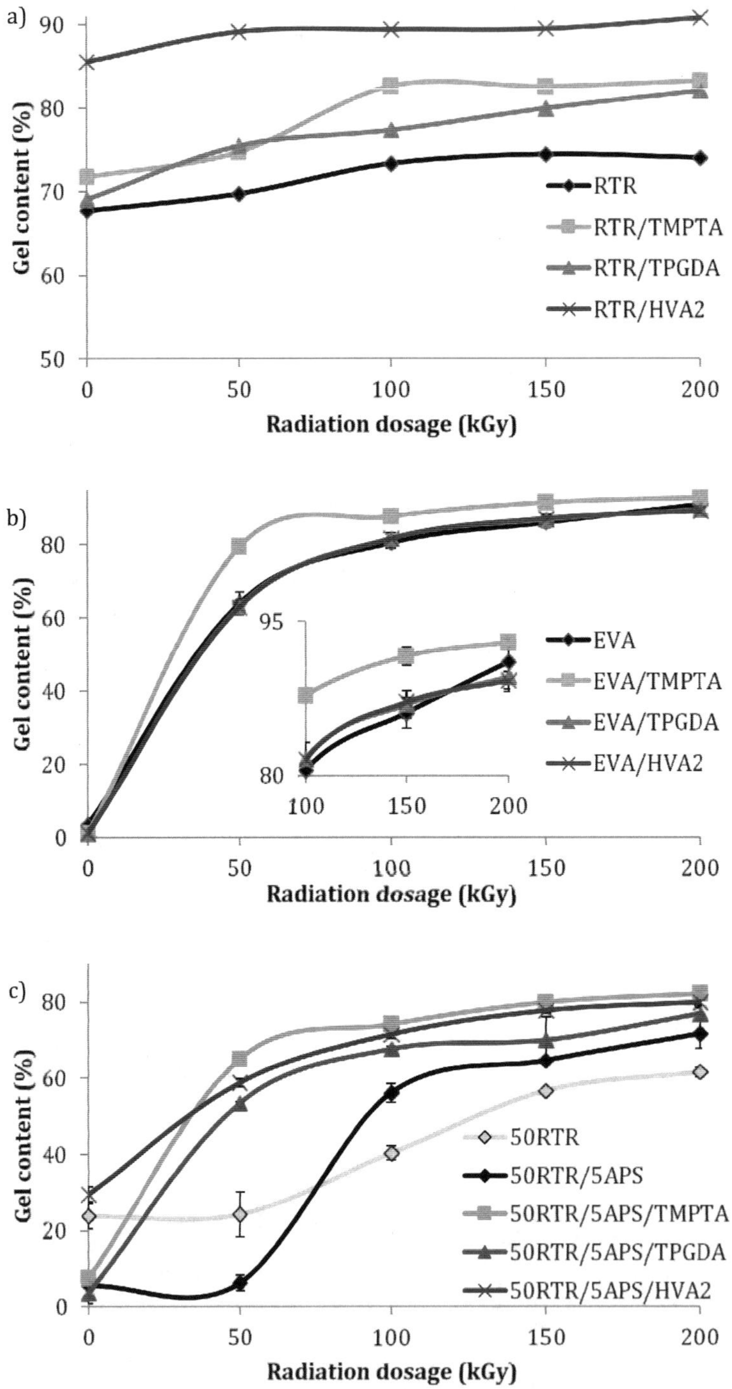

Fig. 2. Gel content values of a) RTR, b) EVA and c) 50RTR/5APS blends as a function of radiation sensitizers and radiation dose

Presence of radiation sensitizers has completely resolved the delay in crosslink formation observed in neat 50RTR and 50RTR/5APS blends. Upon irradiation of the 50RTR/5APS blends, TMPTA was found to yield the highest gel content values, closely followed by HVA2 and TPGDA. Before irradiation (0 kGy), a gel content value of 29.5% was observed in 50RTR/5APS/HVA2 composition. Judging from the gel content analysis of RTR and EVA in the presence of HVA2, the gel value observed in 50RTR/5APS/HVA2 composition could be mostly due to crosslink formation in RTR component of the blend. These observations suggest that the use of radiation sensitizers enhance the crosslink formation in the 50RTR/APS blends.

Charlesby-Pinner equation was used to determine the ratio of chain scission to crosslinking (p_0/q_0) in RTR, EVA and 50RTR/5APS blends in the presence of radiation sensitizers. The p_0/q_0 values have been listed in Table 2. In general, RTR showed the highest p0/q0 ratio, while EVA showed the lowest value and the blends showing an intermediate value. The p_0/q_0 values above 1 indicates the dominance of chain scissioning over crosslinking within the matrix. Presence of TMPTA and TPGDA in RTR decreased, while HVA2 further increased the p_0/q_0 values compared to neat RTR. The p_0/q_0 value of RTR/HVA2 was 1.93 indicating about 2 scissions could be happening per crosslinking in the RTR matrix, which would lead to substantial decrease in the molecular weight of RTR matrix. This further corroborates the reasons for difficulties in moulding RTR/HVA2 samples. Although RTR/TMPTA and RTR/TPGDA recorded a decline, the p_0/q_0 values were still higher than 1, indicating chain scissioning still dominates over crosslinking in the RTR matrix even in the presence of radiation sensitizers.

Table 2. Values p_0/q_0 of RTR, EVA and 50RTR/5APS blends in the presence of radiation sensitizers.

Designation	Neat	TMPTA	TPGDA	HVA2
RTR	1.8724	1.7718	1.7920	1.9255
EVA	0.2473	0.2723	0.2740	0.2805
50RTR/5APS	0.5478	0.6700	0.6960	1.0908

Both EVA and 50RTR/5APS blends displayed a slight increase in p_0/q_0 values in the presence of all three radiation sensitizers. This is to be expected as crosslinking process was found to have already effectively take place in the neat EVA and 50RTR/5APS matrix upon irradiation. Whereas the addition of radiation sensitizers would enhance the crosslinking process to a certain absorbed radiation dosage, upon which, chain scission are deemed to prevail. Presence of radiation sensitizers would allow for optimal crosslinking to be achieved at a lower absorbed dose of irradiation [15-17]. Among all the three radiation sensitizers, HVA2 recorded p_0/q_0 values above 1 for RTR and 50RTR/5APS blends. This indicates the 4phr of HVA2 loading used in this study is more than enough for RTR based system, leading to domination of chain scission. Optimization of HVA2 loading and blending parameters is essential to observe neat crosslink formation in irradiated 50RTR/5APS/HVA2 blends.

4 Conclusion

Formation of crosslink network and enhancement of gel content can be achieved in irradiated RTR and RTR/EVA blend in the presence of radiation sensitizer. TMPTA and TPGDA did perform excellently to increase crosslink network and gel content of RTR/EVA blend. HVA2's role in enhancing crosslink network in RTR/EVA blend could not be confirmed due to excessive loading of HVA2 in the blends.

References

1. S. Ramarad; C.T. Ratnam; M. Khalid; A.C. Luqman. J. Appl. Poly. Sci., **132,** 11 (2015)
2. S. Ramarad; C.T. Ratnam; M. Khalid; A.C. Luqman; S. Hanson. Radiat. Phys. Chem. **130**, 362 (2017)
3. M. Du, B. Guo, D. Jia. J. Poly. Res., **12**, 473 (2005)
4. S.K. Dutta; A. K Bhowmick ; T.K Chaki. Radiat. Phys. Chem., **47**, 913 (1996)
5. Y. Yin ; M. Liu ; X. Zheng ; S.Z. Shen ; P. Deng. J. Appl. Poly. Sci., **127**, 200 (2013)
6. C.T. Ratnam; M. Nasir ; A. Baharin ; K. Zaman. Nucl. Instrum. Methods Phys. Res., B, **171**, 455 (2000)
7. M. Zurina; H. Ismail; C.T. Ratnam. Polym. Test., **27**, 480 (2008)
8. T. Yasin ; S. Khan ; M. Shafiq ; R. Gill. Radiat. Phys. Chem., **106**, 343 (2015)
9. C.T. Ratnam; S. Ramarad; M. Khalid; A.S. Zainal Abidin; L.T. Chuah., 2014. J. Thermoplast. Compos. Mater. 26, 464 (2016)
10. I. Banik ; S.K Dutta ; T.K. Chaki ; A.K. Bhowmick. Polym. J., **40**, 447 (1999)
11. M. Awang ; H. Ismail. Polym. Test., **27**, 321(2008)
12. K. Dijkhuis. PhD Thesis, University of Twente. Enschede, the Netherlands, (2008).
13. K. Dijkhuis; W. Dierkes; P. Sutanto; F. Picchioni; L. Janssen; J. Noordermeer. International Rubber Conference, Maastricht, the Netherlands, (2005)
14. P. Sutanto; F.L. Laksmana; F. Picchioni; L.P.B.M. Janssen. Chem. Eng. Sci., **61**, 6442 (2006)
15. V. Vijayabaskar; S. Bhattacharya; V.K. Tikku; A.K. Bhowmick. 2004. Radiat. Phys. Chem., **71**, 1045 (2004)
16. H.M. Ng; S.T. Bee; Ct. Ratnam; L.T. Sin; Y.Y. Phang; T.T. Tee; A.R. Rahmat. Nucl. Instrum. Methods Phys. Res., B, **319**, 62 (2014)
17. G. Burillo; M. Galicia; M.D.P Carreon; M. Vazquez; E. Adem. 2001. Radiat. Phys. Chem., **60**, 73 (2001)

Mechanical performance and thermal behaviour of Carrageenan Gel with Lavender Essential Oil

Khairul Anuar Mat Amin[1,*], and *Taqiah* Abdul Hadi[1]

Faculty of Science and Marine Environment, Universiti Malaysia Terengganu (UMT), 21030 Kuala Nerus, Terengganu, Malaysia.

Abstract. This study determines the mechanical performance and thermal behaviour of carrageenan (CR) gel with lavender essential oils (LEO). The CR gel was produced via crosslink with calcium ion followed by inclusion of LEO at different concentrations. The compressive strain of CR gel was increased upon addition of LEO could be due to formation of micelles in the gel matrix and therefore restrain the strain and improved the flexibility of the gels. In contrast, the compressive stress of the CR-LEO gels was decreased than CR gels. The water vapour transmission rates (WVTRs) of the CR-LEO2 gel were improved and assumed due to micelles formation and allowed water diffused through the gel's matrix. The thermal behaviour of CR-LEO gels was improved with reducing weight loss (%) of the gel with higher concentration of LEO. CR gel shows weight loss at 91%, and CR-LEO3 gel shows the weight loss at 76%. In conclusion, the inclusion of LEO has shown promising results in tailoring the mechanical performances and thermal activity and can be applied as biomaterials for pharmaceutical applications.

1 Introduction

Carrageenan (CR), an anionic biopolymer produced from marine red algae family has been received great attention to be used in many applications particularly in food industry such as stabilizer and emulsifier, thickening and gelling agents to improve the texture [1]. Not limited to that, it has been used in non-food industry such as printing, cosmetic, textile, and pharmaceutical applications [2, 3]. The gelling strength of CR is depending on the presence of cations such as K^+ and Ca^{2+}, i.e. it will be increased once the number of ions increases until it will reach the optimum level. Calcium ions is a double charged cation and able to form intermolecular ionic bridges between the sulphate groups of CR. Lavender essential oil (LEO) is a natural oil extracted from *Lavandula augustifolia* that contain volatile aroma compound which is concentrated hydrophobic [4]. It is traditionally used for their therapeutic properties and biological activities. For biological activities, it has been reported that LEO exhibit the antioxidant, anticonvulsant, antifungal, antimicrobial activities and a promising candidate to be further explored for pharmaceutical applications [5]. Phytochemical components of LEO which could responsible for all biological activities included linalool, 1,8-cineole, camphor, endo-borneal, α-pinene, camphene and other compounds [4]. Therefore, in this study, the CR gel incorporated LEO will be prepared and the mechanical performance, water vapour transmission rates and thermal behaviour will be examined.

* Corresponding author: kerol@umt.edu.my

2 Materials and Method

2.1 Materials

Carrageenan (product number-C1013), glycerine (product number-G2289), anhydrous calcium chloride, CaCl2 (product number-C5670) and Tween 80 (product number-P1754) were purchased from Sigma Aldrich, St Louis, MO, USA. Lavender essential oil (LEO) was obtained from Living-well, Victoria, Australia. All materials were used as received without any purification.

2.2 Preparation of Carrageenan gel incorporated with LEO

Carrageenan (CR) gel incorporated lavender essential oil (LEO) was prepared via crosslinking with calcium ion. 1% (w/w) of CR was dissolved in deionized water and continuously stirred at 75 °C for 30 minutes followed by addition of 50% (w/w) of glycerine. 30mM of calcium chloride solution was added drop wise in the mixture. The mixture was continuously stirred for 1 hour. Next, the LEO at 1% (w/w), 2% (w/w) and 3% (w/w) were added into the CR gel and further heated at the same temperature for 1 hour. The solubility of the LEO in CR solution was eased with addition of tween-80 (surfactant) at ratio of (1:3) between tween-80 and LEO, respectively. The mixtures were then poured into petri dishes and left for 24 hours at room temperature. The CR gels were named according to the concentration of LEO, for an example, CR gel with 1% (w/w) is named as CR-LEO1.

2.3 Characterization of hydrogel

2.3.1 Mechanical performance.

Compression measurement was obtained using Instron Universal Testing machine (model 3366) with 10kN loads and cross-speed was set at 10mm/min to characterize the mechanical properties of the gels. The samples were cut to 2cm x 2cm and the slopes of the linear part of the stress-strain curve were used to calculate the compressive stress and strain-at-break. The compressive strain at maximum compressive extension and Young's modulus were obtained. The test was repeated triplicates for CR-LEO1, CR-LEO2 and CR-LEO3.

2.3.2 Water Vapour Transmission Rate (WVTR)

The test was carried out by cutting the sample to 3cm x 3cm and placed on top of the mouth of a glass vial with diameter 1.6 cm and contained distilled water. The samples were left in the desiccator containing silica gel (RH~50%) for 3 hours. Then, the samples were weight and the value of WVTR (g m^{-2} h^{-1}) calculated using the equation 2.1:

$$WVTR = (W_i - W_f)/ A \times t \text{ g m}^{-2} \text{ hour}^{-1} \quad (1)$$

where W_i and W_f are initial and final weight of glass vial, respectively and A is the test area of the sample in m2 and t is the time duration.

2.3.3 Thermogravimetric Analysis (TGA)

Thermogravimetric analysis was carried out by using Mettler Toledo TGA/DSC 1. The samples were placed in ceramic crucible to be heated from 25°C to 200°C under nitrogen atmosphere with a nominal gas flow of 30mL/min at heating rate of 10°C/min.

2.3.4 Differential Scanning Calorimetry (DSC)

DSC was carried out by using DSC Q2000 calorimeter. The samples (15mg ± 0.5) were placed in ceramic crucible and nitrogen gas was purged at a flow of 50 mL/min and heating rate at 10°C/min. The heating range was from room temperature to 200°C.

3 Results and Discussion

3.1 Mechanical performance

The compressive strength of CR-LEO gel was assessed via compression test and summarized in Table 1. CR gel without LEO possesses the highest mechanical properties with compressive stress at 3.2 ± 0.1 kPa, compressive strain at 65.7 ± 1.6% and Young's modulus at 60.1 ± 1.9 kPa. Addition of LEO decreased the compressive stress, compressive strain and Young's modulus of CR-LEO1 gel. For an example, the compressive stress, compressive strain and Young's modulus of CR-LEO1 were recorded at 1.9 ± 0.1 kPa, 47.2 ± 2.1% and 40.8 ± 0.8 kPa, respectively. However, slight improvement in compressive strain of CR-LEO2 and CR-LEO3 gels was observed could be due to the emulsion polymerization process [6]. The addition of essential oil could create micelles in the gel matrix and therefore restrain the strain and improved the flexibility of the gels. Besides that, the inclusion of LEO which composed active ingredients such as linalool (contains hydroxyl group) could create hydrogen bonding between CR-LEO and decreased the intermolecular forces along polymer (CR-CR) chains and enhanced the strain of the gels [6].

Table 1. Compressive stress, compressive strain, Young's modulus and water vapour transmission rates (WVTRs) of CR, CR-LEO1, CR-LEO2, and CR-LEO3.

Sample	Compressive stress (kPa)	Compressive strain (%)	Young's modulus (kPa)	WVTR ($g\ m^{-2}\ h^{-1}$)
CR	3.2 ± 0.1	65.7 ± 1.6	60.1 ± 1.9	117 ± 14
CR-LEO1	1.9 ± 0.1	47.2 ± 2.1	40.8 ± 0.8	120 ± 3
CR-LEO2	1.0 ± 0.1	55.4 ± 4.4	18.8 ± 18.8	135 ± 38
CR-LEO3	0.6 ± 0.1	63.2 ± 5.9	18.8 ± 9.4	99 ± 8

3.2 Water Vapour Transmission Rate (WVTR)

Water vapour transmission rates (WVTRs) were used to determine the barrier properties of the sample. The presence of LEO affected the WVTR values of CR gels (Table 1). The inclusion of LEO increased the WVTRs value of CR-LEO2 gel to 135 ± 38 and reduced to 99 ± 8 at higher concentration of LEO (CR-LEO3). The increased WVTR value of CR-

LEO2 gel could be due to formation a lot of micelles in the gel's matrix and allowed the water diffused through the gel as reported in gellan gum with virgin coconut oils [6]. No clear explanation of the decreased WVTRs of CR-LEO3 and further study will be carried out in understanding the mechanism behind it. However, these values still remain within the range of WVTR values (90 – 2,893 g m−2 d−1) as reported for eight commercially available synthetic wound dressings [7].

3.3 Thermogravimetric Analysis (TGA)

Thermogravimetric analysis (TGA) is performed to measure the weight change as a function of temperature, time and atmosphere. This measurement is used to predict the thermal properties of sample due to its ability to measure the sorption or desorption of volatiles compound, decomposition of the compound, oxidation and reduction. Fig. 1 shows the thermogravimetric thermograms (TGA) and derivative thermograms (DTG) of CR, CR-LEO1, CR-LEO2, and CR-LEO3 gels. CR, CR-LEO1 and CR-LEO2 gels show one stage weight loss, while two stages of weight loss was observed for CR-LEO3 gel (Fig. 1(b)). Weight loss below (T≤ 100 °C) is common due to evaporation of the moisture contents in the samples. The second stage of weight loss of CR-LEO3 gel (DTG peak) at 118 °C could be due to evaporation of lavender essential oil. The results are in agreement to previous study reported the evaporation of lavender EO at 125 °C [8]. Lower concentration of lavender (CR-LEO1 and CR-LEO2) does not show a second stage of degradation. Inclusion of lavender essential oil has also decreased the weight loss (%) of the CR-LEO gel. The weight loss of CR gel was examined at 91%, and the CR-LEO1, CR-LEO2 and CR-LEO3 were improved to 87%, 76% and 74%, respectively.

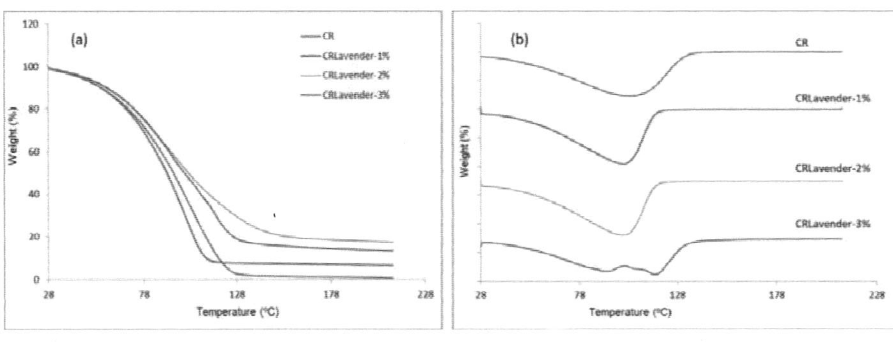

Fig. 1. (a) Thermogravimetric thermograms (TGA) and (b) Derivative thermograms (DTG) of CR, CR-LEO1, CR-LEO2, and CR-LEO3 gels.

3.4 Thermogravimetric Analysis (TGA)

Differential scanning calorimetry (DSC) analysis is used to determine the melting temperature and glass transition of sample. The inclusion of LEO slightly decreased the glass transition (T_g) of the CR-LEO gel (Table 2). Not limited to that, the temperature onset (T_o), temperature melting (T_m) and temperature completion (T_c) were also decreased with addition of LEO. The T_o of CR gel is at 128 °C and the T_c at 148°C (T_c - T_o ~ 20 °C), and for CR-LEO2 gel, the T_o is at 118 °C and the T_c at 129 °C (T_c - T_o ~ 11 °C). The same observation has been reported by previous study using clove essential oil in citrus pectin film [9]. The decreased in thermal activity of CR-LEO gel could be due to a partial separation of CR and LEO, as well as expectation of micelles formation in the matrix of

CR-LEO [6], and reduced the melting points of the gels. Further study is needed to understand the structure of the CR-LEO matrix which effecting the mechanical and thermal behaviour of the gel.

Table 2. Differential scanning calorimetry properties of CR, CR-LEO1, CR-LEO2, and CR-LEO3 gels. Glass transition (T_g), temperature onset (T_o), temperature melting (T_m) and temperature completion (T_c).

	Temperature (°C)				Range (°C)
Sample	T_g	T_o	T_m	T_c	T_c - T_o
CR	32	128	145	148	20
CR-LEO1	32	103	118	121	18
CR-LEO2	31	109	125	129	20
CR-LEO3	31	118	126	129	11

4 Conclusion

The carrageenan (CR) gel with lavender essential oil (LEO) was successfully prepared and the mechanical performance, water vapour transmission rates (WVTRs) and thermal behaviour were examined. The results show that the compression strains of CR-LEO gels were improved expected due to the micelles formation in the gels matrix and formation of hydrogen bond between CR and LEO. In contrast, the compressive stress of the CR-LEO gels was decreased compared to CR gels. The WVTRs values of CR-LEO gels were improved could be due to micelles appearance and permitted the water diffusion through the gel's matrix. The thermal activities of CR-LEO gels were improved with enhanced weight loss of the gels. Lower weight loss (%) was obtained after addition of LEO to 76% (CR-LEO3) compared to CR gels (91%). The mechanical and thermal activities show that the CR-LEO gels can be used as a candidate in pharmaceutical application which offered flexibility of the materials and due to biological activities of LEO, expecting to promote the synergistic affect to the cell growth.

The authors wish to thanks the Ministry of Higher Education, Malaysia for financial assistance under the Fundamental Research Grand Scheme (FRGS/1/2018/STG/07/UMT/02/11), Faculty of Science and Marine Environment and Central Laboratory, Universiti Malaysia Terengganu for providing the facilities for undertaking this work.

References

1. S. Hotchkiss, M. Brooks, R. Campbell, K. Philip, A. Trius. Carrageenans: sources and extraction methods, molecular structure, bioactive properties and health effects, 1st edn. Nova Science Publications Inc., New York, (2016)
2. O. Olatunji, Carrageenans, in: Aquatic Biopolymers, Springer, 2020, pp. 121-144.
3. M.A.Z. Fatin Aliya Kasmi, Aznatul Jannah Abu Bakar and Khairul Anuar Mat Amin. J Pure Appl Microbiol., **14** (2020)
4. K. Smigielski, R. Prusinowska, A. Stobiecka, A. Kunicka-Styczyñska, R. Gruska. Journal of Essential Oil Bearing Plants, **21** (2018)

5. E. Jamróz, L. Juszczak, M. Kucharek. Int. J. Biol. Macromol., **114** (2018)
6. N.A. Ismail, S.F. Mohamad, M.A. Ibrahim, M. Amin, K. Anuar. Advances in Biomaterials, **2014** (2014)
7. P. Wu, A. Fisher, P. Foo, D. Queen, J. Gaylor. Biomaterials, **16** (1995)
8. J. González-Rivera, C. Duce, D. Falconieri, C. Ferrari, L. Ghezzi, A. Piras, M.R. Tine. Innovative Food Science & Emerging Technologies, **33** (2016)
9. T. Nisar, Z.-C. Wang, X. Yang, Y. Tian, M. Iqbal, Y. Guo. Int. J. Biol. Macromol., **106** (2018)

AUTHOR INDEX